PROGRESS IN ELECTRON PROPERTIES OF SOLIDS

PHYSICS AND CHEMISTRY OF MATERIALS WITH LOW-DIMENSIONAL STRUCTURES

VOLUME 10

PROGRESS IN ELECTRON PROPERTIES OF SOLIDS

Festschrift in honour of Franco Bassani

Edited by

E. DONI
C.N.R., Dipartimento di Fisica, Università di Pisa, Pisa, Italy

R. GIRLANDA
Istituto di Struttura della Materia, Università di Messina, Messina, Italy

G. PASTORI PARRAVICINI
Dipartimento di Fisica, Università di Pisa, Pisa, Italy

A. QUATTROPANI
Institut de Physique Théorique, EPFL, Lausanne, Switzerland

KLUWER ACADEMIC PUBLISHERS
DORDRECHT / BOSTON / LONDON

Library of Congress Cataloging in Publication Data

Progress in electron properties of solids.

(Physics and chemistry of materials with low-dimensional structures ; v. 10)
1. Electron transport. 2. Solid state physics.
3. Bassani, F. I. Girlanda, R. II. Bassani, F.
III. Series.
QC176.8.E4P757 1989 530.4'12 89-20063

ISBN 0-7923-0337-7

Published by Kluwer Academic Publishers,
P.O. Box 17, 3300 AA Dordrecht, The Netherlands.

Kluwer Academic Publishers incorporates
the publishing programmes of
D. Reidel, Martinus Nijhoff, Dr W. Junk and MTP Press.

Sold and distributed in the U.S.A. and Canada
by Kluwer Academic Publishers,
101 Philip Drive, Norwell, MA 02061, U.S.A.

In all other countries, sold and distributed
by Kluwer Academic Publishers Group,
P.O. Box 322, 3300 AH Dordrecht, The Netherlands.

Printed on acid-free paper

Printed in The Netherlands

Table of Contents

Preface

This volume on the novelties in the electronic properties of solids appears in occasion of Franco Bassani sixtieth birthday, and is dedicated to honour a scientific activity which has contributed so much of the development of this very active area of research. It is remarkable that this book can cover so large a part of the current research on electronic properties of solids by contributions from Bassani's former students, collaborators at different stages of his scientific life, and physicists from all over the world who have been in close scientific relationship with him. A personal flavour therefore accompanies a number of the papers of this volume, which are both up-to-date reports on present research and original recollections of the early events of modern solid state physics.

The volume begins with a few contributions dealing with theoretical procedures for electronic energy levels, a primary step toward the interpretation of structural and optical properties of extended and confined systems. Other papers concern the interacting state of electrons with light (polaritons) and the effect of the coupling of electrons with lattice vibrations, with emphasis on the thermal behaviour of the electron levels and on such experimental procedures as piezospectroscopy. Electron-lattice interaction in external magnetic field and transport-related properties due to high light excitation are also considered. The impact of synchroton radiation on condensed matter spectroscopy is discussed in a topical contribution, and optical measurements are presented for extended and impurity levels.

Electronic collective properties are to be taken into account to improve or modify the results obtained by the more traditional single-particle description. Excitons, their interactions, also in the presence of lattice coupling, are covered in a few contributions. Also more specific correlation effects in the presence of magnetic fields, in Auger transitions, and as origin of order phenomena are discussed. Collective effects in the theory of glassy and disordered materials complete the survey of this topic of large theoretical and experimental impact.

Other papers are devoted to nonlinear spectroscopy, a field which has grown in the last years as a refined implementation to traditional spectroscopy. The topics presented in the following deal with magnetooptics, coherence effects and multiphoton spectroscopy.

High-T_c superconductivity has by now become a turning point in material research. Theoretical approaches toward the understanding of different classes of superconducting materials are presented together with experimental photoemission results.

Finally, the wide area of two-dimensional electronic systems is largely covered with contributions on the optical properties of surfaces and layered coatings. A few papers are then devoted to the theoretical and experimental analysis of electronic states in superlattices and quantum wells; new theoretical approaches and computational procedures are examined for dealing with electronic levels and electronic transitions in such systems. More specific experimental effects in quantum wells, and theory of core excitons in superlattices are also considered.

In conclusion, we are confident that such a rich and various range of contributions gathered in this volume should constitute an adequate survey of present research on the electronic properties of condensed matter, as well as an appropriate tribute to Franco Bassani scientific achievements in the field.

The Editors

TO FRANCO BASSANI

On the occasion of his sixtieth birthday, October 1989

On October 29, 1989 Professor Franco Bassani, eminent Italian Physicist, celebrates his sixtieth birthday. A small group of his friends and colleagues has decided to mark this occasion, by presenting to him a collection of actual scientific results, elaborated by those physicists all over the world, who owe so much to Bassani's outstanding teaching and research work. The present volume is the tangible result of this decision, and although a severe selection had to be made among the very numerous potential contributors, the editors feel confident that it is representative of Bassani's tremendous influence upon the development of modern solid state physics.

Franco Bassani is born on October 29, 1929 in Milano, the capital city of Lombardia, the most industrious and fertile part of northern Italy. He studies physics at the University of Pavia, where, in November 1952, he obtains his diploma with special mention (laurea con lode). His excellent performances as a student earn him a stipend of the Italian National Research Council, which allows him to carry out his first personal research project on defects in ionic crystals at the University of Milan.

In 1954 he leaves Italy to pursue the same subject for two more years in the research group of Frederick Seitz at the University of Illinois in Urbana, USA. This first expatriation is of considerable importance for Bassani's professional development, since with Seitz, Bardeen and others, Illinois University had become a world famous center for solid state research.

Following his nomination in 1956 he comes back to his native country, where he holds the position of 'professore incaricato' (assistant professor) first at the University of Palermo and then, from 1957 to 1959, at his Alma Mater, the University of Pavia. It is here that he begins to develop the keen interest in the electronic structure of semiconductors and insulators, which constitutes one of the landmarks of Bassani's scientific work.

The dynamic evolution of solid state theory, which he had encountered earlier in Urbana, brings him back to the United States in 1959 as research associate of the Argonne National Laboratory in Argonne, Illinois. He continues his work on the electronic states in solids and makes his decisive contributions to calculations of band structures both by the tight-binding approach and the method of orthogonal plane waves. In this context he proposes a new procedure for band calculations, which has since become known under the name of pseudo-potential method, and he contributes actively to its development. Today this method is a well established tool, used by theoreticians around the world, permitting not only band calculations but also the determination of the total energy of crystals.

Early 1964 Bassani reintegrates Italy after having been nominated full professor at the University of Messina. His arrival there marks the beginning of a most impressive carrier in the Italian University System. Thus, in 1966 he becomes professor for theoretical physics at the University of Pisa, only to move on to Rome University in 1969.

If Rome appears to many as the culmination of a brilliant carrier, it could not entirely satisfy Bassani. Indeed, toward the end of the 1970's life in Rome had become difficult

R. Girlanda et al. (eds.), Progress on Electron Properties of Solids, 1–2.

for the quiet, reserved and somewhat conservative professor, who does not shout with the leftist herds of the students. He takes a leave of absence (1979/80) to the long familiar University of Illinois, devoting himself entirely to his research, which after his second homecoming from the United States also includes studies of the optical behaviour of nonmetallic solids. This second landmark in Bassani's work comprises the first detailed analysis of the optical properties of graphite and germanium and the theoretical interpretation of a welth of experimental data on non-linear spectroscopy.

Even more than his band calculations his optical work brings Bassani in contact with experimentalists and makes him one of the first European solid state theoreticians working close to experiment. Thus, he plays an important role in the design and the interpretation of experiments on the Frascati Synchrotron, whose direction he assumes, while working at the University of Rome.

During his leave of absence from Rome, he is nominated ordinary professor of solid state physics at the most prestigious of all Italian Universities, the Scuola Normale in Pisa, that has been founded in the early 19th century by Napoleon. Here in the Toscana, where so many cultural and historical monuments mix with a most enchanting countryside, the culturally-minded Bassani has finally found in 1980 the peace and restfulness that he treasures so highly and that represents the well deserved reward for an outstanding contribution to modern science.

Bassani's carrier is marked by many displacements inside and outside of Italy. In his peregrinations he was always accompanied by his wife Serenella, who surrounded him with the care and comfort of a loving life companion. It certainly was not always easy for her to cope with the problems arising from Franco's many moves from institute to institute, from country to country. These problems were amplified by the presence of two wonderful children, Clara and Marco, who had to be looked after. It seems adequate, therefore, to say that Serenella, by assuming the heavy responsibility for most practical aspects of every-day life of the Bassani family, has largely contributed to the professional success of her husband. If we owe a lot to Franco for his physics, we also owe a lot to Serenella for making it all possible!

Dear Franco,

We first met at the occasion of your brief visit in summer 1964 at the Zurich laboratories of RCA. I have visited you in all the Italian Universities at which you tought, and in doing so, I have learned to love your country and to appreciate your compatriots. I have helped to bring you and your bright students many a time to the Federal Institute of Technology in Lausanne. Each visit brought an enrichment for me and I would therefore like, at this occasion to express my deep gratitude to what you have done for me and for the Institution at which I teach. And, of course, I add my very best wishes for your 60th birthday and all your future undertakings.

E. Mooser

THEORY OF BANDS, BONDS, AND OPTICAL PROPERTIES OF SEMICONDUCTORS

MARVIN L. COHEN

Department of Physics
University of California
&
Materials and Chemical Sciences Division
Lawrence Berkeley Laboratory
Berkeley, CA 94720
USA

This paper reviews and comments on the major developments in the area of optical properties of semiconductors. The focus is on the interpretation of structure in optical spectra in terms of interband transitions. Franco Bassani contributed pioneering papers in this important field which, in turn, had a large influence on our present understanding of the electronic structure of semiconductors and of solids in general.

1. Introduction

Franco Bassani's expertise covers many subfields of condensed matter physics. His contributions to the theory of impurity states, excitons, energy band theory, optical properties, and a host of other areas are well-known and well-documented by his excellent text [1], his papers, and the creative and important research done by him and his students. In this review, I will concentrate on an area where Bassani's research helped to launch an important subfield of solid state physics.

2. The Early Days

Although quantum mechanics was developed in the 1920's and atomic and molecular spectra were unraveled in the following 20 years, a similar situation did not exist for solid state physics. Optical spectra of metals in the intraband regime had been explored by Drude before the development of quantum mechanics. In Drude's model we visualize the free electrons classically with a mean free path determined by the electron scattering caused by defects or phonons. However, the quantum mechanical picture of solids predicted energy bands. The sharp atomic levels were broadened into bands, and it was assumed that transitions between the broad bands would lead to broad structure in the optical spectra in addition to intraband Drude-like structure.

Although this general view was supported by experimental measurements, there still was structure to be interpreted. It was not sharp on the scale of atomic spectra, but features could be distinguished, and they clearly were caused by interband transitions. The focus for semiconductors was on the fundamental gap, but unfortunately, band structure calculations were not refined sufficiently to predict the sizes of fundamental gaps let alone the entire spectra associated with interband transitions.

This was the situation in the early 1960's, and these are the 'early days' for this

R. Girlanda et al. (eds.), Progress on Electron Properties of Solids, 3–9.

subfield of solid state physics. From a more formal point of view, schemes [2,3] for calculating band structures had been suggested, and the groundwork had been done for proceeding from the OPW method [3] to a pseudopotential approach [4]. The time was ripe for theorists to use the fact that nearly free-electron-like models for metals seemed to give good energy bands when these models were tested. An important advance was the concept of using experimental measurements such as those done with optical probes to give the data for empirical band structures.

The idea is to set up the band structure problem in terms of a potential which could be determined with just a few parameters [5-8]. The parameters, in turn, could be evaluated by comparing theoretical output with experiment. A crucial paper which established the most fruitful path was by Brust, Bassani, and Phillips [7] in 1962. Franco Bassani was at the Argonne National Laboratories, Jim Phillips was at the University of Chicago, and David Brust was a graduate student (as was I) at that time working with Phillips. The scheme was to fit three Fourier coefficients of the *Ge* potential and to calculate the imaginary part of the frequency-dependent dielectric function $\epsilon_2(\omega)$.

Brust traveled back and forth between the University and Argonne and provided the link between Bassani and Phillips. I kept up with the progress but did not get involved until one afternoon when Brust brought back a page of scattered dots from Argonne. This page contained the first realistic interband dielectric function $\epsilon_2(\omega)$ for a solid, but it looked like the results of a dart game. While Brust and Phillips discussed possible improvements, I began doing a three-point averaging by hand, and there was the spectrum.

I hadn't met Franco Bassani during that period, but I was being tutored indirectly by him through Brust. I was in awe of the achievement and of Bassani who then moved on to important research in other fields.

3. Pseudopotentials

Let me now discuss this work from the point of view of the development of pseudopotential theory. The pseudopotential was invented by Fermi [9] in 1934 (Fig. 1) to explain shifts in spectral lines of alkalai vapor arising from foreign gases [10].

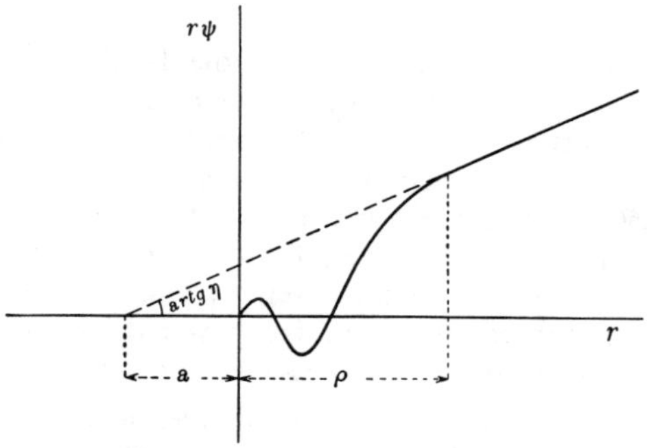

Figure 1. Fermi's approximation to the wavefunction to determine a pseudopotential for alkali atoms.

Hellmann [11] had also used a pseudopotential scheme to describe the energy levels of alkali atoms. The basic concept is to begin with atomic wavefunctions and then develop potentials which reproduce the outermost parts of the wavefunction.

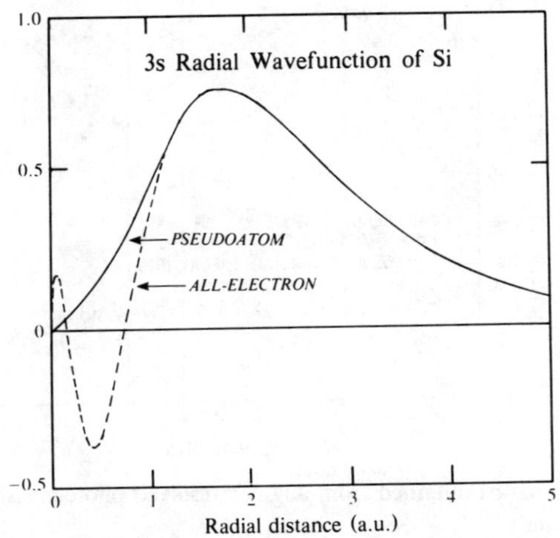

Figure 2. The 3*s* radial wavefunction for *Si* (dashed line) and the pseudowavefunction approximation (solid line).

The above approach is similar to modern schemes [12-17] (Fig. 2) where the pseudopotentials are primarily intended for solids. Since solid state effects are dominated by the outer positions of the electronic wavefunctions, weak potentials are constructed to reproduce this part and eliminate the nodes as shown in Fig. 2. The resulting potential behaves as an attractive Coulomb potential at large distances and is weak or repulsive in the core. This approach and the methods involved are well-documented in the literature, and they form the basis for the approximations made by Brust, Brassani, and Phillips [7].

The empirical approach usually called the Empirical Pseudopotential Method (EPM) assumes that the average potential $V(\mathbf{r})$ felt by an electron arising from electron-core and electron-electron interactions can be expressed in terms of a few Fourier coefficients of an atomic potential $V_a(\mathbf{G})$

$$V(\mathbf{r}) = \sum_{\mathbf{G}} V_a(\mathbf{G}) S(\mathbf{G}) e^{i\mathbf{G}\cdot\mathbf{r}} \tag{1}$$

where the structure factor

$$S(\mathbf{G}) = \frac{1}{N_a} \sum_{\tau} e^{-i\mathbf{G}\cdot\vec{\tau}} \tag{2}$$

where N_a is the number of basis atoms, τ is the basis vector, and $\mathbf{G}$ is the reciprocal lattice vector.

The solution of the Schrödinger equation

$$\left[\frac{p^2}{2m} + V(\mathbf{r})\right] \psi_{n\mathbf{k}}(\mathbf{r}) = E_n(\mathbf{k}) \psi_{n\mathbf{k}}(\mathbf{r}) \tag{3}$$

gives the Bloch wavefunctions for each band n and state $\mathbf{k}$ and the energy eigenvalues. From these results, the $\epsilon_2(\omega)$ and the other optical constants can be computed [1,18].

Comparison between the calculated and observed spectra allowed the determination of the $V_a(\mathbf{G})$ in Eq. (1) by constraining the theoretical results to be consistent with experiment. Dozens of bands structures [18-20] were calculated, and when angular resolved photoemission was perfected, it was found that these band structures were in good

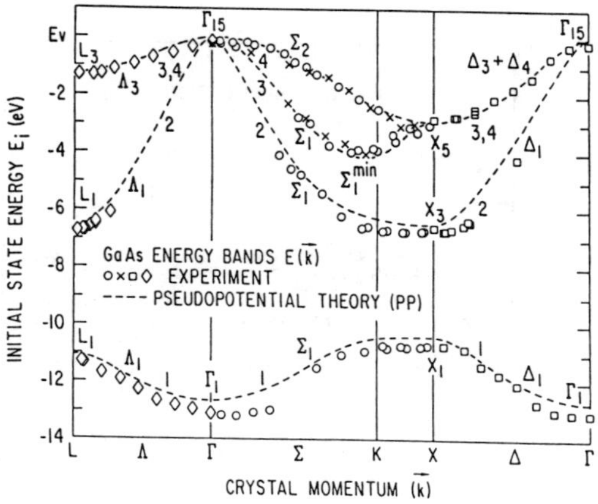

Figure 3. *GaAs* energy bands obtained from angular resolved photoemission experiment and from empirical pseudopotentials.

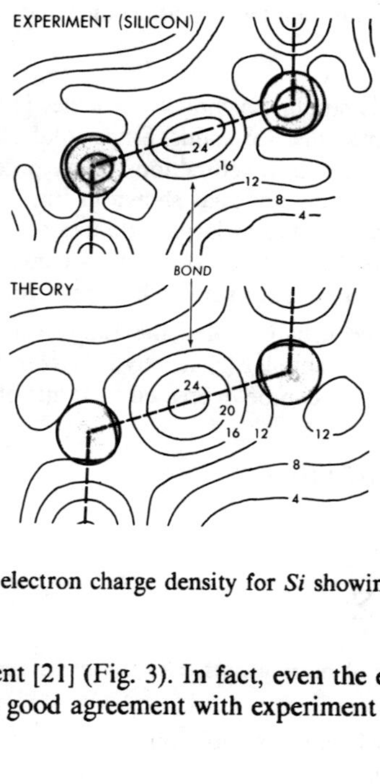

Figure 4. Contour plots of valence electron charge density for *Si* showing the comparison between experiment and theory.

agreement [18-20] with experiment [21] (Fig. 3). In fact, even the electronic charge density from the EPM calculations gave good agreement with experiment [18,22] (Fig. 4).

4. Some Extensions

Since the charge density was reliable and it is known that the principal change in electronic structure near a surface is the redistribution of electronic charge, schemes [23,24] were developed to separate the core part of the pseudopotential from the electron-electron part. Near a surface, the core contribution could be held fixed while the electron-electron contribution changes. This led to detailed calculations of electronic structure at surfaces [23,24] and interfaces [25].

Another important aspect of the charge density is its role in bonding. Phillips and Van Vechten [26] developed a theory relating the spectral features of semiconductors to their bonding. They were able to separate classes of structures and to develop an ionicity scale using dielectric information. By using pseudopotential results, it was possible to compute [27] the charge in the covalent bond and examine this property as a function of the Phillips-Van Vechten ionicity parameter. This scheme illustrated the role of the bond charge in stabilizing covalent structures. A recent scaling extension [28] of the above approach yields the bulk modulus for diamond and zincblende semiconductors.

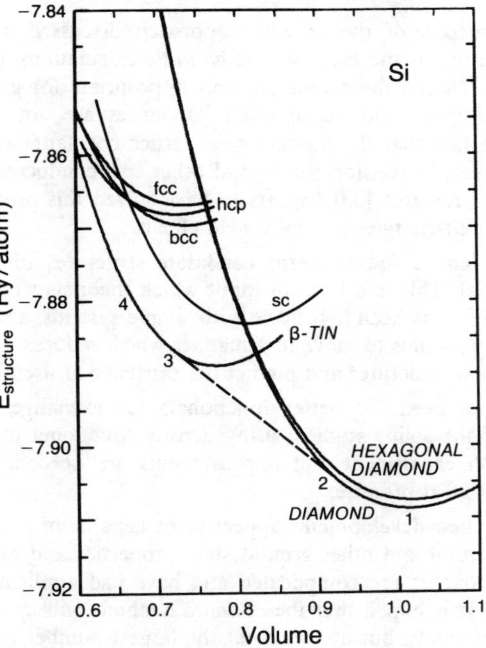

Figure 5. Total energy versus volume for *Si* in various crystal structures. The dashed line is a common tangent between the diamond and β-tin phases (see text).

The above approaches are semi-empirical, and they pointed the way for more first-principles calculations. The calculations were motivated by the consideration of structural properties of semiconductors. Since the approach for studying surfaces already separted the core from the valence contributions of the potential, a scheme was developed to calculate the total energy [29] of the solid for an assumed arrangement of cores while allowing the valence electrons to choose a minimum energy configuration. The calculation was based on a momentum representation [30] of the various energy terms.

A propotype result [31] is given in Fig. 5 where the total energy is plotted as a function of volume for several assumed structures of *Si*. These curves can be used to obtain the lattice constant (given by the volume at the minimum of the curve), the bulk modulus (curvature), and the transition pressure for a solid-solid structural transition. For example, the dashed line illustrates the paths for transforming *Si* from the diamond to the $\beta-Sn$ structure with reduced volume (increased pressure). At point 1, the solid is in the diamond structure, at 2 it begins to transform, at 3 the transformation is complete, and at 4 it is in the $\beta-Sn$ structure.

The total energy structural calculations gave results for structural reconstructions at surfaces, phonon spectra, pressure-induced effects such as structural phase transitions, electron-phonon couplings, and was even responsible for successful predictions of superconductivity [32]. All of this was done from first principles. The calculations require only the atomic number, mass, and candidate structures.

5. The Situation Now

Despite the great success of the *ab initio* approach discussed above, three problems became evident. The first is the fact that the *ab initio* calculations are based on a density functional approach. Hence, the results are only appropriate for ground state properties. Excited states, band gaps, and quasiparticle properties are not treated correctly. For example, despite the fact that the method gives lattice constants and bulk moduli to an incredibly high degree of precision for *Si* and other semiconductors, the band gap is in error by 50%. Recent research [33] appears to have solved this problem by the introduction of quasiparticle self-energies and local field effects.

The second problem is the need for candidate structures to determine the crystal structure of a material. This is a form of input which theorists would like to avoid. Fortunately, good progress has been [34] made with simple systems, and extensions are being examined. By allowing atoms to move in a manner which reduces the total energy, it may be possible to find new structures and predict the existence of useful materials.

Finally, there is a need for better functionals for exchange and correlation. The cohesive energies of the solids studied, using density functional theory, are in relatively good agreement with experiment, but improvements are needed. Recent Monte Carlo techniques [35] show great promise.

At this point, the new developments appear to be capable of giving an *ab initio* theory which will give structural and other ground state properties and excited state properties. However, these approaches are complicated and have had application to only a limited number of crystals. It is hoped that the *ab initio* methods will be simplified and applied to a large number of solids, but at this point the largest number of accurate band structures are those determined with the empirical scheme which Franco Bassani helped to create almost 30 years ago.

This work was supported by National Science Foundation Grant No. NSF88-18404 and by the Director, Office of Energy Research, Office of Basic Energy Sciences, Materials Sciences Division of the US Department of Energy under Contract No. DE-AC03-76SF00098.

References

1. F. Bassani and G. Patori Parravicini, *Electronic States and Optical Transitions in Solids*, Pergamon, Oxford (1975).
2. J.C. Slater, *Phys. Rev.* **51**, 846 (1937).
3. C. Herring, *Phys. Rev.* **57**, 1169 (1940).
4. J.C. Phillips and L. Kleinman, *Phys. Rev.* **116**, 287 (1959).
5. J.C. Phillips, *Phys. Rev.* **112**, 685 (1958).
6. F. Bassani and V. Celli, *J. Phys. Chem. Solids* **20**, 64 (1961).

7. D. Brust, J.C. Phillips and F. Bassani, *Phys. Rev. Lett.* **9**, 94 (1962).
8. D. Brust, M.L. Cohen and J.C. Phillips, *Phys. Rev. Lett.* **9** 389 (1962).
9. E. Fermi, *Nuovo Cimento* **11**, 157 (1934).
10. E. Amaldi and E. Segre, *Nuovo Cimento* **11**, 145 (1934).
11. H.J. Hellmann, *J. Chem. Phys.* **3**, 61 (1935).
12. T. Starkloff and J.D. Joannopoulos, *Phys. Rev. B* **16**, 5212 (1977).
13. A. Zunger and M.L. Cohen, *Phys. Rev. B* **20**, 4082 (1979).
14. D.R. Hamann, M. Schluter and C. Chiang, *Phys. Rev. Lett.* **43**, 1494 (1979).
15. G.P. Kerker, *J. Phys. C* **13**, L189 (1980).
16. S.G. Louie, S. Froyen and M.L. Cohen, *Phys. Rev. B* **26**, 1738 (1982).
17. M.T. Yin and M.L. Cohen, *Phys. Rev. B* **25**, 7403 (1982).
18. M.L. Cohen and J.R. Chelikowsky, *Electronic Structure and Optical Properties of Semiconductors*, Springer Verlag (1988).
19. M.L. Cohen and T.K. Bergstresser, *Phys. Rev.* **141**, 789 (1966).
20. J.R. Chelikowsky and M.L. Cohen, *Phys. Rev. B* **14**, 556 (1976).
21. T.C. Chiang, J.A. Knapp, M. Aono and D.E. Eastman, *Phys. Rev. B* **21**, 3513 (1980).
22. L.W. Yang and P. Coppens, *Solid State Comm.* **15**, 1555 (1974).
23. J.A. Appelbaum and D.R. Haman, *Rev. Mod. Phys* **48**, 3 (1976).
24. M. Schluter, J.R. Chelikowsky, S.G. Louie and M.L. Cohen, *Phys. Rev. B* **12**, 4200 (1975).
25. M.L. Cohen, *Advances in Electronics and Electronic Physics*, Vol. 51, eds. L. Marton and C. Marton, Academic, New York (1980).
26. J.C. Phillips, *Bond and Bands in Semiconductors*, New York, New York (1973).
27. J.P. Walter and M.L. Cohen, *Phys. Rev. B* **2**, 1821 (1970).
28. M.L. Cohen, *Phys. Rev. B* **32**, 7988 (1985).
29. M.L. Cohen, *Phys. Scripta T* **1**, 5 (1982).
30. J. Ihm, A. Zunger and M.L. Cohen, *J. Phys. C* **12**, 4401 (1979).
31. M.T. Yin and M.L. Cohen, *Phys. Rev. B* **26**, 5668 (1982).
32. M.L. Cohen, *Science* **234**, 549 (1986).
33. M.S. Hybertsen and S.G. Louie, *Phys. Rev. B* **34**, 5390 (1986).
34. R. Car and M. Parrinello, *Phys. Rev. Lett.* **55**, 2471 (1985).
35. S. Fahy, X.W. Wang and S.G. Louie, *Phys. Rev. Lett.* **61**, 1631 (1988).

ATOMIC PSEUDO-HAMILTONIANS FOR QUANTUM MONTE CARLO

G.B. BACHELET(1), D.M. CEPERLEY(2), M.G.B. CHIOCCHETTI(1) & L. MITÁŠ(1,3)

(1) *Dipartimento di Fisica*
Università degli Studi di Trento
38050 Povo Trento, Italy

(2) *National Center for Supercomputing Applications and*
Department of Physics
University of Illinois at Urbana-Champaign
Urbana, Illinois, USA

(3) *Centro CNR, Povo Trento, Italy*
and
Institute of Physics, EPRC SAS,
Bratislava, Czechoslovakia

Quantum simulations, successful for simple systems, remain difficult for atoms with core electrons. We discuss the possibility of replacing core electrons by a local yet angular-momentum dependent pseudo-Hamiltonian where momentum operators appear only to the second power, as required by fixed-node Quantum Monte Carlo. We describe two ways of constructing it from norm-conserving local-density atoms: a simpler approach, which works for $s-p$ atoms, and a more sophisticated one, based on a simulated-annealing technique, which fixes *s*, *p* and *d* states. The use of this new pseudo-Hamiltonian in Green's Function Monte Carlo (both fixed-node and release-node) gives accurate electron affinities, ionization and binding energies for second-row atoms and diatomics. This opens the way to quantum simulations of many condensed-matter systems.

1. Introduction

Quantum Monte Carlo (*QMC*) methods, in particular Green's Function Monte Carlo, can achieve remarkable accuracy for total ground state energies including correlation effects that are left out or approximated by other computational methods such as Hartree-Fock or local-density-functional theory, and can treat systems of up to 1000 electrons, far beyond the capabilities of configuration-interaction methods. Examples of such calculations include the ground state properties of liquid and solid ^3He and ^4He [1], the homogeneous electron gas [2] solid hydrogen at high pressures in the molecular and atomic phases [3], and many small molecules [4-6]. However, the presence of atomic cores shows the convergence of the simulation to such an extent that calculations of atoms heavier than neon appear not to be feasible. There are several related problems [7]:

(i) The distances and energies of the 1*s* electrons determine the time step of the simulation and these scale with the atomic number as Z^2

(ii) the statistical errors are proportional to the correlation energy, which, for $Z<20$, increases as $Z^{1.5}$, while the physically relevant valence energies are more or less constant throughout the periodic table; finally

(iii) the computational effort per step of the simulation increases as the total number of electrons cube.

R. Girlanda et al. (eds.), Progress on Electron Properties of Solids, 11–25.

Combining all the effects the total computational effort shows a very strong dependence on the atomic number Z; estimates are between $Z^{5.5}$ [7] and $Z^{6.5}$ [8]. It is then very desirable to eliminate the core electrons completely which will hopefully allow Monte Carlo (MC) simulations of heavier atoms in extended systems.

Pseudopotentials, model potentials or frozen core approximations have a very long history (see e.g. references [9-12]); the state of the art is such that much more serious approximations are made today in solving the correlated atom than in reducing an atom to a pseudoatom. Modern pseudopotentials, born within local-density-functional or Hartree-Fock theory, (see e.g. ref. [13] and references therein) have been successfully tested in the context of variational MC calculations by Fahy et al. [14]; however, being nonlocal (they contain angular-momentum projection operators), they are generally inappropriate for QMC simulations. To extend the concepts of modern pseudopotential theory to QMC simulations (and in general beyond single-particle theory) the development of entirely new theoretical and computational tools is required [15,16]. What we present in this paper is the possibility of totally replacing the core electrons by a local yet angular-momentum dependent atomic Hamiltonian in such a way that the methods developed for QMC are still applicable.

2. Green's Function Monte Carlo

In Green's Function Monte Carlo [17] one begins by sampling an ensemble of configurations R (R is the set of $3N$ coordinates and N is the number of electrons) from some arbitrary probability distribution $f_0(R)$. This probability distribution is then iterated:

$$f_{n+1}(R) = \int G(R,R';t) f_n(R') dR', \tag{1}$$

where n is the iteration number and G the Green's function:

$$G(R,R';t) = \Psi(R)\Psi(R')^{-1} \langle R \,|\, \exp[-t(H-E_T)] \,|\, R' \rangle. \tag{2}$$

Here $\Psi(R)$ is the trial function, H the Hamiltonian, t the time-step (the imaginary or thermal time-step) and E_T the trial energy. At large n, $f_n(R)$ approaches $\Psi(R)\Phi_0(R)$ where $\Phi_0(R)$ is the ground-state wave function. As the trial function approaches the ground-state wave function the variance of the estimate of the ground state vanishes. This is the chief motivation for applying importance sampling and makes QMC simulations of large systems practical.

The main limitation of Green's Function Monte Carlo arises from the need to keep the Green's function non-negative [18], so that it can be sampled as a probability distribution. One can, in fact, treat Green's functions with negative pieces; one simply samples the absolute value of G and carries along the sign as a weight. However, for large values of n this gives rise to an exponentially growing noise. In condensed matter physics, this procedure is known as either the release-node [19] or the transient-estimate [18] method. Such methods are not appropriate for large systems, since the statistical error grows rapidly with the number of particles.

Any fermion system will have a Green's function with negative pieces because the trial function changes sign whenever two electrons with the same spin exchange. The fixed-node approximation [2,4-6] is made in order to treat these systems. There one divides the configuration space into two volumes: a region where $\Psi(R)$ is positive, and a region where $\Psi(R)$ is negative. The fixed-node Hamiltonian equals H for matrix elements within the same region but matrix elements from one region to the other are set to zero.

There are three important properties of this approximation:

(A) The fixed-node Green's functions is non-negative so that the error will decrease inversely proportional to the square root of the number of steps of the random walk.

(B) The fixed-node energy is an upper bound [17] to the true fermion energy: $E_{FN} \geqslant E_0$.

(C) As the nodes of Ψ approach the nodes of Φ_0 (i.e. the region where $\Phi_0\Psi$ is negative vanishes) $E_{FN} - E_0$ vanishes quadratically. Only the position of the nodes is important, not the values of the trial function anywhere else.

The first property allows one to simulate many-fermion systems: the second and third show the method is variational. In fact, in most systems the error introduced by the fixed-node approximation is extremely small. The use of Hartree-Fock *nodes* will give 99% of the correlation energy for typical electronic systems [5,6].

3. Atomic Pseudo-Hamiltonians and Quantum Monte Carlo

We show here that to preserve the three properties discussed at the end of the previous section the Hamiltonian must be local and contain at most second derivatives. If H is a non-local Hamiltonian, and its matrix elements between the two regions are deleted, properties (B) and (C) will not hold in general. For that reason we restrict ourselves to local Hamiltonians only. Such Hamiltonians can be expanded in powers of momentum operators:

$$H = g(R) + \sum_a g_a(R) p_a + \sum_{a,b} g_{ab}(R) p_a p_b + \cdots \tag{3}$$

Now for (A) to hold $G(R, R_0; t)$ must be positive for all R and R_0 in the same region. Suppose we construct the second moment of G about R_0

$$c_{ab}(R_0;t) = \int (R - R_0)_a (R - R_0)_b G(R, R_0; t) dR. \tag{4}$$

This tensor must be positive definite for all trial functions (except if $\Psi(R_0) = 0$), and times t; since it vanishes at zero time (G reduces to a delta function), the derivative of c at zero time must be positive definite. This can be explicitly evaluated to give:

$$\left.\frac{dc_{ab}}{dt}\right|_0 = \Psi^{-1} \left[\frac{\partial^2 H}{\partial p_a \partial p_b} \right] \Psi, \tag{5}$$

where first we take the time derivative of Eq. (4), use the definition of G, Eq. (2), the hermitian property of H, and then commute the R's with the momentum operators. Suppose the highest power of the momentum in H is n. Then $dc/dt\,|_0$ will be proportional to the $(n-2)$th derivative of the trial function and, unless $n = 2$, $dc/dt\,|_0$ (and thus c) cannot be made positive definite for *all* choices of the trial function which, in turn, implies that G will have negative pieces. In conclusion, for G to be positive for all R and R_0 in the same region, it is necessary that $n = 2$ and that the tensor $(\partial^2 H / \partial p_a \partial p_b)$ be positive definite. The above three requirements (A,B,C) and hermiticity lead to the following general form of a Hamiltonian:

$$H = V(R) + \sum_{a,b} p_a g_{ab}(R) p_b \tag{6}$$

where the tensor g_{ab} is positive definite. To simplify further, suppose that the pseudo-Hamiltonian (PH) for a collection of atoms has the same structure as the full Hamiltonian:

$$H = \sum_{i<j} \frac{1}{r_{ij}} - \frac{1}{2}\sum_i \nabla_i^2 + \sum_{i,I} h_I(r_{iI}). \tag{7}$$

Atomic units are used throughout this paper, i and j refer to electrons and I to ions. The most general form of the atomic *PH* with these properties (and spherical symmetry) is:

$$h(r) = -\frac{1}{2}\nabla a(r)\nabla + \frac{b(r)L^2}{2r^2} + v(r), \tag{8}$$

where $a(r)$, $b(r)$ and $v(r)$ are radial functions, and L^2 is the usual angular momentum operator. Since the tensor $(\partial^2 H/\partial p_a \partial p_b)$ must be positive definite we require:

$$a(r)+1>0, \quad a(r)+b(r)+1>0. \tag{9}$$

Notice that if the above requirement is not fulfilled the Hamiltonian is not bounded below. On physical grounds it is rather evident that outside the core of an atom a and b will vanish and $v = -Z_v/r$ (Z_v is the valence charge). By modifying the kinetic energy operator inside the atomic cores we are, in effect, making the electronic mass position and direction dependent. This is to be distinguished from a momentum dependent mass (see e.g. ref. [9]) which would not lead to a suitable Hamiltonian.

We have argued that there is very little freedom in choosing the *PH* once one has accepted conditions A,B,C and that it be a sum of atomic Hamiltonians. One possibility is to keep A,B,C but make the local potential a more general function of R; for example, one could modify the electron-electron interaction inside the core to mimic the effects of core polarizability. We do not consider such a possibility here. Other authors [8,20] have made a local potential by projecting a non-local potential onto a Hartree-Fock trial function. That does not satisfy conditions (B) or (C) since it requires the trial function be accurate everywhere, not just at the node.

4. Atomic Pseudo-Hamiltonians and Norm-Conserving Pseudopotentials

Our goal is to replace the true system (a full-core atom) with a pseudo system (a valence-only atom). The latter should ideally reproduce the behavior of the former outside the core region in a wide variety of situations: not only for a *reference atomic configuration* (e.g. the ground-state valence configuration), but also in a molecular or solid-state environment. In other words the ideal atomic *PH*, though energy independent, is highly transferable: it should closely track the true Hamiltonian over a wide range of energies around the atomic ground state, so that, even if the formation of bonds and bands shifts the energy eigenvalues away from the atomic ground state, the pseudoatom continues to behave like the true, full-core atom.

Hamann, Schlüter and Chiang [21] have shown that, whenever the many-body problem is reduced to a set of self-consistent single-particle equations (for example in local-density-functional or Hartree-Fock theory), the above requirement is met by norm-conserving pseudopotentials, which, for a given valence configuration, identically reproduce the valence states of a full-core atom outside the core: this ability also ensures optimum transferability, because of a remarkable relation between the integral of the wavefunction squared and the energy derivative of the logarithmic derivative of the wavefunction (see e.g. ref. [22]). Norm-conserving pseudopotentials are such that, for each angular momentum l which is relevant to the valence properties of the atom, the bound-state eigenvalue ϵ_l is identical for either pseudo (*PS*) or full-core atom (*FC*), and, which is very important, *outside some core radius* $r>r_c$ the normalized radial wavefunction $r\chi_l$, and thus the logarithmic derivative χ_l'/χ_l, are identical:

$$\chi_l^{PS}(r) \equiv \chi_l^{FC}(r); \quad \left[\frac{\chi_l'(r)}{\chi_l(r)}\right]_{\epsilon=\epsilon_l}^{PS} \equiv \left[\frac{\chi_l'(r)}{\chi_l(r)}\right]_{\epsilon=\epsilon_l}^{FC}. \tag{10}$$

Away from ϵ_l discrepancies are only second order in the energy $\epsilon-\epsilon_l$, because Eq. (10) plus normalization implies that the first energy derivative is identical at ϵ_l [21]

$$\frac{d}{d\epsilon}\left[\frac{\chi_l'(r)}{\chi_l(r)}\right]_{\epsilon=\epsilon_l}^{PS} \equiv \frac{2}{[\chi_l^{PS}(r)]^2}\int_0^r dr'[\chi_l^{PS}(r')]^2 \tag{11}$$

$$\equiv \frac{2}{[\chi_l^{FC}(r)]^2}\int_0^r dr'[\chi_l^{FC}(r')]^2 \equiv \frac{d}{d\epsilon}\left[\frac{\chi_l'(r)}{\chi_l(r)}\right]_{\epsilon=\epsilon_l}^{FC}.$$

As mentioned, the exact fulfillment of the above conditions is relevant only for angular momenta which correspond either to occupied valence orbitals in the atom, or to empty orbitals whose energy is low enough to be potentially involved in the bonding process. This gives angular-momentum dependent nonlocal pseudopotentials, reflecting the physical fact that different partial waves feel the effect of orthogonality to different core shells. For most atoms excellent norm-conserving pseudopotentials are obtained by imposing Eqs. (10,11) to s, p and d pseudowavefunctions (see e.g. ref. [13]); for many atoms it may be sufficient to fix only s and p waves.

Since local-density-functional and Hartree-Fock theories successfully describe many properties of atoms, molecules and solids, it is very likely that the angular-momentum dependence is a general property of transferable *PH*'s and is needed in the exact many-body theory. The fact that the bare-ion pseudopotentials derived from either local-density-functional or Hartree-Fock theory are essentially identical [23] in spite of the different approximations involved in the two approaches further suggests that the norm-conserving pseudopotentials derived in that context represent an excellent approximation to their general many-body counterpart.

In a single-particle theory like local-density-functional theory the form of our *PH* allows us in principle to fix up to three partial waves. The radial Schrödinger equation for a single-particle state of angular momentum l is for our *PH*:

$$\left[-\frac{1}{2}(a+1)\frac{d^2}{dr^2}+(a+b+1)\frac{l(l+1)}{2r^2}- \right. \tag{12}$$

$$\left. -\frac{1}{2}\frac{da}{dr}\left[\frac{d}{dr}-\frac{1}{r}\right]+v+v_{HXC}\right]\chi_l = \epsilon_l\chi_l,$$

where v_{HXC} is the sum of the Hartree and exchange-correlation potentials. We determine the three functions a,b and v by demanding that, in a given valence configuration, the self-consistent valence eigenvalues and radial eigenfunctions which solve Eq. (12) be identical to those of some norm-conserving l-dependent pseudopotential $v_l(r)$ (see e.g. ref. [13]). Substract the pseudopotential Schrödinger equation from Eq. (12): v_{HXC} and ϵ_l drop out and

$$\left[-a\frac{d^2}{dr^2}+(a+b)\frac{l(l+1)}{r^2}-\frac{da}{dr}\left[\frac{d}{dr}-\frac{1}{r}\right]+2(v-v_l)\right]\chi_l = 0. \tag{13}$$

If the *reference atomic configuration* contains s, p and d orbitals, Eq. (13) gives for $l = 0,1,2$ three conditions for the three functions a,b and v, which are thus uniquely determined: one gets a first order differential equation for a and, once that is solved, b and v are obtained in terms of a, da/dr and combinations of pseudopotentials and self-consistent wavefunctions. From their form one recognizes that a,b and v automatically recover the correct behavior outside the core ($a \rightarrow 0$, $b \rightarrow 0$, and $v \rightarrow -Z_v/r$ for $r > r_c$).

It appears that we have a prescription to build up a local norm-conserving Hamiltonian which exactly mimics the action of an l-dependent, nonlocal pseudopotential for the three partial waves s, p and d, and thus excellent for the vast majority of atoms (only lanthanides, rare earths and few other atoms really need an accurate f-wave, and, for these atoms, pseudopotentials are less accurate anyway; see e.g. ref. [13]). Unfortunately upon actual calculations it turns out that the a and b functions resulting from Eq. (13) for $l = 0,1,2$, do not satisfy the conditions of Eq. (9) for many atoms and various types of norm-conserving pseudopotentials. However, as already mentioned, there is a large class of atoms which form $s-p$ bonded molecular and solid-state systems, and for them two partial waves are usually sufficient. As a first attempt we can then concentrate on these atoms, and, instead of fixing three partial waves (s,p and d) we only focus on two of them (s and p). We drop the differential equation for a, and with some trial and error choose the simple analytic fom $a = a_0 \exp[-(r/r_c)^k]$ which vanishes outside the core ($r > r_c$). Now Eq. (13) can be used to fix exactly two partial waves (e.g. s and p). The three parameters a_0, r_c and k are chosen such that

(i) the range of $a(r)$ roughly agrees with the physical extent of the atomic core

(ii) Eq. (9) is satisfied, and

(iii)the d states are reasonably reproduced.

TABLE I

The parameters of the function $a(r) = a_0 \exp[-(r/r_c)^k]$ for the atoms shown, in atomic units (see text).

atom	a_0	r_c	k
Na	0.		
Mg	−0.15	1.00	4
Si	−0.74	1.30	4
Cl	−0.80	1.10	6

Our choice of a_0, r_c and k is shown in Table I for *Na*, *Mg*, *Si* and *Cl*; the radial functions a,b and v are shown in Figure 1 for *Si*. The *Si* logarithmic derivatives, important for transferability, are plotted against energy in Figure 2, and show that the new *PH* is of comparable quality as the original, nonlocal pseudopotential.

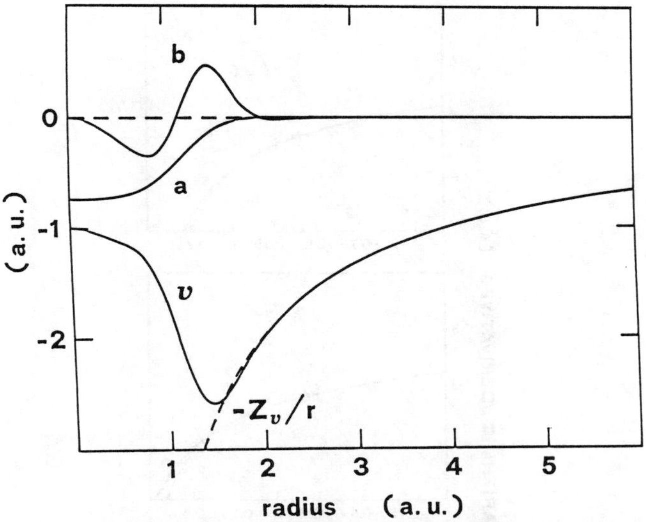

Figure 1. Silicon atom: the two kinetic-energy radial functions $a(r)$ and $b(r)$, and the atomic potential, $v(r)$, which characterize our local pseudo-Hamiltonian Eq. (8). Here we show the silicon pseudo-Hamiltonian which fixes s and p states exactly, and d states only approximately: $a(r)$ has been parametrized and Eq. (13) solved only for $l = 0, 1$ (see Section 4 and Table I). Atomic units are used. A dashed line shows the Coulombic tail $-4/r$.

5. Second-Row $s-p$ Atoms and Molecules

Now we present some tests of our new pseudo-Hamiltonian: a set of *QMC* calculations for second-row atoms and diatomic molecules [15]. Our *MC* calculations have been performed with a simple generalization of the diffusion *MC* algorithm [2,4-6]. This makes an approximation to the Green's function of Eq. (1) which is exact in the limit of small time steps. The more rigorous and efficient, but complicated, Domain Green's Function Monte Carlo algorithm would dispense with this time-step error [1]. In either algorithm there are three basic steps: branching, diffusion and drift. The program for the *PH* runs somewhat slower than one with a constant mass for these small systems but for a large system there will be no appreciable increase in computational time. We have used the same trial wave function employed in other *QMC* calculations [5], a pair-product, Slater-Jastrow function, where the molecular orbitals are linear combinations of the single particle local-density-functional orbitals. The parameters in this trial function are fixed by minimizing the variational energy with the reweighting method [17]. For atoms we use a minimum basis set while for molecules we add a few extra p functions in the bond direction and a Gaussian in the middle of the bond. This type of functions works well if $a(r)$ does not approach -1, but for *Cl* the pair correlation term should include some radial dependence to account for the sizable change of the effective mass in the core.

Variational *MC* (evaluation of the trial function energy), fixed-node diffusion *MC* and release-node *MC* calculations have been performed on the atoms and molecules shown in Table II. Release-node calculations were performed to determine the accuracy of the *PH* and of the fixed-node approximation. It is possible to converge to the exact energy, but that procedure becomes more costly for systems with more electrons. The calculations shown took between 1 minute of *CRAY* time (for *Na*) to 2 hours (for Cl_2).

The accuracy of the *PH* varies. The electron affinities are in perfect agreement with

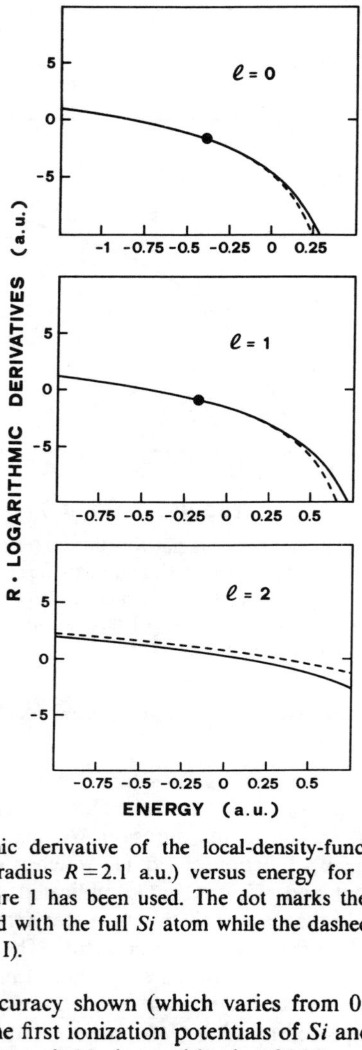

Figure 2. Silicon logarithmic derivative of the local-density-functional radial wavefunction χ_l (evaluated at the covalent radius $R=2.1$ a.u.) versus energy for *s,p* and *d* partial waves. The pseudo-Hamiltonian of Figure 1 has been used. The dot marks the position of the eigenvalue ϵ_l. The solid lines are calculated with the full *Si* atom while the dashed lines are with the *Si PH* (see Eq. (8), Section 4 and Table I).

experiment [24] to the accuracy shown (which varies from $0.07\,mH$ for *Na* to $4\,mH$ for *Cl*). The same holds for the first ionization potentials of *Si* and *Cl*, while the experimental ionization potential of *Na* and *Mg* is outside the *QMC* error bar by $6\,mH$ and $4\,mH$ respectively. The homo-nuclear diatomic binding energies are in excellent agreement with experiment for Si_2, Cl_2 and Mg_2, but 10% too large for *Na*. This discrepancy may be related to the problems pointed out by Louie et al. [29] for atoms with few valence electrons. Figure 3 compares the Si_2 binding energy curve as calculated with $PH-QMC$ and the configuration-interaction method [30].

TABLE II

The results of Bachelet et al. [15] presented in Sections 4 and 5. *LSD*: local-spin-density; *NLPP*: non-local pseudopotentials [13]; *PH*: pseudo-Hamiltonian, Eq. (8) and Table I of this work (see Section 4). *QMC*: Green's Function Monte Carlo, release-node [15]. The experimental values are from Weast [24] for atoms and from Huber and Herzberg [25] for molecules. All energies are in hartrees. E.A.: electron affinity; *I.P.*: ionization potential; E_B: binding energy of the homo-nuclear diatomic molecule (no zero-point motion included) at the experimental bond separation. Shown in parentheses is the error bar: 0.118(3) = 0.118±0.003.

Sodium $Z=11$	*E.A.*	1 *I.P.*	2 *I.P.*	3 *I.P.*	E_B
full core *LSD*		0.1974			-
NLPP + LSD		0.1932			0.0334[a]
PH + LSD		0.1932			-
PH + QMC	0.0204 (7)	0.1951 (1)			0.0310 (6)
experiment	0.0201	0.1889			0.0269
Magnesium $Z=12$					
full core *LSD*		0.2874	0.5635		0.0043[b]
NLPP + LSD		0.2804	0.5535		-
PH + LSD		0.2804	0.5536		-
PH + QMC	not stable	0.2849 (3)	0.5585 (2)		0.0015 (7)
experiment	not stable	0.2810	0.5525		0.00196
Silicon $Z=14$					
full core *LSD*		0.3027	0.5996	1.2339	0.1470[b]
NLPP + LSD		0.3034	0.5993	1.2212	-
PH + LSD		0.3034	0.5989	1.2205	-
PH + QMC	0.051 (1)	0.301 (1)	0.6057 (7)	1.2356 (5)	0.118 (3)
experiment	0.0509	0.2995	0.6007	1.2308	0.119 (4)
Chlorine $Z=17$					
full core *LSD*		0.4888	0.8819	1.4522	-
NLPP + LSD		0.4849	0.8750	1.4537	-
PH + LSD		0.4842	0.8720	1.4487	-
PH + QMC	0.138 (4)	0.475 (4)	0.878 (3)	1.461 (3)	0.08 (1)
experiment	0.1329	0.4765	0.8750	1.4556	0.0924

a: ref. [26]. *b*: ref. [27,28].

6. Further Refinements: *s,p* and *d* Exact

As already mentioned, if one looks for the exact solution of Eq. (13) using available *s*, *p* and *d* norm-conserving atomic wavefunctions (as obtained e.g. following ref. [21] or ref. [31]), then the resulting radial functions $a(r)$, $b(r)$ do not fulfill the conditions of Eq. (9). However, the pseudopotential constructions suggested by Hamann et al. in 1979 [21], or by Kerker in 1980 [31], are only two convenient procedures; many other valid schemes can in principle be conceived. Norm-conserving pseudopotentials are by no means uniquely defined inside the core, and, on the contrary, for each atom there exists a large class of equivalent norm-conserving pseudopotentials whose eigenfunctions fulfill Eqs.

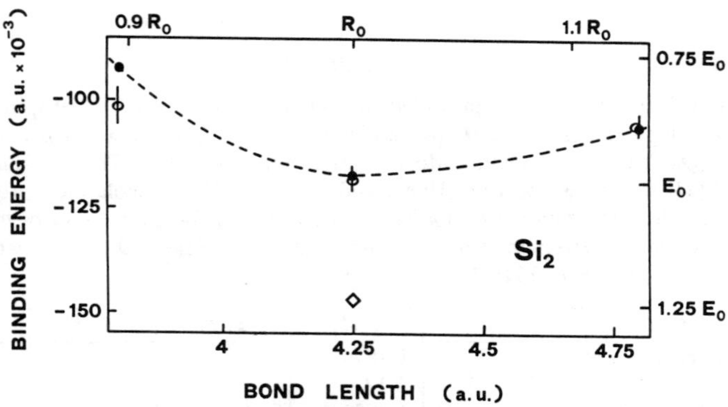

Figure 3. The energy of the Si_2 versus bond length as computed with $QMC-PH$ (symbols with error bars, Bachelet et al. [15] pseudo-Hamiltonian of Figure 1), and with the full-core CI method (dots, ref. [30]). The local-spin-density full-core result is shown as a diamond (ref. [27,28]). The experimental binding energy and equilibrium bond length are E_0 and R_0 respectively (ref. [24]).

(10,11) but differ from each other inside the core. So instead of keeping the existing pseudopotentials and giving up the d orbitals, as successfully done in the previous two sections for $s-p$ atoms, one may well consider a more general construction: starting from a full-core local-density-functional atom one has to find a set of s, p and d peudowavefunctions which not only enjoy the property of 'conserving the norm' Eqs. (10,11), but also, when plugged into Eq. (13), will yield radial functions $a(r)$, $b(r)$ and $v(r)$ which fulfill the conditions of Eq. (9). That such a set of peudowavefunctions exists at all is neither obvious nor easy to derive from our formulae: after expressing the conditions of Eq. (9) in terms of pseudowavefunctions one finds a rather involved nonlinear inverse-scattering problem. But the question is a very interesting one because it crucially affects the possibility of using the local pseudo-Hamiltonian of Eqs. (8,9) for d-metals. Two of us [32] have recently proposed to attack this problem from the numerical side, by a simulated-annealing technique [33-35]. The s,p and d radial wavefunctions are expressed in the Kerker form

$$\chi_l^{PS}(r) = r^{l+1}\exp[p_l(r)] \tag{14}$$

inside the core ($r<r_c$), but, instead of using polynomials or other analytic functions which would considerably limit the variational flexibility, the three optimal p_l functions ($l=0,1,2$) are searched among all the continuous bounded functions (whith continuous first and second derivatives) which fulfill the matching condition in r_c and the condition $p_l' = r^\alpha$, $\alpha \geqslant 1$ in the origin (which ensures that the pseudopotentials v_l and the radial functions a,b and v remain finite in the origin). On the computer these radial functions may be represented on a sufficiently dense regular grid. Both the norm conservation Eq. (11) and the conditions on a and b expressed by Eq. (9) are then imposed via simulated-annealing techniques: first a suitable energy (or cost) function is defined, which by construction is zero when all conditions are satisfied and is otherwise alway positive (C is a positive constant):

$$E_{cost} = E[a+1]+E[a+b+1]+\sum_{l=0,1,2} N_l \tag{15}$$

$$E[f] = C\int_0^{r_c} f(r)(\mathrm{sign}[f(r)]-1)dr \tag{16}$$

$$N_l = \left|\int_0^{r_c} \left\{[\chi_l^{PS}(r)]^2 - [\chi_l^{FC}(r)]^2\right\} dr\right|. \tag{17}$$

Then an appropriate dynamics in the space of the p_l' functions allows them to slowly relax towards the minimum-energy configuration. What is the meaning of the minimum-energy configuration? In principle there are two possibilities. If the minimum energy is $E_{\text{cost}}>0$, then the norm conservation Eq. (11) and the conditions expressed in Eq. (9) are incompatible with each other, and the sought solution does not exist; the corresponding set of three p_l's will only represent some trade off between Eq. (11) and Eq. (9). If, instead, the minimum energy is $E_{\text{cost}} = 0$, then the corresponding p_l's are such that the conditions of Eqs. (9) and (11) are fulfilled, and an exact solution to our problem has been found. In practice, on the computer, we will stop the simulation run as soon as E_{cost} reaches zero within some desired accuracy.

In passing we would like to point out that the power and generality of the proposed simulated-annealing method in the context of pseudopotential approaches is by no means exhausted by the present application. It is possible to find more appropriate representations of the involved functions and/or better strategy for movements in the configuration space; only the most necessary terms were incorporated in the definition of the energy Eqs. (15-17), but other terms can be included too - for example, terms which can increase the smoothness of the pseudo-Hamiltonian functions or which can limit a range of the function values. More generally, a very large variational freedom of the method can be used to further optimize traditional pseudopotentials; for example, to maximize their smoothness.

7. Open Problems for *d*-Metals

After solving the *d*-wave problem (applications to silicon will be presented in the next section) let us mention a severe problem which remains to be faced when studying transition elements, especially those of the first row (*Sc* through *Cu*); this problem, quickly mentioned at the end of Section 5, may also effect $s-p$ atoms with few valence electrons and a shallow underlying core (like *Na*). For these systems an accurate set of norm-conserving *s,p* and *d*-waves for a given valence configuration (the so-called *reference atomic configuration*, for which the pseudopotential and the pseudo-Hamiltonian are constructed) may not be sufficient to ensure optimum transferability, as pointed out by Louie et al.[29]: whenever significant core-valence overlaps occur, the linearization of the core-valence exchange potential, implicit in the 'unscreening' procedure, becomes a questionable approximation (see also ref. [36]). Within local-density-functional theory a cure to this problem was proposed by Louie et al. [29] and successfully applied by Greenside and Schlüter and Moullet *et al.* to first-row transition elements and cesium atoms and dimers, respectively [36]. This cure can be extended to variational *MC* [37] but again not to general *QMC* techniques, and something else has to be invented. One possibility is to give up pseudoatoms and try some frozen-core approximation, as proposed by Hammond et al. [16]; the comparison of their results with ours (the benchmark being the *Si* atom) suggests however that a much larger error bar will result. Another possibility, which two of us are presently testing for *Cu* [32], is to include the shallow cores in the valence shells, as e.g. done by Starkloff and Joannopoulos [38]. This, as expected, works quite well, but the price is the inclusion of eight more electrons per atom (the underlying $s-p$ core), which also will increase the error bar in *QMC* simulations. In few words, this is an open problem which requires new ideas.

8. Results for the Silicon Atom and Conclusions

We illustrate the results obtained by the method presented in Section 6 for the case of silicon. For the atom we have used the ionized configuration Si^{2+} $(3s^1 3p^{0.75} 3d^{0.25})$ with common $r_c = 1.75 a.u.$ After some experience we have realized that the conditions of Eq. (9) are much more difficult to fulfill than the norm conservation Eq. (11). Since then we have started the simulations from configurations which fulfill Eq. (9) but not the norm conservation, and used a value of $C = 1000$ in Eq. (16). In this way during the simulation run $a(r)+1$ and $a(r)+b(r)+1$ remained positive as required by Eq. (9), due to the strong effect of the term Eq. (16) in the energy-cost function, while the norms slowly converged towards their correct values. For the silicon atom we stopped the simulation when the value of E_{cost} was in the range of 10^{-4}, entirely due to small residual discrepancies in the norms.

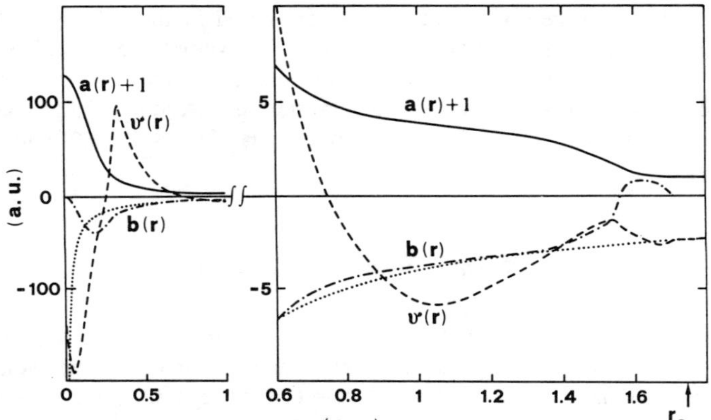

Figure 4. Silicon atom: the two kinetic-energy radial functions $a(r)+1$ (solid line) and $b(r)$ (dash-dotted), and the atomic potential, $v(r)$, which characterize our local pseudo-Hamiltonian Eq. (8). Here we show the silicon pseudo-Hamiltonian which fixes s and p *and* d states exactly, obtained by Mitáš and Bachelet (ref. [32]) by a simulated-annealing technique (see Section 6). Atomic units are used. Here, unlike Figure 1, the coulombic tail $-4/r$ is indicated by a dotted line.

The functions obtained are shown in Figure 4. The functions exhibit larger variations than usual pseudopotentials. However, the corresponding pseudo wavefunctions (Figure 5) are smooth and similar to those produced by 'traditional' norm-conserving pseudopotentials [21,31]. With this new set of a, b and v functions we have performed some preliminary diffusion *MC* simulations for Si^{2+}. This is especially simple because with two valence electrons the spatial part of the wavefunction never changes sign and one does not need any importance sampling; the diffusion *MC* is as simple as in the H_3^+ calculations of Anderson [4], except that a spatially varying mass has to be introduced. The result obtained for the third ionization potential 1.222±0.0221 a.u. is in excellent agreement with the experimental value of 1.230 a.u. [24]. Compared to straight diffusion *MC* there are higher demands in the computer time, essentially because of the rapidly varying mass inside the core.

The *Si* results just presented are evidently preliminary, but rather encouraging: the new local pseudo-Hamiltonian Eq. (8) can reproduce the action of a nonlocal norm-conserving pseudopotentials for all three partial waves in the test case of silicon and gives excellent results for the two-electron Si^{2+} pseudoatom. Tests of this new pseudo-Hamiltonian for less simple physical systems and for various atoms are underway.

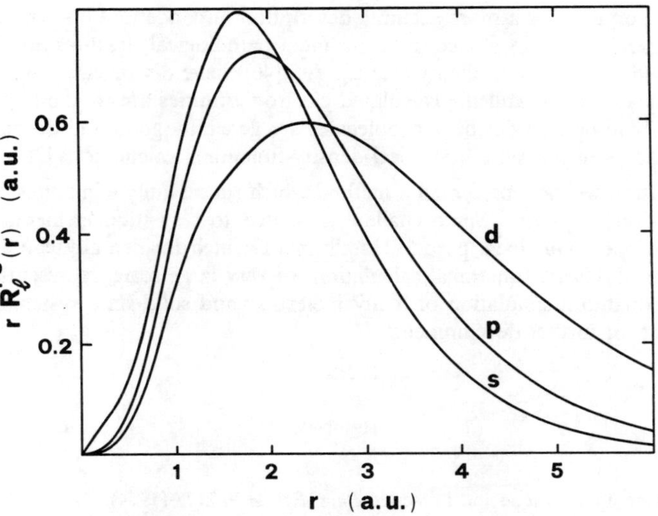

Figure 5. The radial wavefunctions $\chi_l(r) = rR_l(r)$ which correspond to the bound state of the pseudo-Hamiltonian shown in Figure 4 for $l = 0, 1, 2$. Atomic units are used.

A separate test which has its own interest concerns the one-electron excitations of the Si^{3+} pseudoatom. In Table III we compare the results of norm-conserving pseudopotentials, the pseudo-Hamiltonian Eq. (8) obtained in Section 4 (Figure 1), and the pseudo-Hamiltonian obtained here (Figure 5).

TABLE III

The energy of the Si^{3+}. *NLPP*: nonlocal pseudopotentials (ref. [13]); $PH-1$: *s,p* pseudo-Hamiltonian (ref. [15], Sections 4, 5 and Figure 1, this work); $PH-2$: *s,p,d* pseudo-Hamiltonian (ref. [32], Section 6 and Figure 4 of this work). Experimental values are from ref. [39].

	3s	3p	3d	4s	4p	4d	4f	5s
NLPP	-1.659	-1.341	-0.936	-0.781	-0.672	-0.522	-0.501	-0.454
$PH-1$	-1.654	-1.347	-0.851	-0.848	-0.710	-0.494	-0.496	-0.524
$PH-2$	-1.659	-1.348	-0.933	-0.766	-0.661	-0.519	-0.518	-0.443
Exper.	-1.659	-1.333	-0.928	-0.775	-0.664	-0.520	-0.501	-0.450

What we do is to solve the Schrödinger equation for one electron in the field of the bare pseudoion Si^{4+} in the ground state and various excited states, for which detailed experimental results are available [39].

Comparing the various pseudoions considered in Table III it appears that the *s,p,d* pseudo-Hamiltonian ($PH-2$) works much better than the $s-p$ pseudo-Hamiltonian ($PH-1$) for *d*-states, as expected. But besides that a more gneral message comes from Table III: the remarkable agreement with the experiment up to very high excitation energies suggests that bare pseudoions obtained from a local-density-functional atom behave almost as real physical objects; in other words, it confirms that the core-valence interaction may be well represented (and then unscreened) within single-particle approximations. This was also indirectly confirmed by the good results of our $QMC-PH$ calculations for many-electron atoms and molecules (Table II, Figure 3 and this section), which, on the other hand, also showed that an exact description of valence-valence interaction is of crucial importance for a more accurate description of condensed matter systems. By an exact treatment of valence electrons, in fact, pathological features of local-density-functional and Hartree-Fock theory (see e.g. ref. [40]) have disappeared: for example, our negative ions are bound and the calculated electron affinities are in good agreement with experiment; binding energies of our molecules are generally good and do not suffer from the overbinding which results from local-density-functional calculations [26-28].

In conclusion we have presented a method which successfully eliminates core electrons from $s-p$ atomic systems. Some challenges related to transition elements and shallow cores remain open. Our local pseudo-Hamiltonian, which incidentally can also speed up traditional local-density-functional calculations of very large scale, represents a promising tool for the quantum simulation of many molecular and solid-state systems, and a good starting point for further developments.

References

1. M.H. Kalos, D. Levesque and L. Verlet, *Phys. Rev. A* **9**, 2178 (1974).
2. D.M. Ceperley and B.J. Alder, *Phys. Rev. Lett.* **45**, 566 (1980).
3. D.M. Ceperley and B.J. Alder, *Phys. Rev. B* **36**, 2092 (1987).
4. J.B. Anderson, *J. Chem. Phys.* **63**, 1499 (1975).
5. P.J. Reynolds, D.M. Ceperley, B.J. Alder and W.A. Lester, *J. Chem. Phys.* **77**, 5593 (1982).
6. J.W. Moskowitz, K.E. Schmidt, M.A. Lee and M.H. Kalos, *J. Chem. Phys.* **77**, 349 (1982).
7. D.M. Ceperley, *J. Stat. Phys.* **43** 815 (1986).
8. B.L. Hammond, P.J. Reynolds and W.A. Lester, Jr., *J. Chem. Phys.* **87**, 1130 (1987).
9. M.L. Cohen and V. Heine, in: *Solid State Physics* **24**, 38, edited by H.E. Ehrenreich, F. Seitz and D. Turnbull, Academic, New York, 1970.
10. V. Heine and D. Weaire, in: *Solid State Physics* **24**, 249, edited by H.E. Ehrenreich, F. Seitz and D. Turnbull, Academic, New York, 1970.
11. G.F. Bassani and G. Pastori Parravicini, *Electronic States and Optical Transitions in Solids*, Pergamon Press, Oxford, 1974.
12. M. Schlüter and L.J. Sham, *Phys. Today*, Febraury 1982.
13. G.B. Bachelet, D.R. Hamann and M. Schlüter, *Phys. Rev. B* **26**, 4199 (1982).
14. S. Fahy, X.W. Wang and S.G. Louie, *Phys. Rev. Lett.* **61**, 1631 (1988).
15. G.B. Bachelet, D.M. Ceperley and M.G.B. Chiocchetti, *Bull. Am. Phys. Soc.* **33**, 567 (1988); *Phys. Rev. Lett.* **62**, 2088 (1989).
16. B.L. Hammond, P.J. Reynolds and W.A. Lester, Jr., *Phys. Rev. Lett.* **61**, 2312 (1988).
17. D.M. Ceperley and M.H. Kalos, in: *Monte Carlo Methods in Statistical Physics*, edited by K. Binder, Springer-Verlag, Berlin, 1979.
18. K.E. Schmidt and M.H. Kalos, in: *Monte Carlo Methods in Statistical Physics II*, edited by K. Binder, Springer-Verlag, Berlin, 1984.
19. D.M. Ceperley and B.J. Alder, *J. Chem. Phys.* **81**, 5833 (1984).
20. M.M. Hurley P.A. Christiansen, *J. Chem. Phys.* **86**, 1069 (1987).
21. D.R. Hamann, M. Schlüter and C. Chiang, *Phys. Rev. Lett.* **34**, 1494 (1979).
22. L. Landau and E. Lifschitz, *Mécanique Quantique* MIR, Moscow (1966).
23. F. Gygi and A. Baldereschi, *Phys. Rev. B* **34**, 4405 (1986).
24. R.C. Weast (editor in chief), *CRC Handbook of Chemistry and Physics*, 69th edition (1988-

1989), Boca Raton, Florida.

25. K.P. Huber and G. Herzberg, *Molecular Structure and Molecular Spectra. IV. Constants of Diatomic Molecules*, Van Nostrand Reinhold, New York, 1979.
26. J.L. Martins, J. Buttet and R. Car, *Phys. Rev. B* **31**, 1804 (1985).
27. R.O. Jones, *Phys. Rev. A* **32**, 2589 (1985).
28. R.O. Jones, in: *Electronic Structure Calculations*, edited by K.P. Lawley, Wiley, New York, 1987.
29. S.G. Louie, S. Froyen and M.L. Cohen, *Phys. Rev. B* **26**, 1738 (1982).
30. A.D. McClean, B. Liu and G.S. Chander, *J. Chem. Phys.* **80**, 5130 (1984).
31. G.P. Kerker, *J. Phys. C* **13**, L189 (1980).
32. L. Mitáš and G.B. Bachelet, unpublished (1989).
33. S. Kirkpatrick, G.D. Gelatt, Jr. and M.P. Vecchi, *Science* **220**, 671 (1983).
34. V. Černý, Preprint 1982 and *J. Optim. Theory Appl.* **45**, 41 (1985).
35. V. Černý, in: *Proceedings of the IBM Europe Summer School*, Oberlech 1986, and private communication.
36. H.S. Greenside and M. Schlüter, *Phys. Rev. B* **28**, 535 (1983); I. Moullet, W. Andreoni and P. Giannozzi, *J. Chem. Phys.*, in print (1989).
37. S.G. Louie, private communication 1989.
38. Th. Starkloff and J.D. Joannopoulos, *Phys. Rev. B* **16**, 5212 (1977).
39. S. Bashkin and J.O. Stoner, Jr., *Atomic Energy Levels and Grotrian diagrams 1*, North-Holland, Amsterdam, 1975.
40. J. Perdew and A. Zunger, *Phys. Rev. B* **23**, 5048 (1981).

ON THE ELECTRONIC AND STRUCTURAL PROPERTIES OF SMALL CLUSTERS

WANDA ANDREONI

IBM Research Division
Zurich Research Laboratory
CH-8803 Rüschlikon
Switzerland

1. Introduction

A large interest has developed during the last five years on the physics and chemistry of small atomic aggregates in the size range of $2-100$ atoms [1]. This was due to the advent of the laser vaporization technique to produce microclusters of a large variety of materials and to the advance in the direct experimental investigation of size-selected clusters in the gas phase [2]. The progress in cluster physics has provided new clues to surface science, especially in relation with research in catalysis and microelectronics. The study of the clusters in beam has also renewed the hope of obtaining better insight into fundamental questions, such as the evolution of the electronic and structural properties of a material with increasing size of the atomic aggregate, from the molecule to the solid.

The best-know result comes from abundance spectra, which show that for a given material special sizes N exist, at which the aggregate is more abundant: These are called 'magic numbers'. Figures 1 and 2 show the abundance spectrum of *Na* [3] and *Si* [4], respectively, where the higher peaks are labelled. The magic numbers seem to be characteristic of the type of bonding, i.e. they are roughly the same for *Si* and *Ge*, and for *Na* and *K*. They indicate a special stability of the N-atom cluster with respect to others. However, in most cases the reason for this is unclear. Moreover, it is fair to say that the experimental study of the electronic properties of the 'free' microclusters is still in its infancy and that the knowledge of structural properties is practically undeveloped.

Such a situation calls for theoretical approaches which allow one to study both electronic and structural properties in a reliable way. A considerable effort has been made with traditional calculations of quantum chemistry [5] and also of solid-state physics [6] to determine the relevant structures for a number of clusters of up to ~ 10 atoms. However, the standard approaches suffer from severe limitations in the search for the equilibrium structures and rely to a large extent on the intuition of the researcher. Recently Giorgio Pastore, Pietro Ballone, Roberto Car, Michele Parrinello and myself have made some progress in the theoretical approach to cluster physics [7] with the aid of computer simulations using the method introduced by Car and Parrinello [8]. This method allows one to perform molecular dynamics calculations with forces derived from the electronic ground state, calculated within the Born-Oppenheimer approximation and within the Density Functional formalism. The search for the equilibrium structure is made with simulated annealing strategies, which provide a powerful tool for the sampling of the potential energy surface of the cluster. In addition, we have studied the evolution of a number of properties with increasing temperature. This is interesting since the temperature of the cluster is generally unknown from experiment but may affect its behavior [9].

R. Girlanda et al. (eds.), Progress on Electron Properties of Solids, 27–40.

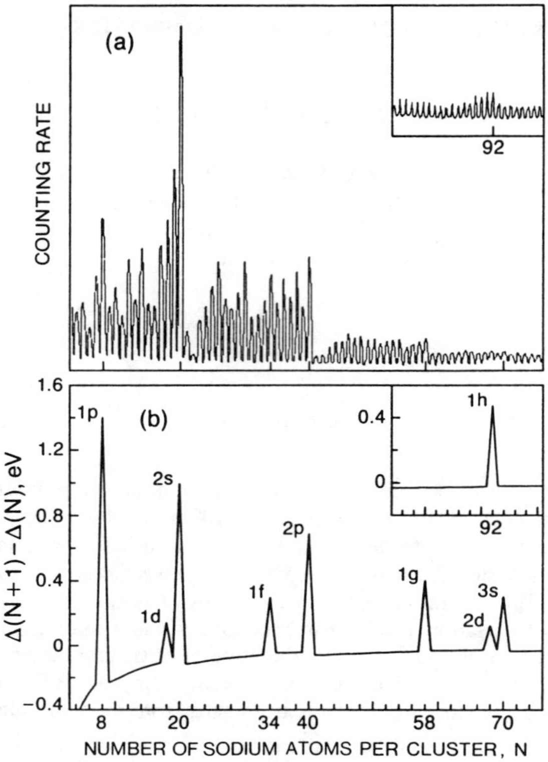

Figure 1. a) Abundance spectrum of sodium clusters and b) results of the jellium-shell model (see text). From [3], © 1984 The American Physical Society.

Our own experience has developed on alkali-metal and semiconductor microclusters. Here, I shall focus on the description of some results on the prototypical clusters, *Na* and *Si*, where I can compare our findings to those of other methods and also to experiment. In the size range up to 10 − 20 atoms, our results suggest some ideas on the evolution of the bonding properties with the size. I shall contrast them with previous views. In particular, I shall discuss the electronic properties of sodium clusters and contrast them with the predictions of the shell model [10]. Such a scheme is particularly suitable for the understanding of the size evolution of the electronic structure. I shall show how it is possible to extend it to clusters other than simple metals, and shall apply it to silicon. For the details of the computational method, I refer to the original articles (Refs. [11-13] and [7]).

2. Alkali-Metal Clusters and the Shell Model

Lithium, sodium and potassium clusters have been studied for many years in a number of laboratories, since their production requires only conventional oven sources [14]. However, a large interest has flourished only after Knight and coworkers [3,10] showed how the abundance spectra can be clearly explained in terms of simple physical models. Figure 1(a) shows the mass spectrum and the existence of a well-defined shell structure

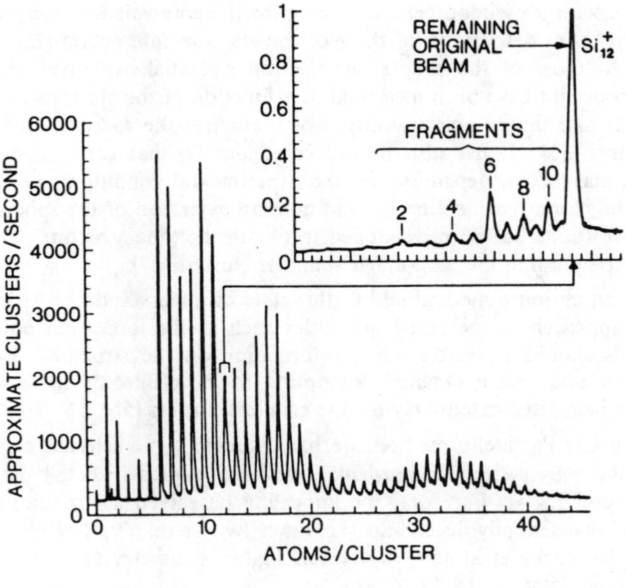

Figure 2. Abundance spectrum of silicon clusters. The insert shows the fragmentation spectrum of Si_{12}^{+}. From Bloomfield et al. [2], © 1985 The American Physical Society.

marked by the magic numbers 8,18,20 etc. Figure 1(b) contains the results of the first simple paradigma introduced to explain these magic numbers and contains the essence of the shell model. It gives the size variation of an energy parameter which contains information about the shell closing and is related to the second derivative of the total energy $E(N)$:

$$\Delta_2 E(N) \equiv \Delta(N+1) - \Delta(N),$$

where

$$\Delta(N) \equiv E(N+1) - E(N). \tag{1}$$

The basic idea of the jellium-shell model is that the valence electron states of sodium clusters are spatially delocalized, as can be expected from the knowledge of the solid, and that they experience a spherically symmetric potential. The former assumption allows one to believe that the electronic structure is to a good approximation independent of the geometry, and that on passing from the N-atom to the $N+1$-atom cluster, the dominant variation in the total energy is the one-electron energy term. This latter assumption allows one then to introduce a simple potential for the electrons in the cluster and to classify the one-electron eigenstates in terms of the angular momentum l. Such a classification naturally gives the shell structure of the eigenvalues, with closing corresponding to a number of electrons $N_{el} = 2, 8, 18, 20$ etc. In the case of an alkali metal, only one valence electron per atom exists, and thus $N = N_{el}$. The agreement with the observed magic numbers is spectacular and shows that the stability of the aggregate is indeed determined by the closing of the electronic shells. This provides a kind of 'giant atom' picture of the cluster [9] and stability criteria analogous to those long familiar in nuclear physics.

More sophisticated calculations of $\Delta_2 E(N)$ (Eq. (1)) have been performed, within the jellium model, with a pseudopotential scheme for the core-valence interaction and with a local density (*LDA*) description of the electron-electron interaction [15]. The results are very similar to those of the simple one-electron potential explained above. By now a number of properties have been measured as a function of the cluster size, such as ionization potentials and dipole polarizability, which confirm the special nature of the magic-number clusters [16]. It has also been pointed out [9] that some modifications of the mass spectra may occur, depending on the experimental conditions, i.e. the nozzle temperature or the carrier gas cooling. In addition, an extension of the spherical shell model to include distortions has been developed to obtain the 'fine structure' of the abundance spectra corresponding to the 'non-magic'-number clusters [17].

What can an ab-initio method add to this state of understanding? Apart from a better quantitative approach to measured quantities such as the ionization potentials, the ab-initio methods should at least provide information on the structural properties, assess their relevance, and give a detailed description of the electronic structure. Traditional methods have been used extensively for the smallest clusters [5b].

Within the Car-Parrinello method, we have studied some selected clusters of sodium and also of the *NaK* system. In particular, we have studied in detail the magic number clusters Na_8 and Na_{20}. For Na_8, the annealing has given a structure which has D_{2d} symmetry and is essentially degenerate (i.e. lower by less than 0.01 eV/atom) with the one proposed by Koutecky et al. [5b], which has higher symmetry (T_d). The former may in fact be seen as a distortion of the latter.

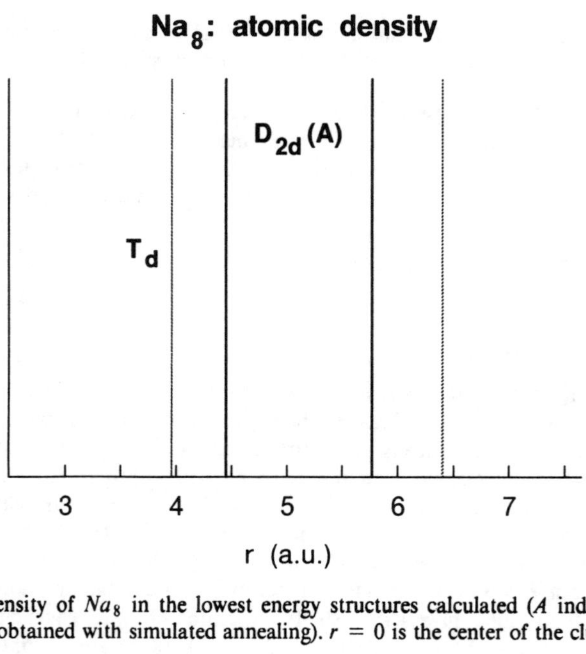

Figure 3. Atomic density of Na_8 in the lowest energy structures calculated (*A* indicates that the D_{2d} modification is obtained with simulated annealing). $r = 0$ is the center of the cluster.

Figure 3 illustrates the atomic density of both structures. Although they are significantly different, in both cases the atoms are arranged in two shells, the average distance from the center is ~5 a.u., the bond angle distribution is dominated by angles around 60° (2/3 are within 90°), the average coordination number is 4.5. Simulated annealing for Na_{20} has given a rather complex structure, which appears in Figure 4.

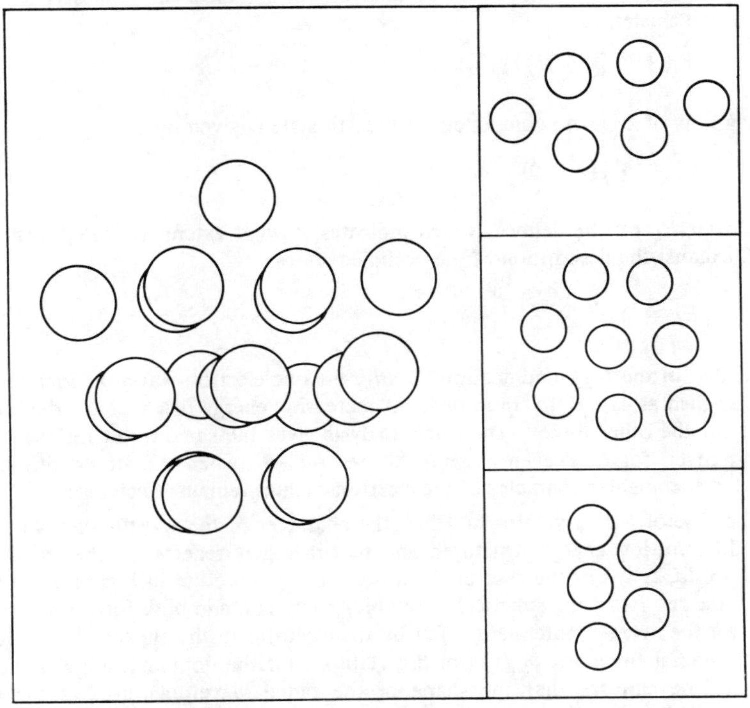

Figure 4. Lowest-energy structure of Na_{20} with, on the right, its decomposition in planes. From [12], © Les Editions de Physique 1989.

It contains a number of interesting features: It 'grows' in layers, as illustrated on the right-hand side of Figure 4, with a type of stacking that, apart from distortions due to the small size, is reminiscent of the hexagonal close packed, has one 12-fold coordinated atom near the center, and is 'compact' (in the sense that the bond angle is dominantly 60°). We have found that a number of local minima exist for Na_{20} within a binding energy range of $0.04\,eV/\text{atom}$. As expected, the potential energy surface around the minima appears to be very flat. This characteristic, which is likely to be typical of the alkali-metal clusters (although more pronounced for Na), also has as a consequence the fact that even at relatively low temperatures the atoms do not oscillate around specific valleys, but rather change their positions significantly. The obvious consequence of this is that traditional static calculations of special geometries for these aggregates are not sufficient to provide an adequate picture.

We have examined the electronic structure with the purpose of assessing the validity of the shell model picture. Generally, one looks at the one-electron energy levels with the idea that the grouping of levels (degeneracy and well-defined gaps) gives a quantitative idea of the validity of the shell model. This however is rather restrictive and may also be disappointing. In fact, let us consider for instance Na_8. A spherical potential would predict a 1s and a 1p threefold degenerate level. In the D_{2d} symmetry, however, the degeneracy of the p-levels is lifted and results in a splitting of a singlet and doublet separated by $\sim 0.4 eV$. This does not mean that the shell model is not correct. I think that a better way of answering this issue is in fact to examine the angular character of the one-electron states $\Psi_i(\mathbf{r})$ which can be analyzed in spherical harmonics centered at the center of the cluster, as

$$\Psi_i(\mathbf{r}) = \sum_{l,m} \psi^i_{lm}(r) Y_{lm}(\Omega). \tag{2a}$$

The weight w^i_l of a given l-component of the i-th state is given by

$$w^i_l = \sum_m \int (\psi^i_{lm}(r))^2 r^2 dr. \tag{2b}$$

A parameter E_L can be defined, which indicates to what extent the components up to a given L exhaust the description of the occupied states:

$$E_L = 1 - \left[\sum_{i_{occ}} \sum_{l=0}^{l=L} w^i_l \right] / N_{el}. \tag{2c}$$

We find that in the D_{2d} modification of Na_8, the one-electron wavefunctions can be perfectly classified as 1s, 1p, 1p, 1p in order of increasing energy ($w_0 = 1$ for the lowest one, $w_1 = 1$ for the other three). The same analysis gives the same result for the T_{2d} structure. Moreover, for Na_8, even at temperatures $\sim 600K$, when the atoms diffuse all over the cluster, the angular character of the electronic states remains unchanged.

In the case of Na_{20}, we also find that the character of the wavefunctions is the same for the different low-energy structures, and that it largely reflects the shell-model prediction (1s, 1p, 1d, 2s). For the case of the lowest energy structure in Figure 4, we report in Figure 5 the results of the spherical harmonics analysis made both for the wavefunctions (b) and for the average potential (a) felt by the electrons in the cluster. In particular, we report the radial functions $\psi_{lm}(r)$ (for a certain m) for the dominant angular momentum $l(w_l \simeq 1)$. One can see that the shape of the radial wavefunctions has the expected behavior and that the electronic potential is essentially spherical. However, in contrast to the case of Na_8, when the temperature increases, thus changing the atomic configurations more drastically, the orbitals at higher energy hybridize. Variation with respect to the predictions of the shell model are invariably small for the 20-atom clusters, i.e. also in K_{20} where the effect of the pseudopotential (and therefore of the structure) is more significant, and even in the $Na_{10}K_{10}$ cluster, where the structure is drastically modified owing to segregation effects which bring the K atoms to the 'surface' of the cluster.

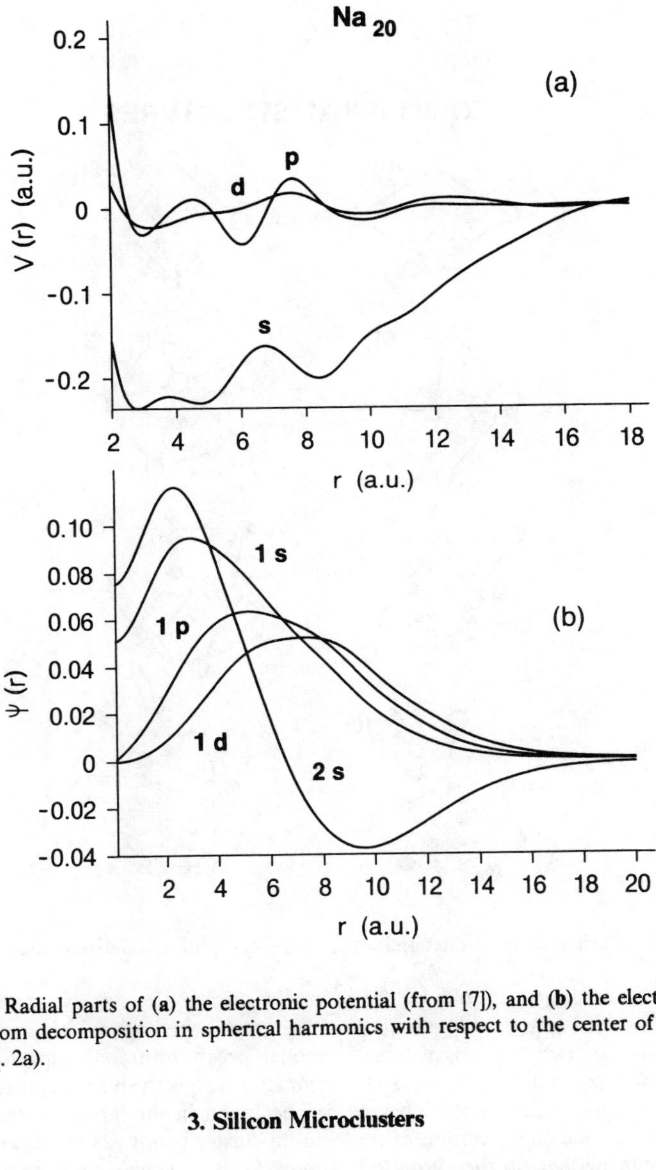

Figure 5. Na_{20}: Radial parts of (**a**) the electronic potential (from [7]), and (**b**) the electronic orbitals, obtained from decomposition in spherical harmonics with respect to the center of the cluster (see text and Eq. 2a).

3. Silicon Microclusters

Recently, a large literature has flourished on silicon microclusters. Beyond the specific interest in cluster physics, they are the benchmark of all the empirical interatomic potentials proposed during the last three years for silicon [18]. As the disappointing outcome of this work has demonstrated, ab-initio calculations are necessary to obtain a correct description of the size effects in silicon. In fact, the chemical bonding changes drastically on passing from the solid to the small clusters. We have found that a crystal fragment is unstable up to the largest size we have been able to calculate, $N = 40$. So far, our most detailed calculations are in the range up to 11. The lowest energy structures, which have come out of the simulated annealing, are displayed in Figure 6.

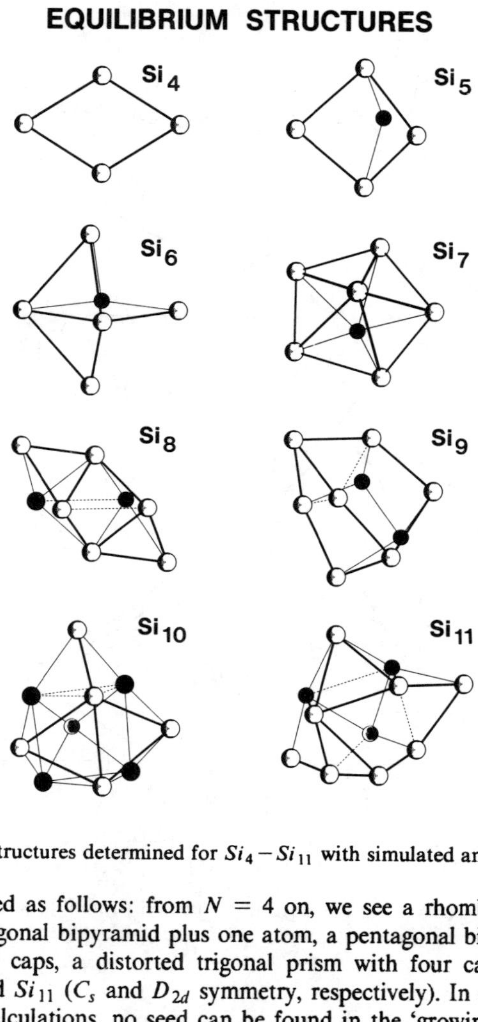

Figure 6. Equilibrium structures determined for $Si_4 - Si_{11}$ with simulated annealing, from [7].

They can be described as follows: from $N = 4$ on, we see a rhombus, a trigonal bipyramid, a distorted trigonal bipyramid plus one atom, a pentagonal bipyramid, a distorted octahedron with two caps, a distorted trigonal prism with four caps for Si_{10}, unique structures for Si_9 and Si_{11} (C_s and D_{2d} symmetry, respectively). In contrast to the predictions of previous calculations, no seed can be found in the 'growing' sequence of these aggregates, i.e. the configuration of the N-atom cluster is not easily related to that of the $N-1$ cluster, neither do they grow by capping faces or simple atom insertions. In comparing the total energy of different relevant structures for a given size, we find that high symmetry is again far from being a good criterion in determining the lowest energy one, while compactness (i.e. smaller average distance from the center of the cluster and dominance of bond angles $<90^{\circ}$) is a good criterion. An interesting case is that of Si_{10}, for which three local minima are shown in Figure 7. The tetracapped trigonal prism (*TTP*) is the result of the annealing. The prism is distorted since the presence of the cap atom makes the two triangular bases inequivalent. The tetracapped octrahedron (*TO*) is the proposal of earlier work [5a] using configuration interaction calculations and corresponds

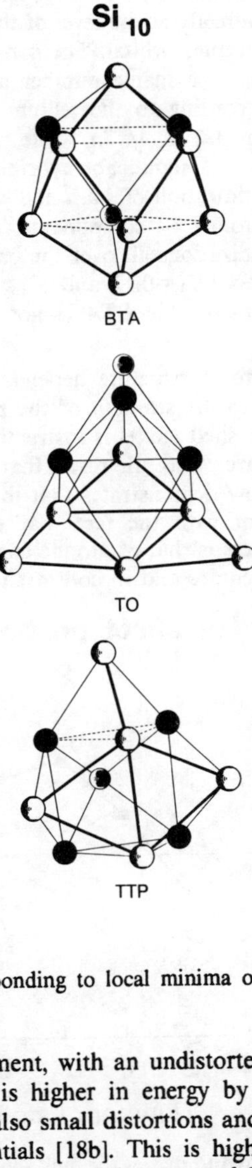

Figure 7. Structures of Si_{10} corresponding to local minima of the potential energy surface (see text).

to a highly symmetrical arrangement, with an undistorted octahedron in the center. We find that in our approach this is higher in energy by $\sim 0.06\, eV/\text{atom}$. The bicapped tetragonal antiprims (BTA) has also small distortions and seems to correspond closely to the prediction of classical potentials [18b]. This is higher in energy than the TTP by $\sim 0.09\, eV/\text{atom}$. By comparison, a crystal fragment of 10 atoms (the 'admantane' cage) is higher by $\sim 0.8\, eV/\text{atom}$ and moreover does not correspond to a local minimum.

The examination of the electron densities of all these clusters shows some peculiar features, such as the presence of a hole in the center of the cluster, with the exception of Si_7. The bonding is less directional than in the bulk, and does not involve hybridization of the sp^3-type. In fact, the tetrahedral angle is not present in the bond angle distributions. Since the jellium model is bound to be a poor approximation for silicon aggregates in any case, nobody has attempted to examine them in terms of a shell model. However, the results of our calculations, when analyzed according to Eqs. (2), give useful information. Si_5 is isoelectronic with Na_{20} but is not a magic number (see Figure 2). The one-electron states do not present a closed shell configuration. However, the lowest energy states can still be classified as $1s$ and $1p$, and three states over the expected five have d-character. Hybridization starts already at the level of the fifth energy level, i.e. at $\sim 1.5\,eV$ below the highest occupied molecular orbital. The contribution of f and g-components is significant, i.e. $E_2 = 0.09$. Si_{10} is a magic number and is isoelectronic with a magic-number Na cluster (Na_{40}). According to the jellium shell model, the sequence of the occupied states should be $1s$, $1p$, $1d$, $2s$, $1f$, $2p$, while the first unoccupied state would be of the $1g$ type. We find that in Si_{10} significant hybridizations exist, but that most of the states can be derived from hybridization of s,p,d and f states, with an extra contribution of 4% of $l = 4$ and only 3% from $l>4$ components ($E_3=0.07$ and $E_4=0.03$). The TO structure implies stronger hybridizations all over the band: Again there is a 7% contribution of $l = 4$ and 5% from $l>4$ ($E_3=0.12$ and $E_4=0.05$). In the BTA structure, the hybridization is weak; however also the BTA is not closed shell, i.e. $E_3 = 0.07$ and $E_4 = 0.03$.

In silicon clusters, the electronic structure depends significantly on the geometry, in contrast to Na clusters, owing to the strength of the pseudopotential. However, I think that the picture provided by the shell model is instructive in understanding how the cluster electron states evolve with size. Also, the result that the contribution from the excited states of the jellium model is low demonstrates that the electron orbitals are moderately localized. This is in agreement with the fact that the dominant structural criterion governing both Si and Na clusters is that of atomic compactness, in analogy with what a 'metallic'-type bonding would require and in contrast to what a covalent bonding would imply [19].

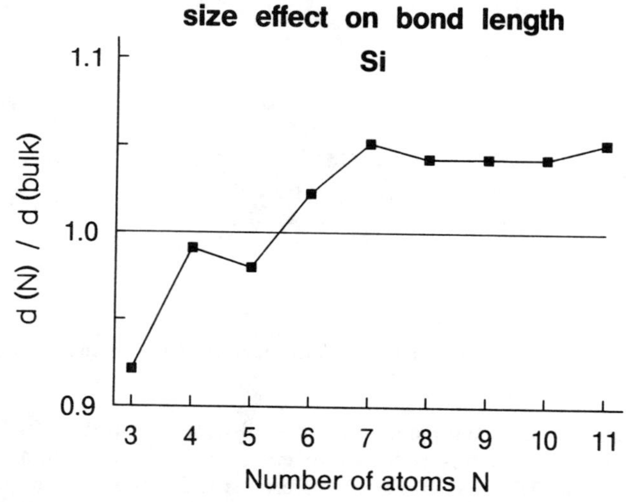

Figure 8. Average bond length referred to that of the bulk solid.

It is also interesting to look at the trends in the structural characteristics as well as in the binding energy with respect to the solid. Figure 8 and 9 illustrate, respectively, the variation of the average bonds length and of the binding energy referred to the values calculated with the same scheme for bulk Si. Apart from the trimer, the variation of the bond length is minor. The fact that from $N = 6$ on, the bond length is larger than in the solid is related to the fact that the average coordination number becomes larger than 4. The binding energy, instead, is dramatically affected by the small size, as expected.

A number of calculations have been made for the fragmentation energies of a given cluster, with the hope of giving an explanation for the magic numbers [20], in the sense

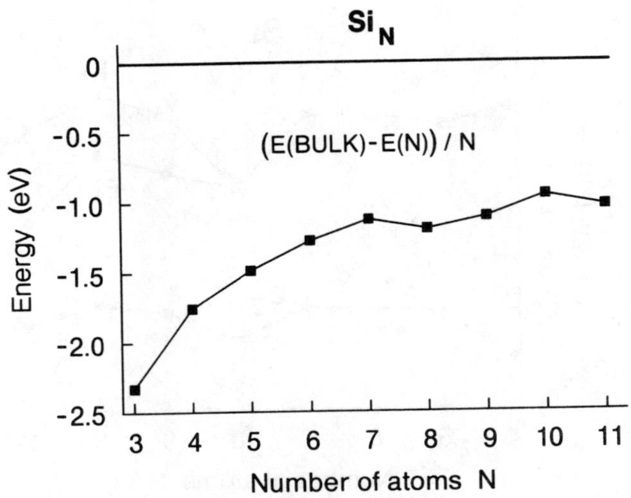

Figure 9. Binding energy referred to that of the bulk solid.

that a larger value implies larger stability. When compared with previous calculations, we find that the better determination of the equilibrium structures made with the simulated annealing, has the effect of smoothing the size dependence of the fragmentation energies. In Figure 10 we report the fragmentation energies

$$F1(N) \equiv E(N) - E(N-1) - E(1) \tag{3a}$$

and

$$F2(N) \equiv E(N) - E(N-M) - E(M), \tag{3b}$$

which correspond to dissociation in the $N-1$ cluster $+1$ atom, and to the most favorable channel, respectively. The two coincide for N up to 8. The qualitative aspect does not change on passing from $F1(N)$ to $F2(N)$, and shows that the existence of the prominent 6 and 10 peaks in the mass spectrum (Figure 2) is not fully explained by this type of approach. This is not surprising, since also thermodynamic and kinetic effects are expected to play a role in the dissociation process. Instead, the particular instability of Si_8 is well reproduced. In the search for a stability criterion, one has often looked at the energy gaps between the occupied and the unoccupied states. This criterion is again based on a naive one-electron picture and is close to that of a shell model. The energy gap is generally strongly underestimated by *LDA* caclulations and is not meaningful per se. However, we believe that the trend with size may be significative. Figure 10 shows that a kind of correlation exists with the fragmentation energy. It changes even more smoothly with the number of atoms and is again relatively narrow for Si_8.

The effects of ionic motion on the structural and electronic properties of these clusters have also been studied [11] for different temperatures. In contrast to the case of sodium, the atoms here tend to oscillate around certain positions up to high temperatures of the order of the melting point of bulk silicon.

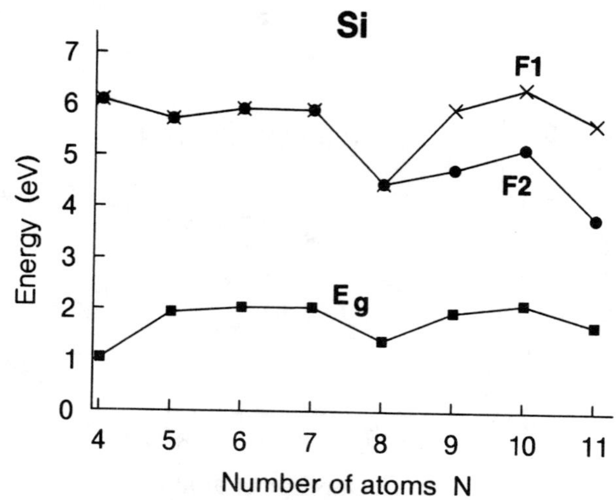

Figure 10. Size variation of fragmentation energies (Eq. 3) and of the lowest energy gap.

4. Conclusions

A systematic and exhaustive theoretical work of either sodium or silicon microclusters is still lacking, especially for the interesting regime of 20-100 atoms. However, insight into ground-state properties, both structural and electronic, starts to deepen. In particular, a number of misleading statements on the size evolution of the equilibrium structures, which originated from the limited reliability of traditional calculations in the search for the ground state, can now be revised. One example is the supposed 'growth' from a common seed [5a,6], or from tetrahedral subunits [5b,21]. Another example is the confusion between the compactness of the atomic arrangement and its high symmetry [5b,21], and the resulting idea that Jahn-Teller distortions present in the small aggregates are in conflict with the former. Clear examples mentioned above are Na_8 and Si_{10}. Moreover, models have been suggested in the search for the critical size at which the transition from close-packed to diamond structure occurs in silicon clusters (see Ref. [19] and references therein). The prediction of the critical size is highly model-dependent. Usually this is derived from comparison of the total energies of crystal fragments of face-centered cubic and diamond crystals. We find that up to $N = 40$ (where the latest model would locate the transition) neither of the two crystal fragments is stable [22].

The jellium-shell model is found to be a good approximation for the description of the electronic structure of the sodium clusters. It is shown to be a useful scheme for the classification of the electron orbitals also in silicon clusters, and especially for the understanding of the evolution of the occupied states with increasing size.

Acknowledgements

It is a pleasure to dedicate this paper to Franco Bassani, who has introduced me to research in solid state physics and has conveyed to me the enthusiasm which has accompanied me during all these years.

References

1. a) Microclusters, Springer Series in Material Sciences, vol. IV, edited by S. Sugano, Y. Nishina and S. Ohnishi, Springer-Verlag, Berlin Heidelberg, 1987.
   b) Elemental and Molecular Clusters, Springer Series in Material Sciences, vol. VI, edited by G. Pacchioni, T.P. Martin and G. Benedek, Springer-Verlag, Berlin Heidelberg, 1988.
2. R.E. Smalley, *Laser Chem.* **2**, 167 (1983); J.B. Hopkins, P.R.R. Langridge-Smith, M.D. Morse and R.E. Smalley, *J. Chem. Phys.* **78**, 1627 (1983); L.A. Bloomfield, R.R. Freeman and W.L. Brown, *Phys. Rev. Lett.* **54**, 2246 (1985); W.D. Rents, A.M. Mujsce, V.E. Bondybey and M.L. Mandich, *J. Chem. Phys.* **86**, 5568 (1987) and references therein; J.L. Elkind, J.M. Alford, F.D. Weiss, R.T. Laaksonen and R.E. Smalley, *ibid.* **87**, (1987) and references therein; O. Cheshnovsky, S.H. Yang, C.L. Pettiette, M.J. Craycraft, Y. Liu and R.E. Smalley, *Chem. Phys. Lett.* **138**, 119 (1987); S.H. Yang, C.L. Pettiette, J. Conceiao, O. Cheshnovsky and R.E. Smalley, *ibid.* **139**, 233 (1987); see also: G. Delacretaz, P. Fayet, J.P. Wolf and L. Wöste, in ref. 1(b), pp. 64-95; W. Begemann, S. Dreihofer, G. Gantefor, H.R. Siekmann, H.K. Meiwes-Boer and H.O. Lutz, *ibid.*, pp. 230-253.
3. W.D. Knight, K. Clemenger, W.A. de Heer, W.A. Saunders, M.Y. Chou and M.L. Cohen, *Phys. Rev. Lett.* **52**, 2141 (1984).
4. See L.A. Bloomfield et al. in ref. [2].
5. a) K. Raghavachari, *J. Chem. Phys.* **84**, 5672 (1986); K. Raghavachari and C.M. Rohlfing, *ibid.* **89**, 2219 (1988).
   b) I. Boustani, W. Pewestorf, P. Fantucci, V. Bonacic-Koutecký and J. Koutecký, *Phys. Rev. B* **35**, 9437 (1987); V. Bonacic-Koutecký, P. Fantucci and J. Koutecký, *ibid.* **37**, 4369 (1988).
   c) G. Pacchioni and J. Koutecký, *J. Chem. Phys.* **84**, 3301 (1986).
   d) M.H. McAdon and W.A. Goddard III, *Phys. Rev. Lett.* **55**, 2563 (1985); C.W. Bauschlicher, Jr., and L.G.M. Pettersson, *J. Chem. Phys.* **84**, 2226 (1986); T.H. Upton, *ibid.* **86**, 7054 (1987); B.K. Rao and P. Jena, *Phys. Rev. B* **37**, 2867 (1988) and references therein.
6. D. Tománek and M.A. Schlüter, *Phys. Rev. B* **36**, 1208 (1987) and references therein.
7. W. Andreoni, Proc. Int'l School of Physics, E. Fermi: *The Chemical Physics of Atomic and Molecular Cluters*, Varenna, Italy, 1988, in press.
8. R. Car and M. Parrinello, *Phys. Rev. Lett.* **55**, 2471 (1985); idem. in: Proc. of the NATO ARW: Simple Molecular Systems at Very High Density, Les Houches, France, NATO ASI Series, Plenum, New York, 1988.
9. W.A. de Heer and W.D. Knight, in ref. 1(b), pp. 45-63.
10. M.L. Cohen, in ref. 1(a), pp. 2-9.
11. P. Ballone, W. Andreoni, R. Car and M. Parrinello, *Phys. Rev. Lett.* **60**, 271 (1988).
12. P. Ballone, W. Andreoni, R. Car and M. Parrinello, *Europhys. Lett.* **8**, 73 (1989).
13. W. Andreoni, G. Pastore, R. Car, M. Parrinello and P. Giannozzi,
    a) Proc. of the NATO ARW: *Band Structure Engineering in Semiconductor Microstructures*, edited by R.A. Abram and M. Jaros, Plenum, New York, in press;
    b) *Proc. of the XIX Int'l Conf. on the Physics of Semiconductors*, Pland, 1988, edited by W. Zawadski, in press.
14. M.M. Kappes, M. Schär and E. Schumacher, *J. Chem. Phys.* **91**, 658 (1987) and references therein.
15. M.Y. Chou, A.N. Cleland and M.L. Cohen, *Solid State Commun.* **52**, 645 (1984); M.Y. Chou and M.L. Cohen, *Phys. Lett.* **113A**, 420 (1986).
16. W.D. Knight, W.A. de Heer, K. Clemenger and W.A. Saunders, *Solid State Commun.* **53**, 445 (1985); W.A. Saunders, K. Clemenger, W.A. de Heer and W.D. Knight, *Phys. Rev. B* **32**, 1366 (1985); W.A. de Heer et al., *Phys. Rev. Lett.* **59**, 1805 (1987).
17. K. Clemenger, *Phys. Rev. B* **32**, 1359 (1985).

18. a) E. Blaisten-Barojas and D. Levesque, *Phys. Rev. B* **34**, 3910 (1986); B.P. Feuston, R.K. Kalia and P. Vashishta, *ibid.* **35**, 6222 (1987); *ibid.* **37**, 6297 (1988); G.A. Antonio, B.P. Feuston, R.K. Kalia and P. Vashishta, *J. Chem. Phys.* **88**, 767 (1988).
b) R. Biswas and D.R. Hamann, *Phys. Rev. B* **36**, 6434 (1987); J. Tersoff, *ibid.* **37**, 6991 (1988), (for the application to microclusters, see ref 11); J.R. Chelikowsky, J.C. Phillips, M. Kamal and M. Strauss, *Phys. Rev. Lett.* **62**, 292 (1989).

19. J.R. Chelikowsky, *Phys. Rev. Lett.* **60**, 2669 (1988).

20. See refs. 5, 6 and K. Raghavachari and C.M. Rohlfing, *Chem. Phys. Lett.* **143**, 428 (1988).

21. J. Koutecký and P. Fantucci, *Z. Physik D* **3**, 147 (1986).

22. W. Andreoni, G. Pastore, R. Car and M. Parrinello, to be published.

ABC THEORY VS. ABC-FREE THEORY OF EXCITON POLARITON

KIKUO CHO

Faculty of Engineering Science
Osaka University
Toyonaka
560 JAPAN

1. Introduction

The purpose of this note is to present an unbiased point of view about the problem of ABC (Additional Boundary Condition), which has been studied for over 30 years as one of the fundamental problems in the exciton physics. The way for presenting the viewpoint is not by following the historical development of ABC theories, but by showing a more general theoretical scheme to treat electromagnetic field in a bounded medium and by bringing it in contrast with the 'first principle' ABC theory. This scheme, which may be called ABC-free theory, can be applied to a wider class of problems than that for ABC theories, and gives equivalent results to those of ABC theories for their common problems. We will not discuss non first principle theories, which are essentially assumptions at macroscopic levels, while the first principle theories derive ABC from microscopic models.

The ABC problem arises from the fact that the dispersion equation for electromagnetic field in a spatially dispersive (bulk) medium, $\epsilon(k,\omega) = (ck/\omega)^2$, has two or more solutions, i.e., polaritons, $\{k=k_j(\omega);\ j=1,2,...\}$. This is due to the wave number (k) dependence of dielectric function $\epsilon(k,\omega)$ which represents the spatially dispersive (or non-local) nature of the medium. The field in a semi-infinite medium is obviously a superposition of these polariton modes (with possible distortion near the boundary of the medium). Then, the coefficients of the linear combination do not seem to be determined uniquely by Maxwell boundary conditions (MBC) alone, since the number of waves on both sides of the boundary exceeds that of MBC. Hence, one needs ABC to fix the field uniquely. This is the usual way to introduce the ABC problem. This problem seemed inevitable, and has been recognized to be very fundamental, because any optical measurements (reflection, transmission, scattering, etc.) dealing with signal intensity cannot be analyzed without solving it. During the last three decades, a large number of theoretical and experimental works have been devoted to the study of the ABC problem together with the multi-branch structure of polariton dispersion. For the details of the problem and reviews, see [1-5].

Apart from historical details of controversial arguments, today's common understanding about the 'first principle' treatment of the ABC problem is as follows: For a given bounded medium with spatial dispersion effect, one should

[a] prepare the eigenvalues and eigenfunctions of the ground and excited states, with proper consideration of quantum mechanical boundary conditions,

[b] calculate polarizability (tensor) $\chi(\mathbf{r},\mathbf{r}';\omega)$ in site representation in terms of the result of [a], and then,

[c] solve the Maxwell equation (with constitutive equation) in terms of this χ;

R. Girlanda et al. (eds.), Progress on Electron Properties of Solids, 41–49.

$$(c/\omega)^2 \,\mathrm{rot}\,\mathrm{rot}\,\mathbf{E}(\mathbf{r},\omega) = \mathbf{E}(\mathbf{r},\omega) + 4\pi \int \chi(\mathbf{r},\mathbf{r}';\omega)\mathbf{E}(\mathbf{r}',\omega)d\mathbf{r}'. \tag{1.1}$$

When the medium is a semi-infinite system, the solution can be written as a linear combination of bulk polariton modes (with possible distortion near the boundary), and at the same time one obtains a (set of) ABC from the condition that such a superposition should satisfy the above equations. There are a few theoretical calculations [6,7] carrying through these steps starting from a simplified model system for the step [a]. Historically, before these calculations appeared, there had been theories [8-10] dealing only with the last step [c] on the basis of some assumed forms of χ for a semi-infinite system. Though the forms of χ used in these works do not have quantum mechanical foundation as to the boundary condition, they did show that the Maxwell equation contains the complete information including ABC.

There are different ways to describe the Maxwell equation and the constitutive equation between polarization and electric field, such as the integral representation used by Birman and Sein [8], and the simultaneous equations for 'coherent electron-hole amplitude' and electric field used by Stahl [11]. Though they may look different, they are essentially equivalent to eq. (1.1). This will be explicitly shown for Stahl's approach in §5.

What we call ABC-free theory is another way to solve eq. (1.1), and it is in principle applicable to, not only semi-infinite medium, but also any medium with arbitrary size, shape and (internal) structure. This method makes use of a general character about the form of $\chi(\mathbf{r},\mathbf{r}';\omega)$: Linear response theory allows us to express χ of an arbitrary system in the form

$$\chi_{ij}(\mathbf{r},\mathbf{r}';\omega) = \sum_{\lambda} \bar{\chi}_{\lambda}(\omega)\rho_{\lambda i}(\mathbf{r})^{*}\rho_{\lambda j}(\mathbf{r}'), \tag{1.2}$$

where λ is the quantum number of the excited states of the system in consideration, and $(i,j) = (x,y,z)$. Some details of this expression will be discussed in §2. The essential point is that χ is a sum of separable functions of **r** and **r**′, which greatly simplifies the treatment of the Maxwell integro-differential equation. Making use of this general character of χ, we can express the solution of the Maxwell equation $\mathbf{E}(\mathbf{r},\omega)$ in terms of the arbitrary constants originating only from the homogeneous part of eq. (1.1);

$$(c/\omega)^2 \,\mathrm{rot}\,\mathrm{rot}\,\mathbf{E}(\mathbf{r},\omega) = \mathbf{E}(\mathbf{r},\omega). \tag{1.3}$$

Namely, the solution is expressed with the minimum number of arbitrary constants which can be fixed uniquely by MBC alone. This situation is parallel to the case of local response, where the field is determined by an equation similar to eq. (1.3) where the field on the right hand side is multiplied by a local dielectric function $\epsilon_L(\omega)$.

The rest of this note is organized as follows: In §2, we briefly argue about the form of $\chi(\mathbf{r},\mathbf{r}';\omega)$. A typical procedure of ABC theories will be shown in §3, and a formal treatment of ABC-free theory is given in §4. Discussions are given in §5, including a proof of equivalence of Stahl's approach to ABC and ABC-free theories and some comments about the future of 'ABC-problem'.

2. Polarizability in Site Representation

We first divide the whole medium into cells of volume V_0. The size of $V_0^{1/3}$ should be chosen to be much smaller than the wavelength of the polariton in consideration, so that the variation of the field within a cell can be neglected. Each cell is specified by a (discrete) vector **R**, and so are the electric field **E** and polarization **P** in each cell. When an external field with frequency ω is applied to a cell at **R**′, the Hamiltonian for the electron-radiation interaction is

$$H' = -\mathbf{P}(\mathbf{R}')\cdot\mathbf{E}(\mathbf{R}')e^{-i\omega t}, \tag{2.1}$$

where **P**(**R**′) is the polarization density operator integrated over the cell at **R**′,

$$\mathbf{P}(\mathbf{R}') = \int_{V_0(\mathbf{R}')} \sum_l e\mathbf{r}_l\delta(r-\mathbf{r}_l)d\mathbf{r}. \tag{2.2}$$

The summation over l runs through all the electrons. Then, the standard procedure of linear response theory [12] leads to the expression of polarizability χ (at $T = 0K$, for simplicity) as

$$\chi_{ij}(\mathbf{R},\mathbf{R}';\omega) = \frac{1}{\hbar V_0^2}\sum_\lambda\left\{\frac{<0|P_i(\mathbf{R})|\lambda><\lambda|P_j(\mathbf{R}')|0>}{\omega_\lambda-\omega-i\gamma}\right. \tag{2.3}$$

$$\left. + \frac{<0|P_j(\mathbf{R}')|\lambda><\lambda|P_i(\mathbf{R})|0>}{\omega_\lambda+\omega+i\gamma}\right\},$$

where λ is the quantum number of the system's eigenstates with eigenenergy $\hbar\omega_\lambda$ measured from the ground state energy, $|0>$ the ground state, γ a phenomenological damping constant. The factor V_0^2 (instead of V_0) is introduced with the understanding that (2.3) is used as an integral kernel in the Maxwell equation (1.1) (though **R** is originally introduced as a discrete variable); correspondingly, we use (**r**,**r**′) instead of (**R**,**R**′) in later sections.

The result (2.3) is quite general: It can be applied to any system where electronic polarization is of main interest. All the informations about the material, size, shape, (possible) internal structure, and boundary condition of the medium are included in $\{|\lambda>,\hbar\omega_\lambda\}$. It is obvious that (2.3) has the form (1.2) generally. The **R**-dependence of $\rho_\lambda(\mathbf{R})$ arises from the structure of the wave function $|\lambda>$: while ρ_λ is localized for an impurity state, it is extended as $\exp[i\mathbf{K}\cdot\mathbf{R}]$ for an exciton state with translational wave vector **K**.

3. ABC Theory

The next step in usual ABC theory is to calculate $\chi(\mathbf{r},\mathbf{r}';\omega)$ in a closed form by carrying out the summation over λ. Its feasibility depends on the microscopic model employed for the calculation of $\{|\lambda>,\hbar\omega_\lambda\}$. The models must specify how they describe the change in the bulk wave function as exciton approaches to the surface of the medium. For large samples, the energy eigenvalue remains to be that of the bulk, and it is specified by internal quantum number n(1s, 2s, etc.) and translational wave vector **K**. The wave function in the bulk region consists of the incoming and outgoing bulk waves specified by (n, **K**). Between these two waves there is a phase shift which is related with the details of the surface potential. In the surface region, there can be a surface localized amplitude of the exciton wave function in addition to the bulk components mentioned above. Therefore, a model for the calculation of $\{|\lambda>,\hbar\omega_\lambda\}$ should specify all the details of the phase shift and surface localized amplitude.

Up to now, there are two examples of such a model calculation for a semi-infinite medium [6,7], and an extension of the latter example to a slab geometry [13-15]. In one case [6], the distortion of the exciton wave function near the surface (surface localized amplitude) is completely neglected, but the phase shift for the reflection of a bulk wave is parametrized in a general form including rigid, free and lossy surfaces. In the other case [7], the surface localized amplitude is explicitly considered, and expressed approximately in terms of a single evanescent wave component for the translational motion. In both cases, the summation over the translational wave vector, K_z (perpendicular component to the surface), can be done rigorously. The resultant $\chi(z,z';\omega)$ consists of a linear combination of the terms

$$\exp\{iq\,|z-z'|\},\quad \exp\{iq(z+z')\}, \tag{3.1}$$

in the case of [6], where q is the z-component of the wave vector of bulk exciton with energy $\hbar\omega$, i.e.,

$$q = \{\frac{2M}{\hbar^2}(\hbar\omega-\hbar\omega_x+i\hbar\gamma-\hbar^2K_{\parallel}^2/2M)\}^{1/2},\quad (\mathrm{Im}[q]>0). \tag{3.2}$$

Here, $\hbar\omega_x$ is the exciton energy at $K = 0$. The term with $(z+z')$-dependence represents the effect of a boundary. In the case of [7], exciton wave has two components $\exp(\pm iqz)$ and $\exp(-Pz)$. Therefore, one gets additional terms in $\chi(z,z';\omega)$ as

$$\exp(\pm iqz-Pz'),\quad \exp(\pm iqz'-Pz),\quad \exp(-Pz-Pz'). \tag{3.3}$$

The solution of the Maxwell equation in terms of the kernel χ with the (z,z')-dependences of the types (3.1) and (3.3) can be obtained either by Fourier transform [6] or by reduction of the original integro-differential equation into a higher order differential equation [7]. In both methods, one finds the solution in the form of a superposition of bulk polariton waves (plus evanescent waves in the case of [7]), and an extra relationship is obtained that determines the relative amplitudes of the bulk polariton waves in the solution. This extra relationship is the ABC obtained from 'the first principle' theory.

The above two examples handle the case of s-polarized light. An extension of the latter model to the case of p-polarized light, where one needs two ABC's, has been done by the same authors [16].

'First principle' ABC theory has also been worked out for a slab [13,15]. This geometry is experimentally quite useful for the study of ABC problem. In this case, the quantum number λ contains a size quantized translational wave vector K. For the model of D'Andrea and Del Sole [7], K is allowed to take the values satisfying [13]

$$Kd-2\tan^{-1}(K/P) = \pi\times(\text{integer}), \tag{3.4}$$

where d is the thickness of the slab and $\exp(-Pd)$ is assumed to be negligibly small. The summation over such a discrete set of K to get a closed form of χ can be done rigorously by rewriting the summation as an integral in a complex plane. The resulting χ contains more terms than (3.1) and (3.3), especially a term like $\exp(-P\,|z-z'|)$, but the Maxwell equation can be solved essentially in a similar way to that used for semi-infinite case: The field is written as a superposition of bulk-like polariton-modes and surface localized amplitudes. The wave numbers of these modes depend on the thickness d for very thin slabs, but for larger values of d, they tend to those of bulk polariton and iP. The coefficients of the superposition satisfy two sets of linear equations, which correspond to the ABC's for the front and back surfaces of the slab.

The inclusion of the evanescent wave $\exp(-Pz)$ is a simplified step to consider the distortion of exciton wave function near the surface. This model allows us a wider class of ABC than that for the case of no distortion of exciton wave function, which leads to the so-called Pekar ABC [6,7]. This enables us to treat the situation, where the intuitive exciton dead layer model of Hopfield and Thomas [17] applies, from a microscopic point of view, as stressed by D'Andrea and Del Sole [7]. At a certain stage, a better description of the distortion of exciton wave function may be required. Since each physically different surface will require a specific surface distortion of exciton wave function, a study of more realistic wave function near the surface would be one of the most important future problems of the ABC theory mentioned above.

4. ABC-free Theory

The polarizability (1.2) as a tensor has the form of a dyadic

$$\chi(\mathbf{r},\mathbf{r}';\omega) = \sum_{\lambda}\bar{\chi}_{\lambda}(\omega)\, \rho_{\lambda}(\mathbf{r})^{*}\, \rho_{\lambda}(\mathbf{r}'). \tag{4.1}$$

In terms of vector potential $\mathbf{A}(\mathbf{r},\omega)$ and Lorentz condition, the Maxwell equation (1.1) can be rewritten as

$$(\nabla^2+k_0^2)\mathbf{A}(\mathbf{r}) = -4\pi\int d\mathbf{r}'\chi(\mathbf{r},\mathbf{r}')\{k_0^2\mathbf{A}(\mathbf{r}')+\nabla'\nabla'\cdot\mathbf{A}(\mathbf{r}')\}, \tag{4.2}$$

where ∇' is the grad operator for $\mathbf{r}'$ and

$$k_0 = \omega/c. \tag{4.3}$$

Substituting (4.1) in (4.2), we get

$$(\nabla^2+k_0^2)\mathbf{A}(r) = -4\pi k_0^2\sum_{\lambda}\bar{\chi}_{\lambda}(\omega)\rho_{\lambda}(\mathbf{r})F_{\lambda}, \tag{4.4}$$

where

$$F_{\lambda} = \int d\mathbf{r}\rho_{\lambda}(\mathbf{r})^{*}\cdot\{\mathbf{A}(\mathbf{r})+(1/k_0^2)\nabla\nabla\cdot\mathbf{A}(\mathbf{r})\}. \tag{4.5}$$

The general solution of eq. (4.4) is the sum of its special solution $\mathbf{A}_s(\mathbf{r})$ and the general solution $\mathbf{A}_h(\mathbf{r})$ of the homogeneous equation

$$(\nabla^2+k_0^2)\mathbf{A}(\mathbf{r}) = 0, \tag{4.6}$$

namely

$$\mathbf{A}_h(\mathbf{r}) = \mathbf{A}_0\, \exp(ik_0\hat{\kappa}\cdot\mathbf{r}), \tag{4.7}$$

$$\mathbf{A}_s(\mathbf{r}) = k_0^2\sum_{\lambda}\bar{\chi}_{\lambda}(\omega)F_{\lambda}\int d\mathbf{r}'\frac{\exp(ik_0|\mathbf{r}-\mathbf{r}'|)}{|\mathbf{r}-\mathbf{r}'|}\rho_{\lambda}(\mathbf{r}'), \tag{4.8}$$

where the constant vector $\mathbf{A}_0$ and the unit vector $\hat{\kappa}$ should be determined according to the geometry of the problem. In the case of a semi-infinite medium (or a slab), the component of $\hat{\kappa}$ parallel to the surface should be conserved. Therefore, $\hat{\kappa}$ can take two values $(\kappa_{\parallel},\pm\kappa)$ for each given value of $\kappa_{\parallel}$. Accordingly, we can write

$$\mathbf{A}_h(r) = \{\mathbf{A}_{0+}\exp(ik_0\kappa z)+\mathbf{A}_{0-}\exp(-ik_0\kappa z)\}\exp(ik_0\kappa_{\parallel}x). \tag{4.9}$$

Substituting $\mathbf{A}$ in (4.5) with $\mathbf{A}_h+\mathbf{A}_s$, (4.8) and (4.9), we get the simultaneous linear equations for $\{F_{\lambda}\}$ as

$$F_{\lambda} = \mathbf{A}_{0+}\cdot\mathbf{N}_{\lambda+}+\mathbf{A}_{0-}\cdot\mathbf{N}_{\lambda-}+\sum_{\nu}D_{\lambda\nu}F_{\nu}, \tag{4.10}$$

where $\mathbf{N}_{\lambda\pm}$ and $D_{\lambda\nu}$ can be calculated from known quantities. The solution of (4.10) is given as

$$F_{\lambda} = \mathbf{A}_{0+}\cdot\sum_{\nu}(1-D)^{-1}_{\lambda\nu}\mathbf{N}_{\nu+}+\mathbf{A}_{0-}\cdot\sum_{\nu}(1-D)^{-1}_{\lambda\nu}\mathbf{N}_{\nu-}. \tag{4.11}$$

From (4.8), (4.9) and (4.11), we get $\mathbf{A} = \mathbf{A}_h + \mathbf{A}_s$ as a linear combination of $\mathbf{A}_{0+}$ and $\mathbf{A}_{0-}$. Since

$$\mathbf{E} = (i/k_0)\{k_0^2\mathbf{A} + \nabla\nabla\cdot\mathbf{A}\}, \tag{4.12}$$

the electric field in the medium is also a linear combination of $\mathbf{A}_{0\pm}$. Namely, the arbitrary constants in the final expression of the field are only those introduced in the homogeneous solution $\mathbf{A}_h(\mathbf{r})$. In other words, the number of the independent waves in the medium is determined only by the homogeneous part of the Maxwell equation (in spite of the existence of an inhomogeneous part in the equation due to the spatial dispersion effect), and this is a common feature to the case of local media where the Maxwell equation can be written without inhomogeneous term. Therefore, the arbitrariness contained in the solution, ($\mathbf{A}_{0\pm}$), can be fixed uniquely by the Maxwell boundary conditions alone. As seen explicitly, there is no need of mentioning ABC at any stage of this solution.

In this way, we can obtain the ABC-free solution of the Maxwell equation for spatially dispersive media [18]. Obviously this solution is equivalent to that with ABC (in the case of semi-infinite medium or slab), since they are the general solution of the same equation. But the ABC-free formalism is more general, since it is applicable to media, where no bulk polaritons are expected, such as very thin films, quantum wells, fine particles etc.

The application of the ABC-free formalism has been done to slab geometries with different models for the surface distortion of exciton wave function, i.e., either without distortion [18] or with the distortion described by a single evanescent wave, $\exp[-Pz]$ [15,19], as in the application of the ABC theory mentioned in the previous section. These examples contain the case of semi-infinite systems by taking the limit of 'thickness $\rightarrow \infty$'. In all these cases, the field in the medium is expressed as a sum $\mathbf{A}_h + \mathbf{A}_s$, (4.9) and (4.8). When the summation over λ (size quantized translational wave vector) contained in (4.8) is carried out, it is shown rigorously that [i] the homogeneous part $\mathbf{A}_h$ is extinguished, [ii] the remaining part consists of bulk (-like) polariton modes and (for the model with surface distortion of wave function) surface localized amplitudes, and [iii] the result is equivalent to that of the ABC theory for the corresponding model [15,18,19].

The case of exciton in a very thin slab, including the case of quantum wells, is also an appropriate system for the ABC-Free formalism. In such systems, exciton states are no more described by the bulk amplitudes plus surface localized part. The eigenvalues and eigenfunctions must be looked for by more appropriate methods. Once they are obtained, however, the formalism leads us directly to physically observable quantities, such as reflectance and transmittance. Two cases have been treated; a simplest case of single level exciton in a quantum well [18], and a more elaborate case connecting quantum well and thin slab regimes [20]. Many other applications of the ABC-free formalism would be possible in view of its general character. One of the important aspects is that the nonlocal nature of the medium is treated from a microscopic level, so that one can appropriately judge whether the system can be regarded as a local medium or not as the sample size becomes smaller than the light wave length.

5. Discussions

A. Another ABC-free theory

Stahl and coworkers have studied the electromagnetic response of electron-hole pair excitations in a seemingly different way [11,21,22]. This formalism consists of the simultaneous equations for electric field $\mathbf{E}(\mathbf{R},\omega)$ and coherent electron-hole amplitude $Y(\mathbf{R},\mathbf{r},\omega)$:

$$\{E_g - \hbar\omega - \frac{\hbar^2}{2M}\nabla_R^2 - \frac{\hbar^2}{2\mu}\nabla_r^2 - V(\mathbf{r})\}Y = \mathbf{M}(\mathbf{r})\cdot\mathbf{E}(\mathbf{R}), \tag{5.1}$$

$$(\epsilon_b\omega^2 + c^2\nabla_R^2)\mathbf{E} = -2\omega^2\int \mathbf{M}(\mathbf{r})Y(\mathbf{R},\mathbf{r})d\mathbf{r}, \tag{5.2}$$

where E_g is the band gap energy, $V(\mathbf{r})$ the interaction potential between electron and hole, $\mathbf{M}(\mathbf{r})$ the transition dipole density, ϵ_b background dielectric constant, $\mathbf{R}$ and $\mathbf{r}$ the center of mass and relative coordinates, respectively, of the electron and hole. Using extra assumptions

$$\mathbf{M}(\mathbf{r}),\ \ V(\mathbf{r}) \propto \delta(\mathbf{r}), \tag{5.3}$$

and reducing the equations to one-dimensional ones, they show that, by requiring boundary conditions to the electron-hole amplitude $Y(R,\mathbf{r})$, one can uniquely determine the reflectivity of a semi-infinite crystal only by means of the usual Maxwell boundary conditions.

Though it is sometimes claimed [21] that this formalism is different from or superior to the usual approaches in terms of the polarizability $\chi(\mathbf{r},\mathbf{r}';\omega)$ from linear response theory, one can show that these two are essentially equivalent: Let us introduce the eigenvalues and eigenfunctions of the operator on the left hand side of (5.1), i.e.,

$$\{E_g - \frac{\hbar^2}{2M}\nabla_R^2 - \frac{\hbar^2}{2\mu}\nabla_r^2 - V(\mathbf{r})\}\psi_\lambda(\mathbf{R},\mathbf{r}) = \hbar\omega_\lambda\psi_\lambda(\mathbf{R},\mathbf{r}). \tag{5.4}$$

Then, the solution linear in $\mathbf{E}$ of eq. (5.1) is obtained as

$$Y(\mathbf{R},\mathbf{r}) = \sum_\lambda \int d\mathbf{r}' \int d\mathbf{R}' \frac{\psi_\lambda(\mathbf{R},\mathbf{r})^*\psi_\lambda(\mathbf{R}',\mathbf{r}')}{\hbar(\omega_\lambda - \omega)}\mathbf{M}(\mathbf{r}')\cdot\mathbf{E}(\mathbf{R}'). \tag{5.5}$$

From (5.2) and (5.5) one gets

$$(\epsilon_b\omega^2 + c^2\nabla_R^2)\mathbf{E}(\mathbf{R}) = -2\omega^2\int d\mathbf{R}' \sum_\lambda \frac{\overline{\mathbf{M}}_\lambda(\mathbf{R})^*\overline{\mathbf{M}}_\lambda(\mathbf{R}')}{\hbar(\omega_\lambda - \omega)}\cdot\mathbf{E}(\mathbf{R}'), \tag{5.6}$$

where

$$\overline{\mathbf{M}}_\lambda(\mathbf{R}) = \int d\mathbf{r}\ \psi_\lambda(\mathbf{R},\mathbf{r})\ \mathbf{M}(\mathbf{r}). \tag{5.7}$$

The Maxwell equation (5.6) corresponds to a system having a dielectric function

$$\epsilon(\mathbf{R},\mathbf{R}';\omega) = \epsilon_b\delta(\mathbf{R}-\mathbf{R}') + \frac{2}{\hbar}\sum_\lambda \frac{\overline{\mathbf{M}}_\lambda(\mathbf{R})^*\overline{\mathbf{M}}_\lambda(\mathbf{R}')}{\omega_\lambda - \omega}. \tag{5.8}$$

If we put the non-resonant term in the polarizability (2.3), i.e., the term proportional to $(\omega_\lambda + \omega + i\gamma)^{-1}$, into a background constant (ϵ_b), the Maxwell equation used in the previous sections is completely equivalent to the present one (5.6). Therefore, the three microscopic approaches,

(a) ABC theory in §3,

(b) ABC-free theory in §4,

(c) Stahl's coherent wave theory,

based on the different mathematical procedures, should lead to physically equivalent results, if the starting model is the same. However, due to the difference in mathematical treatment, the microscopic models of excitons and surfaces tractable in or suitable to each approach could be different. As mentioned in §4, the levels of the models that can be handled by ABC and ABC-free theories, in the case of semi-infinite crystals and slabs, are the same, but ABC-free theory is more general since it can be applied to cases where no bulk polariton is expected as a possible wave in the medium. The models so far treated by Stahl's approach are all restricted to the assumption (5.3), and (under this condition) it is claimed that the method is suitable for treating the effect of continuum polariton [21]. Though such an aspect will be also within the reach of ABC or ABC-free theories, we would not try to judge now the superiority of the one to the other. We should rather let each of them be applied to explore various suitable models, so that we get informations about the correspondence between detailed aspects of models and its observable consequences.

B. Future of 'ABC-problem'

In view of the possibility of treating the Maxwell equation for spatially dispersive medium in several different ways either with or without ABC, it would be necessary to redefine the so-called ABC-problem. A very general definition would be 'problems of electromagnetic response of bounded spatially dispersive media'. From theoretical points of view, the object of the study is to describe the response in terms of microscopic models of medium. Since the response depends not only on the material in consideration but also on the details of surfaces in general, we should establish which effects should lead to observable response. For such a study, ABC is just a means to describe the field in the medium, and not the ultimate goal of the study. From experimental points of view, the study of 'ABC-problem' is to find well reproducible spectra (of any kinds) for bounded spatially dispersive media, and to specify the details of the sample surfaces which give the reproducible results. For the latter purpose, one needs measured data of surface roughness (on atomic scale), surface contamination, band bending and related surface charges, and so on.

It may not be appropriate to use the word 'ABC' to define the above-mentioned problems, since its use could be completely avoided in their treatment. In that case, 'ABC(F)-problems' might be a better naming, where (F) means that the problems can be handled by either ABC or ABC-F(ree) theory. (Since the abbreviation ABC has already got a great popularity all over the world, it will be useful to retain ABC in the naming.)

In the sense mentioned above, there are still a variety of important aspects about the ABC(F) problems to be studied in future from both experimental and theoretical points of view.

Acknowledgments

This article is dedicated to Professor F. Bassani to celebrate his sixtieth birthday. The author is grateful to Professor A. Yoshimori for his support and Mr. H. Ishihara for his collaboration in the works about ABC(F) problems.

References

1. S.I. Pekar, *Sov. Phys. JETP* **6** 785 (1957); *Crystal Optics and Additional Light Waves*, Benjamin/Cummings (1983).

2. V.M. Agranovich and V.L. Ginzburg, *Crystal Optics with Spatial Dispersion and Excitons*, Springer (1984).
3. J.L. Birman, *Excitons*, ed. E.I. Rashba and M.D. Sturge, North-Holland (1982).
4. B. Hönerlage, R. Levy, J.B. Grun, C. Klingshirn and K. Bohnert, *Phys. Report* **124**, 161 (1985).
5. M. Ueta, Springer Series in Solid State Sciences **60**, *Excitonic Processes in Solids* Chap. 3 (1986).
6. R. Zeyher, J.L. Birman and W. Brenig, *Phys. Rev.* **B6**, 4613 (1972).
7. A. D'Andrea and R. Del Sole, *Phys. Rev.* **B25**, 3714 (1982).
8. J.L. Birman and J.J. Sein, *Phys. Rev.* **B6**, 2482 (1972).
9. G.S. Agarwal, D.N. Pattanayak and E. Wolf, *Phys. Rev. Letters* **27**, 1022 (1971).
10. A.A. Maradudin and D.L. Mills, *Phys. Rev.* **B7**, 2787 (1973).
11. A. Stahl, *Phys. Stat. Sol.* **b94**, 221 (1979); ibid. **106**, 575 (1981).
12. R. Kubo, *J. Phys. Soc. Jpn.* **12**, 570 (1957).
13. K. Cho and M. Kawata, *J. Phys. Soc. Jpn.* **54**, 4431 (1985).
14. K. Cho and H. Ishihara, preprint.
15. K. Cho and H. Ishihara, Springer Proceedings in Physics, **25** *Excitons in Confined Systems* (1988).
16. A. D'Andrea and R. Del Sole: *Phys. Rev.* **B38**, 1197 (1988).
17. J.J. Hopfield and G.D. Thomas, *Phys. Rev.* **132**, 563 (1963).
18. K. Cho, *J. Phys. Soc. Jpn.* **55**, 4113 (1986).
19. H. Ishihara and K. Cho, preprint.
20. A. D'Andrea and R. Del Sole, Springer Proceedings in Physics, **25** *Excitons in Confined Systems*, (1988).
21. I. Balslev and A. Stahl, *Phys. Stat. Sol.* **b111**, 531 (1982); A. Stahl and I. Balslev, *Phys. Stat. Sol.* **b113**, 583 (1982).
22. L. Gotthard, *Solid State Commun.* **51**, 975 (1984).

TEMPERATURE DEPENDENCE OF THE BAND STRUCTURE OF SEMICONDUCTORS: ELECTRON-PHONON INTERACTION

MANUEL CARDONA[†] AND SUDHA GOPALAN[*]

[†*] *Max-Planck-Institut für Festkörperforschung*
Heisenbergstrasse 1, D-7000 Stuttgart 80,
Federal Republic of Germany

[*] *The University of Western Ontario*
Department of Physics
London, Ontario, Canada N6A 3K7

Changes of the band structure of semiconductors with temperature arise from a trivial thermal expansion effect plus electron-phonon interaction. The latter can be split up into two contributions: those of the first-order electron-phonon Hamiltonian, to second-order in perturbation theory (self-energy terms), and those of the second-order Hamiltonian taken to first-order (Debye-Waller terms). Results of calculations at important critical points and lines for several diamond and zincblende-type semiconductors are reviewed and compared with recent ellipsometric measurements of shifts and broadenings of interband critical points. The temperature dependence of the infrared dielectric constant ϵ_∞ is considered within the same framework. The problem of screening of the electron interaction with long wavelength acoustic phonons is treated using the concept of dielectric midgap energy. The same concept also leads to an estimate of the temperature dependence of Schottky barriers.

1. Introduction

The calculation of the electron energy bands of semiconductors, pioneered by Franco Bassani [1] is largely a solved problem. Early first principles ('ab initio') calculations, such as those performed with the OPW method [1,2] were hampered by the limited capability of available computers and also by the lack of fast and accurate algorithms to achieve self-consistency. As quantitative discrepancies with a growing experimental data base became clear, 'ab initio' calculations gave way to semiempirical methods, such as the highly successful Empirical Pseudopotential Method (EPM) [3] and the empirical tight binding method (ETBM) [4]. More recently, a return to 'ab initio' techniques has taken place, largely due to three developments: the local density approximation (LDA) to treat exchange and correlation problems [5], the invention of norm-conserving pseudopotentials [6], and Brillouin-zone integration methods based on sampling a few spatial points (Baldereschi points) [7]. The latter provided a fast and accurate algorithm for achieving self-consistency. Other methods based on angular momentum wavefunctions around the cores, such as APW [8] and LMTO [9], have also been successful especially to calculate the single particle band structure of the ground state (which cannot be determined experimentally!) and related bulk properties (such as lattice constant, bulk modulus, etc.). They fail, however, to reproduce excited state properties such as the energy gaps observed in optical and other experiments and the effective masses. This failure can be attributed to a breakdown of the LDA which can be corrected by means of more elaborate calculations of exchange and correlation [10].

R. Girlanda et al. (eds.), Progress on Electron Properties of Solids, 51–64.

Once a band structure is reasonably well understood, it is of interest to investigate its dependence on pressure (both hydrostatic and uniaxial) and temperature with experimental and theoretical methods. The theoretical treatment of the former is relatively easy: once a computational code is available one simply has to perform a calculation with the lattice parameters changed as dictated by the applied stress through the elastic compliance constants. Knowledge of the dependence on stress of internal unit cell parameters, seldom available [11], may sometimes be required. In the case of zincblende-type materials calculations have been recently performed with the pseudopotential [12], ETBM [13], and LMTO [14,15] methods. All these methods yield accurately the dependence on strain of *energy differences* such as excitation gaps. The problem of the pressure dependence of *absolute* energies of a *finite* solid (for an infinite solid such concept is meaningless) is considerably more complex [14].

The problem of the temperature dependence of electronic band states is also more complex. Except for a term induced by thermal expansion, which can be trivially obtained from the dependence on strain just discussed, such dependence is due to electron-phonon interaction and also affects the band structure at $T = 0$ through zero point vibrational amplitudes [16,17]. Because of the perturbation of the translational symmetry produced by the phonons, both energy shifts (real parts of self-energies) and broadenings (imaginary parts) result. We discuss these effects for diamond and zincblende-type semiconductors in the next sections.

2. Electron-Phonon Interaction Effects

The first attempt to estimate shifts in energies of electronic states due to electron-phonon interaction was probably made by Fan [18]. He considered second-order perturbation terms involving the emission and subsequent reabsorption of one phonon, i.e., the second term in the rhs of Fig. 1, with the intermediate state $|\mathbf{k}+\mathbf{Q}, n'\rangle$ belonging to *the same* parabolic band as the state under consideration described by

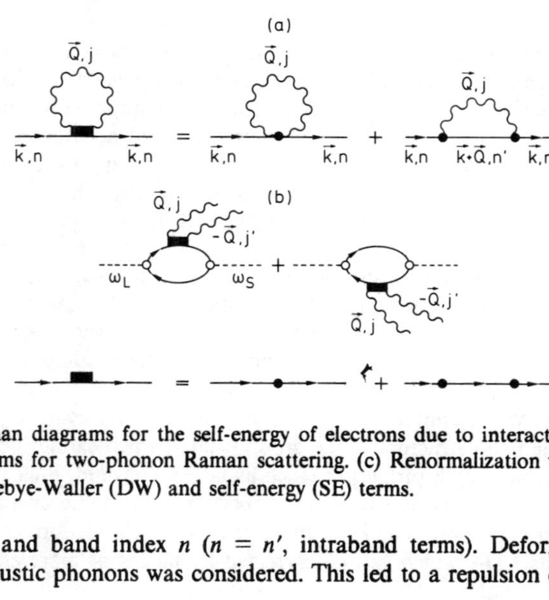

Figure 1. (a) Feynman diagrams for the self-energy of electrons due to interaction with phonons. (b) Feynman diagrams for two-phonon Raman scattering. (c) Renormalization vertex corresponding to the sum of Debye-Waller (DW) and self-energy (SE) terms.

the wavevector $\mathbf{k}$ and band index n ($n = n'$, intraband terms). Deformation potential interaction via acoustic phonons was considered. This led to a repulsion of the band edge by the rest of the band, i.e., to a decrease in the gap with increasing phonon excitation (i.e. temperature, T). At high temperatures this decrease should be linear in T. It was pointed out in [19] that not only intraband scattering was important: interband terms, involving electron-phonon interaction between valence and conduction states, and thus contributing to increases in gaps with T, should also be included. They partly cancel the intraband (Fan) terms. This is the reason for the quantitative success of incomplete theories which only include the first term in the rhs of Fig. 1a, i.e., the so-called Debye-Waller (DW) term [20]. This term is much easier to evaluate than the previous ones; it suffices to perform a pseudopotential calculation with the pseudopotential form factors reduced by Debye-Waller factors ($\exp[-\frac{1}{6}G^2<u^2>]$, where G is the reciprocal lattice vector of the corresponding form factor and $<u^2>$ the average phonon amplitude of the atom under consideration). Thus its evaluation does not require cumbersome Brillouin zone integrations over the Q of the phonons; empirical values of $<u^2>$ obtained from x-ray data can be used [21]. We note that the second terms in the rhs of Fig. 1a (self-energy terms, SE) should also include interband terms belonging to other pieces of the conduction or valence bands, not only band extrema. When the $|\mathbf{k}, n>$ state is not at the lowest band edge, there are some of those states to which it can perform phonon-induced *real* transitions and thus contribute an imaginary part to the self-energy of the $|n, \mathbf{k}>$ electron, i.e., a finite lifetime (energy broadening). We have been engaged over the past few years in a program to measure and calculate shifts and broadenings of energy gaps which appear prominently in the optical spectra of semiconductors [16,17,22-30]. We discuss in what follows some of these calculations and pertinent experimental results.

A question arises in one-electron calculations of SE terms as to whether we should include intermediate states $|\mathbf{k}+\mathbf{Q}, n'>$ which violate the Pauli principle, i.e., which correspond to transitions between occupied and occupied or empty and empty states. Somewhat paradoxically, the answer is yes if we want to apply our calculations to the temperature dependence of *excitation energies*. The reason is that such excitation energies are actually not differences in one-electron energies but rather differences in total (all electrons) energies of the excited minus the ground state. Let us consider the valence electron state participating in a given interband transition (occupied in the *initial* state, empty in the final state). In the *initial* state SE terms representing transitions of that electron to another occupied valence state are forbidden and thus should be excluded from the SE terms. They appear, however, in the final state with a minus sign, as corresponds to a missing electron in the valence band. We thus take care automatically of this contribution to the shift of the transition energy if we compute one-electron SE terms without regard of the Pauli principle: the forbidden term of the initial state which appears with a minus sign in the transition energy, accounts for the SE of the hole in the final state. Proceeding further in this manner, one can see that the difference in one-electron SE's calculated disregarding the Pauli principle give correctly the temperature dependence of excitation energies, i.e., the observable under consideration.

When pseudopotentials and not true potentials are used both SE and DW terms depend on the exact nature of the pseudopotentials which are not uniquely defined. The sum of both, however, corresponds to an energy shift and should, according to the pseudopotential theorem, be uniquely defined. This consideration leads to the so-called acoustic sum rule when applied to acoustic phonons of $Q \rightarrow 0$, i.e., to a uniform translation. Such translation should produce no energy shift, hence the contribution to the DW terms of a phonon labeled by $\mathbf{Q}$ and j, expanded to second order in the phonon displacements, can be rewritten as [27]:

$$\left[\frac{\partial E_{\mathbf{k}n}}{\partial n_{Qj}}\right]_{DW} = \qquad (1)$$

$$-\frac{\hbar}{2N}\sum_{\kappa\kappa' n'}\frac{<\mathbf{k}n|\frac{\partial V}{\partial R_\alpha(\kappa)}|\mathbf{k}n'><\mathbf{k}n'|\frac{\partial V}{\partial R_\beta(\kappa')}|\mathbf{k}n>}{\epsilon_{\mathbf{k}n}-\epsilon_{\mathbf{k}n'}}\times$$

$$\times\left[\frac{1}{M_\kappa\omega_{\mathbf{Q}j}}\epsilon_\alpha(-\mathbf{Q}j\kappa)\epsilon_\beta(\mathbf{Q}j\kappa)+\frac{1}{M_{\kappa'}\omega_{\mathbf{Q}j}}\epsilon_\alpha(-\mathbf{Q}j\kappa')\epsilon_\beta(\mathbf{Q}j\kappa')\right]$$

where N is the number of unit cells, $\epsilon_{\mathbf{k}n}$ represent electronic energies, $R_\alpha(\kappa)$ the α component of the position vector of atom κ, M_κ its mass, $\omega_{\mathbf{Q}j}$ phonon frequencies and E_α their eigenvectors. Equation (1) has a form rather similar to that of the SE energy [27]:

$$\left[\frac{\partial E_{\mathbf{k}n}}{\partial n_{Qj}}\right]_{SE} = \tag{2}$$

$$-\frac{\hbar}{N}\sum_{\kappa\kappa' n'}\frac{<\mathbf{k}n|\frac{\partial V}{\partial R_\alpha(\kappa)}|\mathbf{k}+\mathbf{Q}n'><\mathbf{k}+\mathbf{Q}n'|\frac{\partial V}{\partial R_\beta(\kappa')}|\mathbf{k}n>}{\epsilon_{\mathbf{k}n}-\epsilon_{\mathbf{k}+\mathbf{Q}n'}}\times$$

$$\times e^{-i\mathbf{Q}\cdot(\tau_\kappa-\tau_{\kappa'})}(\mathbf{M}_\kappa\mathbf{M}_{\kappa'}\omega^2_{\mathbf{Q}j})^{-1/2}\epsilon_\alpha(-\mathbf{Q}j\kappa)\epsilon_\beta(\mathbf{Q}j\kappa')$$

except that in general Eq. (1) yields real values while Eq. (2) leads to real and imaginary parts, as already mentioned. For the evaluation of Eqs. (1,2) it is necessary to have a realistic code to calculate the lattice dynamics and a reasonably simple representation of the band structure. We have used for the lattice dynamics bond charge models for diamond structures [31], 14-parameter shell model for zincblende [32]. For the band structure we used an EPM representation without spin-orbit splitting [33].

While for a band structure calculation only pseudopotential form factors $V(G)$ at reciprocal lattice vectors **G** are needed (and fitted to experiment in EPM) in order to evaluate Eqs. (1,2) in the *rigid ion approximation* one needs $V(k)$ at all values of k. Thus the fitted data must be interpolated, and, what is more problematic, extrapolated to $k = 0$. It is known that $V(0)$, i.e. the average value of the potential, is meaningless in an *infinite* crystal. Hence a certain degree of arbitrariness results in the extrapolation of V to $k = 0$, i.e., for the value $V(k)$ with k smaller than the smallest $G \neq 0$. Two rather capricious approaches have been suggested: $V(0) = 0$ [34] and $V(0) = -2E_F/3$ [35]. The latter corresponds to the value of $V(0)$ in a simple metal with Fermi energy E_F (in a semiconductor E_F is usually taken to that of free electrons with the valence electron density). It has been suggested without justification that $V(0) = 0$ gives better values for some electron-phonon interaction parameters. We note, however, that the LA phonons of long wavelength produce a deformation with a hydrostatic component and that the corresponding electron-phonon interaction constant (hydrostatic deformation potential) contains a term proportional to $V(0)$ which, in principle, is undetermined [14]. It must, however, be uniquely defined as it corresponds to an observable (a scattering rate). A method to fix the correct value of this deformation potential is given in [14]. It is based on screening the perturbation produced on a properly defined 'dielectric midgap energy' (DME, the center of the Penn gap) with the static dielectric function. This method yields for the hydrostatic deformation potential of zincblende-type semiconductors a value $a_v \cong 0$ eV, in agreement with that obtained for $V(0) = -2E_F/3$. This may justify the use of this value of $V(0)$. If $V(0) = 0$ is used, on the contrary, we find $a_v \cong -8$ eV, close to that obtained from the LMTO method without screening [14].

Figure 2 shows the temperature dependence of the energies of the E_1 critical points (CP) of GaP, InAs, and GaSb as observed ellipsometrically. These CP's correspond to direct transitions between the top valence and the bottom conduction bands along the {111} directions [30]. These data agree with calculations performed as outlined above,

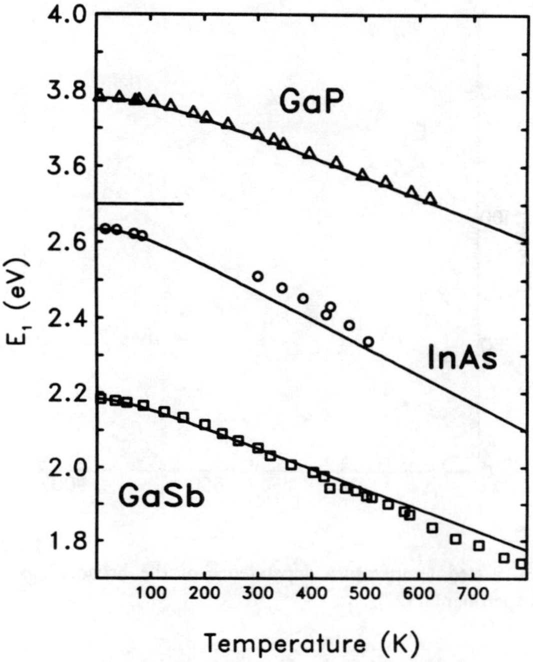

Figure 2. Shift of the E_1-CP energy with temperature for GaP (Δ), InAs (○), and GaSb (□). The solid lines give the results of the calculation for the shift performed as outlined in Sect. 2 [30].

including thermal expansion effect, Debye-Waller term, and real part of the self-energy, also shown in Fig. 2.

In Fig. 3 we show the calculation of the imaginary part of the SE and compare it with experimental data for the E_1 gaps of GaAs [27]. A zero-point broadening of ~20 meV is observed and calculated at ~0 K.

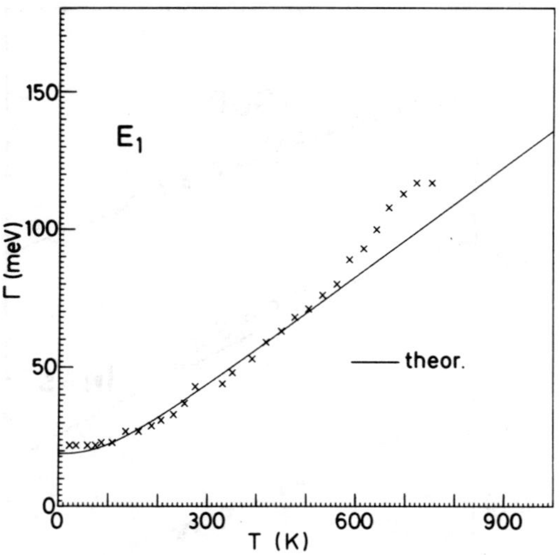

Figure 3. Solid line: calculated temperature dependence of the broadenings of the E_1 gap of GaAs. The points are experimental [27].

3. Second Order Raman Scattering

Feynman diagrams relevant to second order Raman scattering are displayed in Fig. 1b. It can be shown that these diagrams dominate near resonance, i.e., when the laser or scattered frequencies are close to an optically strong interband critical point: diagrams in which one phonon is emitted by the excited electron and the other by the hole are negligible [36]. The similarity between the interaction vertices of Fig. 1a and those of Fig. 1b is rather striking, especially if we limit ourselves to scattering by two phonons of the same branch. Nevertheless the energy denominators for the intermediate state, between interactions with $|\mathbf{Q}, j>$ and $|-\mathbf{Q}, j>$ phonons, are different in Figs. 1a and 1b: in the latter they involve the difference between laser (scattered) frequency and that of excitations of the intermediate states. Very close to resonance, this difference becomes negligibly small and the vertices for second-order Raman and for the SE coincide. The computer code for the calculation of the resonant Raman efficiencies can then be easily obtained through a simple modification of that for the calculation of SE's [36]. Calculations for the E'_0 resonance of Si are compared in Fig. 4 with experimental data (the Γ_1 component of the two-phonon spectra only) [36].

4. Infrared Dielectric Constant ϵ_∞

The calculation of this important macroscopic parameter requires an integration of electronic transition terms, including the appropriate matrix elements of momentum **P**, over the Brillouin zone. For a cubic material: [37,38]

$$\epsilon_\infty = 1 + \frac{8\pi e^2}{3m^2\hbar}\sum_{nn'}\int \frac{d^3\mathbf{k}}{(2\pi)^3}\frac{|<n\mathbf{k}|\mathbf{p}|n'\mathbf{k}>|^3}{\omega_{nn'}^3} \tag{3}$$

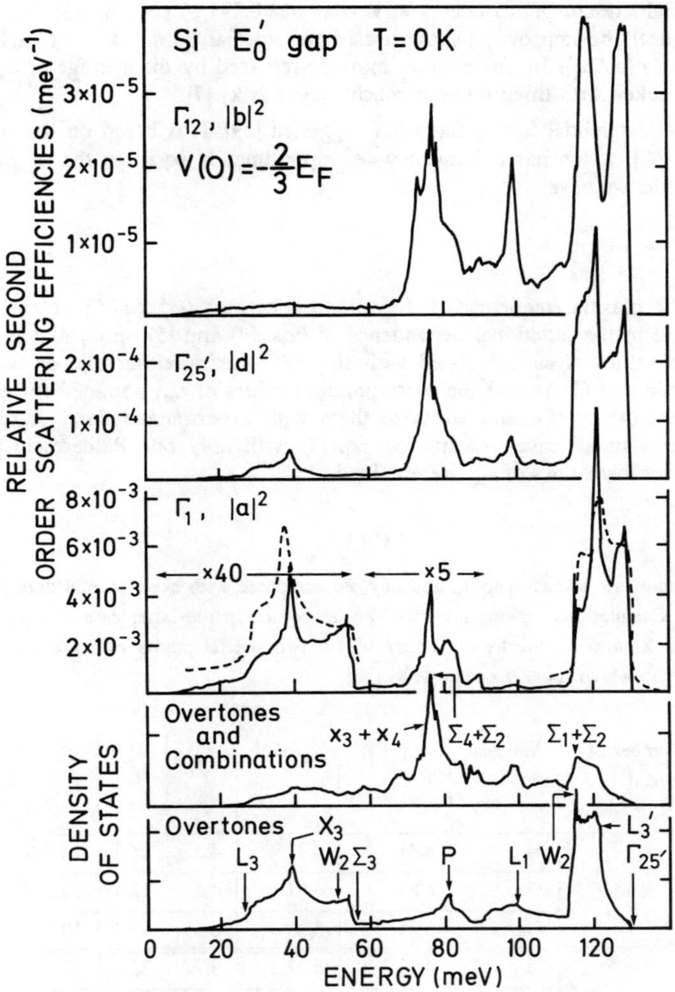

Figure 4. The upper three panels show the calculated Γ_1, $\Gamma_{25'}$, Γ_{12} Raman spectra for Si with the laser photon resonant with the $E_0(\Gamma_{25'}\rightarrow\Gamma_{15})$ gap. The intensity is normalized to the integrated first order intensity. The electron-phonon coupling is derived from an interpolated local pseudopotential with $V(0)=0$. The lower two panels give the density of overtone states ($j=j'$) and the total two phonon density of states (all j,j'). The dashed line represents experiments [36].

where $n\mathbf{k}$ labels an empty and $n'\mathbf{k}$ an occupied electronic state. It has been suggested that the two highest valence bands and the two lowest conduction bands (two because of the two spin directions) may give the main contribution to Eq. (3) while the sampling implied by the integral over $\mathbf{k}$ may be well represented by a small set of special (Baldereschi) points [7]. In the case of a single special point with a gap $\omega_{nn'} = \omega_g$ Eq. (3) can be rewritten as (in atomic units, a_0 in Bohr and ω_g in Hartree = 27.2 eV)

$$\epsilon_\infty = 1 + \frac{256\pi^3}{3\omega_g^3 a_0^5}, \tag{4}$$

where we have assumed that the matrix element of p is $2\pi / a_0$ [38]. The special point for

Eq. (4) (first Baldereschi point) occurs at $\mathbf{k} = (0.622, 0.295, 0)\ (2\pi/a_0)$ [7]. Equation (4) may (or may not!) be improved by the use of two special points: $\mathbf{k}_1 = (3,1,1)\ (\pi/2a)$ and $\mathbf{k}_2 = (1,1,1)\ (\pi/2a)$. In this case ω_g must be replaced by the average of $\omega_g(\mathbf{k}_1)$ and $\omega_g(\mathbf{k}_2)$ with $\mathbf{k}_1$ taken with three times as much weight as $\mathbf{k}_2$ [7].

An even simpler model for ϵ_∞ has been suggested [39]. It is based on the concept of the Penn gap [40], which has a frequency ω_g approximately equal to the ω_g of Eq. (4). Within this model we have:

$$\epsilon_\infty = 1 + \frac{\omega_p^2}{\omega_g^2}, \tag{5}$$

where ω_p is the plasma frequency of the valence electrons ($\omega_p^2 \propto a_0^{-3}$). Note the non-trivial difference in the functional dependence of Eqs. (4) and (5) on ω_g and a_0. We list in Table I the values of ω_g calculated with the EPM taking either one or two special points for Si, Ge and GaAs and the corresponding values of ϵ_∞, obtained for each case, with either Eqs. (4) or (5), and compare them with experimental data. Agreement is surprisingly good in all cases, except for Eq. (4) with only one Baldereschi point, in which case rather low values of ϵ_∞ are obtained.

TABLE I.

Calculated values of the average gap ω_g and of $\epsilon(\infty)$ compared with experimental data [14] for Si, Ge, and GaAs. Calculations for one and two special points, in the latter case two values of ω_g, corresponding to $\mathbf{k}_1$ and $\mathbf{k}_2$, are listed. Under ω_g^b for two special points we have also listed the value of the E_1 gap which roughly occurs at $\mathbf{k}_2$ [38].

Number of special points	Material	$\omega_g(eV)^a$	$\omega_g(eV)^b$	ϵ_∞^c	ϵ_∞^d	ϵ_∞^b
	Si	5.21	4.77	4.3	11.5	12.0
one	Ge	4.83	4.31	4.4	12.3	16.0
	GaAs	5.28	5.20	3.6	9.8	10.9
two	Si	4.09,3.20	4.77,3.4	9.7	20.1	12.0
$\mathbf{k}_1$,weight$\frac{3}{4}$	Ge	3.88,2.19	4.31,2.2	14.9	21.9	16.0
$\mathbf{k}_2$,weight$\frac{1}{4}$	GaAs	4.40,2.87	5.20,3.0	8.6	15.3	10.9

[a]calculated, [b]experimental, [c]Eq. (4), [d]Eq. (5)

5. Dependence of ϵ_∞ on Volume

We can obtain the derivative of ϵ_∞ with respect to volume from Eqs. (4) or (5) provided we know the corresponding derivative of the ω_g's which can be found by varying a_0 in a band structure calculation at the special points. The results of this procedure for $d\omega_g/d\ln V$ and $d\ln\epsilon_\infty/d\ln V$ obtained for Si, Ge and GaAs using either one or two special points are shown in Table II and compared with experimental data. Agreement is reasonable, especially considering the gross simplifications made. The correct sign of $d\ln\epsilon_\infty / d\ln V$ is obtained, except for Si with only one special point (because of a rather delicate cancellation of the effect of V on ω_g and a_0).

TABLE II.

Calculated and measured volume derivative of the infrared dielectric constant of Si, Ge, and GaAs and calculated volume derivatives of the corresponding ω_g's at one and two special points (in eV; in the latter case the values at $\mathbf{k}_1$ and $\mathbf{k}_2$ are given in that order).

Number of special points	Material	$d\ln\epsilon_\infty / d\ln V^a$	$d\ln\epsilon_\infty / d\ln V^b$	$d\ln\epsilon_\infty / d\ln V^c$	$d\omega_g / d\ln V$
	Si	-0.18	-0.02	0.6	-2.55
one	Ge	0.51	0.44	1.6	-3.55
	GaAs	0.64	0.53	0.7	-4.2
two	Si	1.45	0.73	0.6	-2.9,-5,2
$\mathbf{k}_1$,weight$\frac{3}{4}$	Ge	3.6	1.6	1.6	-4.3,-5.7
$\mathbf{k}_2$,weight$\frac{1}{4}$	GaAs	2.7	1.2	0.7	-4.3,-5.8

^{a}Eq. (4), ^{b}Eq. (5), ^{c}Exp, see [38,45]

6. Dependence of ϵ_∞ on Temperature

This dependence also has two components, one due to thermal expansion, which can be evaluated as discussed above, and the other produced by the direct electron-phonon interaction at constant volume. In order to evaluate the latter with Eqs. (4) and (5) we must calculate the temperature dependence of the gaps at the special points in the manner described in Sect. 2. The average linear temperature coefficient of these gaps, calculated in the range 100 - 400 K in which the dependence is fairly linear, is given in Table III for the case of one and two special points, broken up into SE, DW, and total contributions. We also list the corresponding total values of $d\ln\epsilon_\infty / dT$, calculated with Eqs. (4) and (5), and compare them with experimental data. In general the temperature derivatives calculated for one and two special points seem to bracket the experimental results.

TABLE III.

Temperature derivatives of special point gaps and infrared dielectric constants calculated for Si, Ge, and GaAs. The latter are compared with experimental results. For an early calculation of $d\ln\epsilon_\infty / dT$ based only on DW terms see [46].

Number of special points	Material	$d\omega_g / dT^a$ meV/K	$d\omega_g / dT^b$ meV/K	$d\omega_g / dT^c$ meV/K	$d\ln\epsilon_\infty / dT^d$ k^{-1}	$d\ln\epsilon_\infty / dT^e$ K^{-1}	$d\ln\epsilon_\infty / dT^f$ K^{-1}
	Si	-0.20	+0.12	-0.08	4.7×20^{-5}	3.2×10^{-5}	9.4×10^{-5}
one	Ge	-0.25	+0.13	-0.12	7.8×20^{-5}	5.3×10^{-5}	10×10^{-5}
	GaAs	-0.34	+0.23	-0.11	7.3×10^{-5}	4.9×10^{-5}	9×10^{-5}
two	Si	-0.30,-0.39	+0.14,+0.11	-0.16,-0.28	17.1×10^{-5}	9.5×10^{-5}	9.4×10^{-5}
	Ge	-0.331,-0.40	+0.08,+0.03	-0.25,-0.37	44.5×10^{-5}	17.7×10^{-5}	20×10^{-5}
$\mathbf{k}_1,\mathbf{k}_2$	GaAs	-0.42,-0.50	+0.16,+0.10	-0.26,-0.40	33.4×10^{-5}	17.8×10^{-5}	9×10^{-5}

[a]Debye-Waller terms, [b]Self energy terms, [c]DW + SE, [d]Eq. (4), [e]Eq. (5), [f]Experimental, see [38].

7. Temperature Dependence of the Dielectric Midgap Points: Application to Semiconductor Interfaces and Superlattices

When two semiconductors are brought or grown together, charge flows from one to the other at the interface. This occurs even if the two materials have a very small free carrier concentration (low doping and low temperature), in which case the charge transferred extends to only a few monolayers [41]. This charge determines the energy offset between the band structures of both materials. In the case of a metal-metal interface the equivalent transfer is calculated simply by aligning the Fermi levels of both constituents. For each semiconductor there is a characteristic energy, the dielectric midgap energy or DME, which, for the purpose of obtaining the band offsets, plays the same role as the Fermi energy in metals [14,42]. The DME is basically the midpoint of the Penn gap.

In the case of a metal-semiconductor-contact the differences between the Fermi levels and the conduction (n-type material) or valence band (p-type) edges, immediately outside the region of charge transfer at the interface, are called Schottky barriers. One can conjecture that these barriers are obtained by aligning the Fermi energy of the metal with the DME of the semiconductor [43].

The dependence of the DME's on volume can be calculated as described in [14]. It can be used to obtain the dependence of band offsets and Schottky barriers on pressure. Likewise, it is of interest to calculate the dependence of DME's on temperature by the method described in Sect. 2. Together with the dependence of the corresponding band edge it can be used to determine temperature dependences of band offsets and Schottky barriers.

The calculated temperature shifts of the DME's of Si and Ge are displayed in Figs. 5 and 6 together with the calculated shift of the $\Gamma_{25'}$ valence band state and the difference between DME and $\Gamma_{25'}$ shifts, which should correspond to the shift of the p-type Schottky barrier energy (Φ_p).

We note in Figs. 5 and 6 that the $\Gamma_{25'}$ state shifts much more (1.88×10^{-4} eV/K between 300 and 500 K for Si) than the DME (0.35×10^{-4} eV/K), contrary to what happens when changing the volume, in which case both shifts are rather similar (i.e., the

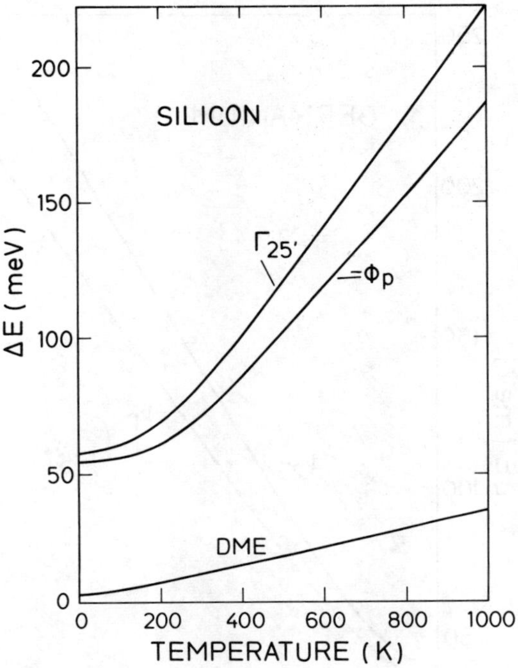

Figure 5. Calculated temperature shift of the dielectric midgap energy (DME) and the top of the valence band ($\Gamma_{25'}$) of silicon and also their difference which corresponds to the shift in the Schottky barrier height $-\Phi_p$.

volume term of $d\Phi_p/dT$ is small). The calculated temperature coefficient of Φ_p is thus -1.5×10^{-4} eV/K for Si and -2.1×10^{-4} eV/K for Ge. Experimental values of the temperature coefficient of Φ_p ranging from -1.4 to -1.7×10^{-4} eV/K are available for Si [44]. Agreement with our estimates is thus excellent.

Acknowledgement

We are greatly indebted to P.B. Allen, who developed some of the original codes to calculate the temperature dependence of electronic states and, to L. Viña, who built our first ellipsometer with very valuable helps of its inventor, D.E. Aspnes. Thanks are also due to the many graduate students and visiting scientists who have participated in the program and whose names can be found in the list of references.

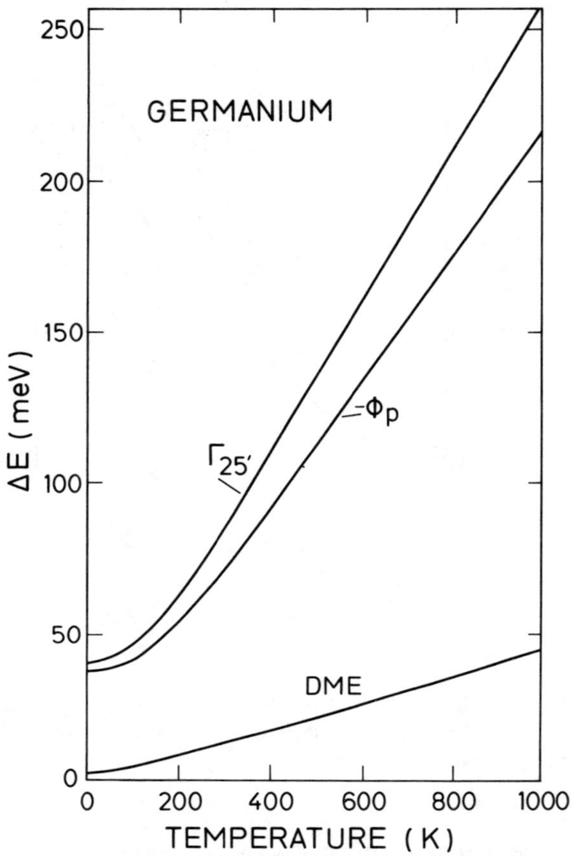

Figure 6. Calculated temperature shift of the dielectric midgap energy (DME) and the top of the valence band ($\Gamma_{25'}$) of germanium and also their difference which corresponds to the shift in the Schottky barrier height $-\Phi_p$.

References

1. F. Bassani, *Phys. Rev.* **108**, 263 (1957); also Semiconductors and Semimetals, Vol. 1, ed. by R.K. Willardson and A.C. Beer, Academic, NY (1966); F. Bassani and M. Yoshimina, *Phys. Rev.* **130**, 20 (1963).
2. F. Herman, *Phys. Rev.* **88**, 1210 (1952); *Physica* **20**, 801 (1954).
3. J.C. Phillips, *Phys. Rev.* **112**, 685 (1966); M.L. Cohen and J.R. Chelikowsky, Electronic Structure and Optical Properties of Semiconductors, Springer Series in Solid-State Sciences 75, Springer, Heidelberg (1988).
4. W.A. Harrison, Electronic Structure and the Properties of Solids; W.H. Freeman, San Francisco (1980); P. Vogl, H.P. Hjalmarson and J.D. Dow, *J. Phys. Chem. Solids* **44**, 365 (1983).
5. P. Hohenberg and W. Kohn, *Phys. Rev.* **136**, B864 (1964); D.M. Ceperly and B.J. Alder, *Phys. Rev. Lett.* **45**, 566 (1980).

6. G.B. Bachelet, D.R. Hamann and M. Schlüter, *Phys. Rev.* **B26**, 4199 (1982).
7. A. Baldereschi, *Phys. Rev.* **B7**, 5212 (1973).
8. B.I. Min, S. Massida and A.J. Freeman, *Phys. Rev.* **B38**, 1970 (1988); H.-Y. Wey and A. Zunger, *Phys. Rev. Lett.* **61**, 1505 (1988).
9. G.B. Bachelet and N.E. Christensen, *Phys. Rev.* **B31**, 879 (1985).
10. M.S. Hybertsen and S.G. Louie, *Phys. Rev. Lett.* **55**, 1418 (1985); W. van der Linden, P. Horsch, *Phys. Rev.* **B37**, 8351 (1988); W. Hanke and L. Sham, to be published.
11. A. Anastassakis and M. Cardona, *Phys. Stat. Sol. (b)* **104**, 589 (1981).
12. K.J. Chang, S. Froyen and M.L. Cohen, *Solid State Commun.* **50**, 105 (1984).
13. A. Blacha, H. Presting and M. Cardona, *Phys. Stat. Sol. (b)* **126**, 11 (1984).
14. M. Cardona and N.E. Christensen, *Phys. Rev.* **B35**, 6182 (1987).
15. N.E. Christensen and L. Brey, *Phys. Rev.* **B**, in press.
16. P. Lautenschlager, P.B. Allen and M. Cardona, *Phys. Rev.* **B31**, 2163 (1985): Si, Ge.
17. S. Gopalan and M. Cardona, *Physica Scripta Reprint Series*, (1988) to be published, GaAs.
18. H.Y. Fan, *Phys. Rev.* **82**, 900 (1951).
19. P.B. Allen and V. Heine, *J. Phys. C* **9**, 2305 (1976).
20. E. Antončik, *Czech. J. Phys.* **5**, 449 (1955).
21. J.S. Reid, *Acta Cryst.* **A39**, 1 (1983).
22. P.B. Allen and M. Cardona, *Phys. Rev.* **B27**, 4760 (1983): Si,Ge.
23. L. Viña, S. Logothetidis and M. Cardona, *Phys. Rev.* **B30**, 1979 (1984): Si,Ge.
24. S. Logothetidis, L. Viña and M. Cardona, *Phys. Rev.* **B31**, 947 (1985): InSb.
25. L. Viña, H. Höchst and M. Cardona, *Phys. Rev.* **B31**, 958 (1985): α-Sn.
26. P. Lautenschlager, P.B. Allen and M. Cardona, *Phys. Rev.* **B33**, 5501 (1986): Si,Ge.
27. P. Lautenschlager, M. Garriga, S. Logothetidis and M. Cardona, *Phys. Rev.* **B35**, 9174 (1987); S. Gopalan, P. Lautenschlager and M. Cardona, *Phys. Rev.* **B35**, 5577 (1987): GaAs.
28. P. Lautenschlager, M. Garriga and M. Cardona, *Phys. Rev.* **B36**, 4813 (1987): InP.
29. P. Lautenschlager, M. Garriga, L. Viña and M. Cardona, *Phys. Rev.* **B36**, 4821 (1987): Si.
30. S. Zollner, S. Gopalan, M. Garriga, J. Humlíček and M. Cardona, Proceedings, ICPS'88, Warsaw: GaP, InAs, GaSb.
31. W. Weber, *Phys. Rev.* **B15**, 4789 (1977).
32. K. Kunc and O.H. Nielsen, *Comp. Phys. Commun.* **16**, 181 (1979); **17**, 413 (1979).
33. M.L. Cohen and T.K. Bergstresser, *Phys. Rev.* **141**, 789 (1966).
34. S. Bednarek and U. Rössler, *Phys. Rev. Lett.* **48**, 1296 (1982).
35. O.J. Glembocki and F.H. Pollak, *Phys. Rev. Lett.* **48**, 413 (1982).
36. M. Cardona and P.B. Allen, *Helv. Phys. Acta* **58**, 307 (1985).
37. F. Bassani, D. Brust and J.C. Phillips, *Phys. Rev. Lett.* **9**, 94 (1962); D. Brust, *Phys. Rev.* **134**, 25 (1964).
38. M. Cardona, Structure and Properties of Solids, ed. by E. Burstein, Academic, New York (1972) p. 514.
39. J.C. Phillips, *Rev. Mod. Phys.* **42**, 290 (1970); J.A. Van Vechten, *Phys. Rev.* **182**, 891 (1969).
40. D. Penn, *Phys. Rev.* **128**, 2093 (1962).
41. N.E. Christensen, *Phys. Rev.* **37**, 4528 (1988); C.G. Van de Walle and R.M. Martin, *Phys. Rev.* **B35**, 8154 (1987).
42. C. Tejedor, F.M. Flores and E. Louis, *J. Phys.* **C10**, 2163 (1977).

43. J. Tersoff, *Surface Sci.* **168**, 275 (1986).

44. M.O. Aboelfotoh and K.N. Tu, *Phys. Rev.* **B33**, 6572 (1986).

45. A.R. Goñi, K. Syassen, K. Strössner and M. Cardona, Proc. Int. Conf. on high Pressure in Semiconductors Physics, Tomaszow (1988), to be published in Semicond. Sci. Technol.

46. P.Y. Yu and M. Cardona, *Phys. Rev.* **B2**, 3193 (1970).

PIEZOSPECTROSCOPY: ELECTRONIC STATES AND PHONONS IN SEMICONDUCTORS AND INSULATORS

A.K. RAMDAS AND S. RODRIGUEZ

Department of Physics
Purdue University
West Lafayette
Indiana 47907
USA

1. Introduction

Spectroscopic investigations have played an essential role in the elucidation of the properties of matter in all states of aggregation. In these studies symmetry considerations become indispensable for the characterization of the quantum mechanical states of the objects under study. These states are classified according to the irreducible representations of the appropriate symmetry group of the system, i.e., they are described by quantum numbers reflecting the values of conserved dynamical quantities.

Professor Bassani has made numerous important contributions to the field of solid state spectroscopy. We mention here, as a single example, his influential book with Pastori Parravicini on 'Electronic States and Optical Transitions in Solids' [1].

The application of external perturbations, modifying the symmetry of the physical states of the system being studied, allows the gleaning of information in a spectroscopic investigation in addition to that obtained in their absence. In many cases these perturbations lift the degeneracies of the energy levels and alter, e.g., the selection rules for optical transitions. Magnetic and electric fields have been widely used as external perturbations in spectroscopic studies [2,3]. The use of stress as an external perturbation had its beginning with the discovery of photoelasticity [4] by Brewster in 1816 but does not appear to have been used in spectroscopic studies until 1946 when Marieé and Mathieu [5] studied the modifications of the Raman spectrum of α-quartz under uniaxial compression. The use of piezospectroscopy in the study of crystals remained infrequent until the early 1960's but, since that time, it has been employed extensively to study electronic and vibrational excitations of solids [6-14].

The effect of uniaxial stress on Raman scattering has recently received considerable attention [15-19].

While the application of uniaxial stress as a tool in spectroscopy is unique to solid state physics, hydrostatic pressure can be employed to study the gaseous, liquid and solid states of matter.

Section II of this paper contains a summary of the results of the theory of elasticity with a view to its application in microscopic investigations of the electronic and vibrational states of crystalline solids. We place special emphasis on symmetry considerations. Section III deals with the general question of the definition of the deformation potentials, i.e., the change in the energy of individual states of a quantum mechanical system upon application of stress. Section IV is concerned with the effect of uniaxial stress on the properties of shallow donors and acceptors in semiconductors. Section V treats a few examples of the effect of stress on the lattice vibrations of crystals as they are revealed in

R. Girlanda et al. (eds.), Progress on Electron Properties of Solids, 65–98.

experiments on inelastic light scattering.

2. Elastic Properties of Solids

2.1. The strain Tensor

We consider a deformable body and a vector **s** joining two points P_0 and P within the solid. The position vectors of P_0 and P measured from an arbitrary origin are designated by $\mathbf{x}_0$ and $\mathbf{x}$ respectively. Assume now that the body is deformed by the action of external forces and that the point whose original position was P occupies the position P' (image of P) defined by a vector $\mathbf{x}'$. Similarly the point P'_0 having position vector $\mathbf{x}'_0$ is the image of P_0. The deformation can be most conveniently characterized by a function $\mathbf{u}(\mathbf{x})$ defined as the displacement of P due to the deformation. Obviously we have

$$\mathbf{x}' = \mathbf{x}+\mathbf{u}(\mathbf{x}). \tag{2.1}$$

Clearly the vector function $\mathbf{u}(\mathbf{x})$ must be single valued and continuous. Now, the change in the vector **s** is

$$\delta\mathbf{s} = \mathbf{s}'-\mathbf{s} = P'_0P'-P_0P = \mathbf{u}(\mathbf{x})-\mathbf{u}(\mathbf{x}_0), \tag{2.2}$$

so that if P and P_0 are sufficiently close to each other we can express $\delta\mathbf{s}$ in the following approximate form

$$\delta\mathbf{s} = (\mathbf{x}-\mathbf{x}_0)\cdot\nabla\mathbf{u} = (\mathbf{s}\cdot\nabla)\mathbf{u}. \tag{2.3}$$

There is no particular advantage to establish the results of the theory of elasticity in vector form so that we take a Cartesian coordinate system whose axes we designate by the suffixes 1, 2, and 3. Thus, the components of the vector **x** are x_1,x_2, and x_3. Then Eq. (2.3) can be rewritten as

$$s'_i = s_i+\delta s_i = s_i+\sum_j s_j\frac{\partial u_i}{\partial x_j}. \tag{2.4}$$

It is convenient to write

$$s'_i = s_i+\frac{1}{2}\sum_j\left[\frac{\partial u_i}{\partial x_j}-\frac{\partial u_j}{\partial x_i}\right]s_j+\frac{1}{2}\sum_j\left[\frac{\partial u_i}{\partial x_j}+\frac{\partial u_j}{\partial x_i}\right]s_j, \tag{2.5}$$

and define

$$\omega_{ij} = \frac{1}{2}\left[\frac{\partial u_i}{\partial x_j}-\frac{\partial u_j}{\partial x_i}\right], \tag{2.6}$$

and

$$\epsilon_{ij} = \frac{1}{2}\left[\frac{\partial u_i}{\partial x_j}+\frac{\partial u_j}{\partial x_i}\right]. \tag{2.7}$$

The nine functions ϵ_{ij} form a tensor called the strain tensor $\underset{\sim}{\epsilon}$. The geometrical properties of this tensor will be the object of later study. First, we describe the mathematical properties of ϵ_{ij} and ω_{ij}. The tensor $\underset{\sim}{\epsilon}$ is symmetric. By this we mean that $\epsilon_{ij} = \epsilon_{ji}$ so that if we arrange the nine numbers ϵ_{ij} in a table it reads

$$\underset{\sim}{\epsilon} = \begin{bmatrix} \epsilon_{11} & \epsilon_{12} & \epsilon_{13} \\ \epsilon_{21} & \epsilon_{22} & \epsilon_{23} \\ \epsilon_{31} & \epsilon_{32} & \epsilon_{33} \end{bmatrix}; \tag{2.8}$$

the arrangement is symmetric under reflection with respect to the diagonal from the upper left corner of the table to its lower right corner. The tensor ω_{ij} is antisymmetric, i.e., it satisfies the condition $\omega_{ij} = -\omega_{ji}$. In particular this implies that the diagonal elements ω_{11}, ω_{22} and ω_{33} vanish. For reasons that will soon become evident we define $\omega_{12} = -\omega_3$, $\omega_{23} = -\omega_1$, and $\omega_{31} = -\omega_2$. Then we write the tensor $\underset{\sim}{\omega}$ in the form

$$\underset{\sim}{\omega} = \begin{bmatrix} 0 & -\omega_3 & \omega_2 \\ \omega_3 & 0 & -\omega_1 \\ -\omega_2 & \omega_1 & 0 \end{bmatrix}. \tag{2.9}$$

It is now possible to write the tensor equation (2.5) in the form

$$\mathbf{s}' = \mathbf{s} + \underset{\sim}{\epsilon}\cdot\mathbf{s} + \vec{\omega}\times\mathbf{s}. \tag{2.10}$$

We see that, if $\underset{\sim}{\epsilon} = 0$, then the deformation is simply a rigid rotation about an axis parallel to the vector $\underset{\sim}{\omega}$ since, then, to first order in $\partial u_i / \partial x_j$ we have $\Sigma_i s_i'^2 = \Sigma_i s_i^2$. From now on we shall be concerned with the case in which $\vec{\omega} = 0$. The components of the tensor $\underset{\sim}{\epsilon}$ acquire a very simple geometrical significance. Let us take first the case in which the only nonzero component of $\underset{\sim}{\epsilon}$ is ϵ_{11}. This corresponds to a situation in which the body is stretched along the 1-axis by the fractional amount

$$\epsilon_{11} = \frac{s'_1 - s_1}{s_1} \tag{2.11}$$

with no change in the other directions. We now consider $\epsilon_{12} = \epsilon_{21} \neq 0$ while all the other components of $\underset{\sim}{\epsilon}$ vanish. If we take two vectors **a** and **b** embedded in the solid and lying parallel to the x_1- and x_2-axes, respectively, before deformation, then after the deformation they become **a**′ and **b**′ where $a'_1 = a$, $a'_2 = \epsilon_{12}a$, $b'_1 = \epsilon_{12}b$, and $b'_2 = b$. The angle θ' between **a**′ and **b**′ becomes

$$\cos\theta' = \frac{\mathbf{a}'\cdot\mathbf{b}'}{|\mathbf{a}'|\,|\mathbf{b}'|} = \frac{2\epsilon_{12}}{1+\epsilon_{12}^2} \approx 2\epsilon_{12}. \tag{2.12}$$

If we define $\alpha_{12} = \frac{\pi}{2} - \theta'$ we find, to first order in small quantities

$$\alpha_{12} \approx 2\epsilon_{12}, \tag{2.13}$$

i.e., the off-diagonal components of $\underset{\sim}{\epsilon}$ give the angular deformation of the solid. A deformation characterized by ϵ_{12} is called a shear strain.

When a body is deformed, the length $|\mathbf{s}|$ of the distance between neighboring points P_0 and $P(|\mathbf{s}| = P_0P)$ becomes

$$|\mathbf{s}'| = |\mathbf{s}|(1 + \sum_{ij}\epsilon_{ij}n_in_j) \tag{2.14}$$

to first order in the strain. Here $n_i (i = 1,2,3)$ are the components of a unit vector parallel to **s**. The quadratic form

$$\sum_{ij}\epsilon_{ij}x_ix_j \tag{2.15}$$

describes a quadric surface. This quantity, being a contraction of a product of two symmetric second rank tensors is invariant with respect to rotations of the coordinate axes. Since ϵ_{ij} is symmetric, the quadric (2.15) can always be reduced to its principal axes by a suitable rotation of the coordinate system.

The fractional change in volume is $\epsilon_{11}'+\epsilon_{22}'+\epsilon_{33}' = \Sigma_i\epsilon_{ii}$. The last equality follows from the invariance of the trace of the strain tensor.

The functions ϵ_{ij} defined by Eq. (2.7) are not independent if it is required that the transformation $\mathbf{x}\to\mathbf{x}' = \mathbf{x}+\mathbf{u}(\mathbf{x})$ be continuous. What are the restrictions to be imposed on the ϵ_{ij} to ensure the existence of a unique $\mathbf{u}(\mathbf{x})$? To answer this question we remark that, in general, the ϵ_{ij} do not determine the u_i. The rotation ω_{ij} must also be specified. The requirement that $\mathbf{u}$ be single-valued and continuous leads to

$$\epsilon_{ij,kl}+\epsilon_{kl,ij}-\epsilon_{ik,jl}-\epsilon_{jl,ik} = 0, \tag{2.16}$$

where $f_{i,j} = \partial f_i/\partial x_j$. This set of equations receives the name of equations of compatibility.

2.2. The Stress Tensor

We consider two types of forces acting within a solid body:

a. Body forces. These are forces like gravitational interactions and the forces exerted by external electric and magnetic fields on bodies carrying net electric charge or electrical current. We call $\mathbf{f}(\mathbf{x})$ the force density at $\mathbf{x}$ so that the total force and torque on the body are

$$\mathbf{F} = \int_{(V)} \mathbf{f}(\mathbf{x})d\mathbf{x}, \tag{2.17}$$

and

$$\mathbf{N} = \int_{(V)} \mathbf{x}\times\mathbf{f}(\mathbf{x})d\mathbf{x} \tag{2.18}$$

where the integrals are carried out over the volume V of the body.

b. Surface forces. These are forces exerted between different parts of the body. We define σ_{ij} as the component along the j-axis of the force per unit area exerted across an element of surface normal to the i-axis. We consider a small volume of the solid in the form of a tetrahedron three of whose faces are parallel to the planes (x_2,x_3), (x_3,x_1) and (x_1,x_2) and thus intersect at right angles at the vertex O. The fourth face is assumed to be a triangle of area ΔS intersecting the three perpendicular edges of the tetrahedron at distances Δx_1, Δx_2 and Δx_3 from their vertex O. We let $\hat{\mathbf{n}}$ be a unit vector normal to this face. The condition of equilibrium of this element of the body is

$$\mathbf{T}_n\Delta S-\frac{1}{2}\mathbf{T}_1\Delta x_2\Delta x_3-\frac{1}{2}\mathbf{T}_2\Delta x_3\Delta x_1-\frac{1}{2}\mathbf{T}_3\Delta x_1\Delta x_2 + \frac{1}{6}\mathbf{f}\Delta x_1\Delta x_2\Delta x_3 = 0 \tag{2.19}$$

where

$$\mathbf{T}_i = \sum_j \sigma_{ij}\hat{\mathbf{x}}_j \quad i,j = 1,2,3, \tag{2.20}$$

and $\mathbf{T}_n$ is the force per unit area on the surface ΔS. Now

$$\Delta S\cos(\hat{\mathbf{n}},\hat{\mathbf{x}}_1) = \Delta S\hat{\mathbf{n}}\cdot\hat{\mathbf{x}}_1 = \frac{1}{2}\Delta x_2\Delta x_3 \tag{2.21}$$

and similar results for the other two components. If, as $\Delta x_1,\Delta x_2,\Delta x_3$ tend to zero the body force remains finite, the last term in Eq. (2.19) can be neglected and we find

$$\mathbf{T}_n = \sum_{ij} \sigma_{ij} \hat{\mathbf{n}} \cdot \hat{\mathbf{x}}_i \hat{\mathbf{x}}_j = \hat{\mathbf{n}} \cdot \underset{\sim}{\sigma} \tag{2.22}$$

where

$$\underset{\sim}{\sigma} = \sum_{ij} \sigma_{ij} \hat{\mathbf{x}}_i \hat{\mathbf{x}}_j \tag{2.23}$$

is the stress tensor.

If the stress tensor contains a single non-vanishing component along its diagonal, say $\sigma_{33} = T$, we say that the body is subjected to uniaxial stress. This is the stress resulting in a long vertical rod loaded by hanging a weight on its end, x_3 being the vertical axis.

If we transform the axes $\hat{\mathbf{x}}_1, \hat{\mathbf{x}}_2, \hat{\mathbf{x}}_3$ to $\hat{\mathbf{e}}_1, \hat{\mathbf{e}}_2, \hat{\mathbf{e}}_3$ the components of the stress tensor become

$$\sigma'_{ij} = \hat{\mathbf{e}}_i \cdot \underset{\sim}{\sigma} \cdot \hat{\mathbf{e}}_j = \sum_{k,l} \hat{\mathbf{e}}_i \cdot \hat{\mathbf{x}}_k \hat{\mathbf{x}}_l \cdot \hat{\mathbf{e}}_j \sigma_{kl}. \tag{2.24}$$

For an uniaxial stress $\underset{\sim}{\sigma} = T \hat{\mathbf{x}}_3 \hat{\mathbf{x}}_3$ this transformation yields

$$\sigma'_{ij} = T \hat{\mathbf{e}}_i \cdot \hat{\mathbf{x}}_3 \hat{\mathbf{e}}_j \cdot \hat{\mathbf{x}}_3. \tag{2.25}$$

The stress tensor of a body subjected to hydrostatic pressure p is

$$\sigma_{ij} = -p\delta_{ij}. \tag{2.26}$$

Clearly, this tensor has the same form in all Cartesian frames of reference.

A stress field for which all components of $\underset{\sim}{\sigma}$ vanish except $\sigma_{11} = -\sigma_{22} = -\sigma$ is a pure shear stress. These forces correspond to a couple exerted along two parallel surfaces as can be immediately demonstrated by a rotation of the coordinate system defined by $\hat{\mathbf{e}}_1 = 2^{-1/2}(\hat{\mathbf{x}}_1 + \hat{\mathbf{x}}_2)$, $\hat{\mathbf{e}}_2 = 2^{-1/2}(-\hat{\mathbf{x}}_1 + \hat{\mathbf{x}}_2)$, $\hat{\mathbf{e}}_3 = \hat{\mathbf{x}}_3$. This transforms the tensor

$$\begin{bmatrix} -\sigma & 0 & 0 \\ 0 & \sigma & 0 \\ 0 & 0 & 0 \end{bmatrix}$$

into

$$\begin{bmatrix} 0 & \sigma & 0 \\ \sigma & 0 & 0 \\ 0 & 0 & 0 \end{bmatrix}$$

proving our statement.

2.3. Equations of Equilibrium

We consider an arbitrary singly connected volume V of the solid bounded by a surface S. The condition of equilibrium is

$$\int_{(V)} \mathbf{f}(\mathbf{x}) d\mathbf{x} + \int_{(S)} \mathbf{T}_n dS = 0. \tag{2.27}$$

The j-component of Eq. (2.27) can be written as

$$\int_{(V)} f_j(\mathbf{x}) d\mathbf{x} + \int_{(S)} \sum_i \sigma_{ij} n_i dS = 0. \tag{2.28}$$

Use of the divergence theorem gives

$$\int_{(V)} \left[f_j(\mathbf{x})+\sum_i \sigma_{ij,i}\right] d\mathbf{x} = 0$$

for every volume V. Thus

$$f_j(\mathbf{x})+\sum_i \sigma_{ij,i} = 0. \tag{2.29}$$

The torque on the volume V has components

$$N_i = \sum_{jk} \epsilon_{ijk} \int_{(V)} x_j f_k d\mathbf{x} + \sum_{jkl} \epsilon_{ijk} \int_{(S)} x_j \sigma_{lk} n_l dS.$$

We now apply the divergence theorem to the second integral and obtain

$$\begin{aligned} N_i &= \sum_{jk} \epsilon_{ijk} \int_{(V)} x_j f_k d\mathbf{x} + \sum_{jkl} \epsilon_{ijk} \int_{(V)} \partial_l (x_j \sigma_{lk}) d\mathbf{x} \\ &= \sum_{jk} \epsilon_{ijk} \left[\int_{(V)} x_j (f_k + \sum_l \sigma_{lk,l}) d\mathbf{x} + \int_{(V)} \sigma_{jk} d\mathbf{x} \right]. \end{aligned}$$

The first integral vanishes by virtue of Eq. (2.29). Thus, since in equilibrium $N_i = 0$ and the result is valid for all volumes we must have

$$\sum_{jk} \epsilon_{ijk} \sigma_{jk} = 0$$

or

$$\sigma_{jk} = \sigma_{kj}. \tag{2.30}$$

Therefore, if there is no net torque, the stress tensor $\underset{\sim}{\sigma}$ is symmetric.

2.4. The Equations of Elasticity

In many materials, when the stresses do not exceed a limit called the elastic limit the components of the stress are linearly related to those of strain. This statement is usually called Hooke's law and is expressed as

$$\sigma_{ij} = \sum_{kl} c_{ijkl} \epsilon_{kl}. \tag{2.31}$$

The quantities c_{ijkl} form a tensor of the fourth rank and are called the elastic constants or elastic stiffness coefficients of the medium. Since $\underset{\sim}{\sigma}$ and $\underset{\sim}{\epsilon}$ are symmetric, we conclude, without loss of generality, that

$$c_{ijkl} = c_{jikl} \tag{2.32}$$

and

$$c_{ijkl} = c_{ijlk}. \tag{2.33}$$

Therefore, there are in general 36 independent components of this tensor. However, in the static case in which an elastic energy can be defined we shall prove that

$$c_{ijkl} = c_{klij} \tag{2.34}$$

thereby reducing the number of independent constants to 21. The free energy per unit volume is

$$F = \frac{1}{2}\sum_{ijkl} c_{ijkl}\epsilon_{ij}\epsilon_{kl}. \tag{2.35}$$

This result is deduced as follows. Within a solid we isolate a volume V bounded by a surface S and increase the stress from zero to σ_{ij} according to $t\sigma_{ij}$ where t varies from zero to unity. The strain also increases from zero to ϵ_{ij} linearly in t. While t changes from t to $t+\Delta t$ the work done on the volume V is

$$\int_{(S)} dS \sum_{ij} t\sigma_{ij} n_i u_j \Delta t = t\Delta t \sum_{ij(V)} \int \frac{\partial}{\partial x_i}(\sigma_{ij}u_j)dx$$
$$= t\Delta t \sum_{ij(V)} \int (\sigma_{ij,i}u_j + \sigma_{ij}\epsilon_{ij})dx.$$

The first term gives the work done against the volume forces. Thus, in the absence of volume forces the elastic energy per unit volume, obtained integrating tdt from $t = 0$ to $t = 1$ is

$$F = \frac{1}{2}\sum_{ij}\sigma_{ij}\epsilon_{ij}. \tag{2.36}$$

Equation (2.35) follows from Eqs. (2.36) and (2.31). The equality (2.34) between the elastic stiffness coefficients follows from the equality of the second derivatives of F with respect to the components of the strain tensor.

It is customary to introduce the following notations

$$\sigma_{11} = \sigma_1, \quad \sigma_{22} = \sigma_2, \quad \sigma_{33} = \sigma_3 \tag{2.37}$$
$$\sigma_{23} = \sigma_4, \quad \sigma_{31} = \sigma_5, \quad \sigma_{12} = \sigma_6$$

and

$$\epsilon_{11} = e_1, \quad \epsilon_{22} = e_2, \quad \epsilon_{33} = e_3 \tag{2.38}$$
$$2\epsilon_{23} = e_4, \quad 2\epsilon_{31} = e_5, \quad 2\epsilon_{12} = e_6.$$

In this notation we write

$$\sigma_i = \sum_{j=1}^{6} c_{ij}e_j \tag{2.39}$$

with $c_{ij} = c_{ji}$. This has the advantage of some simplification often balanced by the disadvantage that, unlike c_{ijkl}, c_{ij} is not a tensor. The free energy reads

$$F = \frac{1}{2}\sum_{j=1}^{6}\sigma_i e_i = \frac{1}{2}\sum_{ij} c_{ij}e_i e_j. \tag{2.40}$$

Equations (2.39) can be solved for e_i in terms of the σ_i's to obtain

$$e_i = \sum_{j=1}^{6} s_{ij}\sigma_j \tag{2.41}$$

where $s_{ij} = s_{ji}$. These quantities are called the elastic compliance coefficients.

We recall that 21 constants are necessary to specify the elastic properties of a medium. However, if the medium possesses rotational and (or) reflection symmetry, relations exist between the constants c_{ij} and hence between the elastic compliance coefficients s_{ij}.

Consider, e.g., that the plane (x_1,x_2) is a mirror plane of the system, i.e., that the transformation $x'_i = x_1$, $x'_2 = x_2$, $x'_3 = -x_3$ leaves the system invariant. The elastic constants must remain invariant under this transformation, i.e., we must obtain $\sigma'_i = \sigma_i$ for $i = 1,2,3,6$ and $-\sigma_i$ for $i = 4,5$. But under reflections in the (x_1,x_2) plane all e_i

remain the same except e_4 and e_5 which change sign. Thus $c_{i4} = c_{i5} = 0$ for $i = 1,2,3,6$. Thus while a triclinic crystal has 21 independent elastic constants, a monoclinic crystal has only 13. Clearly for an orthorhombic crystal $c_{16} = c_{26} = c_{36} = c_{45} = 0$ reducing the number to 9. For a cubic crystal $c_{23} = c_{13} = c_{12}$ and $c_{44} = c_{55} = c_{66}$. Thus, a cubic crystal requires 3 constants for the specification of its elastic properties.

For isotropic bodies only two constants are needed. In fact, the elastic free energy F is a quadratic function of the ϵ_{ij} invariant under the orthogonal group $O(3)$. The only quadratic invariant combinations of the ϵ_{ij} are

$$(\sum_i \epsilon_{ii})^2 \quad \text{and} \quad \sum_{ij} \epsilon_{ij}^2.$$

Thus

$$F = \frac{1}{2}\lambda(\sum_i \epsilon_{ii})^2 + \mu \sum_{ij} \epsilon_{ij}^2. \tag{2.42}$$

The constants λ and μ are called the Lamé coefficients. The equations relating $\underset{\sim}{\sigma}$ and $\underset{\sim}{\epsilon}$ namely

$$\sigma_{ij} = \lambda\delta_{ij}(\epsilon_{11}+\epsilon_{22}+\epsilon_{33})+2\mu\epsilon_{ij} \tag{2.43}$$

follow immediately from Eqs. (2.36) and (2.42).

Comparison with the elastic constants of a cubic crystal shows that $c_{11} = 2\mu+\lambda$, $c_{12} = \lambda$ and $c_{44} = \mu$. Therefore a cubic crystal is elastically isotropic if $c_{11}-c_{12} = 2c_{44}$. This is called the Cauchy relation.

Finally we consider briefly the equation of motion for vibrations in an elastic body. The equation of motion can be deduced from Eq. (2.29) in which we add to $\mathbf{f}(\mathbf{x})$ the density of inertial force $-\rho\,\partial^2\mathbf{u}/\partial t^2$ where ρ is the density of the material. This is nothing but a trivial application of D'Alembert's principle. If $\mathbf{f} = 0$ we have

$$\rho\ddot{\mathbf{u}}_j = \sum_i \partial_i \sigma_{ij}. \tag{2.44}$$

For a cubic crystal we find

$$\rho\ddot{\mathbf{u}}_j = (c_{12}+c_{44})\frac{\partial}{\partial x_j}\nabla\cdot\mathbf{u}+c_{44}\nabla^2 u_j+(c_{11}-c_{12}-2c_{44})\frac{\partial^2 u_j}{\partial x_j^2}. \tag{2.45}$$

We see immediately that the Cauchy relation is a necessary and sufficient condition for elastic isotropy. In that case Eq. (2.45) becomes

$$\begin{aligned}\rho\ddot{\mathbf{u}} &= \mu\nabla^2\mathbf{u}+(\lambda+\mu)\nabla(\nabla\cdot\mathbf{u}) \\ &= (\lambda+2\mu)\nabla^2\mathbf{u}+(\lambda+\mu)\nabla\times(\nabla\times\mathbf{u}).\end{aligned} \tag{2.46}$$

In general there are as many independent elastic constants as there are independent invariant quadratic combinations of the components of the strain tensor. This conclusion follows from consideration of the expansion of the elastic free energy F in terms of the ϵ_{ij} and the requirement that F be invariant under the operations of the point group of the crystal. We use the group-theoretical notation of Koster et al. [20]. For example, in a cubic crystal of symmetry T_d the components of $\underset{\sim}{\epsilon}$ referred to the cubic axes generate the representation $[\Gamma_5\times\Gamma_5] = \Gamma_1+\Gamma_3+\Gamma_5$. Now $\sum_i \epsilon_{ii}$ belongs to Γ_1, $2\epsilon_{33}-\epsilon_{11}-\epsilon_{22}$ and $\sqrt{3}(\epsilon_{11}-\epsilon_{22})$ generate Γ_3 while $\epsilon_{23},\epsilon_{31},\epsilon_{12}$ belong to Γ_5. Use of Unsöld's theorem shows that the only independent quadratic invariants are $(\sum_i \epsilon_{ii})^2$,

$(2\epsilon_{33}-\epsilon_{11}-\epsilon_{22})^2+3(\epsilon_{11}-\epsilon_{22})^2 = 4(\epsilon_{11}^2+\epsilon_{22}^2+\epsilon_{33}^2-\epsilon_{22}\epsilon_{33}-\epsilon_{33}\epsilon_{11}-\epsilon_{11}\epsilon_{22})$ and $\epsilon_{23}^2+\epsilon_{31}^2+\epsilon_{12}^2$. The elastic free energy F is the sum of these three terms multiplied by appropriate coefficients or equivalently,

$$F = \frac{1}{2}c_{11}\sum_i \epsilon_{ii}^2 + c_{12}\sum_{i<j}\epsilon_{ii}\epsilon_{jj} + 2c_{44}\sum_{i<j}\epsilon_{ij}^2 \tag{2.47}$$

$$= \frac{1}{2}(c_{11}-2c_{44})(\sum_i \epsilon_{ii})^2 + c_{44}\sum_{ij}\epsilon_{ij}^2 - (c_{11}-c_{12}-2c_{44})\sum_{i<j}\epsilon_{ii}\epsilon_{jj}.$$

As a second example we consider a crystal of symmetry C_{4v}. In this case we have six independent quadratic invariants in the ϵ_{ij} and thus, six elastic constants are required. The invariants are (the high symmetry axis is labeled 3) $(\epsilon_{11}+\epsilon_{22})^2$, ϵ_{33}^2, $(\epsilon_{11}+\epsilon_{22})\epsilon_{33}$, $(\epsilon_{11}-\epsilon_{22})^2$, ϵ_{12}^2, $\epsilon_{23}^2+\epsilon_{31}^2$.

Under uniaxial stress the symmetry of a crystal is, in general, lowered. The new point symmetry group is the group of elements common to the point group of the undeformed crystal and the symmetry group of the strain quadric. This is called the Curie principle [21]; see also Nye [22].

When a uniform uniaxial compression is applied to a crystal, the strain tensor is represented by a triaxial ellipsoid having symmetry D_{2h} or higher. If two of the principal axes of the strain quadric are equal, this is an ellipsoid of revolution; the appropriate symmetry is then $D_{\infty h}$. If the ellipsoid becomes a sphere, its symmetry group is the orthogonal group $O(3)$. In order to determine the symmetry operations of a strained crystal it is necessary to know the orientation of the strain ellipsoid relative to the crystallographic axes.

An alternative and more convenient procedure is to treat the stress itself as the imposed condition and find the symmetry elements common to the unstrained crystal and to the stress tensor. The equivalence of these two procedures follows from the invariance of the elastic stiffness tensor under the operations of the point group of the undeformed solid. In fact, if R is such a symmetry operation, $P_R\underset{\sim}{\sigma} = P_R(\underset{\sim}{c}\cdot\underset{\sim}{\epsilon}) = (P_R\underset{\sim}{c}P_R^{-1})\cdot(P_R\underset{\sim}{\epsilon}) = \underset{\sim}{c}\cdot(P_R\underset{\sim}{\epsilon})$. Thus if R leaves $\underset{\sim}{\epsilon}$ invariant, it also leaves $\underset{\sim}{\sigma}$ invariant. This procedure is particularly simple to use when the external perturbation is uniaxial compression. In this case the symmetry of the stress tensor is $D_{\infty h}$ with the axis of rotation parallel to the externally applied force.

As an example we consider a crystal of point group symmetry T_d (or a localized center such as a substitutional donor or acceptor in *Si* and *Ge* having, thus, site symmetry T_d). Application of uniaxial stress along <100> reduces the T_d symmetry to D_{2d}. If the force is along a <110> or a <111> direction the symmetry of the strained system is C_{2v} or C_{3v}.

Once the reduced symmetry of the crystal in the presence of the external perturbation is determined, the correlation between the irreducible representations of the two groups can be obtained. From such a correlation, the removal of the degeneracy of a particular energy level can be readily deduced. The polarization characteristics of the stress induced components of an optical transition can also be obtained directly since the energy levels involved in the transition must conform to the irreducible representations of the reduced symmetry. Polarization studies in infrared absorption and inelastic scattering of light permit the determination of the irreducible representation of the states associated with the stress induced components of individual absorption or Raman lines.

2.5. Latent Anisotropy

Consider an optically isotropic cubic crystal like $NaClO_3$. Each individual ClO_3^- ion is clearly anisotropic and the overall isotropy demanded by the cubic symmetry is manifested by the 'orientational degeneracy' of the four ClO_3^- ions per unit cell disposed along the four $<111>$ axes of the structure. The trigonal symmetry and the 'uniaxial' anisotropy of the ClO_3^- ions thus remains 'latent'. Consider the lowest conduction band minima of silicon. These lie along $\mathbf{k} \| <100>$ at a distance about 0.8 of the Brillouin zone edge from $k = 0$ in reciprocal space; these minima conform to the C_{4v} point group. However, optical isotropy is guaranteed by the energetically equivalent minima along *all* the six cubic directions and hence the anisotropy of a single (or individual) minimum remains 'latent'. Anisotropic local centers may be produced in cubic crystals whose anisotropy remains once more 'latent' by the overall distribution of such centers in the body of the crystal, e.g., substitutional *Ni* in *Ge* which displaces along $<100>$ as a result of Jahn-Teller distortion [23,24].

The 'orientational degeneracy' mentioned above can be deduced in a straight-forward fashion. The point group of the constituent, e.g., the site symmetry of the ion in the $NaClO_3$ crystal is a subgroup of the crystallographic point group. The orientational degeneracy, l, is equal to the ratio of the orders of the symmetry group of the crystal (G) and that of the site symmetry (g). For a cubic crystal of symmetry O_h and site symmetry D_{4h}, C_{4v}, C_4, C_3, and C_2, $l = 3, 6, 12, 16$ and 24, respectively. Under ordinary conditions all equivalent positions of a defect in a crystal are populated with equal probability. These remarks apply to 'multivalley' semiconductors and to crystals which have anisotropic constituents.

The 'latent anisotropy' discussed above can be exposed by the application of external fields to the crystals. The external field may affect the different subsets and the associated energy levels differently. For example, a stress with compressive force along a specific cubic direction, **[100]**, will affect the **[100]**, $[\bar{1}00]$ conduction band minima of *Si* differently from those which are along **[010]**, **[010]**, **[001]** and $[00\bar{1}]$. Piezo-spectroscopic effects associated with the removal of latent anisotropy are, therefore, of considerable physical interest.

3. Deformation Potential Theory

While the symmetry arguments outlined above provide qualitative predictions, they do not give quantitative information on the changes experienced by the energy levels of a crystal and on the relative intensities of stress induced transitions, caused by time dependent perturbations (e.g., an incident electromagnetic wave). A deeper insight is obtained by a combined use of symmetry and perturbation theory. This theory was initially developed by Bardeen and Shockley [25] in their study of the mobilities of charge carriers in non-polar semiconductors and insulators. In order to calculate the mean free path of a charge carrier they studied the shifts in the energy bands produced by the thermally excited lattice vibrations. These shifts 'may be used as varying potentials in calculating the behavior of electrons and holes'. They are called deformation potentials.

We consider first the general question of the nature of the energy eigenvalues of an electron in a crystal subjected to stress. To simplify our discussion we study only the case of uniform strain. Our results are applicable when an external uniaxial stress is applied to a flat face of a crystal and to studies of the interaction of electrons with long wavelength phonons.

A point $\mathbf{x} = (x_1,x_2,x_3)$ in a crystal subjected to a uniform strain characterized by a strain tensor $\underset{\sim}{\epsilon}$ is displaced to $\mathbf{x}' = (x'_1,x'_2,x'_3)$ where

$$x'_i = x_i + \sum_j \epsilon_{ij} x_j. \tag{3.1}$$

Each lattice point $\mathbf{n} = \sum_i n_i \mathbf{a}_i$ $(n_i = 0, \pm 1, \pm 2, ...)$ is displaced to

$$(\underset{\sim}{I} + \underset{\sim}{\epsilon})\cdot\mathbf{n} \tag{3.2}$$

after the strain is effected. Our aim is to describe the eigenvalues and eigenfunctions of the conduction and valence electrons after the deformation assuming they are known in the unstrained crystal. Let

$$H = \frac{p^2}{2m} + U(\mathbf{x}) \tag{3.3}$$

be the one-particle Hamiltonian of the unstrained crystal. The periodic potential $U(\mathbf{x})$ is the sum of a scalar potential $V(\mathbf{x})$ and the spin-orbit coupling $(\hbar/4m^2c^2)\vec{\sigma}\cdot(\nabla V \times \mathbf{p})$. Here $\vec{\sigma}$ is the Pauli spin operator. The solutions of the Schrödinger equation

$$H\psi_{\nu k}(\mathbf{x}) = E_\nu(\mathbf{k})\psi_{\nu k}(\mathbf{x}) \tag{3.4}$$

where $\mathbf{k}$, a wave vector in the fundamental Brillouin zone (BZ), and ν, a band index, are assumed known. The functions $\psi_{\nu k}(\mathbf{x})$ are in the familiar Bloch form of a plane wave $\exp(i\mathbf{k}\cdot\mathbf{x})$ modulated by a function $u_{\nu k}(\mathbf{x})$ having the periodicity of the lattice.

In the presence of the strain $\underset{\sim}{\epsilon}$ the one-particle Hamiltonian is of the form

$$H(\underset{\sim}{\epsilon}) = \frac{p^2}{2m} + U(\underset{\sim}{\epsilon},\mathbf{x}). \tag{3.5}$$

We obtain the eigenvalues and eigenfunctions of $H(\underset{\sim}{\epsilon})$ assuming $\underset{\sim}{\epsilon}$ small and using perturbation theory. However, use of $U(\underset{\sim}{\epsilon},\mathbf{x}) - U(\mathbf{x})$ as perturbation is clearly unsatisfactory because even though $\underset{\sim}{\epsilon}$ is small, the relative displacement $\underset{\sim}{\epsilon}\cdot(\mathbf{n}'-\mathbf{n})$ of the lattice points at $\mathbf{n}$ and $\mathbf{n}'$ is large when $|\mathbf{n}'-\mathbf{n}|$ is sufficiently large. Thus, since in ordinary perturbation theory one assumes that the perturbed and unperturbed wave functions have considerable overlap, a modified approach becomes indispensable [26,27]. To solve the Schrödinger equation

$$H(\underset{\sim}{\epsilon})\phi_{\nu k}(\mathbf{x}) = W_\nu(\mathbf{k})\phi_{\nu k}(\mathbf{x}) \tag{3.6}$$

we transform $\mathbf{x}$ by means of the change of variable $\mathbf{x}' = (\underset{\sim}{I} - \underset{\sim}{\epsilon})\cdot\mathbf{x}$. To first order in $\underset{\sim}{\epsilon}$, the variables x'_i have at the point $\mathbf{x}$ in the deformed crystal the same numerical values as the coordinates of that point prior to the deformation. The Hamiltonian $H(\underset{\sim}{\epsilon})$ can now be expressed in terms of the operators $\mathbf{x}'$ and $\mathbf{p}' = (\hbar/i)\partial/\partial\mathbf{x}'$ in the form

$$H(\underset{\sim}{\epsilon}) = \frac{p'^2}{2m} - \frac{1}{m}\mathbf{p}'\cdot\underset{\sim}{\epsilon}\cdot\mathbf{p}' + U\left[\underset{\sim}{\epsilon},(\underset{\sim}{I}+\underset{\sim}{\epsilon})\cdot\mathbf{x}'\right] \tag{3.7}$$

where we keep terms independent of and linear in $\underset{\sim}{\epsilon}$. We observe that the potentials $U(\underset{\sim}{\epsilon},(\underset{\sim}{I}+\underset{\sim}{\epsilon})\cdot\mathbf{x}')$ and $U(\mathbf{x}')$ have their minima at the same values of $\mathbf{x}'$ so that their difference is now truly a small quantity. We define $U_{ij}(\mathbf{x}')$ and $D_{ij}(\mathbf{x}',\mathbf{p}')$ by

$$U\left[\underset{\sim}{\epsilon},(\underset{\sim}{I}+\underset{\sim}{\epsilon})\cdot\mathbf{x}'\right] = U(\mathbf{x}') + \sum_{ij}\epsilon_{ij}U_{ij}(\mathbf{x}') + O(\epsilon^2) \tag{3.8}$$

and

$$D_{ij}(\mathbf{x}',\mathbf{p}') = -\frac{1}{m}p_i'p_j' + U_{ij}(\mathbf{x}'). \tag{3.9}$$

This allows us to write

$$H(\underset{\sim}{\epsilon}) = \frac{p'^2}{2m} + U(\mathbf{x}') + \sum_{ij}\epsilon_{ij}D_{ij}(\mathbf{x}') + \ldots. \tag{3.10}$$

We note that the first two terms in Eq. (3.10), taken together, are formally identical to the Hamiltonian of the crystal in the absence of strain.

Corresponding to $\phi_{\nu k'}(\mathbf{x}')$ of the undeformed crystal, the eigenvalue of (3.10) is

$$W = E_\nu(\mathbf{k}') + \sum_{ij}\epsilon_{ij} < \nu\mathbf{k}' \,|\, D_{ij} \,|\, \nu\mathbf{k}' >. \tag{3.11}$$

The wave function $\phi_{\nu k'}(\mathbf{x}')$ has wave vector $\mathbf{k}'$ in the reduced coordinates and, thus, its wave vector in the BZ of the deformed solid is

$$\mathbf{k} = (\underset{\sim}{I} - \underset{\sim}{\epsilon})\cdot\mathbf{k}'. \tag{3.12}$$

Thus we obtain

$$W_\nu(\mathbf{k}) = E_\nu\left[(\underset{\sim}{I} + \underset{\sim}{\epsilon})\cdot\mathbf{k}\right] + \sum_{ij}\epsilon_{ij} < \nu\mathbf{k} \,|\, D_{ij} \,|\, \nu\mathbf{k} >. \tag{3.13}$$

However, we remember that the BZ of the crystal is modified under stress. The wave vector corresponding to $\mathbf{k}$ in the unstrained crystal is $\mathbf{k}' = (\underset{\sim}{I} - \underset{\sim}{\epsilon})\cdot\mathbf{k}$ so that if we wish to compare energies at the band edges of the strained and unstrained crystal we must compare $E_\nu(\mathbf{k})$ with $W_\nu(\mathbf{k}')$. To first order in $\underset{\sim}{\epsilon}$ this difference in energy is

$$W_\nu(\mathbf{k}') - E_\nu(\mathbf{k}) = \sum_{ij}\epsilon_{ij} < \nu\mathbf{k} \,|\, D_{ij} \,|\, \nu\mathbf{k} >. \tag{3.14}$$

The quantities

$$< \nu\mathbf{k} \,|\, D_{ij} \,|\, \nu\mathbf{k} > = -\frac{1}{m} < \nu\mathbf{k} \,|\, p_i p_j \,|\, \nu\mathbf{k} > + < \nu\mathbf{k} \,|\, U_{ij} \,|\, \nu\mathbf{k} > \tag{3.15}$$

are the deformation potentials.

In the deformable ion model it is assumed that the potential at $(\underset{\sim}{I} + \underset{\sim}{\epsilon})\cdot\mathbf{x}$ in the deformed crystal is equal to that at $\mathbf{x}$ in the undeformed solid. Thus, in this model $U_{ij} = 0$ and the deformation potentials reduce to the first term in the right hand side of Eq. (3.15). In the rigid ion model the result is somewhat more complicated and will not be discussed here [28,27].

Often we deal with the application of stress to systems in which the electron states in the highest occupied band are degenerate. This occurs, e.g., when we study the energy states of electrons at the top of the valence band of a semiconductor such as Si or Ge or in the study of acceptor states in these materials.

As an example we consider a cubic crystal of zinc-blende symmetry (point group T_d) or the acceptor states in a semiconductor having the diamond structure (substitutional site symmetry T_d). The deformation potential is the scalar

$$H' = \sum_{ij}\epsilon_{ij}D_{ij}(\mathbf{x}). \tag{3.16}$$

Since ϵ_{ij} is a symmetric tensor we can take $D_{ij}(\mathbf{x})$ to be symmetric without loss of generality.

We observe that, while the first term on the right hand side of Eq. (3.15) is appropriate for Bloch states in a crystal, for impurity states only the second terms need be considered. In either case the symmetry properties of the deformation potential are the same for identical symmetry groups.

4. Effects of Uniaxial Stress on the Donor and Acceptor Levels in Semiconductors

Under an external stress, the energy band extrema of semiconductors experience shifts and, when the crystal symmetry is lowered, exhibit splittings. Under hydrostatic stress, though the crystal symmetry is preserved, the band extrema shift with respect to one another; thus the particular band extremum relevant for a given set of impurity states can be altered. We restrict our discussion to uniaxial stresses.

The effect of uniaxial stress is described by a deformation potential of the form (3.16). In the experiments we discuss, the stresses applied are well within the elastic limit and the stress-strain relations are given by the standard elastic compliance constants s_{ij}. The quantities D_{ij} are the components of a symmetric second rank tensor operator already discussed in Sec. 3.

We now apply these concepts to donors in the multi-valley semiconductors, *Si* and *Ge*. We recall that the conduction band extrema occur along Δ for *Si* and at *L* for *Ge*. (This nomenclature can be found, for example, in Jones [29].) The shift in the energy of a level at **k** in the Brillouin zone can be written, within the framework of the deformation potential theory, as the expectation value of $H' = \sum_{ij} D_{ij}\epsilon_{ij}$ for the state in question. The interaction H' can conveniently be written as a linear combination of terms invariant under the operations of the group of **k**, i.e., the subgroup of the point group of the crystal leaving **k** invariant. For a level with **k** parallel to **[001]**, using x,y,z as the cubic axes, the only invariants are $(\epsilon_{xx}+\epsilon_{yy})\,(D_{xx}+D_{yy})$ and $\epsilon_{zz}D_{zz}$ so that the change in energy of the level is

$$\Delta E = \Xi_d(\epsilon_{xx}+\epsilon_{yy})+(\Xi_d+\Xi_u)\epsilon_{zz} \tag{4.1}$$

where Ξ_d and Ξ_u are constants called the dilatational and shear deformation potential constants, respectively, and depend only on the band state. In a similar fashion, for *Ge*, in which the conduction band minimum occurs at *L*, with **k** parallel to **[111]**

$$\Delta E = (\Xi_d+\frac{1}{3}\Xi_u)(\epsilon_{xx}+\epsilon_{yy}+\epsilon_{zz})+\frac{2}{3}\Xi_u(\epsilon_{yz}+\epsilon_{zx}+\epsilon_{xy}). \tag{4.2}$$

Equations (4.1) and (4.2) can be represented compactly as

$$\Delta E(\mathbf{k}^{(n)}) = \sum_{ij}\left[\Xi_d\delta_{ij}+\Xi_u\hat{\mathbf{k}}_i^{(n)}\hat{\mathbf{k}}_j^{(n)}\right]\epsilon_{ij} \tag{4.3}$$

where $\hat{\mathbf{k}}^{(n)}$ is a unit vector parallel to $\mathbf{k}^{(n)}$, the position of the nth conduction band minimum in **k**-space.

In the effective mass approximation, the dependence of the donor levels on strain is also given by equation (4.3). If it is assumed that both the dielectric constant and the effective masses characterizing the conduction band minima are unaltered by a small strain, then the energy level scheme of a group *V* donor in *Si* or *Ge* will be unaffected by the stress since the differential equation for the envelope function of the donor is not altered. However, the energy level schemes bearing *different* valley labels will be *shifted* relative to one another by amounts given by Eq. (4.3). Implicit in this deduction is the assumption that Ξ_d and Ξ_u are the same for all the donor levels as well as the conduction band extremum. Experimentally, this turns out to be a valid assumption for the *p*-states. The chemical splitting of the ground states $1s(\Gamma_1+\Gamma_3+\Gamma_5)$ and $1s(\Gamma_1+\Gamma_5)$ of donors in *Si* and *Ge*, respectively, requires the use of a Hamiltonian which includes both chemical splitting and strain; such a calculation has been given by Price [30] for *Ge* and by Wilson and Feher [31] for *Si*.

In order to compare experimental results with the theory it is useful to note that the shift of the 'center of gravity' of the valleys is given by

$$<\Delta E^{(n)}> = (\Xi_d+\frac{1}{3}\Xi_u)(\epsilon_{xx}+\epsilon_{yy}+\epsilon_{zz}). \tag{4.4}$$

Let $\hat{\mathbf{F}}$ be a unit vector along the applied force **F** and T the magnitude of **F** divided by the cross-sectional area of the sample. It can be shown that the shift of a given valley n with respect to the center of gravity is given by

$$\delta E^{(n)} = \pm\Xi_u T(s_{11}-s_{12})[(\hat{\mathbf{k}}^{(n)}\cdot\hat{\mathbf{F}})^2-\frac{1}{3}] \tag{4.5}$$

for *Si*. Similarly for *Ge*,

$$\delta E^{(n)} = \pm\frac{1}{2}\Xi_u T s_{44}[(\hat{\mathbf{k}}^{(n)}\cdot\hat{\mathbf{F}})^2-\frac{1}{3}]. \tag{4.6}$$

Here s_{11}, s_{12} and s_{44} are the elastic compliance constants of the crystal and the + and - signs correspond to tension and compression, respectively.

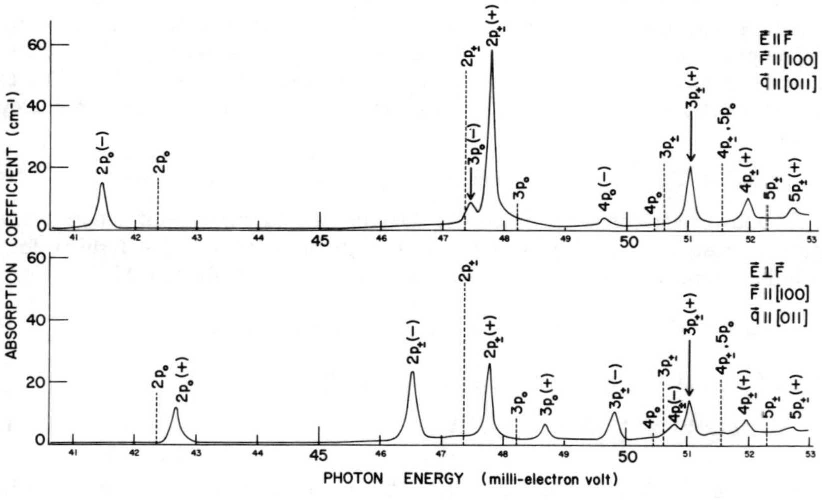

Figure 1. The effect of uniaxial stress on the excitation spectrum of *Si*(*As*) for **F**∥**[100]**; **q** is the direction of light propagation. Measurements were made using liquid helium as coolant. Donor concentration $n\sim3\times10^{15}\mathrm{cm}^{-3}$ (after [32]). (a) **E**∥**F**, **q**∥**[011]**, (b) **E**⊥**F**, **q**∥**[011]**.

In Fig. 1 we show the results of Aggarwal and Ramdas [2] displayed by Ramdas and Rodriguez [33] for the $1s(\Gamma_1)\rightarrow np$ transition in *Si*(*As*) subjected to uniaxial compression along **[100]**; polarized light was used with the electric vector **E** either parallel or

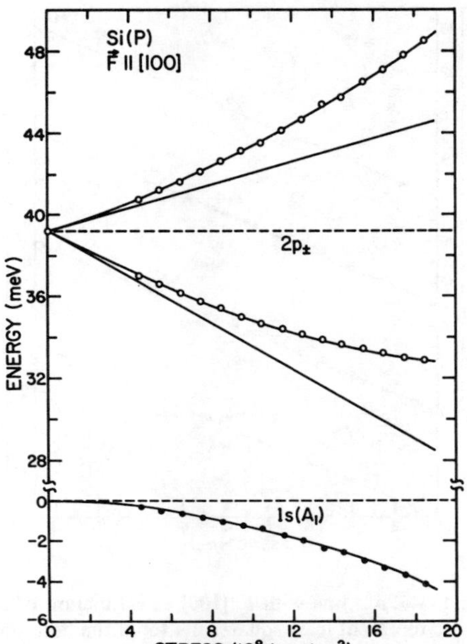

Figure 2. Stress dependence of the energies of the two components of the $1s(A_1) = 1s(\Gamma_1) \rightarrow 2p_{\pm}$ line of $Si(P)$ for **F**||**[100]** (after [34]).

perpendicular to **F**. As can be seen, each excitation line splits into two components, $np(+)$ and $np(-)$; here + and - signs denote the high and low energy components with respect to the zero stress position. In Fig. 2 we show the position of the $1s(\Gamma_1) \rightarrow 2p_{\pm}(+)$ and $1s(\Gamma_1) \rightarrow 2p_{\pm}(-)$ components of $Si(P)$ as a function of stress [34]. The solid straight line are energy shifts of the two components of the $2p_{\pm}$ line from the zero stress position calculated from their energy difference divided in the ratio 2:1. The curved solid lines are due to the shift of the $1s(\Gamma_1)$ ground state with stress. The solid points are computed by taking the difference between the straight lines and the data points. In Fig. 3 we display the splitting $\Delta E_{2p\pm} = E_{2p\pm}(+) - E_{2p\pm}(-)$, as a function of stress for $Si(Sb)$, $Si(As)$, $Si(P)$ and $Si(Mg)$. The linear dependence of $\Delta E_{2p\pm}$ with stress has the same slope for all impurities and yields $\Xi_u = (8.77 \pm 0.07)eV$. A similar study for $Si(Bi)$ by Butler et al. [35] gave $\Xi_u = (8.77 \pm 0.14)eV$. The value for Ξ_u determined in this manner from the piezospectroscopy of donors equals within experimental error that of the conduction band minimum [34]. Jagannath and Ramdas [36] show that the Ξ_u characterizing the energy levels of intersitial lithium donors in Si is equal to that of the conduction band. Tekippe et al. [34] found that the Ξ_u for the $1s(\Gamma_1)$ state is significantly lower than for the np states, decreasing with increasing $6\Delta_c$, the valley-orbit splitting (see [33]); Jagannath and Ramdas [36] report that Ξ_u for the $1s(\Gamma_3 + \Gamma_5)$ state of $Si(Li)$ is identical to that of the np states.

We now turn our attention to the piezospectroscopy of acceptors in Si and Ge. The acceptor states reflect the degeneracy of the $\Gamma_8^+ (p_{3/2})$ and $\Gamma_7^+ (p_{1/2})$ valence band maxima. For **F** in an arbitrary direction, the Γ_8^+ state shifts and splits into two Kramers doublets whereas the Γ_7^+ state merely shifts. In order to exploit symmetry arguments to

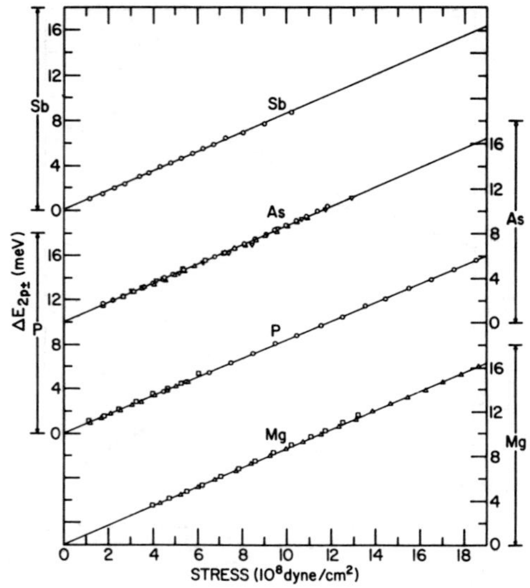

Figure 3. Splitting of the $1s(\Gamma_1)\rightarrow 2p_{\pm}$ line with $\mathbf{F}\|[100]$ as a function of stress for *Sb*, *As*, *P* and *Mg* donors in *Si*. The full lines represent least squares fits to all the data points for a given impurity (after [34]).

the best advantage, we rewrite the strain-induced Hamiltonian H' as the sum of terms involving the irreducible components of the strain tensor. We note that the components of a symmetric second rank tensor generate the representation

$$[\Gamma_5^- \times \Gamma_5^-] = \Gamma_1^+ + \Gamma_3^+ + \Gamma_5^+. \tag{4.7}$$

The combination $\epsilon_{xx}+\epsilon_{yy}+\epsilon_{zz}$ belongs to Γ_1^+, $2\epsilon_{zz}-\epsilon_{xx}-\epsilon_{yy}$ and $\sqrt{3}(\epsilon_{xx}-\epsilon_{yy})$ generate the irreducible representation Γ_3^+ whereas $\epsilon_{yz},\epsilon_{zx},\epsilon_{xy}$ generate Γ_5^+. This permits us to write

$$\begin{aligned} H' = {} & \frac{1}{3}(D_{xx}+D_{yy}+D_{zz})(\epsilon_{xx}+\epsilon_{yy}+\epsilon_{zz}) \\ & +\frac{1}{6}(2D_{zz}-D_{xx}-D_{yy})(2\epsilon_{zz}-\epsilon_{xx}-\epsilon_{yy}) \\ & +\frac{1}{2}(D_{xx}-D_{yy})(\epsilon_{xx}-\epsilon_{yy}) \\ & +2(D_{yz}\epsilon_{yz}+D_{zx}\epsilon_{zx}+D_{xy}\epsilon_{xy}). \end{aligned} \tag{4.8}$$

Within the framework of first order degenerate perturbation theory, it is appropriate to express the submatrices for each multiplet in terms of phenomenological constants, called deformation potential constants. For a Γ_8^+ multiplet the interaction submatrix H' is a linear combination of 16 linearly independent Hermitian matrices. These, in turn, can be constructed from J_x,J_y,J_z the angular momentum matrices for $j = 3/2$ by taking their powers and products [37]. Using Unsöld's theorem, one can show that the only invariant combinations of such matrices which also satisfy time reversal invariance are $\sum_i \epsilon_{ii}\sum_j J_j^2$, $(2\epsilon_{zz}-\epsilon_{xx}-\epsilon_{yy})(2J_z^2-J_x^2-J_y^2)+3(\epsilon_{xx}-\epsilon_{yy})(J_x^2-J_y^2) = 2\sum_i \epsilon_{ii}(3J_i^2-J^2)$ and $\sum_{i<j} \epsilon_{ij}\{J_i,J_j\}$,

where the curly brackets denote an anti-commutator. Thus we can write

$$H' = -a\sum_i \epsilon_{ii} - b\sum_i \epsilon_{ii}(J_i^2 - \frac{1}{3}J^2) - \frac{d}{\sqrt{3}}\sum_{i<j}\epsilon_{ij}\{J_i,J_j\}. \tag{4.9}$$

Here a,b and d are the deformation potentials for the Γ_8^+ multiplet. If the magnitude of the spin-orbit interaction, λ, is comparable to or smaller than the strain energy, it is necessary to consider the Γ_8^+ and Γ_7^+ mutliplets as a single entity. Then the $\Gamma_5^+ \times \Gamma_6^+$ states are described by

$$H' + H_{s.o.} = -a\sum_i \epsilon_{ii} - 3b\sum_i \epsilon_{ii}(I_i^2 - \frac{1}{3}I^2) \tag{4.10}$$

$$-d\sqrt{3}\sum_{i<j}\epsilon_{ij}\{I_i,I_j\} + \frac{1}{3}\lambda \mathbf{I}\cdot\vec{\sigma}.$$

Here σ_i are the 2×2 Pauli spin matrices and I_i are 3×3 angular momentum matrices. The operator $H' + H_{s.o.}$ is to be understood as a 6×6 matrix. If the spin-orbit splitting is large compared to H' it is adequate to use Eq. (4.9).

We will consider the energy states of a bound hole in the rest of our discussion and hence all the terms in the right hand side of equations (4.9) and (4.10) will have the opposite sign. We note that equations (4.9) and (4.10) refer to the valence band of the host crystal and hence have the full symmetry $\overline{O}_h$. It is clear that the lower $\overline{T}_d$ symmetry appropriate for the bound hole does not introduce any changes in the form of these equations. In principle, of course, the deformation potential constants will be different for different bound states. Bir et al. [38] and Suzuki et al. [39] have calculated the deformation potential constants of the ground state of the acceptors in the effective mass approximation. We first consider the case when λ is large. It can be shown [26,40] that the center of gravity of a Γ_8 multiplet of an acceptor is shifted by a $\sum_i \epsilon_{ii}$ whereas it splits into two doublets separated by the energy Δ given by

$$\Delta^2 = 2b^2[(\epsilon_{xx} - \epsilon_{yy})^2 + (\epsilon_{yy} - \epsilon_{zz})^2 + \tag{4.11}$$

$$+ (\epsilon_{zz} - \epsilon_{xx})^2] + 4d^2[\epsilon_{yz}^2 + \epsilon_{zx}^2 + \epsilon_{xy}^2].$$

For uniaxial stress with **F** along an axis with direction cosines n_x, n_y and n_z

$$\Delta^2(n_x,n_y,n_z) = \Delta_{100}^2 + 3(\Delta_{111}^2 - \Delta_{100}^2)K(n_x,n_y,n_z) \tag{4.12}$$

where $\Delta_{100} = 2b(s_{11} - s_{12})T$, $\Delta_{111} = (d/\sqrt{3})s_{44}T$ and $K(n_x,n_y,n_z) = n_y^2n_z^2 + n_z^2n_x^2 + n_x^2n_y^2$; $|\Delta_{100}|$ and $|\Delta_{111}|$ are the energy separations of the doublets for $\mathbf{F}\|\langle 100\rangle$ and $\mathbf{F}\|\langle 111\rangle$, respectively. It is interesting to note that if $|\Delta_{100}| = |\Delta_{111}|$, then $\Delta(n_x,n_y,n_z)$ is the same for all directions of **F**; we refer to this as 'stress isotropy'. The extrema of K occur for $\mathbf{F}\|\langle 111\rangle$, $\langle 110\rangle$ and $\langle 100\rangle$, these being 1/3, 1/4 and 0 respectively. In Fig. 4 the stress induced splitting of the $\Gamma_8 \rightarrow \Gamma_8$, $\Gamma_8 \rightarrow \Gamma_6$ and $\Gamma_8 \rightarrow \Gamma_7$ for $\mathbf{F}\|\langle 111\rangle$ and $\mathbf{F}\|\langle 100\rangle$ are shown. The figure also shows the selection rules for electric dipole transitions. In table I we give the symmetries and eigenvalues of the stress induced sublevels of a Γ_8 state for $\mathbf{F}\|\langle 111\rangle$ and $\langle 100\rangle$. In Fig. 5 the excitation spectrum of *Si*(*B*) under $\mathbf{F}\|\langle 111\rangle$ is shown [43]. As can be seen lines 1 and 2 split into four components as expected for a $\Gamma_8 \rightarrow \Gamma_8$ transition. The $2p'$ line, on the other hand, splits into two components consistent with a $\Gamma_8 \rightarrow \Gamma_6$ transition. The ordering of the

TABLE I.

Symmetries and eigenvalues of stress-induced sublevels of a Γ_8 state for $\mathbf{F}\parallel$<111> and <100>.

Direction of F	State	Eigenvalue*)
<111>	$\Gamma_5+\Gamma_6(\bar{C}_{3v})$	$a(s_{11}+2s_{12})T+(d/2\sqrt{3})s_{44}T$
	$\Gamma_4(\bar{C}_{3v})$	$a(s_{11}+2s_{12})T-(d/2\sqrt{3})s_{44}T$
<100>	$\Gamma_6(\bar{D}_{2d})$	$a(s_{11}+2s_{12})T+b(s_{11}-s_{12})T$
	$\Gamma_7(\bar{D}_{2d})$	$a(s_{11}+2s_{12})T-b(s_{11}-s_{12})T$

ground state sublevels has been deduced on the basis of the polarization and thermal depopulation effects shown in the figure. This, in turn, permitted the ordering of the sublevels of the excited states for lines 1 and 2. The positions of the high and low energy components of the $2p'$ line, $2p'(+)$ and $2p'(-)$, are shown in Fig. 6 for $\mathbf{F}\parallel$<110>, <100> and <111>. In Fig. 7 we show the spacing $2p'(+)-2p'(-) = \Delta$ as a function of stress [40]. The slopes of the stress dependence of Δ give b_0 and d_0, the ground state deformation potential constants. It is interesting to note that, within experimental error $|\Delta_{111}| = |\Delta_{100}| = |\Delta_{110}|$ yielding the stress isotropy referred to earlier. The origin of this stress isotropy of the ground state of $Si(B)$ is not yet explained. The positions of the stress induced components of lines 1 and 2 as a function of stress are shown in Fig. 8. The nonlinear stress dependence exhibited in particular by lines 2.1, 2.3 and 1.4 can be explained on the basis of the hybridization of the $\Gamma_5+\Gamma_6(C_{3v})$ sublevels of the excited states of lines 1 and 2. This hybridization results in unusual stress dependence of the relative intensities of the components. These aspects of the problem are discussed in detail by Chandrasekhar et al. [40].

The piezospectroscopy of acceptors in germanium has been investigated by Jones and Fisher [44,45], Fisher et al. [45], Dickey and Dimmock [47], Rodriguez et al. [48] and Barra et al. [49]. These studies have yielded the symmetry of the ground and excited states of group III acceptors and the group II acceptors Zn and Zn^-. In the case of $Ge(Zn^-)$, hybridization effects analogous to those of lines 1 and 2 in $Si(B)$ were observed for lines G and D [50,51].

The relative intensities of the stress induced components of a given transition can be exploited in making symmetry assignments for the sublevels of the initial and final states [48]. This proves to be valuable when polarization features alone do not give a unique assignment. This approach is valid only in the absence of significant hybridization with adjacent states. The relative intensities of the stress induced components of an absorption line are obtained by calculating the matrix elements of the electric dipole moment operator $\mathbf{d}$ for the different components of the multiplets. For a transition between two Γ_8 states the matrix elements of $\mathbf{d}$ with respect to states diagonalizing J_z are linear

*)Following Pikus and Bir [41], the eigenvalues are those of the bound hole; Kleiner and Roth [42] use electron energy. The definitions of the deformation potential constants of the latter are related to those of the former by $D_u = -\frac{3}{2}b$ and $D'_u = -\frac{1}{2}\sqrt{3}\,d$.

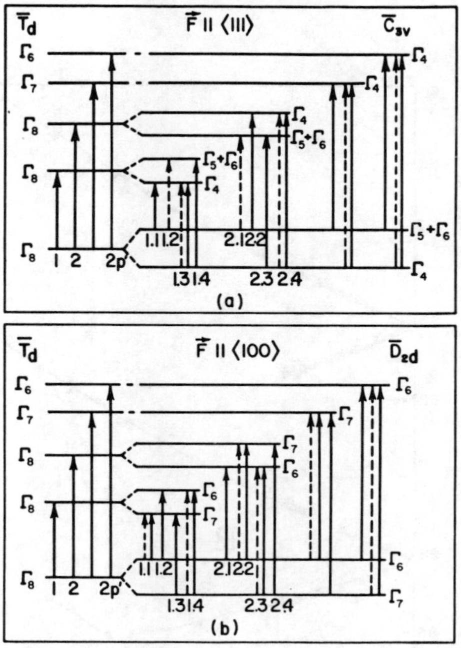

Figure 4. The allowed transitions from a Γ_8 ground state to Γ_6, Γ_7 and Γ_8 excited states of the double group $\bar{T}_d$ with compressive force **F** along <111> (a) or <100> (b). The site symmetry is C_{3v} and D_{2d} for the <111> and <100> compressions, respectively; the designations next to the levels denote the irreducible representations of the appropriate double point group following the notation of Koster et al. [20]. The solid and dashed lines in the stress-induced components correspond to incident radiation polarized perpendicular to and parallel to the applied force, respectively (after [32], Fig. 43; line 2.3 for **F**∥<111>: $\Gamma_4 \rightarrow \Gamma_{5+6}$ occurs only in the perpendicular polarization; the present figure corrects an error in this reference).

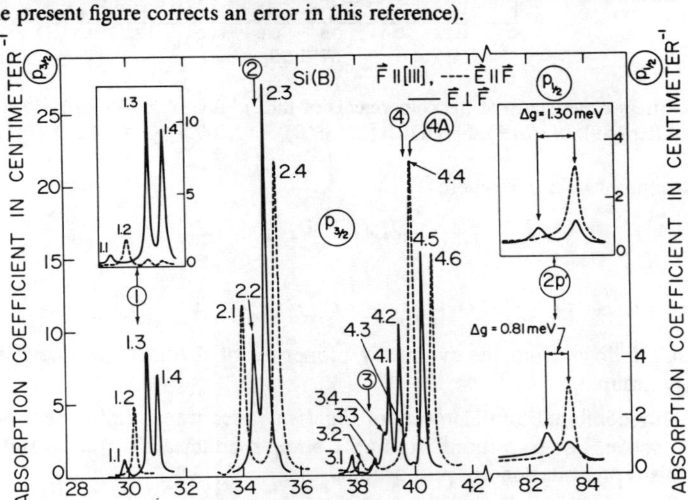

Figure 5. The effect of uniaxial stress on excitation lines $1-4A$ and $2p'$ of $Si(B)$; **F**∥<111>. The two insets show the effect of a larger stress on lines 1 and 2′. Liquid helium used as coolant. $p(300K) = 1.5 \times 10^{15} \text{cm}^{-3}$ (after [43]). ---, **E**∥**F**—, **E**;⊥**F**.

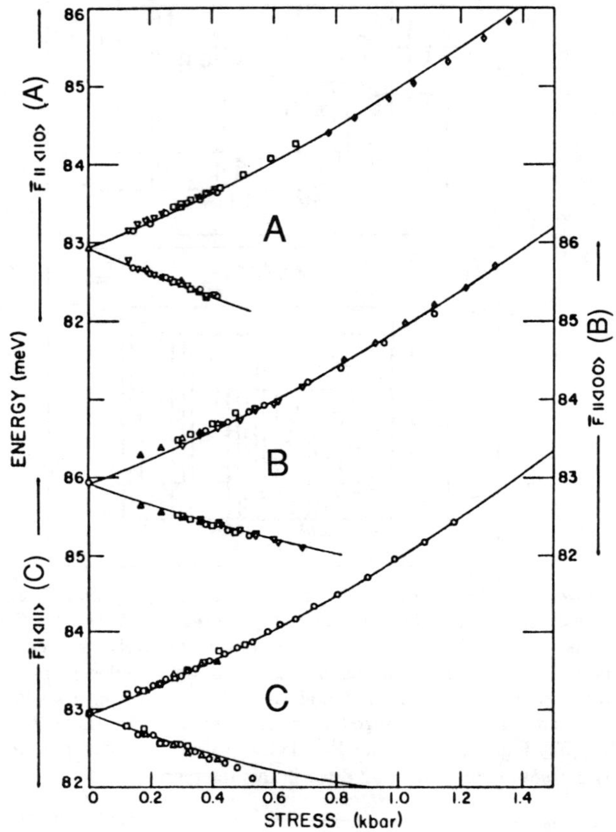

Figure 6. Stress dependence of the components of the $2p'$ line of $Si(B)$ for $\mathbf{F}\|$<111>, <100> or <110> (after [40]). A:$Si(B)$, $\mathbf{F}\|$<110>; B:$Si(B)$, $\mathbf{F}\|$<100>; C:$Si(B)$, $\mathbf{F}\|$<111>.

combinations of **U** and **V** where

$$U_x = \{J_y,J_z\}, \quad U_y = \{J_z,J_x\}, \quad U_z = \{J_x,J_y\}, \tag{4.13}$$

and

$$V_x = \{J_x,(J_y^2-J_z^2)\}, \; V_y = \{J_y,(J_z^2-J_x^2)\}, \; V_z = \{J_z,(J_x^2-J_y^2)\}. \tag{4.14}$$

This result follows from the symmetry properties of **d** under the operations of the site symmetry group.

The only combination of powers of the **J** matrices transforming as a polar vector are **U** and **V** above. For transitions from a Γ_8 state to either a Γ_6 or a Γ_7 state, the matrices of **d** in this representation are proportional to

$$\begin{bmatrix} -\sqrt{3}(\hat{\mathbf{x}}+i\hat{\mathbf{y}}) & 2\hat{\mathbf{z}} & \hat{\mathbf{x}}-i\hat{\mathbf{y}} & 0 \\ 0 & -(\hat{\mathbf{x}}+i\hat{\mathbf{y}}) & 2\hat{\mathbf{z}} & \sqrt{3}(\hat{\mathbf{x}}-i\hat{\mathbf{y}}) \end{bmatrix}, \tag{4.15}$$

and

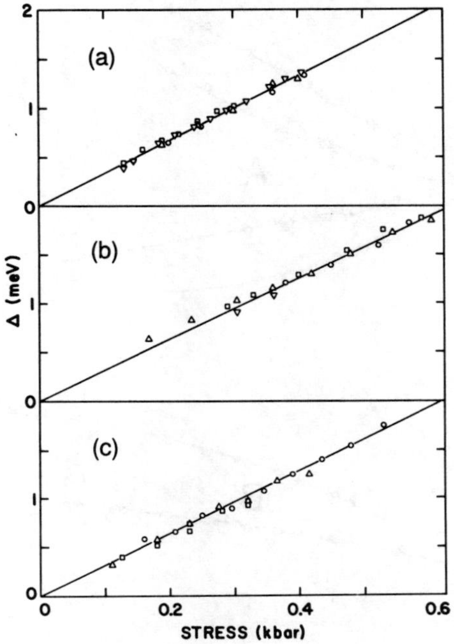

Figure 7. Stress dependence of $\Delta = E[2p'(+)] - E[2p'(-)]$ for F‖<111>, <100> or <110> of $Si(B)$ (after [40]). (a) $Si(B)$, F‖<110>; (b) $Si(B)$, F‖<100>; (c) $Si(B)$, F‖<111>.

$$\begin{bmatrix} \hat{\mathbf{x}}-i\hat{\mathbf{y}} & 0 & \sqrt{3}(\hat{\mathbf{x}}+i\hat{\mathbf{y}}) & 2\hat{\mathbf{z}} \\ 2\hat{\mathbf{z}} & -\sqrt{3}(\hat{\mathbf{x}}-i\hat{\mathbf{y}}) & 0 & -(\hat{\mathbf{x}}+i\hat{\mathbf{y}}) \end{bmatrix}, \tag{4.16}$$

respectively. Here $\hat{\mathbf{x}}, \hat{\mathbf{y}}, \hat{\mathbf{z}}$ are unit vectors along the cubic axes. Thus, in order to obtain the relative intensities of the stress induced components associated with a $\Gamma_8 \rightarrow \Gamma_8$ transition it is enough to obtain two independent, non-zero, matrix elements of **d** while for $\Gamma_8 \rightarrow \Gamma_6$ and $\Gamma_8 \rightarrow \Gamma_7$ only one such quantity suffices. Needless to say, one must obtain, in each case the transformation from the states diagonalizing J_z to the eigenfunctions of the stress Hamiltonian. More details can be found in the work of Kartheuser et al. [33].

5. Phonons in Crystals under Uniaxial Stress

In addition to the translational and point group symmetries of a crystal lattice the equations of motion of the atoms in a solid are invariant under time reversal when the external forces acting on the atoms are independent of their velocities. We consider the dynamical properties of a perfect crystal containing f atoms per primitive cell. We designate the lattice sites by the translation vectors **n** which are linear combinations with integral coefficients of three primitive non-coplanar vectors. The state of motion of the lattice is specified by giving the $3Nf$ Cartesian components $u_{n\alpha}(\alpha = 1,2,...3f)$ of the displacements of the atoms of the primitive cell located at **n** for all N cells constituting the solid. To first order in the displacements the equations of motion are

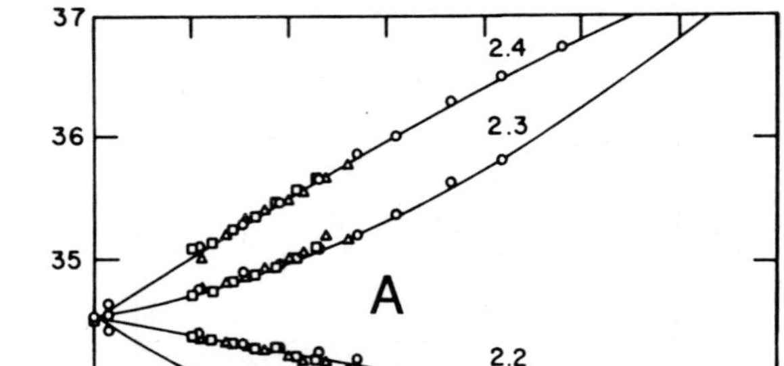

Figure 8. Stress dependence of the energies of the components of line 1 and 2 in $Si(B)$ for $\mathbf{F}\| <111>$ (after [40]).

$$M_\alpha \ddot{\mathbf{u}}_{n\alpha} = -\sum_{n'\alpha'} C_{\alpha\alpha'}(\mathbf{n}-\mathbf{n}')u_{n'\alpha'}. \tag{5.1}$$

The force constants $C_{\alpha\alpha'}(\mathbf{n}-\mathbf{n}')$ are real coefficients, symmetric in $n\alpha$ and $n'\alpha'$ depending only on $\mathbf{n}-\mathbf{n}'$ to conform to the translational symmetry of the structure; M_α is the mass of the atom associated with the αth degree of freedom. The functions

$$u_{n\alpha} = (NM_\alpha)^{-1/2} e_\alpha \exp[i(\mathbf{q}\cdot\mathbf{n}-\omega t)] \tag{5.2}$$

are solutions of Eqs. (5.1) provided ω and the $3f$-dimensional vector $\underset{\sim}{e}$ satisfy the eigenvalue equation

$$\sum_{\alpha'} \overline{C}_{\alpha\alpha'}(\mathbf{q}) e_{\alpha'} = \omega^2 e_\alpha. \tag{5.3}$$

Here $\overline{C}_{\alpha\alpha'}(\mathbf{q})$ is a kind of Fourier transform of the force constants defined by

$$\overline{C}_{\alpha\alpha'}(\mathbf{q}) = \sum_n (M_\alpha M_{\alpha'})^{-1/2} C_{\alpha\alpha'}(\mathbf{n}) e^{-i\mathbf{q}\cdot\mathbf{n}}. \tag{5.4}$$

For convenience we have taken Born-von Karman periodic boundary conditions with a basic period containing N primitive cells. This condition, together with the redundancy of $\mathbf{q}$ vectors differing by vectors of the reciprocal lattice, restricts the values of $\mathbf{q}$ to a finely spaced set of N vectors within the fundamental Brillouin zone of the crystal. The sum over **n** in Eq. (5.4) is limited to one Born-von Karman period. It is easy to show that the $3f \times 3f$ matrices $\bar{C}_{\alpha\alpha'}(\mathbf{q})$ are Hermitian so that for each **q** there are $3f$ eigenvectors $\underset{\sim}{e}$ and $3f$ eigenvalues $\omega_s^2(\mathbf{q})$, $(s = 1,2,...3f)$ some of which may be degenerate. Since the solutions (5.2) are complex, we have constructed $6f$ real linearly independent solutions of the equations of motion for each of the N possible choices of **q**.

Symmetry arguments based on the point group of the structure allow us to determine the possible degeneracies of $\omega_s^2(\mathbf{q})$ at particular values of **q**. For example, for $\mathbf{q} = 0$, $\bar{C}_{\alpha\alpha'}(0)$ is invariant under all operations of the point group, each operation being a permutation of the labels α leaving the matrix invariant. This allows us to classify the normal modes of vibration according to their symmetry properties thereby establishing their degeneracies.

Since we are interested here in first order Raman scattering we study only the nature of the phonons near $\mathbf{q} = 0$ [52]. The dynamical matrix $\bar{C}_{\alpha\alpha'}(\mathbf{q})$ can be written as

$$\bar{C}(\mathbf{q}) = \bar{C}(0) + \mathbf{q}\cdot\mathbf{S} + ..., \tag{5.5}$$

where S is a $3f \times 3f$ Hermitian matrix. Time reversal symmetry is expressed by the condition

$$\bar{C}(\mathbf{q}) = \bar{C}^*(-\mathbf{q}). \tag{5.6}$$

Thus

$$\mathbf{S} + \mathbf{S}^* = 0 \tag{5.7}$$

so that **S** is antisymmetric and purely imaginary. If the system has a center of inversion the permutation of the α indices must be such that $\bar{C}(-\mathbf{q}) = \bar{C}(\mathbf{q})$ so that $\mathbf{S} = 0$. Thus, all optical phonons in crystals possessing a center of inversion vary quadratically with **q** near $\mathbf{q} = 0$. For acoustic phonons we have, of course, $\omega^2 \propto q^2$ so that ω is linear in **q** in all cases.

In general **S** is a vector matrix of $3f$ dimensions. However, if we deal with a three fold degenerate optical phonon it is enough to consider a 3×3 submatrix of S corresponding to the degenerate phonons at $\mathbf{q} = 0$. Then we can write **S** as a linear term in **I** whose components are

$$I_x = \begin{bmatrix} 0 & 0 & 0 \\ 0 & 0 & -i \\ 0 & i & 0 \end{bmatrix}, \quad I_y = \begin{bmatrix} 0 & 0 & i \\ 0 & 0 & 0 \\ -i & 0 & 0 \end{bmatrix}, \quad I_z = \begin{bmatrix} 0 & -i & 0 \\ i & 0 & 0 \\ 0 & 0 & 0 \end{bmatrix}. \tag{5.8}$$

These are the only 3×3 matrices (except for a trivial real multiplicative factor) satisfying the restrictions on **S**. Thus, we write the submatrix of $\bar{C}(\mathbf{q})$ near $\mathbf{q} = 0$ as $\omega_0^2 + 2a\omega_0\mathbf{q}\cdot\mathbf{I} + ...$ The square roots of the eigenvalues of this matrix are $\omega_0 + aqm + ...(m = 1,0,-1)$ for small values of **q**. We observe that **S** transforms as a pseudovector under the operations of the orthogonal group in three dimensions. Since $\mathbf{q}\cdot\mathbf{S}$ must be invariant when **q** is also transformed together with **S** we conclude that $\mathbf{S} \neq 0$ only for crystals having no improper rotations in their point groups. An interesting case where these results are applicable is that of $Bi_{12}GeO_{20}$. This crystal has a body-centered cubic structure containing one formula unit per primitive cell. Its point group symmetry is T. We refer to the work of Imaino et al. [53] for details.

As examples, we consider the cases of calcium fluoride and α-quartz. Calcium fluoride (CaF_2) crystallizes in the face-centered cubic structure O_h^7 with a calcium atom at (0,0,0) and fluorine atoms at $\pm(1/4,1/4,1/4)$. The zone center optical phonons generate the representation $\Gamma_4^- + \Gamma_5^+$ of O_h, each component being 3-fold degenerate.

Alpha-quartz is a trigonal crystal with $D_3^4(D_3^6)$ symmetry with a primitive cell containing three SiO_2 units. At $\mathbf{q} = 0$, it has four non-degenerate optical phonons belonging to the totally symmetric representation Γ_1, four non-degenerate optical phonons belonging to the antisymmetric representation Γ_2 and eight optical phonons belonging to Γ_3, the doubly degenerate irreducible representation of D_3.

In view of the above discussion we expect no linear $\mathbf{q}$ dependence in the optical phonons of CaF_2. On the other hand, we shall see that this is possible for the two fold degenerate (zone center) optical phonons in α-quartz. We recall that a Raman transition in a centro-symmetric system occurs only if parity is conserved. For example, the excitation of a phonon belonging to Γ_4^- in CaF_2 changes the parity of the crystal and hence it is not Raman active. For Raman scattering involving phonons, far from a resonance, the appropriate polarizability tensor is symmetric. Thus its components transform under the operations of the point group of the crystal according to the representation generated by the symmetric direct product of the polar vector representation by itself. For O_h, the polar vector representation is Γ_4^- and its symmetric direct product by itself is

$$[\Gamma_4^- \times \Gamma_4^-] = \Gamma_1^+ + \Gamma_3^+ + \Gamma_5^+ .$$

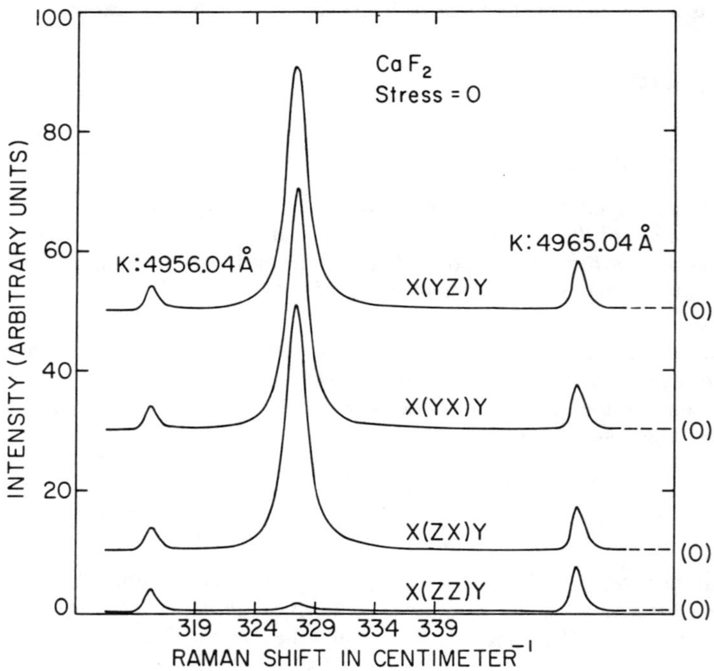

Figure 9. Zero-stress Raman spectrum of CaF_2 and its polarization characteristics; $T \sim 15K$. The gain $G = 0.5$ corresponds to 2000 counts/s for full scale deflection on the ratemeter. $\hat{\mathbf{x}} \| [\mathbf{100}]$, $\hat{\mathbf{y}} \| [\mathbf{010}]$ and $\hat{\mathbf{z}} \| [\mathbf{001}]$. On either side of the Raman line are calibration lines from a potassium discharge lamp (after [19]).

Therefore, the triply degenerate phonon Γ_5^+ is Raman active and the Raman spectrum of CaF_2 has a single Stokes (anti-Stokes) line. The phonon Γ_4^- is infra-red active. The Raman spectrum of CaF_2 is shown in Fig. 9, where the incident light is along one of the cubic axes, $\hat{\mathbf{x}}$, and the scattered radiation is collected along a second cubic axis $\hat{\mathbf{y}}$. The first letter in the parenthesis indicating the scattering geometry is the direction of polarization of the incident light and the second that of the scattered one.

In order to explain the polarization features of the Raman spectra of phonons, we determine the form of the Raman scattering tensor $\underset{\sim}{\alpha}$, assumed to be symmetric away from resonance, by reducing the representation it generates into its irreducible components. In O_h, denoting the cubic axes by $\hat{x}$, $\hat{y}$ and $\hat{z}$, $\alpha_{xx}+\alpha_{yy}+\alpha_{zz}$ belongs to Γ_1^+, $2\alpha_{zz}-\alpha_{xx}-\alpha_{yy}$ and $\sqrt{3}(\alpha_{xx}-\alpha_{yy})$ to Γ_3^+ and α_{yz}, α_{zx} and α_{xy} to Γ_5^+. Thus for CaF_2 the polarizability tensor for a Γ_5^+ optical phonon has only off-diagonal nonvanishing matrix elements referred to the cubic axes; furthermore, these elements are equal. The experimental results in Fig. 9 illustrate these features.

For the point group D_3 characterizing α-quartz, $\frac{1}{2}(\alpha_{xx}+\alpha_{yy})$ and α_{zz} belong to the totally symmetric representation Γ_1, while the two pairs $[(1/2)(\alpha_{xx}-\alpha_{yy}), -\alpha_{xy}]$ and $[\alpha_{yz}, -\alpha_{zx}]$ belong to the two-dimensional representation Γ_3. The first component in each of these pairs corresponds to the component X of a polar vector along one of the binary axes of the crystal, while the second transforms as the Y component with $\hat{y}$ being normal to the binary axis $\hat{x}$ and to the trigonal axis $\hat{z}$. The representation Γ_α generated by the components of $\underset{\sim}{\alpha}$ reduces to twice Γ_1 and twice Γ_3.

Next we use a theorem in group theory which we quote: The inner product of two vectors (or quantum-mechanical states if we are dealing with a dynamical system) is zero unless the vectors belong to the same row of identical irreducible unitary representations. We denote by X, Y, Z states transforming as the components of a polar vector along the axes $\hat{x}, \hat{y}, \hat{z}$ defined above for α-quartz. Z belongs to Γ_2 and the pair (X, Y) to Γ_3. Then $<X|Z> = 0$ and $<X|Y> = 0$. In the same way $<X|(1/2)(X^2-Y^2)> \neq 0$, while $<X|XY> = 0$. To find if a transition is allowed in Raman scattering, we obtain the matrix elements $<\psi_f|\hat{\epsilon}'\cdot\underset{\sim}{\alpha}\cdot\hat{\epsilon}|\psi_i>$. The initial state ψ_i, is the ground state ψ_0 of the crystal with no phonons present. Thus it belongs to Γ_1 and is denoted by $\psi_0(\Gamma_1)$. If ψ_f is also a Γ_1 state, the electric susceptibility tensor is of the form

$$[\underset{\sim}{\alpha}]_{\Gamma_1\to\Gamma_1} = \begin{pmatrix} p & 0 & 0 \\ 0 & p & 0 \\ 0 & 0 & q \end{pmatrix}$$

where p and q are real constants. In a similar way we find

$$[\underset{\sim}{\alpha}]_{\Gamma_1\to X} = \begin{pmatrix} r & 0 & 0 \\ 0 & -r & s \\ 0 & s & 0 \end{pmatrix}$$

and

$$[\underset{\sim}{\alpha}]_{\Gamma_1\to Y} = \begin{pmatrix} 0 & -r & -s \\ -r & 0 & 0 \\ -s & 0 & 0 \end{pmatrix},$$

r and s being suitable real constants. Clearly

$$[\underset{\sim}{\alpha}]_{\Gamma_1\to\Gamma_2} = 0.$$

From these results we deduce that the excitation (de-excitation) of a Γ_1 phonon occurs in the Raman spectrum only in the (xx), (yy) and (zz) polarizations, the former two having equal intensity. For scattering by Γ_3 phonons the (zz) intensity vanishes and off-diagonal components such as (xy) occur. The first-order Raman spectrum of α-quartz should exhibit 12 Stokes lines. Fig. 10 shows the Raman spectrum. We note that 17 lines are present. This is due to the splitting of some of the doubly degenerate lines into transverse optical (TO) and longitudinal optical (LO) phonons owing to the long-range polarization associated with the latter.

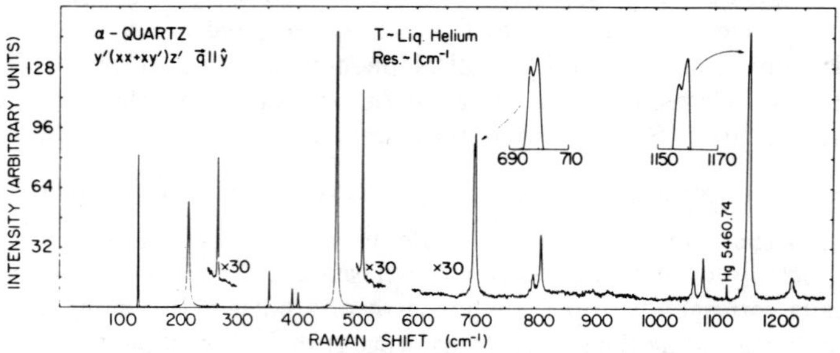

Figure 10. The Raman spectrum of α-quartz for scattering geometry $y'(xx+xy')z'$ with $\mathbf{q}\|\hat{\mathbf{y}}$. Liquid helium used as coolant. The spectrum was excited with 1.5 W of 5145Å Ar^+ radiation and recorded at $1cm^{-1}$ resolution. Note the LO-TO splittings of the 695 and $1160cm^{-1}\Gamma_3$ lines are just resolved. The vertical scale has been expanded by a factor of 30 for the 263 and $509cm^{-1}$ lines and for the spectrum beyond $600cm^{-1}$ (see Fig. 12 for the definition of $\hat{\mathbf{z}}'$ and $\hat{\mathbf{y}}'$) (after [54]).

We explore now the dispersion relation of the optical phonons as a function of wave vector $\mathbf{q}$ near $\mathbf{q} = 0$. We consider phonon states at $\mathbf{q} = 0$ generating the representation $\Gamma_2+\Gamma_3$ of D_3. This corresponds to both vector and pseudovector optical phonons. The dynamical submatrix associated with such vibrations is [55,56,18]

$$\omega_0^2 I_z^2+\omega_2^2(1-I_z^2)+(2B\omega_0/\hbar)\mathbf{q}\cdot\mathbf{I} \tag{5.9}$$

in terms of a suitable real parameter B. We note that when $\mathbf{q} = 0$, the eigenvalue ω_0^2 and ω_2^2 are double and single, respectively belonging to Γ_3 and Γ_2 states. Solution of the secular equation for these phonons gives

$$\omega = \omega_0 \pm Bq_z/\hbar+O(q^2) \tag{5.10}$$

and

$$\omega = \omega_2+O(q^2).$$

Fig. 11 shows the $128cm^{-1}$ Raman line of α-quartz at zero stress recorded with a triple-passed Fabry-Perot interferometer. The upper and lower parts correspond to $\bar{z}'(xx+xz')\bar{y}'$ and $z'(xx+xz')\bar{y}'$, yielding $\mathbf{q}\|\hat{\mathbf{z}}$ and $\mathbf{q}\|\hat{\mathbf{y}}$, as can be seen in Fig. 12. Note $\hat{\mathbf{x}},\hat{\mathbf{y}}$ and $\hat{\mathbf{z}}$ form a right-handed coordinate system with $\hat{\mathbf{x}}$ along a binary axis C_2,$\hat{\mathbf{z}}$ along the trigonal axis C_3,$\hat{\mathbf{y}}' = (1/\sqrt{2})(\hat{\mathbf{y}}-\hat{\mathbf{z}})$ and $\hat{\mathbf{z}}' = (1/\sqrt{2})(\hat{\mathbf{y}}+\hat{\mathbf{z}})$. As can be observed, the $128cm^{-1}$ line exhibits a splitting for $\mathbf{q}\|\hat{\mathbf{z}}$ and not for $\mathbf{q}\|\hat{\mathbf{y}}$. These results are in agreement with the predictions $E_{1,2} = E_0 \pm Bq_z$ for $\mathbf{q}\|\hat{\mathbf{z}}$ and $E_{1,2} = E_0$ for $\mathbf{q}\|\hat{\mathbf{y}}$. The splitting of $(0.115\pm0.003)cm^{-1}$ for $\mathbf{q}\|\hat{\mathbf{z}}$ obtained in the right-angle scattering geometry with 5145Å exciting radiation corresponds to $B = 2.14\cdot10^{-7}$ expressing the energies in cm^{-1}.

We consider now the effect of uniaxial stress on the Γ_5^+ optical phonons in CaF_2 [19]. Under compression along **[001]** the new group is D_{4h} and the Γ_5^+ levels split into a singlet $\Gamma_4^+(D_{4h})$ and a doublet $\Gamma_5^+(D_{4h})$. If the uniaxial compression is applied along **[111]** the new group is D_{3d}; the Γ_5^+ phonons split into a singlet $\Gamma_1^+(D_{3d})$ and a doublet $\Gamma_3^+(D_{3d})$. Finally if the force is along **[110]** the reduced symmetry is D_{2h} and the triply

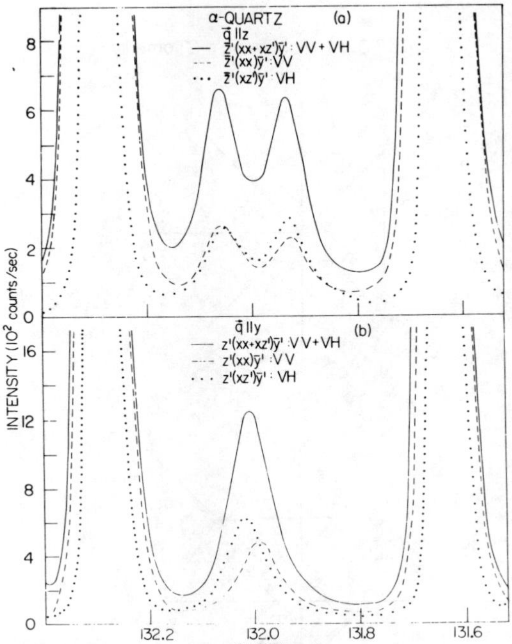

Figure 11. Effect of phonon propagation direction on the $128cm^{-1}$ Raman line of α-quartz. Liquid helium used as coolant. The abscissa refers to the Stokes Raman lines in the center of the figure excited with the 5145Å Ar^+ line. The features occurring on the right and left of the Raman lines are the unshifted laser radiation appearing with an interference order difference of 203 and 204, respectively, referred to the Raman lines (after [57]).

degenerate Γ_5^+ phonon splits into three singlets: $\Gamma_1^+(D_{2h})$, $\Gamma_2^+(D_{2h})$ and $\Gamma_5^+(D_{2h})$.

The effect of uniaxial stress is described by a deformation potential whose theory is completely analogous to that for electrons except that we deal with single valued representations of the point group of the system. The dynamical matrix $\overline{C}(0)$ is modified by addition of a deformation potential matrix

$$D = -a\sum_i \epsilon_{ii}(I_i^2 - \frac{1}{2}I^2) + b\sum_i \epsilon_{ii}I_i^2 - 2c\sum_{i<j}\epsilon_{ij}\{I_i, I_j\}. \tag{5.11}$$

In Fig. 13 the Raman spectrum of CaF_2 is shown in $x(zx)y$ and $x(yx)y$ polarizations for **F**‖**[001]**. There are two stress-induced components with a splitting of $\sim 6.0cm^{-1}$ for a stress of 6.32 kbar. The component at the higher frequency is observed in the $x(yx)y$ polarization whereas the component which shifts very little from the zero-stress position, denoted by the arrow, is observed in the $x(zx)y$ polarization. It should be noted that these polarizations are consistent with those predicted for the new point group D_{4h}.

Fig. 14 shows the polarization characteristic of the stress-induced components for **F**‖**[111]**. Again the Γ_5^+ line splits into two components. The spectrum under applied stress was obtained in this case with incident light which had polarization components along **[111]** as well as **[112]** directions. From the polarization characteristics it is clear that the higher frequency component is the singlet $\Gamma_1^+(D_{3d})$. The lower frequency component is the doublet having symmetry $\Gamma_3^+(D_{3d})$. These results and additional data for

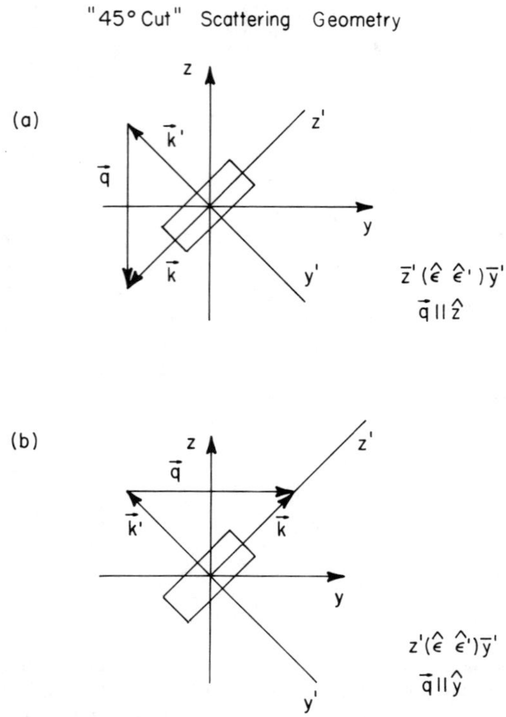

Figure 12. "45 degree cut" scattering geometry. The incident light is along $\mp\hat{\mathbf{z}}'$ in a) and b), respectively, where $\hat{\mathbf{z}}'$ bisects the angle between the trigonal axis along $\hat{\mathbf{z}}$ and $\hat{\mathbf{y}}$, an axis normal to both the trigonal and binary axis, a) $\bar{z}'(\hat{\epsilon}\hat{\epsilon}')\bar{y}'$, $\mathbf{q}\|\hat{\mathbf{z}}$; b) $z'(\hat{\epsilon}\hat{\epsilon}')\bar{y}'$, $\mathbf{q}\|\hat{\mathbf{y}}$ (after [54]).

$\mathbf{F}\|[\mathbf{110}]$ [58] shown in Fig. 15 yield the following deformation potentials for the Raman line of CaF_2 (at $327.3cm^{-1}$): $a = -1287cm^{-1}$, $b = -271cm^{-1}$ and $c = -103cm^{-1}$.

We now turn to the effect of uniaxial stress on the optical phonon of α-quartz. The expansion of the deformation potential for a crystal having D_3 symmetry is [57,54]

$$D = \frac{1}{2}(D_{xx}+D_{yy})(\epsilon_{xx}+\epsilon_{yy})+D_{zz}\epsilon_{zz}+ \tag{5.12}$$
$$\frac{1}{2}\left[(D_{xx}-D_{yy})(\epsilon_{xx}-\epsilon_{yy})+4D_{xy}\epsilon_{xy}\right]+2D_{yz}\epsilon_{yz}+2D_{zx}\epsilon_{zx}.$$

The corresponding deformation potential matrix for a Γ_3 optical phonon, including the effect of linear $\mathbf{q}$ discussed above, can be expressed directly in terms of energy by extracting the square root of the dynamical matrix, expanding it in powers of $\mathbf{q}$ and $\underset{\sim}{\epsilon}$ and keeping only first powers in these quantities. This can be done because the submatrix for $\mathbf{q} = 0$ is positive definite. For a doubly degenerate Γ_3 phonon (at $\mathbf{q} = 0$) the energies of the phonons are obtained by diagonalization of the matrix

$$H = H_0+a(\epsilon_{xx}+\epsilon_{yy})+b\epsilon_{zz}+\mathbf{h}\cdot\vec{\sigma}, \tag{5.13}$$

where

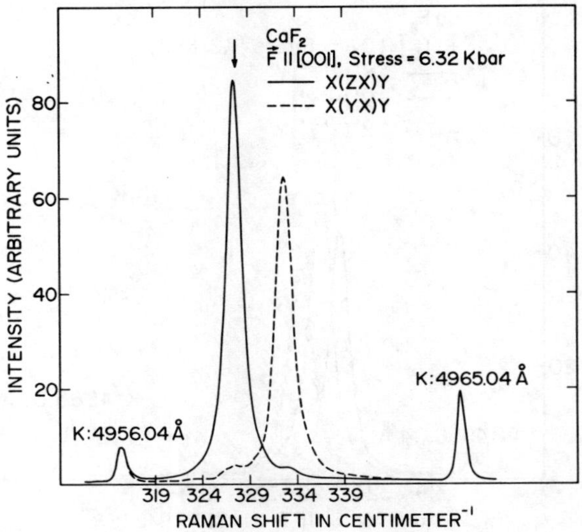

Figure 13. Polarization characteristics of the stress-induced components in CaF_2 for **F**||**[001]**, at a stress of 6.32 *kbar*; $T \sim 15K$. The arrow denotes the zero-stress position of the line. $\hat{x}$||**[100]**, $\hat{\mathbf{y}}$||**[010]**, $\hat{\mathbf{z}}$||**[001]** (after [19]).

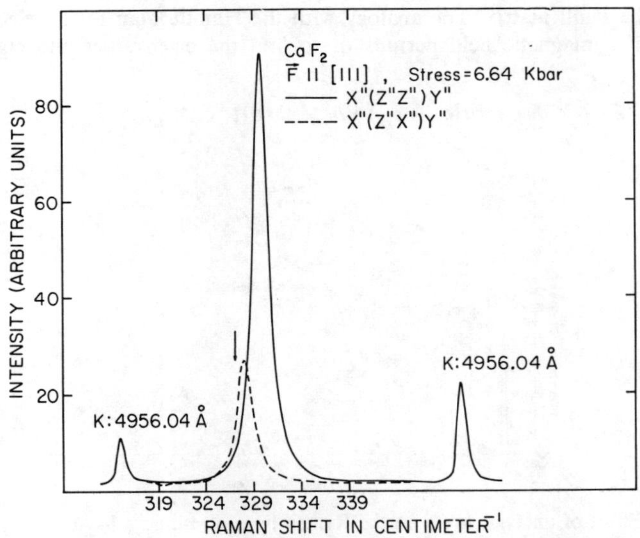

Figure 14. Polarization characteristics of the stress-induced components in CaF_2 for **F**||**[111]**, at a stress of 6.64 *kbar*; $T \sim 15K$. The arrow denotes the zero-stress position of the line. $\hat{x}''$||**[1$\bar{1}$0]**, $\hat{y}''$||**[11$\bar{2}$]**, $\hat{z}''$||**[111]** (after [19]).

$$h_x = 2c\epsilon_{xy} + d\epsilon_{zx}, \tag{5.14}$$

$$h_y = Bq_z, \tag{5.15}$$

$$h_z = c(\epsilon_{xx} - \epsilon_{yy}) + d\epsilon_{yz}, \tag{5.16}$$

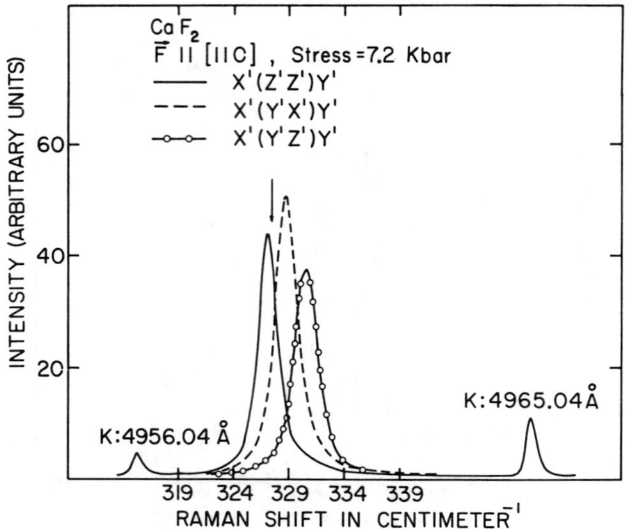

Figure 15. Polarization characteristics of the stress-induced components in CaF_2 for **F**||**[110]**, at a stress of 7.2 *kbar*; $T{\sim}15K$. The arrow denotes the zero-stress position of the line. $\hat{\mathbf{x}}'$||**[$\bar{1}$10]**, $\hat{\mathbf{y}}'$||**[001]**, $\hat{\mathbf{z}}'$||**[110]** (after [19]).

and $\vec{\sigma}$ is the Pauli matrix. The analogy with the Hamiltonian for an electron spin in the presence of a magnetic field permits us to find the eigenvalues and eigenvectors of H. They are

$$E_{1,2} = \hbar\omega_0 + [a(\epsilon_{xx} + \epsilon_{yy}) + b\epsilon_{zz}] \pm |\mathbf{h}|. \tag{5.17}$$

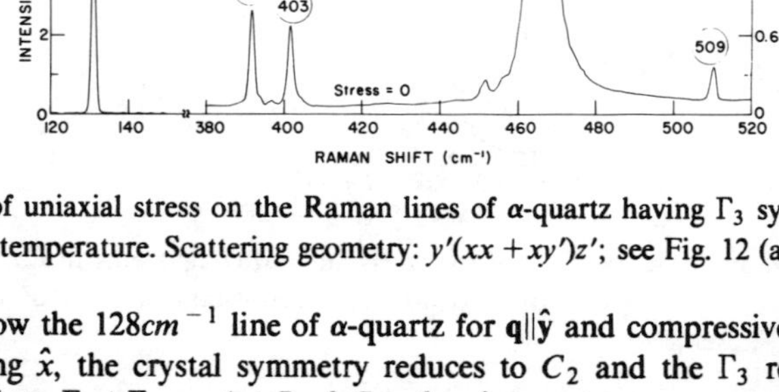

Figure 16. Effect of uniaxial stress on the Raman lines of α-quartz having Γ_3 symmetry. $\mathbf{F}||\hat{\mathbf{x}}$ and $\mathbf{q}||\hat{\mathbf{y}}$, liquid-helium temperature. Scattering geometry: $y'(xx + xy')z'$; see Fig. 12 (after [54]).

In Fig. 16 we show the $128 cm^{-1}$ line of α-quartz for $\mathbf{q}||\hat{\mathbf{y}}$ and compressive force $\mathbf{F}||\hat{\mathbf{x}}$. For compression along $\hat{x}$, the crystal symmetry reduces to C_2 and the Γ_3 representation of D_3 decomposes into $\Gamma_1 + \Gamma_2 = A + B$ of C_2; the eigenvectors corresponding to $\Gamma_1 = A$ and $\Gamma_2 = B$ are X and Y, respectively, and the components in Fig. 16 are labeled accordingly. The splitting $2|\mathbf{h}|$ for $\mathbf{q}||\hat{\mathbf{y}}$ and $\mathbf{F}||\hat{\mathbf{x}}$ is shown in Fig. 17 as a function of compressive stress. The splitting for $\mathbf{q}||\hat{\mathbf{z}}$ and $\mathbf{F}||\hat{\mathbf{x}}$ is given by

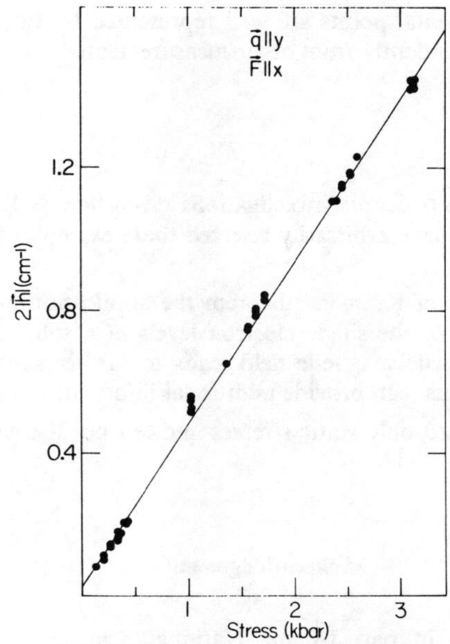

Figure 17. Splitting of the $128cm^{-1}$, $2|\mathbf{h}|$, as a function of uniaxial stress for compressive force **F** along $\hat{\mathbf{x}}$ and phonon propagation direction **q** along $\hat{\mathbf{y}}$. The straight line represents a least-squares fit to the experimental data (after [57]).

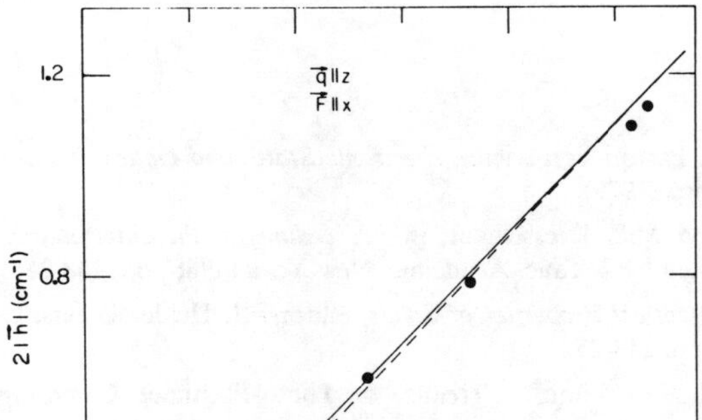

Figure 18. Splitting of the $128cm^{-1}$ line, $2|\mathbf{h}|$, as a function of uniaxial stress for $\mathbf{F}\|\hat{\mathbf{x}}$ and $\mathbf{q}\|\hat{\mathbf{z}}$. The solid line represents a theoretical fit to the data according to Eq. (5.18). The dashed straight line passing through the origin has the slope obtained for $\mathbf{F}\|\hat{\mathbf{x}}$ and $\mathbf{q}\|\hat{\mathbf{y}}$ (after [57]).

$$2|\mathbf{h}| = 2(B^2 q_x^2 + c'^2 T^2)^{1/2}. \tag{5.18}$$

The spacing between the two components of the $128cm^{-1}$ line for this geometry is shown in Fig. 18 and the experimental points are well represented by Eq. (5.18) using values of B and c' determined independently from other measurements.

Concluding Remarks

The exposition of piezospectroscopic investigations given here is limited to a few examples from a vast field. We have arbitrarily selected these examples from the authors' previous work.

We note that, by virtue of Kramers' theorem the application of stress does not split completely the degeneracy of the single electron levels of a solid. Judicious use of both uniaxial stress and an external magnetic field leads to further splitting of the electronic states of the system and, thus, can provide additional information on their nature.

Finally, we have discussed only static stresses and left out the whole area of piezomodulation spectroscopy.

Acknowledgements

This work was supported in part by the National Science Foundation (Grant No. DMR86-16787).

The authors wish to thank Y.R. Lee, G.C. La Rocca and M. Villeret for a critical reading of the manuscript.

References

1. F. Bassani and G. Pastori Parravicini, *Electronic States and Optical Transitions in Solids*, Pergamon Press, Oxford (1975).
2. G. Dresselhaus and M.S. Dresselhaus, in *Proceedings of the International School of Physics 'Enrico Fermi'*, edited by J. Tauc, Academic, New York (1966) pp. 198-256.
3. B.O. Seraphin, in *Optical Properties of Solids*, editor E.I. Heidemenakis, Gordon and Breach, New York (1970) pp. 213-252.
4. E.G. Coker and L.N.G. Filon, A Treatise on Photo-Elasticity, Cambridge University Press (1957) p. 8.
5. M. Marieé and J.P. Mathieu, *Compt. Rend.* **233**, 147 (1946).
6. P. Fisher and A.K. Ramdas, in *Physics of the Solid State*, edited by S. Balakrishna, M. Krishnamurthi and B. Ramachandra Rao, Academic Press, New York (1969) pp. 149-182.
7. I. Balsev, in *Semiconductors and Semimentals*, edited by R.K. Willardson and A.C. Beer (Academic Press, New York) **9**, 403 (1972).
8. I.G. Nolt and A.J. Sievers, *Phys. Rev.* **174**, 1004 (1968).
9. M. Cardona, *Modulation Spectroscopy*, supplement 11 of *Solid State Physics*, edited by F. Seitz, D. Turnbull and H. Ehrenreich, Academic Press, New York (1968).
10. J.F. Asell and M. Nicol, *J. Chem. Phys.* **49**, 5395 (1968).
11. S.S. Mitra, O. Brafman, W.B. Daniels and R.K. Crawford, *Phys. Rev.* **186**, 942 (1970).
12. B.A. Weinstein and G.J. Piermarini, *Phys. Rev. B* **12**, 1172 (1975).
13. G.S. Hobson and E.G.S. Paige, *Proc. Phys. Soc.* **88**, 437 (1966).
14. R.L. Aggarwal and A.K. Ramdas, *Phys. Rev.* **137**, A602 (1965).

15. E. Anastassakis, A. Pinczuk, E. Burstein, F.H. Pollak and M. Cardona, *Solid State Commun.* **8**, 133 (1970).
16. V.J. Tekippe and A.K. Ramdas, *Phys. Letters* **35A**, 143 (1971).
17. F. Cerdeira, C.J. Buchenauer, F.H. Pollak and M. Cardona, *Phys. Rev. B* **5**, 580 (1972).
18. V.J. Tekippe, A.K. Ramdas and S. Rodriguez, *Phys. Rev. B* **8**, 706 (1973).
19. S. Venugopalan and A.K. Ramdas, *Phys. Rev. B* **8**, 717 (1973).
20. G.F. Koster, J.O. Dimmock, R.G. Wheeler and H. Statz, *Properties of the Thirty-two Point Groups*, MIT Press, Cambridge, Mass. (1963).
21. P. Curie, *Collected Works,* Gauthier-Villars, Paris.
22. J.F. Nye, *Physical Properties of Crystals*, Clarendon Press, Oxford (1957) pp. 245-249.
23. P.P. Feofilov and A.A. Kaplyanskii, *Uspekhi* **5**, 79 (1962).
24. A.A. Kaplyanskii, *J. de Physique Colloque* C4-28, C4-39 (1967).
25. J. Bardeen and W. Shockley, *Phys. Rev.* **80**, 72 (1950).
26. G.L. Bir and G.E. Pikus, *Symmetry and Strain-induced Effects in Semiconductors*, Wiley, New York (1974) pp. 295-309.
27. E.P. Kartheuser and S. Rodriguez, *Phys. Rev. B* **33**, 772 (1986).
28. F.S. Khan and P.B. Allen, *Phys. Rev. B* **29**, 3341 (1984).
29. H. Jones, *The Theory of Brillouin Zones and Electronic States in Crystals*, 2nd ed., Amsterdam, North-Holland Publishing Co. (1975).
30. P.J. Price, *Phys. Rev.* **104**, 1223 (1956).
31. D.K. Wilson and G. Feher, *Phys. Rev.* **124**, 1068 (1961).
32. R.L. Aggarwal and A.K. Ramdas, unpublished (1965).
33. A.K. Ramdas and S. Rodriguez, *Reports on Progress in Physics* **44**, 1297 (1981).
34. V.J. Tekippe, H.R. Chandrasekhar, P. Fisher and A.K. Ramdas, *Phys. Rev. B* **6**, 2348 (1972).
35. N.R. Butler, P. Fisher and A.K. Ramdas, *Phys. Rev. B* **12**, 3200 (1975).
36. C. Jagannath and A.K. Ramdas, *Phys. Rev. B* **23**, 4426 (1981).
37. E.P. Kartheuser, S. Rodriguez and P. Fisher, *Phys. Stat. Solidi b* **64**, 11 (1974).
38. G.L. Bir, E.I. Butikov and G.E. Pikus, *J. Phys. Chem. Solids* **24**, 1467 (1963).
39. K. Suzuki, M. Okazaki and H. Hasegawa, *J. Phys. Soc. Japan* **19**, 930 (1964).
40. H.R. Chandrasekhar, P. Fisher, A.K. Ramdas and S. Rodriguez, *Phys. Rev. B* **8**, 3836 (1973).
41. G.E. Pikus and G.L. Bir, *Soviet Phys. - Solid State* **1**, 1502 (1960).
42. W.H. Kleiner and L.M. Roth, *Phys. Rev. Lett.* **2**, 334 (1959).
43. A. Onton, P. Fisher and A.K. Ramdas, *Phys. Rev.* **163**, 686 (1967).
44. R.L. Jones and P. Fisher, *Solid State Commun.* **2**, 369 (1964).
45. R.L. Jones and P. Fisher, *Phys. Rev. B* **2**, 2016 (1970).
46. P. Fisher, R.L. Jones, A. Onton and A.K. Ramdas, *J. Phys. Soc. Japan* **21**, Suppl., 224 (1966).
47. D.H. Dickey and J.O. Dimmock, *J. Phys. Chem. Solids* **28**, 529 (1967).
48. S. Rodriguez, P. Fisher and F. Barra, *Phys. Rev. B* **5**, 2219 (1972).
49. F. Barra, P. Fisher and S. Rodriguez, *Phys. Rev. B* **7**, 5285 (1973).
50. N.R. Butler and P. Fisher, *Phys. Rev. B* **13**, 5465 (1976).
51. K.J. Duff, P. Fisher and N.R. Butler, *Aust. J. Phys.* **33**, 73 (1980).
52. S. Rodriguez and A.K. Ramdas, in *Highlights of Condensed Matter Theory, Proceedings of the International school of Physics 'Enrico Fermi',* edited by F. Bassani, F. Fumi and M.P. Tosi, Course LXXXIX, North-Holland, Amsterdam (1985) pp. 369-420.
53. W. Imaino, A.K. Ramdas and S. Rodriguez, *Phys. Rev. B* **22**, 5679 (1980).

54. R.J. Briggs and A.K. Ramdas, *Phys. Rev. B* **16**, 3815 (1977).
55. A.S. Pine and G. Dresselhaus, *Phys. Rev.* **188**, 1489 (1969).
56. A.S. Pine and G. Dresselhaus, *Phys. Rev. B* **4**, 356 (1971).
57. M.H. Grimsditch, A.K. Ramdas, S. Rodriguez and V.J. Tekippe, *Phys. Rev. B* **15**, 5869 (1977).
58. S. Venugopalan, Ph.D. Thesis, Purdue University, unpublished (1973).

POLARONIC EFFECTS ON FREE AND HARMONICALLY BOUND ELECTRONS IN MAGNETIC FIELD

G. IADONISI, V. CATAUDELLA* AND D. NINNO

Dipartimento di Scienze Fisiche
Università di Napoli
Pad 19, Mostra d'Oltremare
I-80125 Napoli
Italy

* *Department of Theoretical Physics*
Umea University
S-90187 Umea, Sweden

We study the ground state energy of a three-dimensional polaron in a magnetic field. Our theory gives energy values, which, for small field coincide with those of the free polaron and in the strong field limit with those expected by the decoupling between the electronic and vibrational wavefunctions.

We calculate also the effective mass of the polaron in its motion along the direction of the field when this is weak or strong.

We compare our results with those given by the perturbation theory and the path-integral generalization of the polaron problem in a magnetic field. The result is that both the perturbation theory and the path-integral technique do not give a satisfactory description of the polarons in magnetic field.

1. Introduction

The usual approaches to the polaron problem in the Fröhlich model [1] are perturbative, variational or in the scheme of the Feynman path-integral method. Perturbative calculations can be used only when the electron-phonon coupling constant α is much less than one [2]; the variational scheme uses an Ansatz appropriate to the value of α [3,4]. Finally the Feynman path-integral formulation of the polaron problem [5] allows to make an unique Ansatz to study, for any α, the self-energy, the effective mass and the mobility of the free polaron.

When a magnetic field acts on the polaron, it is not possible to make a mathematical generalization of the Feynman procedure, although this has been done on physical grounds by Peeters and Devreese [6,7].

Their results, in the limit of small α and for any value of the magnetic field, are in agreement with those obtained when the electron-phonon interaction is treated perturbatively [8,10]. Nevertheless, both the path-integral and the perturbative calculations give, in the limit of strong magnetic field, energy values lower than those predicted by the decoupling of the electronic motion with respect to the vibrational ones [11]. This non-adiabatic behaviour is physically justified by the fact that, when the magnetic field is strong, the polarization cannot follow the electronic motion. The incorrect results given by the perturbative and path-integral approaches makes doubtfull their validity.

In this work we present a new variational solution of the three-dimensional polaron

R. Girlanda et al. (eds.), Progress on Electron Properties of Solids, 99–110.

problem in a magnetic field, extending a theory already developed for the two-dimensional case [12]. We are able to study not only the ground state, but also a set of excited states, although we discuss in this work only results to the ground state.

Our theory is valid for any value of α and any magnetic field. The results obtained are coincident with the perturbative ones in the weak field limit. However, when the magnetic field increases, our ground state energies are higher than those calculated perturbatively and in the limit of strong magnetic field, we find the same results as obtained in the non-adiabatic approximation [11]. We explain this fact showing that the average number of virtual phonons associated to the ground state, for a fixed α, is an increasing function of the magnetic field.

When α becomes larger, our energy values are larger than those obtained with the path-integral method. This occurs because for zero magnetic field the Feynman polaron self-energy [5] is correctly lower than those given by Lee, Low and Pines theory [4] and for strong magnetic field, our values go to the correct non-adiabatic results, which are higher than those given by the extension of the path-integral method.

We use the conservation of the components of the total linear momentum and of the total angular momentum in the direction of the magnetic field, i.e. the sum of the linear and angular momentum of the electron and of the phonons. For any fixed value of the angular momentum, we have expressions which permit to find the effective mass of the polaron in its motion along the direction of the magnetic field. We discuss this point and present explicit calculations only for weak and strong magnetic field.

Finally we study the problem of two-dimensional electronic oscillator in a perpendicular magnetic field and interacting with the longitudinal optical phonons via the Fröhlich interaction. This is not an academic problem, because recently quantum dots have been technically realized in a heterostructure, i.e. electrons bound in all directions in a microstructure of dimension less then 1 μm. The electronic potential is harmonic; its frequency ω_0 depends on the dimension of the dot and on the Fermi energy. Transitions between localized electronic states have been observed [13,17].

We show that this problem is very similar to the free polaron in a magnetic field. Also in this case we set up the states and give results of calculations for the ground state. Since ω_0 can be comparable with the phonon frequency, we obtain that the effective number of virtual phonons associate to the state is large and the perturbation theory cannot be applied, even though the material has a small electron-phonon coupling constant. Therefore the ground state energy is different from that of the free harmonic oscillator even when the magnetic field is zero.

The outline of this paper is as follows: in sect. 2 we present our variational theory of the free and harmonically bound polaron in a magnetic field; in sect. 3 we discuss the analytical asymptotic limits of our solution for weak and strong magnetic field; in sect. 4 the numerical calculations for the ground state are presented and in sect. 5 we discuss the results.

2. Mathematical Approach

2.1. Three-Dimensional Free Polaron in Magnetic Field

The hamiltonian of a polaron in a uniform magnetic field in the Fröhlich model [1] is

$$H = \frac{1}{2m}(\mathbf{p}+\frac{e}{c}\mathbf{A})^2 + \sum_{\mathbf{k}} \hbar\omega_L a_{\mathbf{k}}^\dagger a_{\mathbf{k}} + \sum_{\mathbf{k}} (V_k e^{i\mathbf{k}\cdot\mathbf{r}} a_{\mathbf{k}} + h.c.), \tag{1}$$

where **A** is the vector potential given by

$$\mathbf{A} = \left[-\frac{1}{2}By, \frac{1}{2}Bx, 0 \right],$$

V_k the electron-phonon interaction term

$$V_k = -i\frac{\hbar\omega_L}{k}\left[\frac{\hbar}{2m\omega_L}\right]^{\frac{1}{4}}\left[\frac{4\pi\alpha}{V}\right]^{\frac{1}{2}}$$

with

$$\alpha = \frac{e^2}{\hbar\epsilon}\left[\frac{m}{2\hbar m_L}\right]^{\frac{1}{2}},$$

and

$$\frac{1}{\epsilon} = \frac{1}{\epsilon_\infty} - \frac{1}{\epsilon_0}.$$

Here ω_L, V, ϵ_0, ϵ_∞, m indicate the longitudinal phonon frequency, the volume of the system, the static and high frequency dielectric constant and the conduction band mass of the electron. $a_{\mathbf{k}}(a_{\mathbf{k}}^\dagger)$ indicate the destruction (creation) operators of phonons in the free wave representation.

The hamiltonian (1) commutes with

$$P_z = p_z + \hbar\sum_{\mathbf{k}} k_z a_{\mathbf{k}}^\dagger a_{\mathbf{k}}. \tag{2}$$

Denoting by $\hbar Q_z$ the eigenvalue of P_z, the unitary transformation

$$U_1 = \exp\left[i\left[Q_z - \sum_{\mathbf{k}} k_z a_{\mathbf{k}}^\dagger a_{\mathbf{k}} \right] z \right] \tag{3}$$

on H, gives

$$H_1 = U_1^{-1} H U_1 = \frac{1}{2m}\left[\left[p_x + \frac{e}{c}A_x\right]^2 + \left[p_y + \frac{e}{c}A_y\right]^2\right] + \tag{4}$$

$$+\frac{\hbar^2}{2m}\left[Q_z - \sum_{\mathbf{k}} k_z a_{\mathbf{k}}^\dagger a_{\mathbf{k}}\right]^2 +$$

$$+\sum_{\mathbf{k}} \hbar\omega_L a_{\mathbf{k}}^\dagger a_{\mathbf{k}} + \sum_{\mathbf{k}} \left[V_k e^{i\vec{\chi}\vec{\rho}} a_{\mathbf{k}} + h.c.\right]$$

with $\vec{\chi} = (k_x, k_y)$ and $\vec{\rho} = (x,y)$. Introducing the new operators

$$a_{\chi,n,k_z} = \sqrt{\chi}\left[\frac{V}{(2\pi)^3}\right]^{\frac{1}{2}} \int_0^{2\pi} \phi_n(\varphi_{\vec{\chi}}) a_{\mathbf{k}} d\varphi_{\vec{\chi}}, \tag{5}$$

$$a_{\chi,n,k_z}^\dagger = \sqrt{\chi}\left[\frac{V}{(2\pi)^3}\right]^{\frac{1}{2}} \int_0^{2\pi} \phi_n^*(\varphi_{\vec{\chi}}) a_{\mathbf{k}}^\dagger d\varphi_{\vec{\chi}}$$

with $\phi_n(\varphi) = (2\pi)^{-\frac{1}{2}} e^{in\varphi}$, we obtain for H_1 (from now on we use as units for the length and the energy the polaron radius $\left(\frac{\hbar}{2m\omega_L}\right)^{\frac{1}{2}}$ and the phonon energy $\hbar\omega_L$ respectively),

$$H_1 = p_\rho^2 + \left[p_\varphi + \frac{\lambda}{4}\rho\right]^2 \tag{6}$$
$$+ \left[Q_z - \int_{-\infty}^{\infty} dk_z \int_0^{\infty} d\chi \sum_n k_z a^\dagger_{\chi,n,k_z} a_{\chi,n,k_z}\right]^2$$
$$+ \int_{-\infty}^{\infty} dk_z \int_0^{\infty} d\chi \sum_n a^\dagger_{\chi,n,k_z} a_{\chi,n,k_z} +$$
$$\int_{-\infty}^{\infty} dk_z \int_0^{\infty} d\chi \sum_n [T_{kn}(\rho)\varphi_n^*(\varphi) a_{\chi,n,k_z} + h.c.],$$

where

$$T_{kn}(\rho) = 2\pi \left[\frac{V}{(2\pi)^3}\right]^{\frac{1}{2}} \sqrt{\chi}\, V_k i^n J_n(\chi\rho)$$

$\lambda = \dfrac{eB}{mc\omega_L}$, $p_\rho = -i\dfrac{\partial}{\partial\rho}$, $p_\varphi = -\dfrac{i}{\rho}\dfrac{\partial}{\partial\varphi}$ and J_n are the Bessel functions. The operator

$$L_z = -i\frac{\partial}{\partial\varphi} + \int_{-\infty}^{\infty} dk_z \int_0^{\infty} \sum_n n a^\dagger_{\chi,n,k_z} a_{\chi,n,k_z} \tag{7}$$

commutes with H_1 and P_z, so that the eigenfunctions of H can be caracterized by the quantum numbers Q_z and M, eigenvalues of P_z and L_z respectively.

To apply the variational procedure for calculating eigenvalues and eigenstates of H_1, we choose as trial functions

$$|\Psi\rangle = e^{S(\vec{\rho})}|0\rangle\Phi(\vec{\rho}) \tag{8}$$

where

$$S(\vec{\rho}) = \int_{-\infty}^{\infty} dk_z \int_0^{\infty} d\chi \sum_n (f_{\chi,n,k_z}(\vec{\rho})^* a_{\chi,n,k_z} - f_{\chi,n,k_z}(\vec{\rho}) a^\dagger_{\chi,n,k_z}).$$

This form of the states (coherent states [18]) has been used with good results in the polaron [4] and in the polaronic exciton theory [19].

To solve our problem the functions $\Phi(\vec{\rho})$ and $f_{\chi,n,k_z}(\vec{\rho})$ are taken dependent on the coordinate $\vec{\rho}$ and are calculated variationally. It is easy to show that $|\Psi\rangle$ is an eigenstate of L_z with eigenvalue M if $f_{\chi,n,k_z}(\vec{\rho})$ and $\Phi(\vec{\rho})$ have the form

$$f_{\chi,n,k_z}(\vec{\rho}) = F_{\chi,n,k_z}(\rho)\phi_n(\varphi) \tag{9}$$
$$\Phi(\vec{\rho}) = R_M(\rho)e^{iM\phi}. \tag{10}$$

We obtain for $\langle\Psi|H_1|\Psi\rangle$

$$\langle\Psi|H_1|\Psi\rangle = \tag{11}$$
$$\langle\Phi| \left[(p_\rho + j_\rho)^2 + (p_\varphi + j_\varphi)^2 + \frac{\lambda\rho(p_\varphi + j_\varphi)}{2} + \frac{(\lambda\rho)^2}{16}\right.$$
$$+ (Q_z - K_z)^2 + \int_{-\infty}^{\infty} dk_z \int_0^{\infty} d\chi \sum_n \left[(1 + k_z^2)|f_{\chi,n,k_z}(\vec{\rho})|^2\right.$$

$$+ \quad |\nabla f_{\chi,n,k_z}(\vec{\rho})|^2 - (T_{nk}(\rho)\phi_n^*(\varphi)f_{\chi,n,k_z}(\vec{\rho}) + c.c.)\Big]\Big] |\Phi>$$

with $\nabla = (\frac{\partial}{\partial\rho}, \frac{1}{\rho}\frac{\partial}{\partial\varphi})$,

$$\mathbf{j} = \frac{i}{2}\int_{-\infty}^{\infty} dk_z \int_0^{\infty} d\chi \sum_n (f_{\chi,n,k_z}(\vec{\rho})\nabla f_{\chi,n,k_z}(\vec{\rho})^* - c.c.)$$

and

$$K_z = \int_{-\infty}^{\infty} dk_z \int_0^{\infty} d\chi \sum_n k_z |f_{\chi,n,k_z}(\vec{\rho})|^2.$$

To find a stationary state we must minimize eq. (11) with respect to $f_{\chi,n,k_z}(\vec{\rho})$ and Φ. The functional variation with respect to $f_{\chi,n,k_z}(\vec{\rho})$ gives

$$-\nabla^2 f_{\chi,n,k_z}(\vec{\rho}) - \nabla f_{\chi,n,k_z}(\vec{\rho})\cdot\frac{\nabla|\Phi|^2}{|\Phi|^2} + (1+k_z^2)f_{\chi,n,k_z}(\vec{\rho}) \tag{12}$$

$$-2(Q_z - K_z)k_z f_{\chi,n,k_z}(\vec{\rho}) = T_{nk}^*(\rho)\phi_n(\varphi),$$

where

$$\nabla^2 = \frac{1}{\rho}\frac{\partial}{\partial\rho}\left[\rho\frac{\partial}{\partial\rho}\right] + \frac{1}{\rho^2}\frac{\partial^2}{\partial\varphi^2}.$$

We have neglected the terms related to **j**, because, for symmetry, the solution of eq. (12) gives $\mathbf{j} = 0$. Since eq. (12) contains a non-linear term, it cannot be solved exactly; we overcome this difficulty substituting K_z with $<\Phi|K_z|\Phi>$. This quantity, for symmetry, must be zero when $Q_z = 0$; we put then $<\Phi|K_z|\Phi> = \eta Q_z$, with η to be calculated self-consistently. Finally, using eqs. (9) and (10) for $f_{\chi,n,k_z}(\vec{\rho})$ and Φ, we obtain

$$\frac{d^2}{d\rho^2}F_{\chi,n,k_x} + \left[\frac{1}{\rho} + \frac{1}{R_M^2}\frac{d}{d\rho}R_M^2\right]\frac{d}{d\rho}F_{\chi,n,k_z} - \tag{13}$$

$$\left[\frac{n^2}{\rho^2} + \gamma^2\right]F_{\chi,n,k_z} = -T_{nk}(\rho),$$

where

$$\gamma^2 = 1 + k_z^2 - 2k_z Q_z(1-\eta).$$

It is well known that neglecting the electron-phonon interaction the states $R_M(\rho)$ of an electron in an uniform magnetic field have the form [20]

$$R_M(\rho) = e^{-\frac{\rho^2}{4\beta^2}}\rho^{|M|}U_{l,M}(\rho), \tag{14}$$

where $U_{l,M}(\rho)$ are the generalized Laguerre polinomial of order l. The eigenvalues are

$$\lambda\left[l + \frac{M+|M|+1}{2}\right] + Q_z^2.$$

Now, for our problem, we take for the variational form of $R_M(\rho)$ the same expression as (14) with $l = 0$ and β as a variational parameter; in this way we find, for any M, the states of lower energy. Eq. (13) can then be exactly solved. We find

$$F_{\chi,n,k_z}(\rho) = \frac{\beta\Gamma(a)}{\sqrt{2}\,\Gamma(b)}\left[\frac{\rho^2}{2\beta^2}\right]^s \left[\phi\left[a,b,\frac{\rho^2}{2\beta^2}\right]\right. \tag{15}$$

$$\int_\rho^\infty dt\left[\frac{t^2}{2\beta^2}\right]^\nu e^{-\frac{t^2}{2\beta^2}} T^*_{nk}(t)\psi\left[a,b,\frac{t^2}{2\beta^2}\right]$$

$$\left. + \psi\left[a,b,\frac{\rho^2}{2\beta^2}\right]\int_0^\rho dt\left[\frac{t^2}{2\beta^2}\right]^\nu e^{-\frac{t^2}{2\beta^2}} T^*_{nk}(t)\,\phi\left[a,b,\frac{t^2}{2\beta^2}\right]\right],$$

where $\phi(a,b,x)$ and $\psi(a,b,x)$ are the regular and irregular confluent hypergeometric function, $\Gamma(x)$ is the gamma function and

$$a = \frac{1}{2}(-|M| + \sqrt{M^2+n^2+\gamma^2\beta^2}),$$

$$b = 1+\sqrt{M^2+n^2},$$

$$\nu = \frac{1}{2}(1+|M| + \sqrt{M^2+n^2}),$$

$$s = \frac{1}{2}(-|M| + \sqrt{M^2+n^2}).$$

With the function F_{χ,n,k_z} given by (15), we minimize eq. (11) with respect to β, for any fixed value of Q_z, M and B. In such a way we obtain the total energy of the system; its dependence on Q_z permits to calculate the effective polaron mass in the direction of the magnetic field.

It is evident that the average radius of the states (14) decreases when the magnetic field increases; this means that the limit of F_{χ,n,k_z} for β small or large gives the distribution of the phonons when the magnetic field is strong or weak.

In the next section we study these limits.

2.2. Two-Dimensional Harmonic Polaron in a Magnetic Field

The hamiltonian H_1 (eq. (6)) can also be used to describe a two-dimensional polaron in a magnetic field with the following modifications: a) we set $k_z = 0$ and eliminate the integration on k_z; b) in eq. (5) we substitute $\frac{V}{(2\pi)^3}$ with $\frac{S}{(2\pi)^2}$, where S is the normalization area; c) the quantity V_k, which in the three-dimensional case and in our units has the value $-\frac{i}{k}(\frac{4\pi\alpha}{V})^{1/2}$, is substituted by $-\frac{i}{\sqrt{\chi}}(\frac{2\pi\alpha}{S})^{1/2}$. Furthermore, for an electron harmonically bounded the hamiltonian takes the form

$$H_2 = p_\rho^2 + \left[p_\varphi + \frac{\lambda\rho}{4}\right]^2 + \frac{\lambda_1^2\rho^2}{4} + \int d\chi\sum_n[a^\dagger_{\chi,n}a_{\chi,n} + \tag{16}$$

$$(T_{\chi,n}(\rho)\phi^*_n(\varphi)a_{\chi,n} + h.c.)],$$

where $\lambda_1 = \frac{\omega_0}{\omega_L}$. The expression of the average energy with respect to the trial state (8) and the minimization with respect to $F_{\chi,n}$ are written easily from eqs. (11) and (12). Finally we find that the differential equation for $F_{\chi,n}$ is the same as eq. (13) and the solution is given by expression (15) with $\gamma = 1$.

The difference between this problem and that of the free two-dimensional polaron lies in the fact that the value of β which minimizes the energy is different. In particular, in this case, for the ground state, when the magnetic field goes to zero, β remains finite and its value is near to $\frac{1}{2\lambda_1}$; for increasing field the value of β decreases as in the case of the free polaron. We conclude that, when λ_1 is large respect to one, the electron-phonon interaction cannot be treated perturbatively even if the material has a small α.

3. Distribution Function of Phonons for Weak and Strong Magnetic Field

To find the expression of F_{χ,n,k_z} for the ground state ($M=0$) in the limit of weak magnetic field ($\omega_c \ll \omega_L$), we note that the radius of the electronic state must be very large with respect to the polaron radius, so that $\beta \to \infty$. using then the asymptotic formulae

$$\phi(a,b,x) \sim \Gamma(b)(ax)^{-\frac{|n|}{2}} I_{|n|}(2\sqrt{ax}),$$

$$\psi(a,b,x) \sim \frac{2}{\Gamma(1+a-b)}(ax)^{-\frac{|n|}{2}} K_{|n|}(2\sqrt{ax})$$

we can calculate all the integrals in eq. (15); the final result for F_{χ,n,k_z} is

$$F_{\chi,n,k_z} = \left[\frac{V}{2\pi}\right]^{\frac{1}{2}} \frac{V_k^*(-i)^n J_n(\chi\rho)}{1+\chi^2+k_z^2-2k_z Q_z(1-\eta)}. \tag{17}$$

The above expression is the n-th component, in the basis of the angular momentum, of the planar phonon distribution of a free polaron, moving along the z-axis with total momentum Q_z. Consequently, the effective mass coincides with that of the free polaron.

On the other hand, when the magnetic field becomes very strong ($\omega_c \gg \omega_L$), the radius of the electron state becomes smaller than the polaron radius and $\beta \ll 1$. From eq. (15) we see then that only the term $n = 0$ gives a finite contribution. In this case $a = \frac{\gamma^2\beta^2}{2}$, $b = 1$, $\nu = \frac{1}{2}$ and $s = 0$ and the hypergeometric functions $\phi(a,b,x)$ and $\psi(a,b,x)$ become indipendent on x and assume the value one. $F_{\chi,0,k_z}$ is then independent on ρ and has the expression

$$F_{\chi,0,k_z} = 2\pi\sqrt{\chi}\frac{i}{k}\left[\frac{2\alpha}{(2\pi)^2}\right]^{\frac{1}{2}} \frac{e^{-\frac{\chi^2\beta^2}{2}}}{1+k_z^2-2k_z Q_z(1-\eta)}. \tag{18}$$

With this quantity we calculate from eq. (11) the total energy

$$E = \frac{1}{2\beta^2} + \frac{(\beta\lambda)^2}{8} + (1-\eta)^2 Q_z^2 \tag{19}$$
$$+ \frac{\alpha}{\pi}\int_{-\infty}^{\infty} dk_z \int_0^{\infty} d\chi e^{-\chi^2\beta^2} \frac{\chi}{(\chi^2+k_z^2)\gamma^2}\left[\frac{1+k_z^2}{\gamma^2}-2\right].$$

To find the self-energy and the effective mass, we must consider an expansion of eq. (19) in powers of Q_z. At $Q_z = 0$ we have

$$E_0 = \frac{1}{2\beta^2} + \frac{(\beta\lambda)^2}{8} - \frac{\alpha}{\pi\beta}\int_0^{\infty} dx \frac{1}{1+\frac{x^2}{\beta^2}} e^{x^2} E_1(x^2), \tag{20}$$

where

$$E_1(x) = \int_x^\infty dt \frac{e^{-t}}{t}.$$

The self-energy is a complicated decreasing function of β, which must be calculated numerically.

We can also calculate the polaron effective mass; we see that the electronic mass is increased by the factor

$$R = 1 + \frac{4\alpha}{\pi\beta^3} \int_0^\infty dt \frac{t^2 e^{t^2} E_1(t^2)}{(1+\frac{t^2}{\beta^2})^3}. \tag{21}$$

Eq. (21) shows that the polaron effective mass increases on decreasing β, that is, increasing the field strength. The actual value of the polaron effective mass must be calculated numerically.

4. Numerical Results

For the free polaron in magnetic field we have minimized the average energy (eq. (11)) as function of β, using expression (15) for F_{χ,n,k_z} with $M = 0$ and $Q_z = 0$. In Table I we show the results of our calculations for increasing magnetic field.
We compare, for small α, our energies
with those obtained perturbatively and with the Peeters and Devreese theory [6]. When α increases the comparison is done only with path-integral calculations. For strong magnetic field we give also the results obtained in the limit of decoupled electronic and ionic motions. We see that our energies are higher than those given by the perturbation and path-integral theories and tend to the non-adiabatic limit. We have not calculated the polaron effective mass as a function of the magnetic field because Q_z enters in the parameter a of the confluent hypergeometric function and we cannot give an analitycal form of the energy as function of Q_z. The calculations must be done numerically and we shall discuss this point in a future work.

In the previous section we have calculated the polaron mass in the limit of weak and strong magnetic field. In the first case we have found the free polaron mass whereas in the second case Table II shows the dependence of the polaron mass on strong magnetic field. For what concerns the problem of an electron in a quantum dot, the ground state energies can be obtained from those of a free two-dimensional polaron in a magnetic field, defining an effective field through $\lambda^* = \lambda + 4\lambda_1$. In Table III we report the energies calculated in a previous work [12] relative to the two-dimensional polaron in magnetic field.

TABLE I.

We show the ground-state energies of the three-dimensional polaron for $\alpha = 0.1, 1.0, 3.0$ and for different values of the dimensionless magnetic field λ. The column headed E_f are the energies calculated with the path-integral method and are taken from [6]; column E_p lists the results of second-order perturbation theory; column E_v gives the results of our variational theory and column E_a those obtained in the non-adiabetic limit (eq. (20)). The last column shows the values of the average number $<n>$ of phonons for unitary volume. The energies E_a are calculated using the values of β obtained by the minimization of eq. (11).

λ	E_f	E_p	E_v	E_a	$<n>$
		$\alpha = 0.1$			
0.4	-	0.097	0.100	-	0.052
0.6	-	0.195	0.199	-	0.053
0.8	-	0.294	0.298	-	0.053
1.0	-	0.392	0.397	0.453	0.054
1.2	-	0.491	0.496	0.550	0.055
1.4	-	0.590	0.595	0.647	0.056
1.6	-	0.688	0.694	0.744	0.057
1.8	-	0.787	0.794	0.842	0.057
2.0	-	0.886	0.893	0.939	0.058
3.0	-	1.380	1.390	1.430	0.061
5.0	-	2.371	2.385	2.417	0.067
10.0	-	4.855	4.876	4.896	0.077
		$\alpha = 1.0$			
0.4	-0.845	-0.833	-0.79	-	0.52
0.6	-0.760	-0.748	-0.71	-	0.53
0.8	-0.675	-0.663	-0.62	-	0.54
1.0	-0.589	-0.577	-0.53	0.032	0.54
1.2	-0.503	-0.491	-0.44	0.102	0.55
1.4	-0.416	-0.405	-0.35	0.164	0.56
1.6	-0.329	-0.318	-0.26	0.233	0.57
1.8	-0.241	-0.230	-0.16	0.314	0.57
2.0	-0.153	-0.142	-0.07	0.393	0.58
3.0	0.291	0.303	0.40	0.782	0.62
5.0	1.199	1.213	1.35	1.668	0.67
10.0	3.530	3.555	3.76	3.955	0.78
		$\alpha = 3.0$			
0.4	-3.025	-2.898	-2.78	-	1.55
0.6	-2.969	-2.844	-2.74	-	1.59
0.8	-2.912	-2.789	-2.67	-	1.62
1.0	-2.853	-2.732	-2.59	-0.987	1.64
1.2	-2.793	-2.674	-2.54	-0.963	1.67
1.4	-2.732	-2.615	-2.46	-0.942	1.71
1.6	-2.669	-2 .553	-2.39	-0.911	1.72
1.8	-2.606	-2.491	-2.29	-0.858	1.72

2.0	-2.542	-2.427	-2.23	-0.820	1.83
3.0	-2.207	-2.092	-1.83	-0.671	1.87
5.0	-1.490	-1.360	-0.95	0.003	2.01
10.0	0.457	0.664	1.26	1.823	2.37

TABLE II.

We show the three-dimensional polaron effective mass for $\alpha = .1, 1., 3.$ in its motion along the direction of the field in the strong field regime. The values of β used are those which minimize eq. (11). In column 1 we give the values of λ; in column 2,3,4 the values of the ratio R_1, R_2, R_3 of the polaron mass with respect to the band mass. The free polaron ratio, for the above values of α are 1.017,1.167 and 1.5 respectively.

λ	R_1	R_2	R_3
1.0	1.010	1.101	1.322
1.2	1.011	1.111	1.356
1.4	1.012	1.123	1.402
1.6	1.013	1.134	1.422
1.8	1.014	1.141	1.424
2.0	1.015	1.149	1.543
3.0	1.019	1.197	1.592
5.0	1.025	1.247	1.740
10.0	1.035	1.363	2.117

5. Conclusions

We have shown that the motion of a three-dimensional polaron in a magnetic field can be well described using the conservation of the linear and angular momentum of the electron and phonons in the direction of the magnetic field. For any fixed value of Q_z, we can set up the variational wavefunction of smallest energy for any fixed M. The trial wavefunctions, having a form of coherent states in the phonon operators, give energies depending on the unique parameter β.

We have shown, for the ground state, that the position-dependent phonon distribution function goes to the free polaron expression when $\beta \to \infty$ and to the adiabatic limit when $\beta \to 0$. The energies, calculated for $Q_z = 0$ and for several values of the electron-phonon coupling constant, are compared with those calculated within second order perturbation theory and with the extension of the polaron path-integral theory when a magnetic field is present. Our values, for small α, are coincident with the perturbative ones. However, when the magnetic field increases, our ground state energies are larger than the perturbative ones. The average number of virtually excited phonons associated to the ground state increases with increasing the magnetic field and this explains the failure of the perturbative theory.

Our results are not in agreement with those of Peeters and Devreese [6], specially in the strong magnetic field regime. Our conclusion is that there are doubts on the validity of the path-integral approach, particularly because it is not well based mathematically.

TABLE III.

A table showing the ground-state energies of the two-dimensional polaron for $\alpha = 0.1, 1.0, 4.0$ and for different values of the dimensionless magnetic field λ. The column headed E_f are the energies calculated with the path-integral method and are taken from [7]; column E_p lists the results of second-order perturbation theory and column E_v gives the results of our variational theory. For $\alpha = 4.0$ we list the variational and path-integral energies only. The column headed E_a are the energies calculated in the two-dimensional non-adiabatic approximation not reported here and $<n>$ is the average number of virtually excited phonons.

λ	E_f	E_p	E_v	E_a	$<n>$
			$\alpha = 0.1$		
0.1	-0.10951	-0.10905	-0.108	-	0.0787
0.2	-0.06149	-0.06105	-0.057	-	0.0804
0.4	0.03451	0.03492	0.044	-	0.0816
0.6	0.13049	0.13086	0.141	-	0.0874
0.8	0.22647	0.22680	0.238	-	0.0901
1.0	0.32247	0.32275	0.336	0.4373	0.0943
1.5	0.56261	0.56280	0.579	0.6732	0.1019
2.0	0.80300	0.80313	0.823	0.9114	0.1097
4.0	1.76759	1.76763	1.801	1.8747	0.1362
10.0	4.68267	4.68267	4.755	4.8018	0.2000
20.0	-	9.57685	0.707	9.7197	0.2782
			$\alpha = 1.0$		
0.1	-1.59202	-1.54055	-1.52	-	0.791
0.2	-1.56080	-1.51049	-1.45	-	0.812
0.4	-1.49815	-1.45091	-1.38	-	0.840
0.6	-1.43509	-1.39140	-1.32	-	0.876
0.8	-1.37152	-1.33201	-1.23	-	0.909
1.0	-1.30739	-1.27245	-1.14	-0.1266	0.943
1.5	-1.14516	-1.12204	-0.99	-0.0175	1.020
2.0	-0.98330	-0.96870	-0.82	0.1138	1.097
4.0	-0.32748	-0.32374	-0.02	0.7467	1.362
10.0	1.82627	1.82671	2.56	3.0183	2.003
20.0	-	5.76846	7.08	7.1975	2.701
			$\alpha = 4.0$		
0.1	-8.2067	-	-6.1	-	3.163
0.2	-8.2057	-	-6.1	-	3.280
0.4	-8.2031	-	-6.1	-	3.371
0.6	-8.1996	-	-6.1	-	3.517
0.8	-8.1952	-	-6.0	-	3.635
1.0	-8.1899	-	-6.0	-2.0066	3.773
1.5	-8.1728	-	-6.0	-2.3199	4.080
2.0	-8.1502	-	-6.0	-2.5449	4.390
4.0	-8.0090	-	-5.9	-3.0132	5.448
10.0	-7.7004	-	-4.8	-2.9266	8.014
20.0	-	-	-2.4	-1.2099	11.129

We have also discussed the problem of the polaron effective mass in its motion along the z-axis and have calculated it in the limit of weak and strong magnetic field.

Finally we have shown that the motion of an electron under an harmonic external potential and in a magnetic field can be reduced, for the ground state, to the problem of a polaron in an effective magnetic field. We conclude that if the frequency of the harmonic potential is larger than the phonon frequency, the electron-phonon interaction cannot be treated perturbatively even if the Fröhlich coupling constant is small.

Since the experimental informations on the ciclotronic mass comes out from transitions between different states, a complete theory of polarons in a magnetic field requires the calculation of the excited states.

We shall study this point in a future work.

References

1. H. Fröhlich, *Adv. Phys.* **3**, 325 (1954).
2. For a review of this type of calculations, see: J. Appel. in *Solid State Phys.*, F. Seitz, D. Turnbull and H. Ehrenreich eds, **21**, 193 (1968).
3. S.I. Pekar, *Z. Eksp. Teor. Fiz.* **16**, 341, (1946); see also ref. [2].
4. T.D. Lee, F. Low and D. Pines, *Phys. Rev.* **B90**, 297 (1953); see also ref. [2].
5. R.P. Feynman, *Rev. Mod. Phys.* **20**, 367 (1949); *Phys. Rev.* **97**, 660 (1955).
6. F.M. Peeters and J.T. Devreese, *Phys. Rev.* **B25**, 7281 (1982); **B25**, 7302 (1982).
7. W. Xiaoguang, F.M. Peeters and J.T. Devreese, *Phys. Rev.* **B32**, 7964 (1985).
8. D.M. Larsen, *Phys. Rev.* **B30**, 4807 (1984); **33**, 799 (1986).
9. S. Das Sarma, *Phys. Rev. Lett.* **52**, 859 (1984).
10. F.M. Peeters and J.T. Devreese, *Phys. Rev.* **B31**, 3689 (1985).
11. D.M. Larsen, *Phys. Rev.* **B32**, 2657 (1985).
12. D. Ninno and G. Iadonisi, *Phys. Rev. B.*, in press (1989).
13. M.A. Reed, R.T. Bate, K. Bradshaw, W.M. Duncan, W.R. Frensley, J.W. Lee and H.D. Shih, *Journ. Vac. Technol. B* **4**, 358 (1986).
14. K. Kash, A. Scherer, J.W. Worlock, H.G. Craighead and M.C. Tamargo, *Appl. Phys. Lett.* **49**, 1043 (1986).
15. J. Cibert, P.M. Petroff, G.J. Dolan, S.J. Pearton, A.C. Gossard and J.H. English, *Appl. Phys. Lett* **49**, 1275 (1986).
16. H. Tunken, G.J. Dolan, M.B. Panish and S.N.G. Chu, *Appl. Phys. Lett.* **50**, 413 (1987).
17. M.A. Reed, J.N. Randall, R.J. Aggarwal, R.J. Matyi, T.M. Moore and A.E. Wetsel, *Phys. Rev. Lett.* **60** 535 (1968).
18. W.H. Louisell, *Quantum Statistical Properties of Radiation*, Wiley, New York (1973).
19. For a short review of this problem see: G. Iadonisi, *La Rivista del Nuovo Cimento* **7**, 11 (1984); G. Iadonisi, F. Bassani and G. Strinati, *Phys. Stat. Solidi. b*, in press (1989).
20. L.D. Landau and E.M. Lifshitz, *Quantum Mechanics. Non-relativistic Theory*, Pergamon Press (1965), pag. 424.

SPECTROSCOPY WITH SYNCHROTRON RADIATION

A. BALZAROTTI AND U.M. GRASSANO
Dipartimento di Fisica,
Università di Roma II,
Via E. Carnevale,
00175-Roma, Italy

M. PIACENTINI

Dipartimento di Energetica,
Università di Roma "La Sapienza",
Via Scarpa 14,
00161-Roma, Italy

R. ROSEI

Dipartimento di Fisica,
Università di Trieste,
34100 - Trieste
and
Sincrotrone Trieste,
Padriciano 99,
34012 - Trieste, Italy

We describe the results of spectroscopic investigations performed on solid materials using synchrotron radiation at the Italian Synchrotron Radiation Facility PULS showing the importance of such a realization. A perspective for future developments using a third generation storage ring is also briefly outlined.

1. Introduction

Synchrotron radiation (SR) has many features such as a continuous spectral distribution ranging from visible to hard *x*-rays with high intensity, spatial and time structure and high degree of polarization which make it a very attractive tool for spectroscopic purposes.

Too numerous and different are the applications of SR in the study of the optical properties of solids to be possible to summarize them in a short paper. All the standard optical techniques as absorption spectroscopy, reflectivity, luminescence excitation and emission spectroscopy and many others have been performed in the vacuum ultraviolet (VUV) and *x*-ray regions with the SR available at the different storage rings.

We describe here two novel optical techniques and the results of a few optical investigations performed in the last few years at the SR facility PULS using the 1.5 *GeV* ADONE storage ring located at the Frascati National Laboratories of the Istituto Nazionale di Fisica Nucleare.

Several studies described below have been undertaken under the guide and suggestion of Prof. Franco Bassani. His interest for the optical properties of materials is well known

R. Girlanda et al. (eds.), Progress on Electron Properties of Solids, 111–124.

and documented by the numerous contributions he gave to the field. For many years Prof. Franco Bassani headed the PULS project at Frascati. More recently, he made efforts in support of a new facility dedicated to SR under construction at Trieste, of which he chairs the Scientific Program Advisory Committee.

This paper is dedicated to Prof. Franco Bassani in the occasion of his 60th birthday as a recognition of the great impulse that he gave to the Italian SR research.

2. Two-Photon Spectroscopy with SR

Two-photon absorption was theoretically calculated by Goeppert Mayer [1] in 1931 and became possible after the discovery of the laser [2], i.e. of a light source so strong that the small, second order effect could be detected. Various combinations of fixed wavelength lasers, tunable dye lasers, flash lamps have been used over the past twenty years to study two-photon spectra of atoms and solids obtaining new and interesting results [3,4]. Often a new picture of the energy levels of the solids emerged from the spectra due to the different selection rules that are complementary to those involved in the one-photon transitions (even-even transitions in the two-photon absorption versus even-odd transitions in one-photon absorption). The available lasers and light sources restricted the energy range of two-photon experiments at energies smaller than 7 eV and therefore interesting classes of materials such as the alkali and alkaline-earth chlorides and fluorides could not be studied. A breakthrough of this limitation came from the use of SR for providing tunable photons (Fig. 1). In this way nonlinear spectroscopy in the far ultraviolet region became a reality [5]. In order to observe real two-photon absorption, and not simply a two step process, where one photon populates an intermediate state and the second photon induces from it a second transition, the temporal overlapping of the two photon pulses must be carefully controlled. Commonly two-photon experiments were performed using for the tunable photon a flash lamp with pulse duration approximately two orders of magnitude longer than that of laser pulses, thus providing a quasi stationary signal, on top of which one has to measure the changes induced by the laser pulses. In the present case the width of the light pulses from the ADONE storage ring is about 4 ns and that of the used Nd-YAG laser is 15 ns. Therefore proper gating of the laser pulses was necessary to synchronize both pulses and moreover to reveal, as a reference and thus out of coincidence with the laser, the light pulses originating from the same electron bunch after an integer number of orbits in the storage ring [6].

The experiments performed on KCl and NaCl [5], shown in Fig. 2, have shown the importance of the realization. However the energy interval was limited by the laser photon energy and the low intensity of the ADONE synchrotron light pulses introduced some scattering in the data points. A proposal to extend and improve these measurements with a much more intense VUV light source has been presented to the scientific committee of Trieste storage ring whose construction has just begun.

3. Time-resolved spectroscopy with a phase-sensitive fluorometer

The measurements of emission lifetimes and of time resolved luminescence are of great importance for the understanding of the optical cycles of the excited electrons. These studies equally developed in crystalline solids, in liquids, in biological systems, in simple organic molecules, are generally performed in the time domain using, as exciting source, a short light pulse from a laser (whose duration can be nowadays as short as 10 fs) and streak camera or pulse and probe detection system with picosecond resolution. An old technique recently revived measures the same quantities in the frequency domain [7].

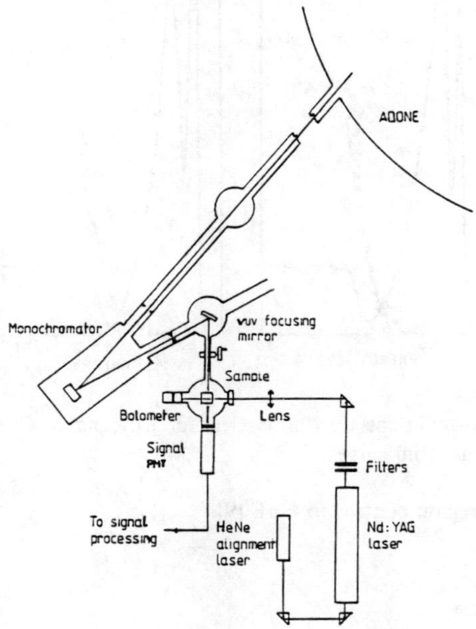

Figure 1. Optical layout of the experimental set-up for two-photon absorption using the SR emitted by the ADONE storage ring.

If one excites a fluorescent sample with a light beam whose intensity is sinusoidally modulated at the frequency ν, also the luminescence intensity is modulated at the same frequency, but shows a phase shift ϕ with respect to the excitation. Moreover, the modulation μ, defined as the ratio between the a.c. and the d.c. parts of the signal, is smaller for the luminescence signal than for the pumping source. The phase shift ϕ and the modulation ratio $M = \mu_{\text{emission}} / \mu_{\text{source}}$ contain all the information about the emission response function $I(t)$ of the system under study. Even in the case of multiple, independent exponential decays, the lifetimes can be extracted from the measurements of ϕ and M at different modulation frequencies [7].

The interest of using SR as the exciting beam stems from the fact that it is a perfectly stable and already modulated light source. One of the problems encountered in building phase fluorometers is the difficulty of modulating the light source at different high frequencies (of the order of 100 Mhz in order to measure 10 ps). The light of a storage ring, being a high repetition rate pulse light source, has a large number of modulation frequencies simultaneously present - all the harmonics of the fundamental revolution frequency ν_0 of the electrons inside the storage ring. The intensity of the harmonics decreases slowly and appreciable signal can be obtained performing the analysis at frequencies up to 80 MHz. The phase fluorometer (Fig. 3) has been used in Frascati mainly in studies of biological molecules [8] and, in the field of color centres, in the measurements of the

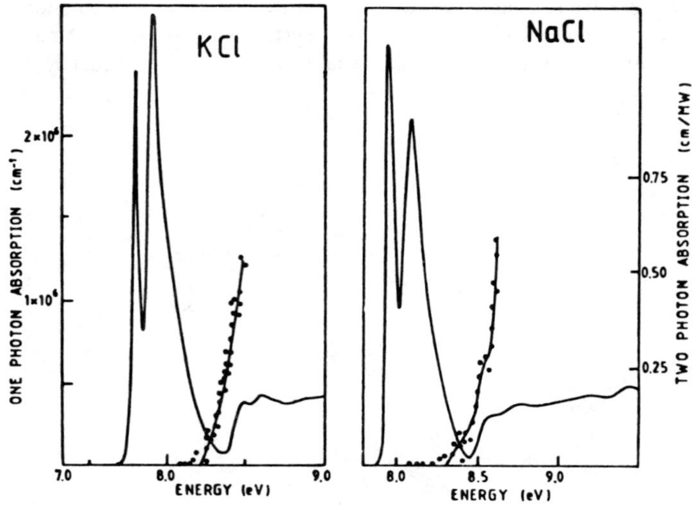

Figure 2. Two-photon absorption spectra (full circles) for *KCl* and *NaCl* at 20*K* compared with one-photon absorption spectra (full curves).

lifetimes of *F* and *F*-aggregate centres in NaF [9].

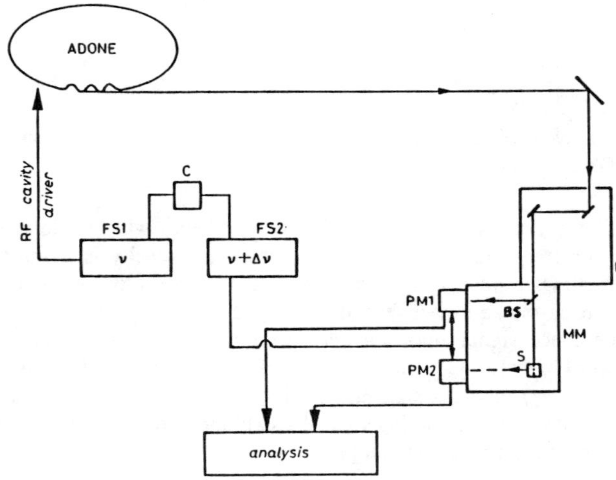

Figure 3. Experimental set-up for lifetime measurements. M monochromator, *MM* measuring module, *BS* beam splitting, *S* sample, *PM* 1 and *PM* 2 photomultipliers, *FS* 1 and *FS* 2 frequency synthetizers, *C* crystal oscillator.

The interest of the present technique, which does not resolve times shorter than 1 ps and therefore is slower than the measurement in the time domain, lies in the possible shift of the excitation wavelength in all the visible and UV range of the spectrum, tunability not easily obtained with present laser sources. Improvements in the Frascati storage ring currently under realization will allow excitation with stronger pulses and also in the VUV up to 10 eV. This will allow to perform routinely lifetime experiments and to explore entirely new problems, for example in large gap materials.

4. Luminescence studies in the vacuum ultraviolet

The SR emitted by ADONE has been used as excitation source for the luminescence studies. The SR is dispersed with an evacuable normal incidence monochromator, that limits the spectral range available at the sample to the 6-30 eV region. The luminescence, collected at 90° with respect to the exciting beam from the front surface of the samples, is analyzed by means of a second monochromator and processed by a single photon counting system. The spectral range covered by this second monochromator is from 2 eV to 6 eV. The SR being a pulsed source allows time-resolved measurements. In the case of our experimental system a time resolution better than 0.1 ns can be achieved with the coincidence single photon counting technique.

Several, interesting researches can be performed with such an experimental apparatus, as one can immediately conceive. We have started by studying the relaxation of the electronic excited states towards the luminescence centres in both doped and pure alkali halides. Doped alkali halides, such as KI:TI^+, are well known phosphors employed, for example, for detecting $x-$ or γ-rays and energetic particles. The primary ionizing excitation produces electron-hole pairs in the host crystal, while the luminescence occurs either in the emission bands of the impurity atoms or as the intrinsic luminescence of the host crystal, depending upon temperature and excitation energy. These processes have been already investigated in detail by many authors who used either electrons or $x-$rays or γ-rays or low-energy photons ($h\nu$<7 eV) for exciting the samples [10]. Thus the general features of the energy transfer from the excited host crystal to the recombination centres are rather well understood. However, there are still several aspects, such as the temperature and excitation energy dependences, that deserve further investigation.

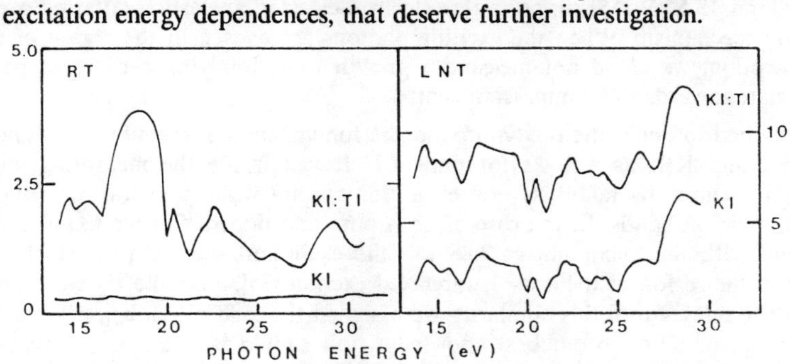

Figure 4. Left panel: Room temperature VUV excitation spectra of pure *KI* (in correspondence of the π emission) and of $KI:Tl^+$ (A_x emission). Right panel: Liquid nitrogen temperature VUV excitation spectra of the *KI* and $KI:Tl^+$ emission at 3.44 eV (intrinsic π band). The pure *KI* spectrum overlaps the $KI:Tl^+$ one and it has been shifted downwards for clarity.

We have measured from 14 eV to 30 eV the excitation spectra of the A_x and A_T luminescence bands of Tl^+ in KI:Tl^+ crystals at several temperatures as well as the excitation spectra of the intrinsic luminescence (the σ and π bands) of both pure and doped crystals [11,12]. At temperatures above the critical temperature, T_q, for thermal quenching of the intrinsic luminescence, the excitation is transfered to the impurity centre, while below it only the intrinsic luminescence could be detected. In the case of KI, T_q = 100 K and the different regime was already manifest at 80 K, as shown in Fig. 4. For KBr, $T_q = 32\,K$ and at 80 K in KBr:Tl^+ the emission is still concentrated in the Tl^+ bands. Our excitation spectra of the Tl^+ bands are very similar to the excitation spectrum of the π luminescence in pure KBr measured at 5K by Yanagihara et al. [13]. At low temperatures, when an electron is excited in the conduction band, the crystal lattice relaxes almost immediately near the hole, which becomes self-trapped forming a V_k centre [a (halogen)$_2^+$ molecular ion whose axis is oriented along a <110> direction]. The V_k centres are stable and cannot migrate through the sample. The excited electrons are then captured by the V_k centres, forming excited self-trapped excitons, which decay radiatively with the emission of the well-known σ and π bands. Thus only the intrinsic luminescence can be excited in both pure and doped crystals at low temperatures. Above T_q, the V_k centre becomes unstable; the hole is released and can migrate near a Tl^+ site, where it is trapped. The hole is transfered to the Tl^+ ion, which becomes a Tl^{++} ion. One of the excited electrons can be captured by one of these Tl^{++} ions and a $(Tl^{++})^*$ excited ion is formed. The $(Tl^{++})^*$ ion decays radiatively emitting the A_x and A_T emission bands, as if it were excited directly. The entire process described above, i.e. photon absorption, energy transfer and radiative decay, requires some time. The absorption and the emission processes occur with their own characteristic lifetimes, which do not depend on the energy of the incident photons. Instead the transfer time may depend on the excited electron energy through the mechanisms by which the electrons loose their energy and thermalize, namely electron-electron and electron-lattice interactions. We measured the lifetime of the σ band in pure KI at several temperatures between 15 K and 77 K as a function of the excitation photon energy [14]. The σ luminescence lifetime of 2.3$\pm$0.07 ns did not depend on temperature and on exciting energy in the explored region. Instead we found a growth time of 260$\pm$70 ps for 13.5 eV excitation energy only, which does not depend on temperature. We attributed the observed growth time of the σ luminescence to the fact that in KI electrons excited by 13.5 eV photons have enough energy for exciting bound electron-hole pairs. A depression detected in the yield of the photoemitted electrons around 13.5 eV [15] supports this electron-electron anelastic scattering process as a relaxation mechanism. When the exciting photons are chosen in the region of core exciton absorption, we could not measure a growth time, implying a different path of the excited state towards the luminescent centre.

We studied in detail the region around the fundamental absorption threshold for Tl^+ doped KI and KBr as well as for pure KI cleaved inside the measuring chamber in ultra-high vacuum, by taking several excitation spectra while sampling with high resolution the emission bands. In the case of both pure and doped KBr we found a new emission band with the maximum at 2.88 eV, the excitation spectrum of which is sharply peaking in the region of the $n = 1$ unrelaxed exciton [16], alike the E_x band observed in the alkali iodides. From this similarity we assigned the 2.88 eV emission band to the E_x luminescence in KBr. To our best knowledge, this should be the first observation of the E_x band in an alkali halide not belonging to the iodide family.

In $KI:Tl^+$ the excitation of the C band produces at 20 K the A_x band at 2.9 eV, the A_T band at 3.7 eV and a third band at 4.5 eV. From absorption measurements the C band appears as a triplet due to the dynamical Jahn-Teller interaction [17]. The excitation in the different components bands (C_1 at 5.28 eV, C_2 at 5.31 eV and C_3 at 5.37 eV) produces marked differences of the intensities of the emission bands [16]. For example, as shown in Fig. 5, excitation in the lower energy component C_1 yields a stronger emission in the 4.5 eV luminescence band, while excitation in the C_3 band increases the relative strength of the A_x and A_T emission bands. It seems therefore that the electrons excited into the C band sublevels split by the Jahn-Teller interaction, relax along different paths. Excitation of the C_3 band increases the probability of relaxation into the A band

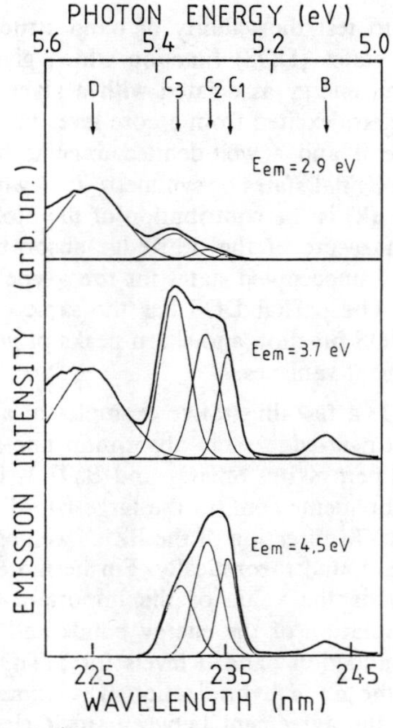

Figure 5. Excitation spectra of $KI{:}Tl^{+}$ at 20 K. The emission energies correspond, from top to bottom, to the A_x band, the A_T band and the 4.5 eV band. The thin lines are best-fit Gaussian curves used for sinthetize the experimental spectra.

configurational curve, leading to stronger A_x and A_T emissions.

This research program is continuing towards other materials and towards the simultaneous investigation of different relaxation mechanisms, such as defect formation and photon-stimulated desorption.

5. Core level spectroscopy

Optical excitations in the high-energy range by the use of a tunable and intense source of soft x-ray photons like SR provide a unique tool to investigate the electronic properties and the local structure of many solid materials [18]. Here, we give a short account of the major achievements obtained in recent years at Frascati through the optical excitation of the core levels of several solid systems.

The experiments have been performed using the x-rays emitted by the ADONE storage ring operated at a typical electron current of 50 mA. A 1.5 mrad fan of light was collected by the entrance slit of a channel-cut Si (220)/(111) single crystal monochromator at about 17 m from the tangent point with an average resolution of 10^{-4}. The spectral purity of the light, as determined by a Si (Li) detector, was rather good, containing less than 1% of second harmonic and 0.01% of third harmonic at 5 keV. Absorption and fluorescence measurements in both the near-edge region and far above the edge were performed to obtain information mostly related to the lowest unoccupied electronic levels and to the interatomic distances and coordination numbers, respectively.

A quantity of interest to test the validity of band structure calculations is the projected (partial) density of states (DOS) function which gives the number of electronic states per unit cell and unit energy associated with a given atomic orbital of symmetry $\nu\{l,m\}$. When the electrons are excited from a core level E_l having negligeable dispersion through the Brillouin zone Ω and a well-defined angular momentum l, the absorption coefficient $\mu(\omega)$ picks out the final states of symmetry $l\pm1$ which are coupled to the initial state of symmetry l. If $A_{f,\nu}(\mathbf{k})$ is the contribution of the atomic orbital forming the final crystal wavefunction, the measure of the near-edge absorption coefficient gives directly the weighted final density of unoccupied states for the ν-type symmetry orbital on the j-th atom and for the band f. The partial DOS has the same critical behaviour (Van Hove singularities) as the final DOS function and sharp peaks of $\mu(\omega)$ arise mostly from regions of Ω where the energy gradient vanishes.

Fig. 6 and Fig. 7 provide a few illustrative examples of absorption data showing how informative high-resolution near-edge x-ray absorption fine structure (NEXAFS) spectra can be. For d-band cubic perovskites $SrTiO_3$ and $BaTiO_3$ in which Ti is 6-fold coordinated to oxygen, the measurements confirm the large two-dimensional character of the p conduction band along the XM direction of the BZ as well as the P_0 and P_2 singularities occurring at the band edges found theoretically. Furthermore, the detailed analysis of the structural features [19] yields the value of the important two-centre $(pd\pi)$ and $(pd\sigma)$ integrals used in the computation of the energy bands and indicates that the cristalline $t_{2g}-e_g$ splitting of the conduction band d levels for $SrTiO_3$ is smaller than previously assumed, suggesting that the $p-d$ hybridization was underestimated in band-structure calculations. Additionally, the agreement between single-electron DOS calculations and experimental data rules out large excitonic corrections at the edge. In TiFe [20], an important Asano-type transition-metal alloy, and in transition metal carbides and nitrides [21] the partial p-like DOS of titanium and iron but not the total or the d-like DOS describe rather well the lowest conduction band features at the Ti K- and Fe K-edges. The comparison between the binding energy position of the L_3 core level with respect to the Fermi energy determined from absorption and x-ray photoemission spectra and the theoretical DOS provide information on the electronic screening mechanism of valence electrons following the ejection of a core electron.

The extended x-ray absorption fine structure (EXAFS) technique has proven to be a rather accurate technique to determine interatomic distances and coordination numbers in a large variety of different systems [22]. EXAFS is basically caused by the interference of the outgoing photoelectron wave with the portion of it which results from the back-scattering of the atoms surrounding the absorbing atom. In distinction to NEXAFS, which yields information on high-order correlation function, the EXAFS technique is sensitive to pair correlations. Among the broad spectrum of possible applications of this technique we select two special cases in which peculiar structural information on atomic scale has been achieved in amorphous and disordered semiconductors.

We consider first the structural change as a function of deposition temperature from amorphous to crystalline phase of Ge films grown by thermal evaporation in vacuum over <111> Si substrates [23]. X-ray or electron diffraction analysis do not show structural differences with the method of preparation nor with such parameters as substrate temperature, rate of deposition, pressure, etc. over distances less than 15-20 Å. As a consequence it has been proven a difficult test to ascertain how one arrives at the long range disorder characteristic of the amorphous state. The EXAFS analysis, on the contrary, provides a clear indication that the transition from the amorphous to crystalline state is a continuous process so that no transition temperature T_c can be defined, opposite to what was usually assumed in the literature. This can be easily seen in Fig. 8, where

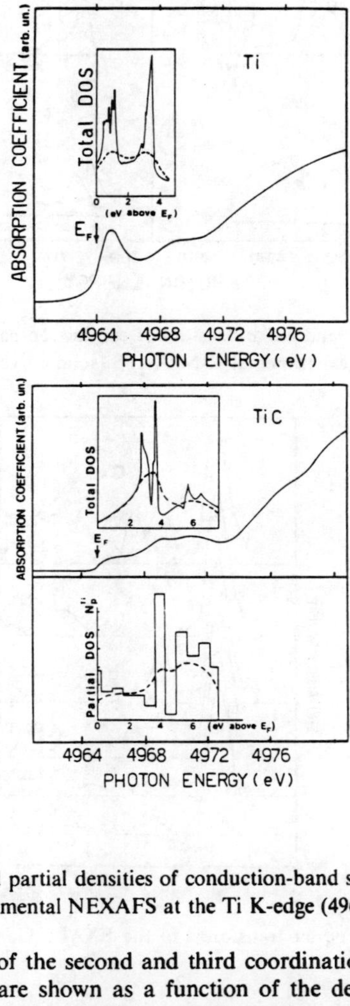

Figure 6. Theoretical total and partial densities of conduction-band states of Ti metal and TiC (insets) compared with the experimental NEXAFS at the Ti K-edge (4964 eV).

the measured contribution of the second and third coordination shells obtained from the K-shell absorption of Ge, are shown as a function of the deposition temperature. The gradual structural evolution toward the crystalline order occurring in the 130-300 °C temperature range accounts for the continuous change of some macroscopic properties observed in this temperature interval. A detailed analysis of the Debye-Waller factor allows to conclude in favor of the existence of ordered microstructures (clustering) at medium-range level as opposed to a random topological disorder which decreases with increasing deposition temperature.

The experimental determination of the microscopic structure and of the thermodynamic properties of semiconducting alloys is another example in which the full potential of the EXAFS technique is exploited. In the past, the local structure of substitutional $A_xB_{1-x}C$ ternary (pseudobinary) alloys, was supposed to possess, in the majority of

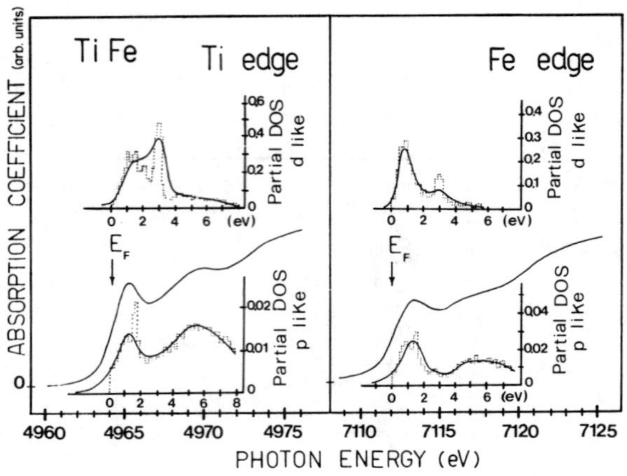

Figure 7. Calculated partial *d*- and *p*-like densities of conduction-band states of TiFe at Ti and Fe edges, (insets) compared with experimental K-NEXAFS (solid curves).

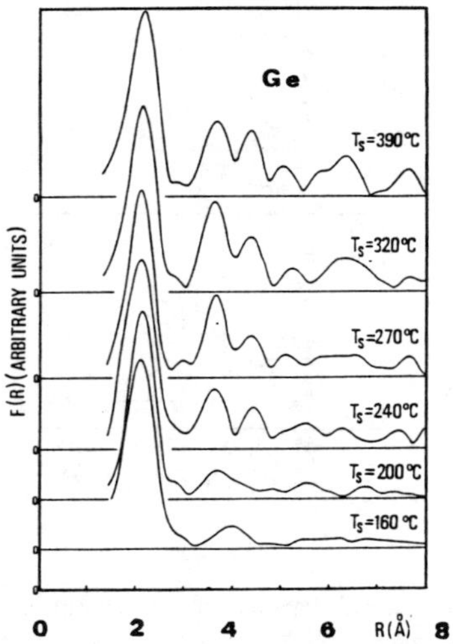

Figure 8. Magnitude of the Fourier transforms of the EXAFS Ge K-shell absorption at different deposition temperatures.

cases, a well-defined zincblende-like structure in which one of the two fcc sublattices (anion and cation) is chemically disordered. This assumption, based on structural determinations by *x*-ray diffraction, was consistent with the observation of a linear behaviour of the alloy lattice constant with composition known as Vegard's law. Such a simplified picture was shown to be in disagreement with the distribution of nearest-neighbors distances measured by EXAFS in a large number of ternary alloys [24,25]. The EXAFS measurements indicate a bimodal distribution of bond lengths which stay almost unchanged when the alloy composition is varied.

This result is rich of important consequences. A microscopic model of the alloy lattice structure, based on five basic ordered structures, can be introduced and a complete calculation of the formation energies of these structures is possible using both phenomenological [24] and state-of-the-art density functional total energy methods [26]. Furthermore, a detailed description of the coherent solid-solid phase diagrams (Fig. 9) and of the main thermodynamic properties of these alloys can be accomplished [27].

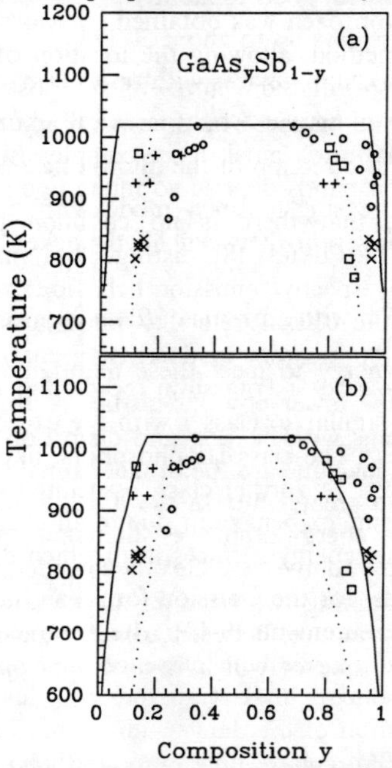

Figure 9. Calculated phase diagram of $GaAs_ySb_{1-y}$a) including and b) neglecting second nearest-neighbours relaxation. Solid line: Miscibility gap; dotted line: spinodal curve. Experimental data form several sources are also shown (symbols).

Using a novel method to compute the free energy of mixing [28], we have determined the excess probability distribution of the ordered configurations, the interaction and cluster parameters of the alloy, the distribution of nearest-neighbors distances and the degree of energy relaxation taking place during the alloy formation. The present knowledge of the microscopic structure of semiconducting alloys, as provided by EXAFS, is put now on much more firm ground than before and makes it possible to understand recent observations as superlattice formation and the influence of the substrate in the epitaxial growth of these alloys.

6. Future Perspectives

The pioneering work performed at Frascati with the 1.1 GeV electronsynchroton [29] and, subsequently, with the ADONE storage ring, settled the basis of future developments in Italy. At present an ultra-high brightness VUV/soft *x*-ray machine (ELETTRA) dedicated to SR researches is under construction at Basovizza near Trieste. A workshop of the potential users of the facility held in May 1987 attracted nearly 200 participants mainly from Italian academic and industrial organizations. A large number of proposals were put forward in the fields of physics, chemistry, biology, medicine, and industrial applications [30]. The required perofrmance of ELETTRA to meet the needs of these users may be summarized as follows:

i) radiation from insertion devices (wigglers and undulators) covering the range from 10 eV to 1-2 keV with a peak brilliance of $10^{18} - 10^{19}$ photons/s/mm^2/0.1% bw between 0.1 keV and 1 keV with good tunability.

ii) radiation from multipole wigglers up to about 30 keV.

iii) large number of possible insertion devices with up to 6 m straight sections.

iv) 10 to 12 openings for bending magnet beamlines with apertures of at least 20 mrad.

v) high stability of position and direction of the photon beams.

vi) possibility to operate in the single bunch mode with bunch length in the range 100-200 ps and with a filling ratio between filled bucket and adjacent ones greater than 10^3.

vii) beam lifetime longer than 6 h with currents of 200-400 mA in the multi bunch mode.

The optimum storage ring energy to meet these requirements is 1.5-2 GeV. Advantages of the lower energy are the lower power densities of the undulator radiation and therefore potentially less problems with heating and distortion of beamline components. Arguments in favor of a 2 GeV machine are the broader tunability range and higher photon energies of undulators and a greater flux in the medium hard *x*-rays from multipole wigglers. Moreover the higher energy improves the stability of the electron beam. Presently a 1.5 GeV machine upgradable to 2 GeV is considered. The storage ring lattice consists of 12 periods each containing two bending magnets in a double-bend achromat or 'expanded Chasman-Green' structure. Between adjacent quadrupoles there are 12 long straight sections each 6 m long, one of which is used for injection leaving 11 sections available for insertion devices.

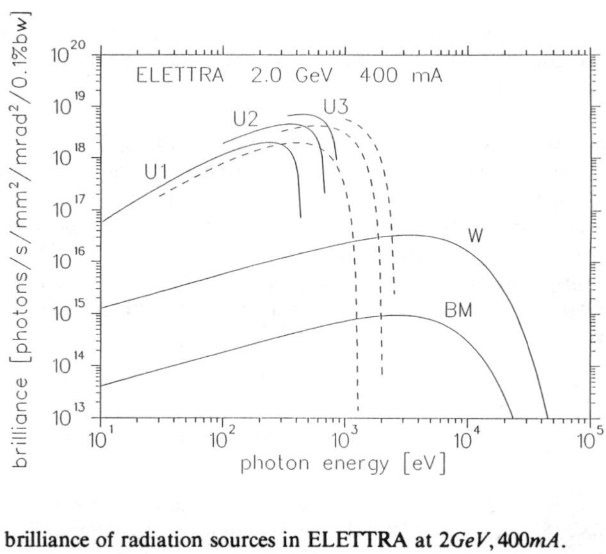

Figure 10. Spectral brilliance of radiation sources in ELETTRA at $2GeV, 400mA$.

Provisions are made to take one bending magnet beamline from each achromat also. The most important parameters of ELETTRA are given in Table I.

In Fig. 10 we present the spectral brilliance obtainable from a bending magnet and from several representative insertion devices, calculated for a 2.0 GeV, 400 mA machine.

TABLE I.
Main parameters of the storage ring ELETTRA.

Circumference (m)	259.2	
Number of long straight sections	12	
Length for insertion devices (m)	6	
Maximum beam current, multibunch (mA)	400	
Number of bunches	432	
Energy (GeV)	2	1.5
Natural emittance (10^{-9} m rad)	7.4	4.2
Maximum single bunch current (mA)	9.8	8.5
Rms bunch length at 400 mA (ps)	55	49
Rms natural bunch length (ps)	18	12
Insertion devices electron beam size (mm)	$\sigma_x = 0.228$	$\sigma_y = 0.044$
Insertion devices el. beam divergence (mrad)	$\sigma'_x = 0.033$	$\sigma'_y = 0.017$

Multipole wigglers ($\approx$ 70 poles) extend the useful spectral range to about 40 keV. Three undulators have been chosen with a minimum photon energy in the first harmonic of 10 eV (U1) and 100 eV (U2) and of 1 keV (U3) in the third harmonic.

To fully utilize the high performance offered by insertion devices in the low emittance ring ELETTRA, the design of the beamlines must face new problems such as the presence of high power densities on the optical elements. For VUV and soft x-ray beamlines grazing incidence mirrors are usually the first optical elements and a study of the thermally induced errors has been carried out using finite element analysis [31]. The first results of this study show that the thermal effects are significant.

Another problem that must be considered is the quality of the optical elements in terms of the figure accuracy. An average slope error of 3 arcsec on aspherical optical surfaces is a major limitation to the optical performance of monochromators for undulator radiation [32] and designs with plane and spherical surfaces are considered for focussing and monochromatization.

References

1. M. Goeppert-Mayer, *Ann. Phys.* **9**, 273 (1931).
2. W. Kaiser and C.G. Garret, *Phys. Rev. Lett.* **7**, 229 (1961).
3. D. Fröhlich and M. Sondergeld, *J. Phys. E* **10**, 761 (1977).
4. A. Quattropani and R. Girlanda, *Rivista Nuovo Cim.* **6**, 1 (1983).
5. R. Pizzoferrato, M. Casalboni, R. Francini, U.M. Grassano, F. Antonangeli, M. Piacentini, N. Zema and F. Bassani, *Europhys. Lett.* **2**, 571 (1986).
6. R. Pizzoferrato and M. Casalboni, *J. Phys. E* **20**, 896 (1987).
7. R.D. Spencer and G. Weber, *Ann. N.Y. Acad. Sci.* **158**, 361 (1969).
8. E. Gratton, D.M. Jameson, N. Rosato and G. Weber, *Rev. Sci. Instrum.* **55**, 486 (1984).
9. U.M. Grassano, M. Piacentini and N. Zema, *Nuovo Cim.* **7D**, 379 (1986).
10. N. Itoh, *Adv. Phys.* **31**, 491 (1982); *Cryst. Latt. Def. Amorph. Mat.* **12**, 103 (1985).

11. F. Antonangeli, F. Fermi, U.M. Grassano, M. Piacentini, A. Scacco and N. Zema, *Solid State Commun.* **49**, 323 (1984).
12. F. Antonangeli, F. Fermi, U.M. Grassano, M. Piacentini, A. Scacco and N. Zema, *J. Luminescence* **31&32**, 314 (1984).
13. M. Yanagihara, Y. Kondo and H. Kanzaki, *J. Phys. Soc. Jpn* **52**, 4397 (1983).
14. F. Antonangeli, A. Chiari, F. Fermi, U.M Grassano, M. Piacentini, A. Scacco and N. Zema, *Synchrotron Radiation at Frascati*, Atti di Conferenze, Compositori Bologna (Italy), Ed. S. Mobilio, F. Patella, S. Stipcich **5**, 177 (1987).
15. S.N. Ivanov, E.R. Il'mas, Ch.B. Lushchik and V.V. Mikhailin, *Sov. Phys. Solid State* **15**, 1053 (1973).
16. F. Antonangeli, M. Piacentini, N. Zema, U.M Grassano and A. Scacco, *Cryst. Latt. Def. Amorph, Mat.* **16**, 323 (1987).
17. P.H. Yster and C.J. Delbecq, *J. Chem. Phys.* **21**, 892 (1953).
18. See, e.g., A. Balzarotti and M. Piacentini, in *Electronic Structure and Electronic Transitions in Layered Materials*, Ed. V. Grasso, Reidel, Dordrecht (1986).
19. A. Balzarotti, F. Comin, L. Incoccia, M. Piacentini, S. Mobilio and A. Savoia, *Solid State Commun.* **35**, 145 (1980).
20. N. Motta, M. De Crescenzi and A. Balzarotti, *Phys. Rev. B* **27**, 4712 (1983).
21. A. Balzarotti, M. De Crescenzi and L. Incoccia, *Phys. Rev. B* **25**, 6349 (1982).
22. T.M. Hayes and J.B. Boyce, *Solid State Physics*, Ed. F. Seitz, D. Turnbull and H. Ehrenreich, Academic, New York (1982).
23. F. Evangelisti, M.G. Proietti, A. Balzarotti, F. Comin, L. Incoccia and S. Mobilio, *Solid State Commun.* **37**, 413 (1981).
24. A. Balzarotti, M.T. Czyzyk, A. Kisiel, N. Motta, M. Podgorny and M. Zimnal-Starnawska, *Phys. Rev. B* **B30**, 2295 (1984); *ibidem* **31**, 7526 (1985).
25. J.C. Mikkelsen, Jr. and J.B. Boyce, *Phys. Rev. B* **28**, 7130 (1983); *ibidem* **31**, 6903 (1985).
26. A. Qteish, N. Motta and A. Balzarotti, *Phys. Rev. B.* **39**, 5987 (1989).
27. P. Letardi, N. Motta and A. Balzarotti, *J. Phys. C: Solid State Phys.* **20**, 2853 (1987); A. Balzarotti, *Physica* **146B**, 150 (1987).
28. A. Balzarotti, M.T. Czyzyk, A. Kisiel, P. Letardi, N. Motta, M. Podgorny, M. Zinmal-Starnawska, *Advances in Solid State Physics* **25**, 689 (1985); *Z. Phys. B-Cond. Matter* **62**, 153 (1986).
29. A. Balzarotti, M. Piacentini and M. Grandolfo, *Lett. Nuovo Cim.* **3**, 15 (1969).
30. Proc. Workshop on Scientific and Technological Applications of Synchroton Radiation, R. Rosei and L. Fonda Ed., Miramare, Trieste (1987).
31. C. Lenardi, C. Vecile, R. Vitali and R. Rosei, *Int. Rep. TRSP-IUS* (unpublished).
32. H.A. Padmore, *Proc. SPIE* **733**, 253 (1987); H.P. Hogrefe, M.R. Howells and E. Hoyer, ibidem 274 (1987).

THEORETICAL STUDIES OF THE TRANSIENT RESPONSE OF HIGHLY PHOTOEXCITED CARRIERS IN SEMICONDUCTORS

P. KOCEVAR

Insitut für Theoretische Physik
Universität Graz
Universitätsplatz 5
A-8010 Graz
Austria

We review our present theoretical understanding of ultrafast relaxation processes in highly photoexcited electron-hole plasmas in polar semiconductors, with special emphasis on the picosecond and subpicosecond spectroscopy of bulk *GaAs* and *GaAs*-based quantum-well structures.

1. Introduction

The recent years have seen a continuously growing experimental and theoretical effort to clarify the various routes of energy transfer during and after photoexcitation of semiconductor surfaces by ultrashort laser pulses of high intensity. From a fundamental point of view this research is motivated by the ever increasing sophistication of picosecond and subpicosecond laser spectroscopy, which at present allows to study the fastest relaxation phenomena in solids. But the study of the transient dynamics of highly excited carriers in semiconductors is also of substantial technological interest, as these processes will ultimately determine and limit the speed of most ultrafast electronic and electrooptic semiconductor devices.

Originally it was the so called 'laser-annealing debate' of the early eighties about the experimental finding in Silicon of an unexpectedly slow power transfer from a laser-pulse excited electron-hole plasma into the semiconductor lattice which had triggered much of the present 'hot-carrier' research [1,2,3]. These findings for nanosecond pulses were corroborated by earlier picosecond transmission measurements in *GaAs* [4,5], which had also given evidence for a strongly reduced energy relaxation of the photoexcited carrier plasma. In view of the extremely short intercarrier scattering times in these experiments the differential spectra were analysed by use of Fermi-Dirac (FD) and Maxwell-Boltzmann (MB) distributions for the electrons and holes with a common time-dependent carrier temperature as supported by the correspondingly near-exponential frequency dependence of the spectra. The emission of optical phonons as the dominant cooling mechanism for the highly energetic plasma after its rapid initial thermalisation also seemed to be well established: in both *Si* and *GaAs* through the optical deformation potential coupling of longitudinal (*LO*) and transverse (*TO*) optical phonons to holes, and in *GaAs* in addition through the still stronger polar-optical (*po*) coupling of *LO* phonons to electrons and holes.

Nevertheless the surprising finding was that the carrier-cooling rates, as obtained from standard hot-electron theory [6] and appropriately modified for the free-carrier screening

R. Girlanda et al. (eds.), Progress on Electron Properties of Solids, 125–138.

of the long-range po couplings, were at least twice as high as the experimental rates, resulting in plasma temperatures far below the experimental ones. The discrepancy was attributed by the authors to possible effects of *LO*-phonon amplification and a correspondingly reduced cooling efficiency of the *LO* phonons for the carriers. This hypothesis was later qualitatively confirmed by detailed transport calculations for the coupled systems of hot carriers and nonequilibrium phonons [7,8] and independently established by both experimental [9] and theoretical [10,11] work on picosecond hot-carrier luminescence.

Besides the hot-phonon effect some further mechanisms were and are still debated which, depending on sample and excitation characteristics, could also noticeably impede the rapid power transfer into the lattice: (i) strong free-carrier screening of the *LO* phonons without pronounced phonon amplification, (ii) the reduction of the initial excitation density by carrier diffusion and (iii) retardation of the cooling of the electrons in the central (Γ) conduction-band minimum through transitory accumulation of electrons in higher band valleys (see Figure 1).

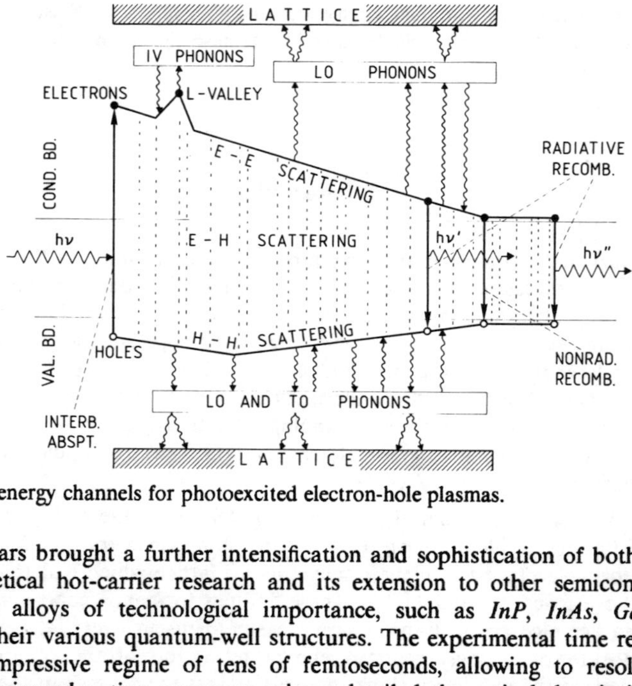

Figure 1. Main energy channels for photoexcited electron-hole plasmas.

The last years brought a further intensification and sophistication of both experimental and theoretical hot-carrier research and its extension to other semiconductors and semiconductor alloys of technological importance, such as *InP*, *InAs*, *GaP*, *GaAlAs*, *InGaAs*, and their various quantum-well structures. The experimental time resolution has reached the impressive regime of tens of femtoseconds, allowing to resolve even the fastest intercarrier relaxation processes, whose detailed theoretical description has only recently become possible through refined direct [12] and Monte-Carlo solutions [13,14,15] of the transport equations.

At present the most spectacular progress is being achieved in the study of the detailed carrier dynamics in single and multiple quantum-well structures by way of intersubband hot-carrier spectroscopy. In many cases spectacular effects of size quantisation and related dimensionality effects are seen in the excitation, dissipation, diffusion, and recombination properties of hot carriers. Moreover, the spatial discretisation within the individual quantum wells has opened the possibility of time- and space-resolved studies of diffusion and recombination [16] and of d.c. transport [17] of photoexcited carriers. This research has many obvious implications for our understanding of the carrier dynamics in both bulk and layered materials and also for the development of prospective new electronic [18,19] and electrooptic [19,20,21,22] devices. In the remaining sections we describe some of these recent and some of the earlier studies in more detail.

2. Transient Carrier-Phonon Dynamics

This section gives a short account of the interdependent effects of phonon disturbances and free-carrier screening of the po couplings. Effects of nonequilibrium optical phonons were for many years considered a backwater in hot-electron physics. They had already been predicted in connection with laser-pulse excitation of Germanium [23] and still earlier in connection with transient high-d.c. field transport in polar semiconductors [24,25]. Moreover, the creation of phonon disturbances by intense and short laser pulses had for many years been the prerequisite of excite-and-probe Raman spectroscopy of optical phonons [26,27], which of course involved the same electronic phonon-excitation mechanism as the laser-pulse spectroscopy of hot carriers in our present examples.

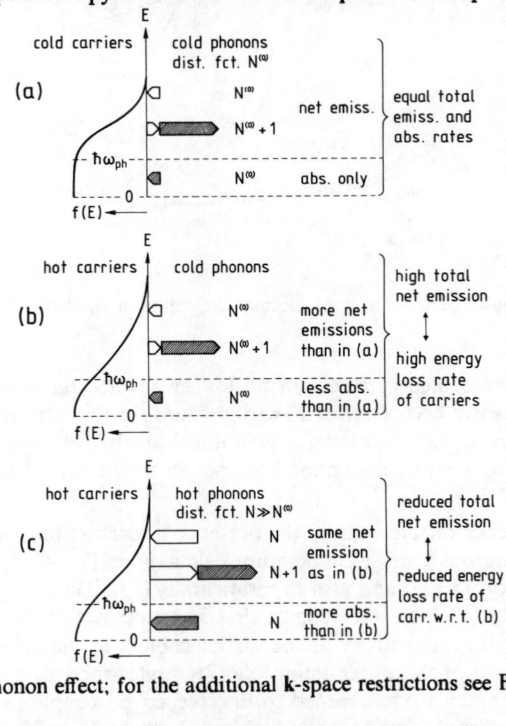

Figure 2. The hot-phonon effect; for the additional **k**-space restrictions see Figure 3.

Figure 2 summarises the essentials of the hot-phonon effect. For a completely isolated carrier-phonon system in equilibrium the total rates for phonon emission and absorption are equal, because the net (i.e. the spontaneous) emission rate of the 'active' regime of carriers with energies above the phonon-emission threshold $\hbar\omega_{ph}$ is exactly balanced by the absorption rate of the 'passive' regime of carriers with energies below $\hbar\omega_{ph}$ (Figure 2a). The balance is offset by the laser excitation which heats the carriers by transferring carriers from lower to higher energies. This increases the net phonon emission and reduces the phonon absorption near the band minimum (Figure 2b), resulting in a non-vanishing total emission rate and amplification of the phonons. As consequence of the phonon amplification the absorption rate increases, whereas the net emission rate of the 'active' regime, as determined by the spontaneous emission, remains unchanged (Figure 2c). So the cooling efficiency of the phonons is reduced by the *relative* increase of the phonon absorptions and would even eventually vanish, were it not for the nonelectronic phonon losses which keep the phonon distribution function below its value for a mutual carrier-phonon equilibrium.

However, for the reabsorption of the initially amplified phonons by the 'passive' carriers, the schematic picture of Figure 2 misses important **k**-space restrictions due to the energy and momentum conservation for each carrier-phonon ($c-ph$) process. Consider the cooling of the carriers after a short-pulse excitation as they relax from their injection energy E_{exc} to lower energies. The nearer a carrier approaches its band extremum the more increases the minimum $|\mathbf{q}|$ for a $c-ph$ process, as shown in Figure 3 for the Γ- and *L*-electrons in *GaAs*.

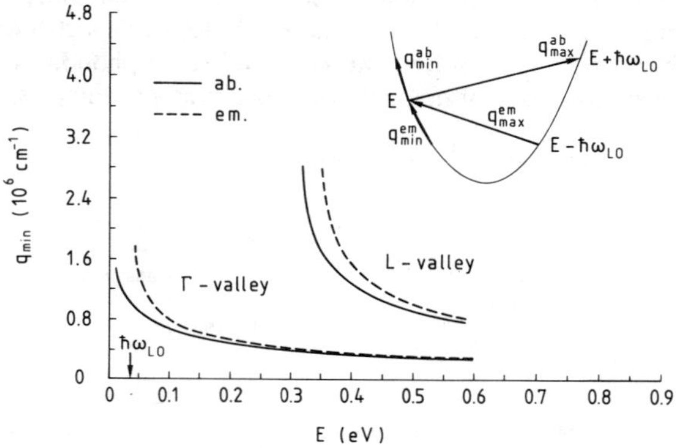

Figure 3. **k**-space anlaysis of the transient electron-*LO* phonon dynamics in *GaAs* for parabolic valleys.

So the majority of the cooled carriers can no longer absorb the small-$|\mathbf{q}|$ *LO* phonons which had predominantly been emitted at earlier times. In this way these **k**-space restrictions not only influence the hot-carrier relaxation, but also the electronic reabsorption contributions to the apparent *LO*-phonon lifetimes in time-resolved Raman spectroscopy [28,29].

A negligible influence of screening in the presence of such strong phonon disturbances has been found in various carrier-temperature calculations [11,30,31], in calculations of high-d.c. field transport [30,32] and also experimentally [33]. These findings seem to question the practicality of involved screening models whenever hot-phonon effects are noticeable. The reason for this insensitivity of the carrier cooling to the screening and therefore to the detailed strength of the po couplings can be understood as a further hot-phonon phenomenon. By changing from screened to unscreened po couplings the scattering rate of *LO* phonons is increased. This causes a higher *LO*-phonon amplification, which in turn increases the relative proportion of *LO* reabsorptions. So, although the total number of *LO*-phonon scatterings is increased, their net cooling efficiency is reduced, so that the total energy-loss rate of the carriers remains practically unchanged.

Nevertheless, for some years many workers in the field tried to explain the slow plasma cooling only as effect of the free-carrier screening of the po couplings, because the resulting decrease of the *LO*-phonon emission rates was of the right order of magnitude. To put these claims into the right perspective, one has first to establish the range of experimental conditions, for which substantial nonequilibrium-phonon effects should be expected [8]. As the phonon disturbance is proportional to the hot-carrier concentration there is obviously a lower limit to detectable hot-phonon effects, typically at relatively low plasma densities around and below a few $10^{16} cm^{-3}$. So one should expect the experimental cooling rates for the lowest excitation densities to be reproducible by conventional calculations for phonon equilibrium, as was indeed found to be the case in one of the above-mentioned claims [34]. But naturally for such low densities, the influence of screening should be rather small as well. There is also an upper density limit for nonequilibrium-phonon effects: at densities around and above $10^{20} cm^{-3}$ the po couplings are so heavily screened that practically no polar optical scatterings occur.

To summarize the preceding discussion we can state that today the cause of the ineffectiveness of the polar optical interaction for the energy relaxation of highly photoexcited carriers in polar semiconductors seems to be understood: either, at moderately high plasma densities typically between 10^{17} and $5.10^{19} cm^{-3}$, it is *LO*-phonon amplification or, at still higher densities, the complete screening of the couplings. This leaves, at least for high excitation levels, the intrinsically weaker short-range optical deformation-potential couplings as the dominant power-dissipation channel for the carriers [7,8] and might explain the general finding that both polar and nonpolar materials seem to behave very similarly under extreme laser-pulse excitation [35].

3. Transport Models

Semiclassical transport theory as a bookkeeping of the mean particle occupation numbers of quantum states is best suited for the analysis of the photoexcitation and relaxation processes of carriers and phonons. As long as the space-time scales involved allow a kinetic description in terms of free particle trajectories between instantaneous scattering processes with quantum-mechanical scattering cross sections, the semiclassical Boltzmann equations for electrons, holes and phonons should in principle provide an appropriate microscopic description. However, it seems that such a Boltzmann theory for materials like *GaAs* with their strong polar couplings is pushed to its limit set by the Heisenberg principle for the energy uncertainty $(\Delta E)_c$ and the lifetime τ_c of a carrier state. Indeed the po-scattering frequency τ_c^{-1} is of the order of the *LO*-phonon frequencies $\omega_{LO} = (\Delta E)_c / \hbar$, so that $(\Delta E)_c \cdot \tau_c \approx \hbar$, which marks the ultimate borderline for the concept of an instantaneous scattering event for both carriers and phonons. To navigate in such treacherous waters near or even outside the conventional limits of Boltzmann transport is typical for hot-electron research, but most workers in the field find encouragement by the general empirical finding that the Boltzmann equation in solid-state and plasma physics works much better than expected on formal transport-theoretical grounds.

The framework of semiclassical transport theory implies the notion of particles: in our case of carriers with effective mass m, crystal momentum $\hbar\mathbf{k}$ and kinetic energy $E(\mathbf{k})$, and of phonons as ballistic particles with crystal momentum $\hbar\mathbf{q}$ and energies $\hbar\omega(\mathbf{q})$. In *GaAs*-like materials the dominant scattering mechanisms for the highly excited carriers (near-zone-center Γ-electrons and holes, *L*-valley electrons) are intercarrier $(e-e, h-h, e-h)$ scattering and the interaction with near-zone-center *LO* and *TO* phonons and with the large-$\mathbf{q}$ phonons involved in electronic $\Gamma - L$ valley transfers. The intervalley (iv) phonons can be assumed to be in thermal equilibrium, as their distribution covers a region of $\mathbf{q}$-space much larger than that of the electronically active *LO* and *TO* modes. The *LO*- and *TO*-phonon dynamics consists of the various $c-ph$ and phonon-phonon $(ph-ph)$ interactions. The $ph-ph$ processes tend to thermalise the excited phonons by causing their decay into pairs of electronically inactive near-zone-boundary phonons. Thus these decay modes constitute the primary energy sink for the photoexcited carrier-phonon system and initiate, through their own rapid cascading into phonons of still lower frequency, the eventual transformation of the electronic excitation energy into lattice heat (Figure 1).

As far as the different $c-ph$ couplings are concerned it suffices to recall the q^{-2} dependence of the transition probabilities for unscreened po scattering of LO phonons. For the case of a maximal, i.e. static, free-carrier screening of these long-range couplings, the 'bare' transition probabilities are modified to $q^2(q^2+q_s^2)^{-2}$, producing a small-q cutoff at the Thomas-Fermi (TF) screening parameter q_s of the hot carriers. It is important to evaluate q_s for the *nonequilibrium* carriers, as their screening efficiency is certainly much smaller than that of the equilibrium carriers in the usual linear-response theory. This will make q_s time-dependent through its dependence on the continuously changing concentrations and mean energies of the different types of carriers. Strictly speaking, because of the finite frequency of the LO phonons, the free-carrier screening of the po couplings should be treated as both nonlinear and dynamical through the use of the frequency- and wavevector-dependent dielectric function $\epsilon(\mathbf{q},\omega_{LO})$ of the nonequilibrium electrons and holes. But in any case the screening obtained from any one of the very elaborate dynamical and nonlinear response theories [36,37,38] must fall somewhere between the two easily evaluated extreme cases of 'hot' TF screening and unscreened couplings. A comparison of the results obtained for these two extreme cases should therefore suffice to predict the maximal screening effects on the hot-carrier cooling, with the possible exception of antiscreening phenomena. Such plasma-LO phonon resonance effects are predicted from linear response theory and are to be expected under excitation conditions for which the effective plasma frequency of carriers approaches the LO-phonon frequency [39]. However, their influence on the carrier cooling has been found to be negligible [37].

Most of the generally accepted transport models have for simplicity neglected the space dependence of the carrier and phonon distributions. As near-zone-center LO and TO modes have a very small frequency dispersion (ie. group velocity) and lifetimes of the order of picoseconds, their spatial propagation can be safely neglected. Moreover, carrier diffusion out of the optically active surface layer can in a first approximation be treated as a volume effect by adding a particle-loss term to the nonradiative bimolecular recombination [11].

According to the foregoing physical picture for the main routes of energy transfer within and out of the photoexcited carrier-phonon system, one has to solve the coupled time-dependent Boltzmann equations for the carrier and phonon distribution functions $f(\mathbf{k},t)$ and $N(\mathbf{q},t)$. We first discuss the carrier equation

$$\frac{df}{dt} = \left[\frac{df}{dt}\right]_{exc} + \left[\frac{df}{dt}\right]_{c-ph} + \left[\frac{df}{dt}\right]_{c-c} + \left[\frac{df}{dt}\right]_{c-c'} + \left[\frac{df}{dt}\right]_{loss}, \tag{1}$$

where the carrier and phonon indices denote $e(\Gamma)$, $e(L),h$ and LO,TO, which in quantum-well systems should be supplemented by additional subband or mode labels in cases of size quantisation of carriers [40] and geometrical confinement or folding effects on the phonons [41,42]. Eq. (1) and the Boltzmann equation for the phonons are coupled by their respective $c-ph$ collision integrals, each containing the statistical weights $N+1$ or N for phonon emissions or absorptions as well as the factor $f(1-f')$ as probability for the occupation of the initial and non-occupation of the final carrier state per $c-ph$ process. The meaning of the various contributions to (1) is obvious from the notation, except that $c-c$ and $c-c'$ distinguish between equivalent (such as $e_\Gamma - e_\Gamma$ or $h-h$) and inequivalent (e.g. $e-h$) carrier-carrier scattering.

The excitation term in (1) contains the nonexcitonic interband photogeneration of $e-h$ pairs plus possible small contributions of free-carrier absorptions and Auger processes at high plasma densities. The dominant interband term depends on the temporal and spectral intensity profile $I_\nu(t)$ of the laser pulse, often assumed to be exponential or gaussian in time, and on the illumination area. One first corrects the energy flux density of the laser pulse for losses along the beam path (e.g. at the windows of the cryostat) and for the reflection at the sample surface, leaving typically one third of the original pulse intensity to enter the sample. By use of the quantum yield (≈ 1) and absorption coefficient $\alpha_\nu^{(j)}(j=1,2)$ of the material one then obtains the $e-h$ pair generation rates for the experimentally interesting cases of one- or two-photon absorption as $\propto \alpha_\nu^{(1)} I_\nu(t)/h\nu$ or $\propto \alpha_\nu^{(2)} I_\nu^2(t)/2h\nu$ [9,12].

For a simple parabolic bandstructure with band gap E_g the kinetic energies of the photogenerated electron and hole per absorbed photon (of energy $h\nu > E_g$) are $E_{exc}^{(e)} = (h\nu - E_g)m_h/(m_e+m_h)$ and $E_{exc}^{(h)} = (h\nu - E_g)m_e/(m_e+m_h)$ with straightforward corrections for nonparabolicity and warping [43] or for size quantisation in quantum well structures [44]. For high carrier densities n_c, effects of band filling and band-gap renormalisation are easily incorporated through the Pauli-exclusion factor $1-f_e(E_{exc}^{(e)})-f_h(E_{exc}^{(h)})$ [8,12] and by use of the empirically known change of E_g with n_c [12,43].

Two further carrier-heating processes are free-carrier absorption and Auger recombination, both at high densities. Their rates can be written as $\approx \alpha_{fca}$ and $\approx \alpha_{aug} n_c^2$, with α_{fca} and α_{aug} as known material constants. One should note here that these and all other optical material parameters are in most cases only known from low-excitation measurements, but nevertheless also used for the high excitation levels of our present concern. For special band structures (e.g. in *GaSb*) and under favourable excitation conditions Auger processes can be of major importance [45].

The particle losses in (1) are dominated by nonradiative ('bimolecular') recombination of band-edge electrons and holes, including Auger recombination at high densities. The general use of frequency-integrated recombination rates $\approx B_{rec} n_c$ is probably too much of an oversimplification. The differential rates as inferred from the time- and frequency-resolved luminescence spectra are found to strongly depend on the energies of the recombining carriers. In *GaAs*, for example, the increase of $B_{rec}(\nu)$ about one order of magnitude for increasing emitted photon energies [9] obviously reflects the rapid downward relaxation of the carriers towards their respective band minima and possibly also the faster spatial diffusion of more energetic carriers out of the optically active absorption layer [46].

We now turn to the phonon Boltzmann equation

$$dN(\mathbf{q},t)/dt = -\sum_{couplings} \{(N(\mathbf{q},t)-N_c(\mathbf{q},t))/\tau_c\} - (N(\mathbf{q},t)-N^{(0)})/\tau_{ph} \qquad (2)$$

for the rate of change of the phonon distribution through the $c-ph$ and $ph-ph$ processes. The first term describes the $c-ph$ scatterings and is automatically in the form of a linear relaxation, with N_c and the relaxation time τ_c defined by the $\mathbf{k}$-space integrals of the statistical weight $f(1-f')$ of the participating carrier states times the energy-conserving δ-function for phonon emissions and absorptions. The $ph-c$ rate τ_c^{-1} pushes the phonons towards a mutual equilibrium with the carriers, as expressed by the distribution function N_c being a functional of the instantaneous carrier distribution. In general the $ph-ph$ collision integral contains as leading term the cubic lattice anharmonicity which is of third order in the phonon occupancies N of the interacting modes [47]. But for our near-zone-center LO and TO modes an enormous simplification is achieved through the fact that their decay products belong to electronically inactive and rapidly decaying modes. This allows one to take their distributions as undisturbed and thus to express $(dN/dt)_{ph-ph}$ in terms of a well-defined and experimentally measurable relaxation time τ_{ph} as $-(N(\mathbf{q},t)-N^{(0)})/\tau_{ph}$, showing the tendency of the decays to relax $N(\mathbf{q},t)$ towards the thermal-equilibrium Planck distribution $N^{(0)}$ and thus to thermalise the excited phonons. For *GaAs* the generally accepted values for $\tau_{LO}(\approx\tau_{TO})$ are $8ps$ for $T\rightarrow 0$ and $3ps$ at room temperature [30].

In view of the spectroscopical evidence for the change from one 'bare' *LO*-phonon branch to two branches of mixed *LO* phonon-plasmon ($ph-pl$) modes at carrier densities for which their plasma frequency approaches ω_{LO}, substantial efforts are being made to modify, at least for high excitation levels, the above-formulated bare carrier-*LO* phonon dynamics by rewriting eqs. (1) and (2) in terms of hybridising plasmon and *LO* modes and dynamically screened polar couplings [10,37]. But this straightforward procedure might raise more problems than it sets out to solve. First its substantiation by many-body techniques would also require a renormalisation of the free electrons and holes and of their $c-c$ couplings. Second the decay rate of the $ph-pl$ excitation quanta cannot automatically be assumed to play the same role as the bare *LO*-phonon decay rates which act, in the simple formulation, as the rates for the energy transfer out of the excited $c-ph$ system into the lattice. The reason is that the plasmon part of each mixed mode would always decay back into single carrier excitations. For the question of the power transfer from the laser into the lattice it is essential that this transfer only involves the zone-boundary phonons from the decay of the bare phonon part of each $ph-pl$ excitation. As two hybrid $ph-pl$ modes just share the weight of one original bare *LO* phonon we expect the formation of mixed $ph-pl$ modes to be irrelevant for the net energy flux into the lattice. The plasmon admixture of each excitation acts as vehicle for an intermediate reshuffling of electronic energy between collective and single-particle constituents of the excited carrier system. So any type of plasmon formation could have a transitory influence on the spectroscopically determined time evolution of the single-particle energies of electrons and holes. In any case, a consistent theory of these collective phenomena would have to go beyond the strategy of just replacing each *LO* phonon by two dynamically screened $ph-pl$ excitations with lifetimes treated in analogy to τ_{LO} and therefore as relaxation times for purely nonelectronic losses.

4. The Carrier-Temperature and Monte-Carlo Approaches

The two most widely used approaches for solving eqs. (1) and (2) are the carrier-temperature model and the ensemble Monte-Carlo technique. As in the pioneering times of hot-electron theories for nonlinear d.c. transport [6], the first successful attempts used the concept of a carrier temperature. Under the assumption that the $c-c$ scatterings are sufficiently frequent to establish a heated *FD* distribution f_{FD} for each type of photogenerated carrier, the carrier temperatures T_c and chemical potentials μ_c are determined as functions of time from the instantaneous concentrations and mean kinetic energies of the carriers during and after the excitation pulse. These are obtained from rate equations for the carrier number and energy in the presence of the phonon disturbances which in turn are determined by a consistent solution of (2). Due to the relatively simple functional form of f_{FD} this solution is straightforward, because the $ph-c$ collision integrals can be solved in closed form, with N_c (naturally) obtained as the Planck distribution at $T_c(t)$ and with τ_c as explicit function of $T_c(t)$ and $\mu_c(t)$ for each $c-ph$ coupling [7,8,10]. After inserting the resulting momentary phonon-induced energy losses of the carriers

$$(dE/dt)_{c-ph} = -\sum_{couplings}\sum_{modes}\hbar\omega(dN/dt)_{ph-c} = \tag{3}$$

$$= \sum_{couplings}\sum_{modes}\hbar\omega(N-N_c)/\tau_c$$

into the total rate equation for the carrier energy, the latter can be stepwise integrated in parallel with (2) and with the rate equation for the carrier number, starting with the equilibrium situation just before the onset of the photoexcitation. In this way one obtains the various carrier temperatures, carrier populations and optical phonon distributions as functions of time during and after the laser pulse.

The advantage of this approach is the direct correspondence of theory and experimental analysis, because almost all of the early time-resolved absorption or luminescence experiments were evaluated in terms of *FD* or *MB* carrier distributions with time-dependent carrier temperatures. Before the advent of experimental subpicosecond resolution the even simpler model of only one common temperature for both electrons and holes had been used in in the theoretical [7,8,10] and phenomenological description of the experiments [5], because a picosecond timescale was thought to be sufficiently long to ensure the complete internal thermalisation of the plasma via $e-h$ collisions. But it was soon recognised in two-temperature (2*T*) calculations for $T_e \neq T_h$ [43,48,49] that for frequently encountered plasma densities the internal thermalisation of electrons and holes can take several picoseconds, well within the resolution of many picosecond experiments.

The important practical aspect of the 2*T* approach lies in the fact that the analysis of the differential spectra by use of two *FD* distributions reveals the same single-exponential behaviour as for the 1*T* model: because of the large mass difference between electrons and holes the dominant exponential involves T_e only. This has the important consequence that the experimental plasma temperature as determined by a 1*T* analysis is always T_e [38], especially at those earlier times, at which the initially more energetic electrons (as $E_{exc}^{(e)} >> E_{exc}^{(h)}$) are still much 'hotter' than the holes. So the use of separate electron and hole temperatures and of an extra rate equation for the energy exchange between the electrons and holes is essential for the time span between the internal thermalisation of each carrier species, taking typically a few tenths of a picosecond, and the mutual thermalisation. The regime of mutual thermalisation was first estimated through an analysis for a picosecond laser pulse of sufficiently low frequency to create 'cold' holes with energies below the threshold for *LO*-phonon emission [43]. For a clearer demonstration the phonons were assumed to be in equilibrium to exclude any effects of phonon reabsorption. The authors found a noticeable heating of the holes by $e-h$ collisions during the first few picoseconds, revealing an inexpectedly long mutual thermalisation phase for the electrons and holes. For times beyond this complete thermalisation regime the computationally simpler 1*T* model provides the adequate description of the $c-ph$ dynamics [48], as demonstrated in the eventual convergence of T_e and T_h in Figure 4, taken from a 2*T* analysis including hot phonons [49].

Typical 1*T* calculations are illustrated in Figures 5 and 6. The upper T_c-curve of Figure 5 shows satisfactory agreement between the theory [30] and the experimentally determined plasma temperatures during and after a 25*ps* excitation pulse, as derived from time-resolved differential luminescence spectra [9]. Much too low carrier temperatures are found for phonon equilibrium (i.e. vanishing τ_{ph}), indicating a nonequilibrium-phonon induced retardation of the carrier cooling. The upper part of the figure shows indeed a strong *LO* and *TO* phonon buildup. The amplification of the *LO* phonons by the *po* interaction is much stronger than the amplification of the *TO* phonons by the weaker optical deformation-potential coupling: the N^{LO} curve is much nearer to its 'ineffectivity' limit N_c^{LO} for mutual $c-ph$ equilibrium (compare eq. (2)). The already mentioned negligible influence of screening in the presence of such strong *LO*-phonon disturbances is demonstrated by the two upper T_c-curves for hot-carrier screened and unscreened *po* couplings. An expected reduction of the phonon amplification and of the resulting retardation of the carrier cooling for faster phonon thermalisation is seen from a comparison of Figure 5 with the analogous results of Figure 6 for a *CdS* sample of much

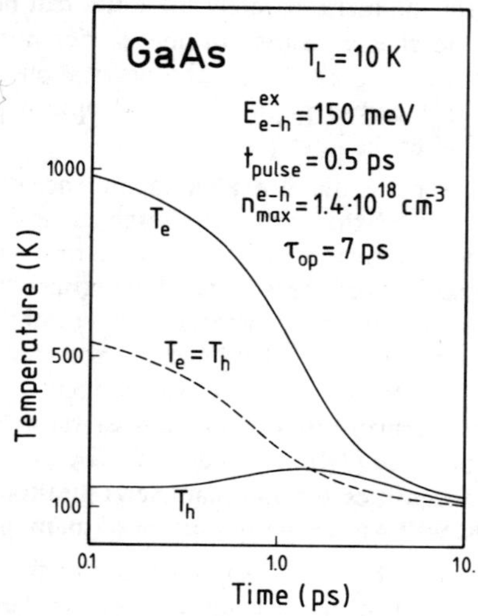

Figure 4. T_c-analysis of the mutual electron-hole thermalisation in *GaAs* after a 0.5 picosecond excitation pulse (from ref. [49]).

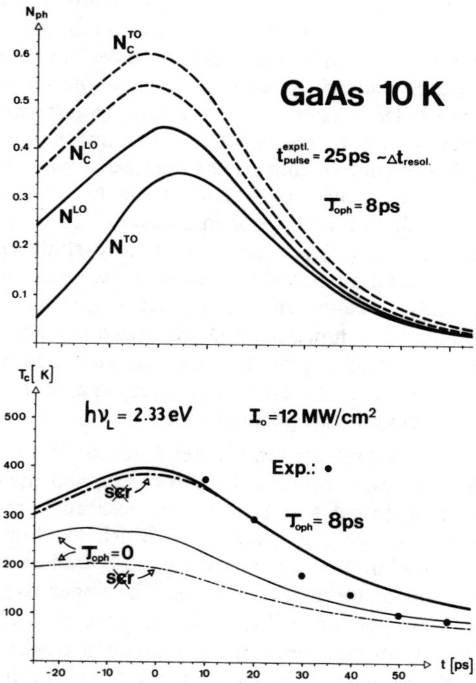

Figure 5. T_c-analysis of hot-carrier luminescence in *GaAs* during and after a 25 picosecond excitation pulse (from ref. [11]).

shorter *LO*-phonon lifetime under comparable excitation conditions.

A practically important aspect of carrier-temperature models concerns the use of the equilibrium notion of a particle temperature for the extremely nonequilibrium situations

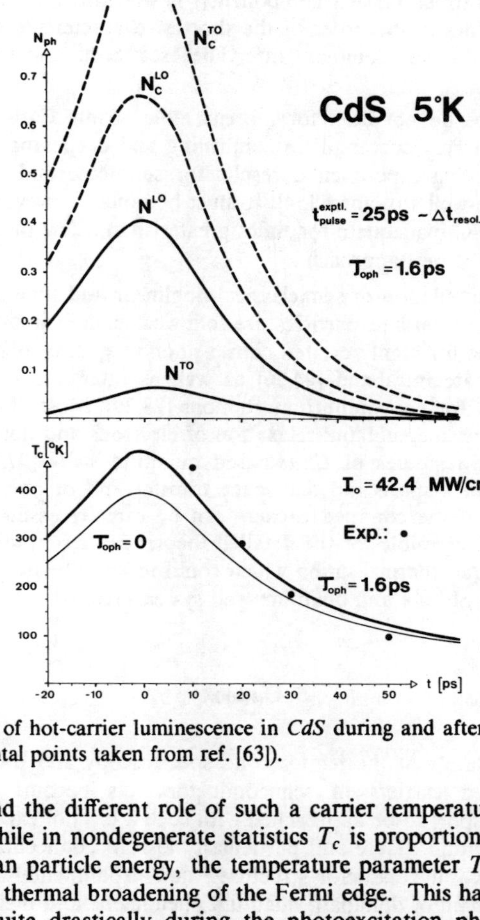

Figure 6. T_c-analysis of hot-carrier luminescence in *CdS* during and after a 25 picosecond excitation pulse (experimental points taken from ref. [63]).

of our examples and the different role of such a carrier temperature for degeneracy and nondegeneracy. While in nondegenerate statistics T_c is proportional to and thus a direct measure of the mean particle energy, the temperature parameter T_c of a *FD* distribution is a measure of the thermal broadening of the Fermi edge. This has the consequence that T_c can decrease quite drastically during the photoexcitation phase, just to fulfill the requirement of fixed energy per carrier under the condition of an increasing carrier density (i.e. increasing chemical potential). So this 'generation cooling' [8,11] would also occur in the absence of any genuine energy-loss processes such as phonon emissions, and could be misinterpreted as energy relaxation. The analogous phenomenon of a spurious 'recombination heating' [11] could similarly be taken as a sign of a reduced power dissipation during the recombination phase after the excitation pulse. Although both effects and their counterparts for nondegenerate statistics are well understood and easily estimated [50], they somewhat obscure the relevant relaxation dynamics during the photoexcitation and on the longer (nanosecond) timescales of the recombination regime [45,51]. This same drawback is of course already inherent in the presentation of experimental results in terms of plasma temperatures obtained by fitting differential spectra through *FD* or *MB* distributions.

The carrier-temperature model has several serious limitations. First the analytical solution of the $ph-c$ collision integrals in (2) requires simple parabolic bandstructures and thus misses effects of valence-band warping and conduction-band nonparabolicity. Second and more severely, the T_c models are not able to describe the now experimentally accessible femtosecond relaxation phenomena before and during the internal thermalisation of the various plasma components; it will also fail in cases, where these thermalisations take times of the order of the shortest characteristic times for the $c-ph$ dynamics, e.g. in situations of dominant intervalley scattering and low plasma densities [52,53].

All these restrictions do not exist for the ensemble Monte Carlo (EMC) techniques which have lately been very successful in confirming and deepening the older T_c-model results and in reproducing experimental results for subpicosecond phenomena in bulk materials and quantum-well structures [54]. It must be noted, however, that its numerical complexity renders EMC inadequate for times greater than a few picoseconds, for which the T_c-model remains the best approach.

In a typical EMC simulation of semiclassical nonlinear and time-dependent transport several thousand or ten thousand particles are followed in their individual histories [55]. To adapt this technique for highly excited carrier-phonon systems one had to devise new algorithms to incorporate intraband [44,56] as well as interband [13,15] carrier-carrier scattering and to allow for nonequilibrium phonons [28,29]. Most of these heavy simulations concentrate on the intersubband relaxation of electrons and holes as obtained from experimental picosecond studies of *GaAs*-based quantum wells [57,58,59], because the various contributions of **k**-space and real-space transfer and of phonon disturbances to the energy relaxation of the confined carriers can be directly assessed. Moreover, such simulations will be indispensible for the detailed theoretical description of ultrashort carrier dephasing and carrier thermalisation which could recently be experimentally resolved in femtosecond studies of bulk and quantum-well systems [60,61].

5. Outlook

In summary we can state that the femtosecond spectroscopy and theoretical description of highly photoexcited carriers in semiconductors has become the first topic of condensed-matter transport to be approached with such a level of experimental resolution and theoretical expenditure. Here and particularly also in connection with ballistic d.c. transport in ultrasmall semiconductor structures, new experimental refinements are just touching upon the domain of dominant quantum phenomena and their practical exploitation in submicron devices.

Looking into the future we can thus foresee another fruitful period of close interplay between experiment and solid-state transport theory. However, the traditional semiclassical transport picture of our foregoing discussion, in which intuition has always been and still is of great use, will necessarily be replaced by the much more abstract framework of quantum-transport theory.

References

1. Ellen Yoffa, in *Laser and Electron Beam Processing of Material*, Eds. C.W. White and P.S. Pearcy, Acad. Press, 1980, p. 59.
2. H. Kurz, *Physica B* **117**, **118**, 1010 (1983).
3. A. Compaan, *Proc. 3rd Trieste ICTP-IUPAP Semic. Symp. on High Excitation and Short Pulse Phenomena*, Ed. M.H. Pilkuhn, North Holland, 1985, p. 425.
4. R.F. Leheny, J. Shah, R.L. Fork, C.V. Shank and A. Migus, *Solid St. Comm.* **31**, 809 (1979).

5. J. Shah, *J. de Physique* **52**, C7-445 (1981).
6. E. Conwell, *Solid State Physics, Suppl.* **9**, Eds. F. Seitz, D. Turnbull, H. Ehrenreich, Acad. Press, 1967.
7. P. Kocevar, W. Pötz and W. Porod, *Physica B* **117, 118**, 220 (1983).
8. W. Pötz and P. Kocevar, *Phys. Rev.* **B28**, 7040 (1983).
9. W. Graudszus, Ph.D.Thesis, Univ. Stuttgart 1985, unpublished, see also ref. [1].
10. J. Collet, A. Cornet, M. Pugnet and T. Amand, *Solid St. Comm.* **42**, 883 (1982).
11. P. Kocevar, *Physica* **134B**, 401 (1985).
12. J. Collet and T. Amand, *J. Phys. Chem. Solids* **47**, 153 (1986); J. Collet, J.L. Oudar and T. Amand, *Phys. Rev.* **B34**, 5443 (1986).
13. M.A. Osman, U. Ravaioli, R. Joshi, W. Pötz and D.K. Ferry, *Proc. 18th Int. Conf. Phys. Semicond.*, Ed. O. Engstrom, World Scientific Press, Singapore, 1986, p. 1311.
14. S. Goodnick and P. Lugli, *Appl. Phys. Lett.* **51**, 584 (1987).
15. S. Goodnick and P. Lugli, *Phys. Rev.* **B38**, 10135 (1988).
16. D.J. Westland, D. Mihailovic, J.F. Ryan and M.D. Scott, *Surf. Sci.* **196**, 399 (1988).
17. R.A. Höpfel, J. Shah, P.A. Wolff and A.C. Gossard, *Phys. Rev.* **B37**, 6941 (1988).
18. N. Yokoyama, K. Imamura, H. Ohnishi, T. Mori, S. Muto and A. Shibatomi, *Solid St. Electronics* **31**, 577 (1988); B.F. Levine *ibid.* p. 583.
19. F. Capasso, *Physica Scripta* **T19**, 199 (1987).
20. Th. Pfeiffer, J. Kuhl, M. Serenyi, H. Habermeier, E.O. Göbel, L. Palmetshofer and A. Axmann, *Phys. Scripta* **T13**, 100 (1986).
21. For an updated overview see ref. [62].
22. S.T. Eng, *Physica Scripta* **T19**, 205 (1987).
23. H. van Driel, *Phys. Rev.* **B19**, 5928 (1979).
24. P. Kocevar, *J. Phys.* **C5**, 3349 (1972); *Acta phys. Austriaca* **37**, 259 (1973).
25. P. Kocevar, in *Physics of Nonlinear Transport in Semiconductors*, Eds. D.K. Ferry, J.R. Barker and C. Jacoboni, Plenum Press, New York, 1980, p. 401.
26. J. Shah, R.C.C. Leite and J.F. Scott, *Solid St. Comm.* **8**, 1089 (1970).
27. Aurea Vasconellos and R. Luzzi, *Solid St. Comm.* **32**, 1219 (1979).
28. P. Lugli, C. Jacoboni, L. Reggiani and P. Kocevar, *Appl. Phys. Lett.* **50**, 1251 (1987); see also P. Lugli, C. Jacoboni, L. Reggiani, and P. Kocevar, SPIE vol. 793 on *Ultrafast Lasers Probe Phenomena in Bulk and Microstructure Semiconductors*, 1987, p. 102.
29. P. Lugli, P. Bordone, M. Rieger, P. Kocevar and S.M. Goodnick, *Phys. Rev.* **B39**, 7852 (1989).
30. P. Kocevar, Festkörperprobleme, *Advances in Solid State Physics* **27**, 197 (1987).
31. A.C.S. Algarte, *Phys. Rev.* **B28**, 2388 (1985).
32. M. Rieger, P. Kocevar, P. Lugli, P. Bordone, L. Reggiani and S.M. Goodnick, *Phys. Rev.* **B39**, 7866 (1989).
33. K. Leo, W.W. Rühle and K. Ploog, *Phys. Rev.* **B38**, 1947 (1988).
34. C.H. Yang, J.M. Carlson-Swindle, S.A. Lyon and J.M. Worlock, *Phys. Rev. Lett.* **54**, 2045 (1986).
35. N. Bloembergen, priv. comm.
36. J.B. Barker and D. Lowe, *J. de Physique* **42**, C7-293 (1981).
37. S. Das Sarma, J.K. Jain and R. Jalabert, *Phys. Rev.* **B37**, 6290 (1988).
38. J.H. Collet, *Phys. Rev.* **B39**, 7659 (1989).
39. S. Doniach, *Proc. Phys. Soc.* **73**, 849 (1959).
40. T. Ando, A.B. Fowler and F. Stern, *Rev. Mod. Phys.* **54**, 437 (1982).

41. B.K. Ridley, *J. Phys.* **C15**, 5899 (1982).
42. W. Cai, C.M. Marchetti and M. Lax, *Phys. Rev.* **B35**, 1369 (1987).
43. M. Ashe and O.G. Sarbei, *Phys. Stat. Sol. (b)* **126**, 607 (1984).
44. S. Goodnick and P. Lugli, *Phys. Rev.* **B37**, 2578 (1988).
45. P.A. Snow, D.J. Westland, J.F. Ryan, T. Kerr, H. Munekata and L.L. Chang, in ref. [62] p. 595.
46. T. Kuhn and G. Mahler, *Phys. Rev.* **B35**, 2827 (1987).
47. P.G. Klemens, *Solid St. Phys.* **7**, 1 (1958).
48. M. Asche and O.G. Sarbei, *Phys. Stat. Sol. (b)* **141**, 487 (1987).
49. W. Pötz, *Phys. Rev.* **B36**, 5016 (1987).
50. D. Bimberg and J. Mycielski, *Phys. Rev.* **B31**, 5490 (1985); *J. Phys.* **C19**, 2363 (1986).
51. K. Leo and W.W. Rühle, *Solid St. Comm.* **62**, 659 (1987).
52. J. Shah, B. Devaud, T.C. Damen, W.T. Tsang, A.C. Gossard and P. Lugli, *Phys. Rev. Lett.* **59**, 2222 (1987).
53. S. Zollner, S. Gopalan and M. Cardona, *Appl. Phys. Lett.* (1989) to be published.
54. D.K. Ferry, M.A. Osman, R. Joshi and M. Kann, *Solid-St. Electronics* **31**, 401 (1988).
55. C. Jacoboni and L. Reggiani, *Rev. Mod. Phys.* **55**, 645 (1983).
56. R. Brunetti, C. Jacoboni, A. Matulionis and V. Dienys, *Physica* **134B**, 369 (1985).
57. D.Y. Oberli, D.R. Wake, M.V. Klein, J. Klem, T. Henderson and H. Morkoc, *Phys. Rev. Lett.* **59**, 696 (1987).
58. A. Seilmeier, H.J. Hübner, G. Abstreiter, G. Weimann and W. Schlapp, *Phys. Rev. Lett.* **59**, 1345 (1987).
59. A.M. de Paula, R.A. Taylor, C.W.W. Bradley, A.J. Turberfield and R.F. Ryan, in ref. [62].
60. P.C. Becker, H.L. Fragnito, C.H. Brito Cruz, R.L. Fork, J.E. Cunningham, J.E. Henry and C.V. Shank, *Phys. Rev. Lett.* **61**, 1647 (1988).
61. W.H. Knox, D.S. Chemla, G. Livescu, J.E. Cunningham and J.E. Henry, *Phys. Rev. Lett.* **61**, 1290 (1988).
62. Proc. 4th Int. Conf. on Superlattices, Microstructures and Microdevices, Trieste 1988, Ed. J. Dow, Superlattices and Microstructures Vol. 5 (1989).
63. H. Saito and E.O. Göbel, *Phys. Rev.* **B31**, 2360 (1985).

OPTICAL SPECTROSCOPY IN TRANSITION-METAL DICHALCOGENIDES

A. BORGHESI, L. NOSENZO AND A. STELLA

Università di Pavia, Italy

G. GUIZZETTI

Università di Modena, Italy

F. LÉVY

EPFL Lausanne, Switzerland

The optical spectroscopy of layered and in particular of transition metal dichalcogenides is an interesting and stimulating subject which has played a major role in the recent developments of solid state physics. A detailed investigation of the optical structures due to interband transitions has been carried out by means of reflectance and thermoreflectance. Pure compounds and mixed systems have been studied in some detail. Among the latter ones, a particular attention has been devoted to $Ti_{1-x}Hf_xSe_2$, which changes from semimetallic to semiconducting as a function of composition and shows an anomalous evolution of the energy position of its main spectral features versus lattice and other structural parameters. The effect of phase transitions like charge density waves on the optical response, and in particular on the plasma resonance, free carrier absorption, energy gap etc., is also studied and discussed.

1. Introduction

Layered compounds have been extensively studied both theoretically and experimentally after some pioneering works in the sixties [1,2]. A few properties of such materials, which have been discovered and clarified in the seventies [3,4], have shown particularly interesting and updated aspects connected to some of the most promising developments of solid state physics.

In this respect, graphite and transition metal dichalcogenides (*TMD*) should be primarily mentioned, where intercalation [5,6] and two-dimensional [4] phenomena like charge density waves (*CDW*) phase transitions take place.

Intercalation is achieved through the insertion of atomic or molecular layers of guest chemical species between layers of a host material, characterized by intraplanar binding forces larger than the interplanar ones. Thus intercalation compounds are very similar to artificially structured materials like semiconductor superlattices. In both cases an additional periodicity related to a large unit cell containing multiples of the primitive unit cells of the constituents is established in the direction perpendicular to the layer planes (z-direction). When the thicknesses of the layers of the two constituents of a semiconductor heterojunction superlattice become very small (i.e. of the order of a few lattice constants), the electronic structure of the system is essentially characterized by free electron states in zone folded energy bands, exactly like in layered intercalated compounds (in

R. Girlanda et al. (eds.), Progress on Electron Properties of Solids, 139–151.

particular graphite intercalated compounds).

In addition to these cases of superlattice periodicity perpendicular to the layer planes, the creation of larger real space dimensions of supercells, and therefore smaller reciprocal space dimensions with folding of the bands inside the smaller Brillouin Zones (*B.Z.*) may also take place in the layers. This typically happens, for instance, with the charge density waves (*CDW*) phase transition in $2H-TaSe_2$ at $\approx 110K$ and the concomitant formation of a two-dimensional new periodicity characterized by a 3×3 commensurate superlattice in the plane [7]. The effect of such two-dimensional phenomena on plasmons, free carrier behaviour, interband transitions, etc., has been studied by means of several techniques (electrical, optical, neutron and *X*-ray diffraction, etc.).

One remarkable feature associated to such a phase transition, besides the 'remapping' of the electronic states in the $\frac{1}{3}\times\frac{1}{3}$ reduced *B.Z.*, is represented by the opening of a gap at the Fermi level of the order of $0.2\,eV$. It is worthwhile noting that the inspection of small gaps in the electronic density of states and the study of their relevance on the fundamental properties of specific solids have become an important issue of the spectroscopy in the infrared, particularly in connection with the developments of new materials besides *TMD*: metallic and intermetallic superconductors, high T_c superconductors, rare earth compounds having evidence of heavy fermions, etc.

In this paper we shall focus our attention on some features of *TMD*'s optical spectra related to intra- and inter-band transitions, to *CDW* and super-lattice transitions, and to anomalies in the behaviour of mixed compounds.

2. Structural and Electronic Properties of *TMD*'s

It is well known that *TMD*'s are layered crystals with formula MX_2, where M is a transition-metal atom from the groups $-IVB$, $-VB$ or VIB of the Periodic Table and X is a chalcogen atom (S, Se, Te). The layer structure is built up from sandwiches $X-M-X$, loosely bound by weaker atomic forces, in which each metal atom is 6-fold coordinated with chalcogen atoms placed in hexagonally close packed planes.

The relative alignment of M-atom planes within a single $X-M-X$ sandwich and the stacking of sandwiches determine several phototypes, whose structural and crystallographic details are largely available in many specific or review papers [1,8,9].

As for what concerns the dichalcogenides with $X = S, Se$, here studied (Figure 1), the IVB metals $(M = Ti, Zr, Hf)$ are octahedrally coordinated in the $1T$ polytype; among the VB metals $(M = Nb, Ta)$, while Nb has trigonal prismatic coordination in the $2H$ polytype, Ta presents both the $1T$ and $2H$ polytypes (the integer indicates the number of sandwiches $X-M-X$ per unit cell, while T and H denote trigonal and hexagonal symmetry, respectively).

In particular some compounds of this family, like $1T-TiSe_2$, $1T-TaS_2$ and $2H-TaSe_2$, undergo structural phase transitions by lowering the temperature with the appearance of superlattices [6,10-12] as already mentioned in the introduction.

The literature reports a considerable number of works concerning the calculation of the electronic structure of *TMD*'s. Almost all of these compounds have been examined with different approaches: *APW* method [8], layer method [13], semi-empirical *LCAO* [14-16], first-principles *LCAO* [17], self-consistent *OPW* [18] with relativistic corrections [19], self-consistent *LMTO* [20]. However new calculations, applicable to all compounds of the family are desirable: in particular calculations for the optical spectra of *TMD*'s

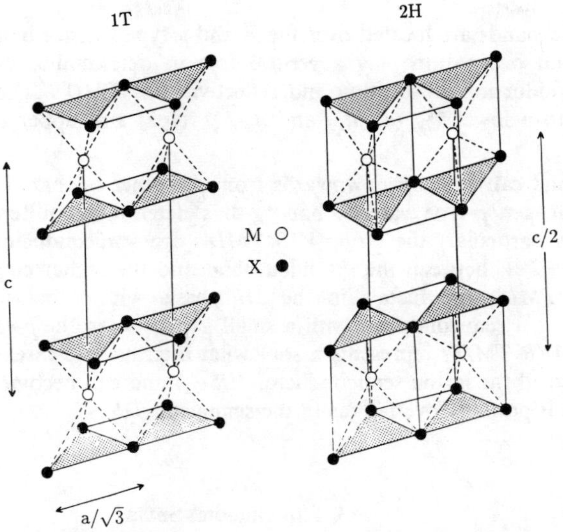

Figure 1. Crystal structures of *TMD*'s both in octahedral (1*T* polytype) and trigonal prismatic (2*H* polytype) coordination.

which start from calculated energy bands and treat the momentum matrix rigorously [21].

While theoretical results agree on the general features of the energy band structure, some discrepancies exist in the finer details, e.g. the optical gap values and their attribution to transitions at specific high-symmetry points of the *B.Z.* Although the electronic structures are considerably more complicated for the 2*H* than for the 1*T* polytype, mainly due to the double number of atoms in the unit cell, the main features of both phases can be summarized as follows [1], with reference to the Figure 2.:

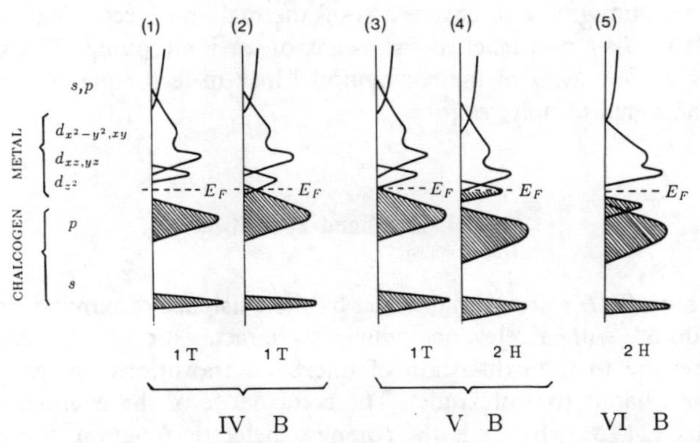

Figure 2. Simplified picture of the electronic energy level diagram for group-*IV*, $-V$, and $-VI$ *TMD*'s.

a) The valence bands of *TMD*'s are separated in two groups, one at lower energy and the other at higher energy, mainly derived from $s-$ and $p-$orbitals of the chalcogen, respectively.

b) The metal d-bands are located over the s- and p-type valence bands and their position, splitting and occupation play a central role in determining the physical properties (such as conduction, absorption and reflectivity) of *TMD*'s. The five d-bands can be divided into a lower t_{2g} (d_{z^2},d_{xy} and $d_{x^2-y^2}$ type) and upper e_g group (d_{xz} and d_{yz} type).

c) The d_{z^2} band can split off downwards from the other t_{2g} bands and overlap with the upper chalcogen p-type valence bands, thus determining different properties of the crystals. In particular: the group-*VIB* *TMD*'s are semiconductors, with a gap of the order of $1-2\,eV$ between the d_{z^2} filled band and the higher empty t_{2g} bands; all the group-*VB* *TMD*'s are metals in the $2H$ phase, with a half-filled d_{z^2} band, while $1T-TaS_2$ is a semiconductor with a small gap between the p-valence and d_{z^2} bands; the group-*IVB* *TMD*'s represents a somewhat intermediate case, since the $p-d$ gap is generally small (as in the semiconductor $HfSe_2$) and can become negative, i.e. the two sets of bands partially overlap (as in the semimetal $TiSe_2$).

3. Experimental Setup

The samples studied in the present work were single crystals, grown by iodine chemical transport reaction. Shortly before the experiment they were cleaved or their surface was carefully cleaned by rubbing them on a cloth drenched with methyl alcohol.

Reflectance (R) measurements, at near normal incidence, were performed with a Perkin Elmer 330 spectrophotometer from 0.5 to $6\,eV$. The experimental apparatus used to obtain, in the $1-9\,eV$ range, thermoreflectance (*TR*) spectra, i.e., to determine wavelength-dependent relative variation $\Delta R/R$ of the reflectivity R produced by a temperature modulation ΔT, has been discussed in detail in a previous paper (see, e.g., ref. [22]). Here we recall only that a temperature modulation $\Delta T \approx 1K$ was obtained at $1.5Hz$ by indirect heating, using a $1\,ohm$ germanium heater, fastened to the copper sample holder. The crystals of $3\times3\times0.2\,mm^3$ were gently pressed against the heater, using a thin layer of vacuum grease to ensure a good thermal and mechanical contact. A working pressure $p \approx 10^{-8}\,Torr$ was reached by means of an ionic pump. The modulation signal ΔR was detected by means of the conventional lock-in technique and integrated in order to improve the signal-to-noise ratio.

4. Interband Transitions

The R spectra of *TMD*'s are characterized by the existence of minima, sometimes referred to a as 'windows', with a relevance which is characteristic of each crystal of the series [23]. They are due to the exhaustion of interband transitions involving specific valence and conduction bands (or subbands). The coincidence of the energies of R minima and $\mathrm{Im}\{\epsilon^{-1}\}$ peaks [24,25], where ϵ is the complex dielectric function, suggests that they can be regarded as indicative of (partial) plasma resonances. Superimposed on the broad bands separated by such minima, a number of fine structures appear, due to interband transitions at high-symmetry points of the *B.Z.*

Thermoreflectance gives a different response in two cases: the structures due to interband transitions yield weak features, whereas the R minima cause strong dispersionlike signals. The lineshape analysis of the *TR* signal shows that in this family of compounds the broadening is the predominant modulation mechanism. Actually, typical values of

$3\times10^{-5}\,eV/K$ and $1.5\times10^{-4}\,eV/K$ are obtained for the 'shift' and 'broadening' thermal coefficients of the optical structures, respectively. So, in order to determine the photon energies of the structures, the positive peak of the *TR* signal due to plasmons [26] and the minimum of the *TR* signal due to interband transitions are usually chosen.

A) Crystals with octahedral coordination

In a previous work [27] the optical spectra of crystals of the series $MX_2(M=Ti,Zr,Hf$ and $X=S,Se)$ have been analyzed in the $1-9\,eV$ range by means of *TR*. The energies of the structures as a function of the lattice parameters *a* and *c* show a general behaviour [23,28]: the energy gaps show a roughly linear variation versus *c* for the sequences $TiS_2-TiSe_2-TiTe_2$; $TiS_2-ZrS_2-HfS_2$ and $TiSe_2-ZrSe_2-HfSe_2$. As for the dependence on the lattice parameter *a*, only the sequence $TiS_2-TiSe_2-TiTe_2$ shows a linear variation. The experimental spectra have been compared with theoretical curves of reflectivity obtained from a Hartree-Fock-Slater band calculation [29], obtaining a quite satisfactory agreement, summarized in Table I.

TABLE I.

Energies (expressed in *eV*) of the structures of *TR* interband spectra. Energies labeled with an asterisk are indicative of partial plasma resonances.

TiS_2		1.48	1.72		2.00	2.83*	3.32	3.75		5.15			7.05	
ZrS_2	2.40	2.59	2.71	2.95	3.25	3.78*	4.70	5.05	5.58	5.80	6.25	7.20	7.70	
HfS_2	2.85	3.04	3.22	3.36	3.60	4.44*	5.10	5.50	6.10		6.80	7.50	7.95	
$TiSe_2$				1.34	1.55	2.50*	2.70	3.00	3.45	4.32	4.72	5.36	6.20	7.25*
$ZrSe_2$	1.65	1.95	2.15	2.30	2.49	3.43*	4.30	4.66	4.95	5.40	5.90	6.80	7.10	
$HfSe_2$	2.08	2.26	2.52	2.74	3.10	3.75*		4.75		5.55	6.25	6.98	8.00	
TaS_2			1.40	1.95		3.55*	3.85	4.55	5.60					
NbS_2		2.00*	2.40	2.95	3.50	4.05*	5.10	6.95	7.75					
$NbSe_2$			1.55	2.15	2.55	3.05*	3.90	4.35	7.00	7.90				

TR measurements have been extended to $1T-TaS_2$ crystals (Figure 3), giving indication in the $1-9\,eV$ interval of the existence of two regions of interband transitions, separated by a partial plasma frequency at $3.5\,eV$. This energy is higher than that of $2.6\,eV$ previously derived [30] from electron energy loss spectra (*EELS*). The weak modulations of the above two regions are reported in Table I.

B) Crystals with trigonal prismatic coordination

In Figure 4 *TR* spectra of metallic $2H-NbS_2$ and $2H-NbSe_2$ are shown. In the case of $NbSe_2$ a direct comparison with reflectivity is possible. Starting from low energy, the reflectivity [30,31] decreases from the high infrared value, typical of a metal, to a strong minimum at $\approx 1\,eV$, indicative of a free carrier plasma resonance. A corresponding peak is found at $0.9\,eV$ in the derived *EELS* spectra [30]. This resonance, which overlaps with the edge of the interband transitions, is responsible for the positive signal in the low energy region of the *TR* spectrum. The *R* minimum at $3.05\,eV$ (*EELS* peak at $2.9\,eV$ [30]) gives rise to a *TR* positive peak at the same energy. It corresponds to a partial plasma resonance which marks the exhaustion of a first important region of interband transitions, characterized by structures at 1.55, 2.15 and $2.55\,eV$ in the *TR* spectrum. The next interband transition region shows *TR* peaks at 3.9, 4.35, 5.1 and $7\,eV$. No clear indication

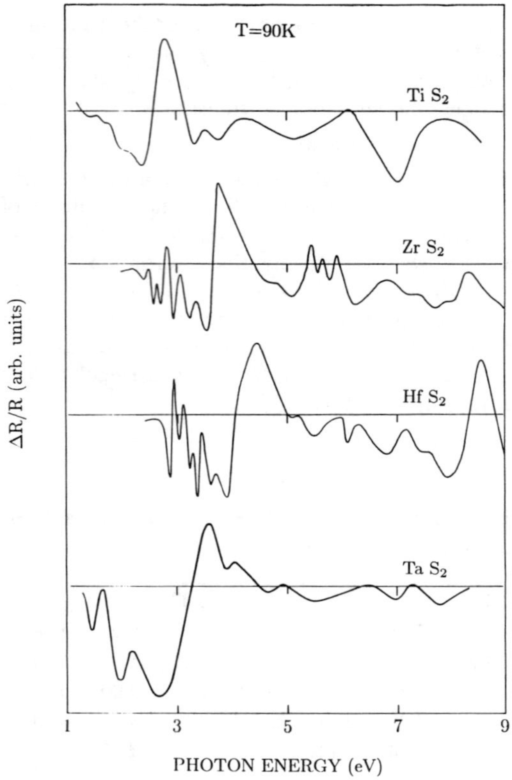

Figure 3. Thermoreflectance spectra at 90*K* for TiS_2, ZrS_2, HfS_2, and TaS_2 crystals.

is found in the *TR* spectrum of the 7.5*eV EELS* peak attributed [30] to the exhaustion of the interband transitions of 9 electrons in the valence band and generally observed in the *EELS* spectra.

TR spectrum of $NbSe_2$ between 1 an 7 *eV* is very similar to that of NbS_2, except for a rigid translation of $\approx 1\,eV$ to lower energies going from *S* to *Se*. A shift of the low energies interband transitions of the same order ($\approx 0.8\,eV$) and in the same direction is generally found in octahedral MX_2 crystals ($M = Ti, Zr, Hf$), going from *S* to *Se*. The corresponding shift of the onset of interband transitions can be responsible for the shift of the free electron plasma frequency from 1 to 2 *eV* in $NbSe_2$, as indicated by the positive peak at this energy in the *TR* spectrum.

5. Optical Evidence of *CDW* Transitions

Some phenomena strictly related to the phase transitions typical of these crystals and briefly referred to in the introduction are quite peculiar and deserve a comment.

Let us confine ourselves to consider a few cases, characterized by the formation of superlattices when cooling down the samples below a given temperature T_c.

a) In the metallic $2H - TaSe_2$ the *CDW* phase transition leads to the formation of a two-dimensional 3×3 superlattice below 110*K*, and also to the opening of relatively large

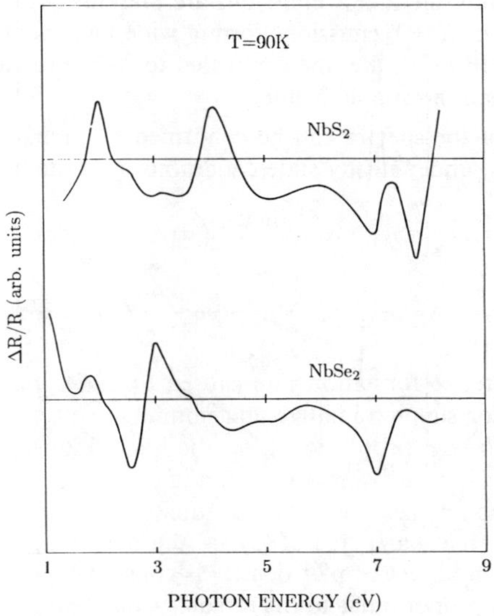

Figure 4. Thermoreflectance spectra at $90K$ for NbS_2 and $NbSe_2$.

gaps ($\approx 0.2-0.5\,eV$) at the Fermi level in order to meet the requirement of minimizing the Coulomb energy of the system in the new configuration below T_c.

Optical absorption is expected to decrease upon heating above T_c because of the closing of the CDW gaps, while the response due to free electron and interband transitions should increase, in connection with the reappearence of the electronic level distribution in the normal phase. This is actually what has been observed, as shown in Figure 5, where the thermal modulation of the optical conductivity is reported, as deduced from TR spectra.

The CDW phase transition is also the origin of a strong perturbation on the plasma resonance spectra around $1\,eV$. The TR curve gives a direct proof of this effect: the gradual variation of the lineshape, which shows very nicely the 'precursor' character of the transition, has been analyzed in term of anomalous shifts and broadenings. Their peculiarity is a consequence of the formation of the CDW gaps.

As seen in the par. 2, the semimetallic $1T-TiSe_2$ is characterized by an overlap between conduction and valence bands of the order of $0.2\,eV$. A gradual uncrossing of the bands takes place by adding known amounts of Hf. As a matter of fact $HfSe_2$ is a semiconductor with a positive gap of $1.13\,eV$.

In the mixed system $Ti_{1-x}Hf_xSe_2$ the gap varies gradually from $-0.2\,eV$ to $1.13\,eV$ as a function of x. A structural phase transition gives rise to a $2\times2\times2$ superlattice, and therefore to a folding of the bands to a smaller $\frac{1}{2}\times\frac{1}{2}\times\frac{1}{2}$ $B.Z.$ In connection to this, in principle it is possible to observe indirect transitions to become direct by cooling the sample below T_c, which decreases from $200K$ for $TiSe_2$ to zero when $x = 0.01$.

Let us consider, for a given stoichiometry a well defined structure, i.e. the negative dip located at energies which are $\approx 0.2\,eV$ lower at $200K$ with respect to $100K$. Figure 6 shows that the dip centered at $\approx 0.4\,eV$ at $100K$ shift to lower energies ($\approx 0.2eV$) at $200K$, due to the opening of a gap at the Fermi level driven by the onset of the phase transition

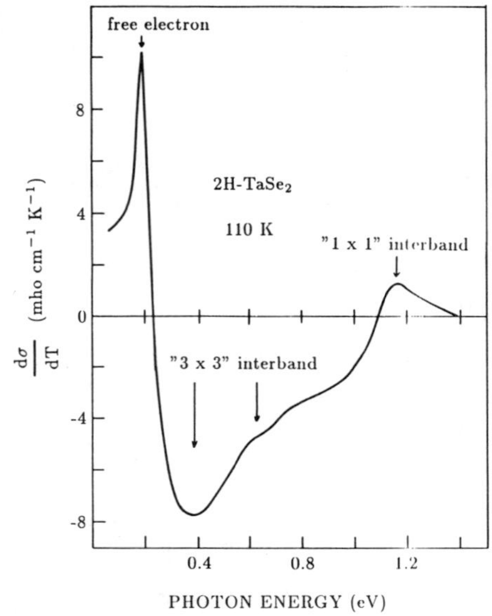

Figure 5. Thermal modulation of the conductivity of $2H-TaSe_2$, extracted from thermoreflectance. Note that optical absorption at the *CDW* gaps decreases upon heating, while the gaps due to the free electrons and interband transitions increase.

[22]. The enhancement of $\Delta R / R$ by a factor of 4-5 is strictly determined by an indirect transition $L_1^{\cdot} - \Gamma_3^{\cdot\cdot}$ which becomes direct below T_c (inset in Figure 6).

6. Optical Properties of Mixed *TMD*'s

The *TR* derivative technique applied to *TMD* mixed compounds yield information on the phase transition. In particular, visible and UV *TR* spectra allow: i) to obtain a detailed classification of the optical spectra of the compounds, with particular attention to their dependence on simple parameters such as bond length or lattice constants; ii) to do that not only by considering the various combinations of *M* and *X* atoms in MX_2 compounds, but also, in a mixed compound like $Ti_{1-x}Hf_xSe_2$, by smoothly driving it from semimetal to semiconductor; iii) to observe the evolution of the optical spectra of mixed crystals also in connection with the behaviour of some structural parameters and to give a precise picture of 'anomalies' and possible bowing effects; iv) to better relate the optical transitions to the band calculations.

For these reasons we have investigated the mixed crystals $Ti_{1-x}Hf_xSe_2$ with $0 \leqslant x \leqslant 1$ by *TR* from 1 to $9\,eV$. With reference to a previous theoretical work [17] we have studied, as the composition x changes, the evolution of the optical spectra versus the variation of the lattice parameters a, c, and of the structural parameter z, which is related to the $M-X$ distance d by the formula

$$d = (\frac{1}{3}a^2 + z^2c^2)^{1/2} \tag{1}$$

and which has a leading role in controlling the semimetallic-semiconductor transition.

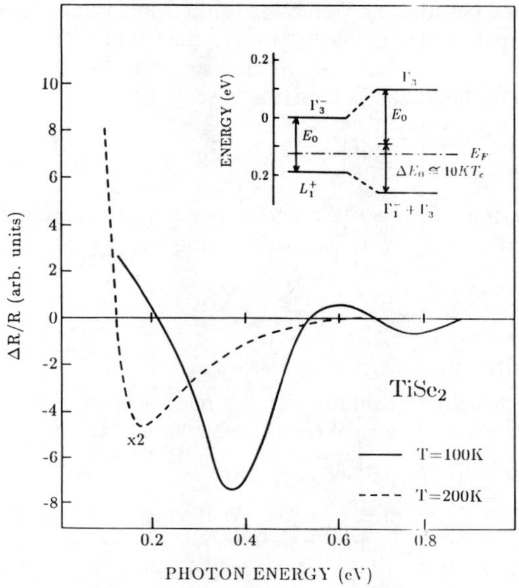

Figure 6. Low-energy interband *TR* spectra at 100*K* and 200*K*. The inset shows the band scheme near the Fermi level below and above T_c. $L_1^+ - \Gamma_3^-$ as indirect in the 1×1×1 phase and direct in the 2×2×2 phase. ΔE_0 is the gap opened by the phase transition.

As for the lattice parameter values, *X*-ray measurements show that a, c and d grow linearly as a function of x, starting from $TiSe_2$ (a = 3.54Å, c = 6.01Å, and d = 2.53Å): the slope gives, within 1%, the correct extrapolation to the values typical of $HfSe_2$ (a = 3.75Å, c = 6.16Å, and d = 2.66Å). On the other hand the z value, calculated on the basis of Eq. 1, remains practically constant (0.2547±0.0005). *EXAFS* measurements indicate that in this alloy the nearest-neighbour bond length versus x has a clear bimodal distribution [32]. This behaviour is expected for highly mismatched alloys (the atomic size mismatch parameter δ, defined in ref. [33], is ≈6% in $Ti_{1-x}Hf_xSe_2$) as a way to relax the elastic energy accumulated in the mixed sublattice [34].

On the basis of these results, a discussion can now be made on the evolution of the structures in the *TR* spectra due to interband transitions. *TR* spectra are reported in Figure 7, while the variation of the energy of some *TR* structures versus the composition parameter x is shown in Figure 8. It can be stressed that the structures at 1.34 *eV* and 1.55 *eV* in $TiSe_2$ merge into those at 2.08 *eV* and 2.52 *eV* in $HfSe_2$. The partial plasma resonance located at 2.5 *eV* in the $TiSe_2$ disappears in the mixed crystals for $x>0.7$ and reappears in $HfSe_2$ as a double structure at 2.3 *eV* and 3.75 *eV*. This is due to the coupled effect of the shift of the above mentioned transitions and of the redistribution of oscillator strengths which causes an enhancement of the transition at 2.08 *eV* in $HfSe_2$. As a matter fact, an enhancement of low energies interband transitions is expected to take place when passing from metal to insulator, as a consequence of the sum rules.

Furthermore, the reduced electronic screening effect in non metallic $HfSe_2$ yields well resolved optical structures and prevents the possibility of establishing a direct correspondence between the structures of the richer spectrum of $HfSe_2$ and of $TiSe_2$. As for the possible bimodal distributions of bonds, we note that no clearcut evidence can be deduced directly from the optical data reported here and reference to *EXAFS*

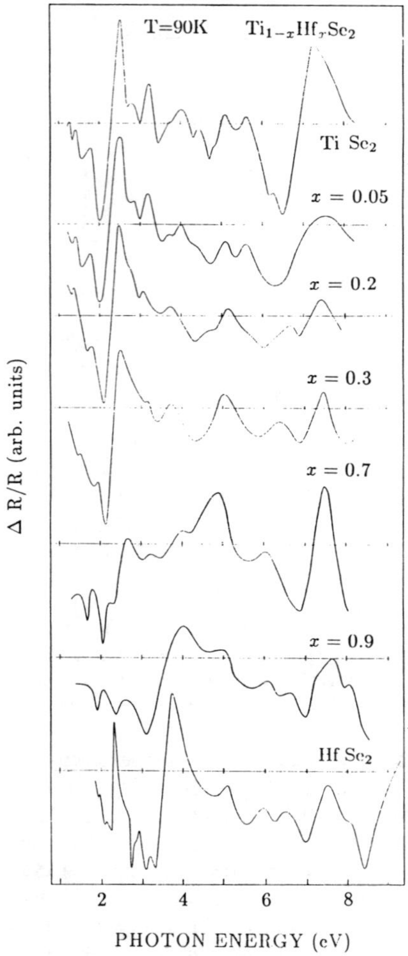

Figure 7. Thermoreflectance spectra for $Ti_{1-x}Hf_xSe_2$ crystals.

measurements should be made in order to reach more conclusive statements in this respect.

At intermediate values of x, the width of the optical structures is expected to be larger than at $x = 0$ or $x = 1$, due to compositional and structural disorder. In this series of mixed crystals, the gradual switching from the metallic to the insulating phase accompanied by the above mentioned shrinking of the linewidths masks the expected behaviour versus x.

The analysis of Figure 8 gives evidence of a bowing effect in this mixed system. Such effect, already mentioned in a previous work [22], is here more clearly identified on lower energy structures owing to the availability of samples with values of x also between 0.5 and 1.0.

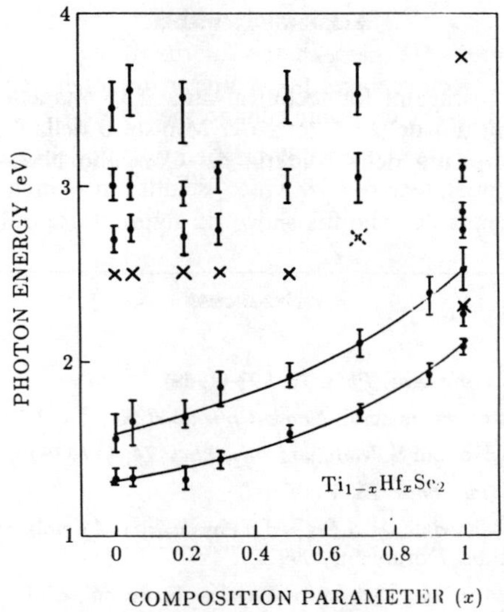

Figure 8. Energies of optical structures due to interband transitions versus composition for $Ti_{1-x}Hf_xSe_2$ crystals (points). Crosses indicate the energies of partial plasma resonances.

Conclusions

This family of compounds is characterized by a wide spectrum of interesting and peculiar properties of quite different nature, which span from exciton confinement in thin semiconducting layers [35] to the temperature evolution of *CDW* [3,12] phase transition in metallic systems. The variety of the phenomena observed in *TMD* is such that on one side it offers the possibility of comparison with other materials on specific issues, and on the other hand stimulated the authors to focus their attention on a few specific aspects, which have been dealt with in this paper and can be briefly summarized as follows:

1) A detailed picture of the main features in the energy band scheme has been obtained through thermoreflectance measurements at low temperature, yielding high resolution spectra of the optical structures.
2) The effect of phase transitions (*CDW*, etc.) on the optical spectra has been nicely evidenced in some significant cases. In particular, the modification of the plasma resonance induced by the *CDW* transitions in $2H-TaSe_2$ and the enhancement of optical structures due to interband transitions becoming direct with the $2\times2\times2$ superlattice of the low-temperature phase ($T\leqslant200K$) in $1T-TiSe_2$ have been observed and analyzed.
3) Anomalies and peculiarities in the interband spectra of mixed systems as a function of composition and structural parameters have been investigated in some detail. Such data, compared with *EXAFS* measurements, have proved to give new information about bond distribution in mixed systems like $Ti_{1-x}Hf_xSe_2$ and are indicative of bowing effects.

Acknowledgements

We wish to thank M. Moscardini for technical aid. The financial support from Centro Interuniversitario di Struttura della Materia del Ministero della Pubblica Istruzione and Gruppo Nazionale di Struttura della Materia del Consiglio Nazionale delle Ricerche is gratefully acknowledged.

References

1. J.A. Wilson and A.D. Yoffe, *Adv. Phys.* **18**, 193 (1969).
2. F. Bassani and G. Pastori Parravicini, *Nuovo Cimento B* **50**, 95 (1967).
3. J.A. Wilson, F.J. Di Salvo and S. Mahajan, *Adv. Phys.* **24**, 117 (1974).
4. A.D. Yoffe, *Adv. Sol. State Phys.* **13**, 1 (1973).
5. Intercalated Layered Materials, F. Lévy, ed., Physics and Chemistry of Materials with Layered Structures, D. Reidel, Dordrecht, 1979.
6. M.S. Dresselhaus, in Proc. Inst. School of Physics 'E. Fermi', CVI Corso, Lerici, 1988; Proc. Int. School of Erice, 1986.
7. D.E. Moncton, J.D. Axe and F.J. Di Salvo, *Phys. Rev. B* **16**, 801 (1977).
8. L.F. Mattheis, *Phys. Rev. B* **8**, 3719 (1973).
9. B.L. Evans, in Physics and Chemistry of Materials with Layered Structures, Vol. 4, ed. by P.A. Lee, D. Reidel, Dordrecht, 1976).
10. F.J. Di Salvo, D.E. Moncton and J.V. Waszcak, *Phys. Rev. B* **14**, 4321 (1976).
11. W.Y. Liang, G. Lucovsky, J.C. Mikkelsen and R.H. Friend, *Phylos. Mag. B* **39**, 133 (1979).
12. C. Campagnoli, A. Gustinetti, A. Stella and E. Tosatti, *Phys. Rev. Lett.* **38**, 95 (1977).
13. G. Wexler and A.M. Woolley, *J. Phys. C* **9**, 1185 (1976); *ibid* **10**, 2601 (1977).
14. N.J. Doran, B. Ricco, D.J. Titterington and G. Wexler, *J. Phys. C* **11**, 685 (1978).
15. R.B. Murray, R.A. Bromley and A.D. Yoffe, *J. Phys. C* **5**, 746 (1972).
16. N.V. Smith, S.D. Kevan and F.J. Di Salvo, *J. Phys. C* **18**, 3175 (1985).
17. A. Zunger and A.J. Freeman, *Phys. Rev. Lett.* **17**, 1155 (1978).
18. H.M. Isomaki and J. von Boehm, *Phys. Lett. A* **89**, 89 (1982).
19. J. von Boehm and H.M. Isomaki, *J. Phys. C* **15**, L733 (1982).
20. G.Y. Guo and W.Y. Liang, *J. Phys. C* **20**, 4315 (1987).
21. J. von Boehm and H.M. Isomaki, *Phys. Rev. B* **10**, 5798 (1982); *ibid.* **10**, 5807 (1982).
22. A. Borghesi, G. Guizzetti, L. Nosenzo, E. Reguzzoni, A. Stella and F. Lévy, *Nuovo Cimento D* **4**, 141 (1984).
23. D.L. Greenaway and R. Nitsche, *J. Phys. Chem. Solids* **26**, 1445 (1965).
24. M.G. Bell and W.Y. Liang, *Adv. Phys.* **25**, 53 (1976).
25. A. Couget, L. Martin and F. Pradal, *C.R. Acad. Sci. B* **272**, 626 (1971).
26. M. Cardona, in Modulation Spectroscopy, Academic, New York, 1969.
27. A. Borghesi, Chen Chen-jia, G. Guizzetti, L. Nosenzo, E. Reguzzoni and A. Stella, *Phys. Rev. B* **33**, 2422 (1986).
28. S.C. Bayliss and W.Y. Liang, *J. Phys. C* **15**, 1283 (1982).
29. H.M. Isomaki and J. von Boehm, *Phys. Rev. B* **26**, 5807 (1982).

30. A.R. Beal, H.P. Hughes and W.Y. Liang, *J. Phys. C* **8**, 4236 (1975).

31. W.Y. Liang, *J. Phys. C* **6**, 551 (1973).

32. A. Borghesi, G. Guizzetti, L. Nosenzo, A. Stella, A. Balzarotti, N. Motta and F. Lévy, Proc. 18th ICPS, ed. by O. Engstrom, World Scientific, Singapore, 1987, p. 1815.

33. A. Zunger and J.E. Jaffe, *Phys. Rev. Lett.* **51**, 662 (1983).

34. A. Balzarotti, M.T. Czyzyk, A. Kisiel, N. Motta, M. Podgorny and M. Zimnal-Starnawska, *Phys. Rev. B* **30**, 2295 (1984); and *Phys. Rev. B* **31**, 7526 (1985).

35. F. Consadori and R.F. Frindt, *Phys. Rev. B* **2**, 4893 (1970).

LIFETIME MEASUREMENTS OF $KBr:Tl^+$ BY A MULTIFREQUENCY PHASE FLUOROMETER

V. GRASSO AND G. SAITTA

Istituto di Struttura della Materia
Università degli Studi di Messina
I-98100 Messina
Italy

The first measurements of emission lifetime in $KBr:Tl^+$ phosphor, using a cross correlation phase and modulation fluorometer, are reported. The decay signal studied at room temperature shows a complex decay with two time constants, a long one ($\tau = 193.21 ns$), together with a very low one ($\tau = 0.288 ns$) never previously reported. No experimental evidence for an intermediate lifetime was found.

1. Introduction

The alkali-halides phosphors, containing Tl^+-type impurity ions have three absorption bands called A, B and C bands in order of increasing energy. The nature of the transitions responsible for the absorptions is well understood. These spectra are due to the electronic transitions from the $(ns)^2$ ground state to the $(nsnp)$ multiple split by the exchange and by the spin-orbit interactions. One of the reasons for the interest, in general, in heavy metal ion impurities with (s^2) electronic configuration incorporated in alkali-halides, is the possibility of testing in a simple system the theory of interaction between an ion and the surrounding medium. Furthermore, a complete understanding of the properties of these systems could produce practical results in the application of these phosphors to luminescent device and detectors. Hystorically, the interpretation by Seitz of their optical spectra provided a simple atomic model for the electronic states of the impurity centre; such a model has been the basis for most of the following elaborations. However, despite its initial simplicity, more experimental measurements were refined, the more one discovered new complications and complexities. In fact, as well summarized and widely reviewed by Ranfagni et al. [1], it turns out that the electronic systems containing Tl^+-like impurities show an overlapping of the effects of spin orbit coupling, electron phonon coupling and a strong Jahn Teller effect.

According to a model proposed by Toyozawa and Inoue [2] and extended by Fukuda [3] and Cho [4] the observed optical structures are due to the dynamic Jahn-Teller distortion of the cubic crystalline field caused by tetragonal and trigonal lattice vibrations. In fact, although the ground electronic state of the Tl^+-like impurity has the same symmetry as the point defect, thus revealing no degeneracy, however when the surrounding lattice is distorted from its equilibrium position, the defect symmetry lowers, giving rise to the splitting of the degenerate excited states.

In the basic Seitz [5] model the A and C absorption bands, typical in this type of phosphors, are assigned to the ${}^1A_{1g} \rightarrow {}^3T_{1u}$ and ${}^1A_{1g} \rightarrow {}^1T_{1u}$ transitions respectively, while the B band is connected with the transitions ${}^1A_{1g} \rightarrow {}^3E_u + {}^3T_{2u}$ levels allowed by coupling of these levels to non-totally symmetric lattice vibrations.

The spectral features of the above absorption bands are essentially determined by

R. Girlanda et al. (eds.), Progress on Electron Properties of Solids, 153–159.

three parameters: W_0, the one-electron energy difference between the excited and the ground states; G, the exchange energy; and ζ the spin-orbit coupling energy. In a first approximation, the experimental data are in a fairly good agreement with a theoretical description of the system adopting a reasonable semiempirical molecular orbital technique [6]. A more satisfactory approach was obtained by Moretti et al. [7] in the framework of the Green's function method, allowing the location of the impurity levels with respect to the valence and the conduction bands of the host crystal. In particular the results showed that all the levels of arrival for the *A*,*B* and *C* bands are comprised within the gap. Unfortunately, the situation is much more complex in the case of the emission, since the features of the emission bands depend so much upon the host lattice, temperature, exciton wavelength, etc., that there is no common pattern of behaviour. A classification scheme, proposed by Fukuda [8], distinguishes among three principal emission classes on the basis of the respective emission behaviour with temperature, upon *A*-band excitation. In particular the class I refers to phosphors that show two different emission bands called A_T and A_X in order of decreasing energy, only below the liquid-nitrogen temperature. This is the case of Potassium Iodide doped with thallium. The class II, that comprises $KBr{:}Tl^+$, is similar to class I with the exception that the A_T and A_X bands persist at all temperatures. The class III phosphors, like $KCl{:}Tl^+$ shows a single emission band at all temperatures. A further class, associated with divalent ions such as S_n^{++} and P_b^{++}, show a more complex behaviour due to all charge compensating vacancy that affects the adiabatic potential energy surfaces from which the emission takes place.

Literature reports decay model for the emission for alkali halides doped with thallium studied by means of lifetime measurements [9-13]. The observed complex decay with two or three components was explained with the presence of a metastable trap ${}^3A_{1u}$ level, responsible of a long-life component, underlying the ${}^3T_{1u}$ level related to a fast component. However, the interpretation of the data coming from experiments is not yet certain since different phosphors seem behave differently and also some reported lifetime is, at least, in the limits of the experimental apparatus performance. Thus, in this work we study the emission of $KBr{:}Tl^+$ by exciting the samples in the *A* absorption band by means of a more refined experimental technique.

Experimental Section

a) Material

The $KBr{:}Tl^+$ single crystals were grown by the standard Kyropoulus method under nitrogen atmosphere and quenced on a copper block. The optical absorption in the region of the *A*-band was measured with a Perkin-Elmer mod. 330 Spectrophotometer, to select samples with an absorbance less then 0.1 to avoid the distorsion of the frequency distribution due to the reabsorption of fluorescence light.

b)Phase and Modulation Fluorometry

Fluorescence lifetime were measured by means of a multifrequency phase and modulation fluorometer (I.S.S. GREG 200) using a scatter solution as reference [14]. The fluorescence is excited by light with an intensity modulated sinusoidally at high frequencies, typically in the *MHz* range. Also, the intensity of the luminescence signal is modulated at the same frequency, but the finite persistence of the excited state will lead to a phase delay ϕ and demodulation μ (defined as the ratio between the a.c. and d.c. part of the signal) of the fluorescence relative to the excitation see Figure 1. Measurements of the phase delay and modulation ratio $M = (\mu_{emission})/(\mu_{excitation})$ provide the basis of two

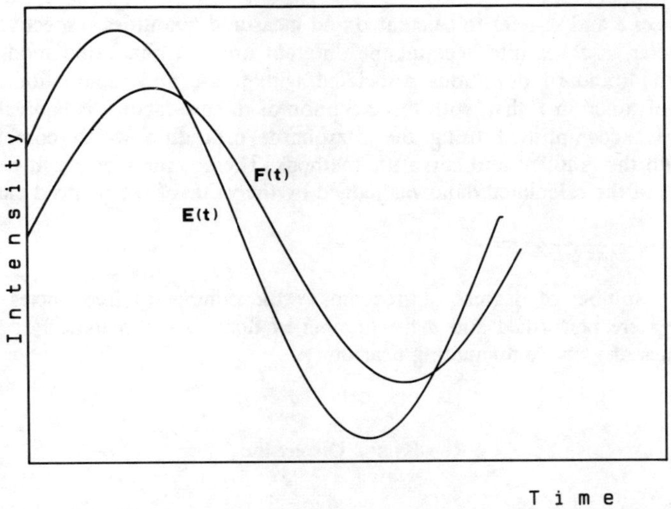

Figure 1. Schematic representation of the excitation $E(t)$ and fluorescence $F(t)$ waveforms. Fluorescence is delayed and demodulated with respect to excitation.

independent determinations of the fluorescence lifetime, i.e. phase (τ^P) and modulation (τ^M) lifetimes [16-17].

$$\phi = tg^{-1}(\omega\tau^P), \quad M = [1+(\omega\tau^M)^2]^{-1/2}. \tag{1}$$

The case of multiple exponential decays is not so easily solvable and the knowledge of both ϕ and M at different frequencies becomes important. In this case relations (1) take the general expression given by:

$$\phi = tg^{-1}(S/G), \quad M = (S^2+G^2)^{-1/2} \tag{2}$$

where

$$S = \sum_i f_i M_i \sin\phi_i, \quad G = \sum_i f_i M_i \cos\phi_i \tag{3}$$

with ϕ_i and M_i given by equation (1). In such system f_i is the fractional intensity of the i-th component ($\Sigma_i f_i = 1$) and τ^P and τ^M are frequency dependent. The function S and G are the sine and cosine Fourier transforms of the emission response function $I(t)$.

$$I(t) = \sum_i \alpha_i \exp(-t/\tau_i) \tag{4}$$

$$S(\omega) = \frac{\int_0^\infty I(t)\sin(\omega t)dt}{\int_0^\infty I(t)dt} \tag{5}$$

$$G(\omega) = \frac{\int_0^\infty I(t)\cos(\omega t)dt}{\int_0^\infty I(t)dt} \tag{6}$$

where α_i is connected to f_i by $f_i = \tau_i\alpha_i$. Thus, by measuring ϕ and M at different modulation frequencies of the incident beam and using a fitting procedure, based on a non-linear least-square analysis that use expressions (2) with τ_i and f_i as free parameters, it is possible to deduce lifetimes. In practice, we minimize the function

$$F = \sum_i \left\{ (1-W)\left[\frac{\phi_i^c - \phi_i^m}{\sigma_p}\right]^2 + KW\left[\frac{M_i^c - M_i^m}{\sigma_m}\right]^2 \right\} \tag{7}$$

where the indices c and m refer to calculated and measured quantities respectively, W is a weight parameter, K takes into account the different units of phase and modulation, σ_p and σ_m are the standard deviations associated with phase and modulation data. The minimization of function F that, with the exception of the pre-factors, is equivalent to the chi-square, was accomplished using the Marquardt procedure which combines good features of both the gradient and curvature methods. The goodness of the fit between the experimental and the calculated data was judged by the value of the reduced chi-square

$$\chi_r^2 = \frac{1}{2N-p}\chi^2 = \frac{\chi^2}{\nu} \tag{8}$$

where ν is the number of degrees of freedom, N the number of frequencies for which measurements were performed and p the number of floating parameters. For all experiments the values of χ_r^2 were fluctuating near unity.

Results and Discussion

An example of the results of ϕ and M obtained for excitation at 255*nm* is in Figure 2.

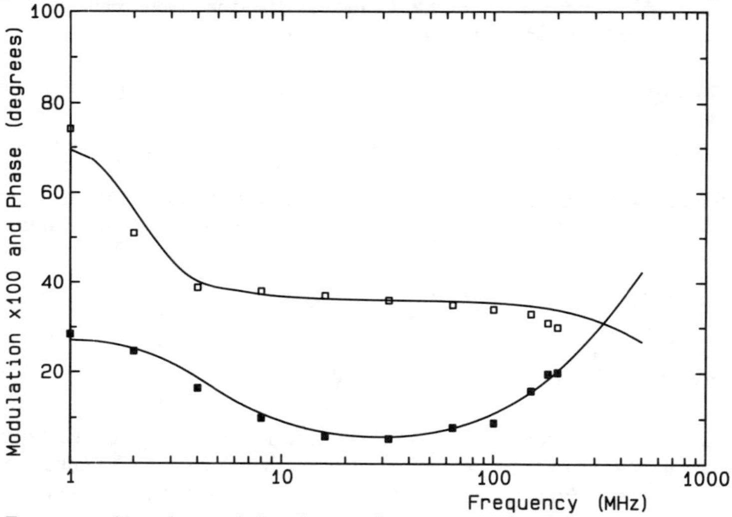

Figure 2. Frequency dependence of the phase (full squares) and modulation data (open squares) and best fit curves that gives two decay times $\tau_1 = 0.288ns$ and $\tau_2 = 193.21ns$.

The solid line is the best-fit curve from relations (2) with the two lifetimes τ_1 and τ_2 and f_1 used as adjustable parameters. The results are consistent with a short decay time $\tau_1 = 0.288ns$ and with a much longer decay time $\tau_2 = 193.21ns$ with an amount of the long component of the order of 64%. The decay time τ_2 agrees with the longest lifetime observed by Benci et al. [18], that reported a complex decay for $KBr:Tl^+$, with three time constants at temperatures less than 150 *K*. In particular, their results show a long lifetime present in the whole temperature range studied (83 to 300*K*), together with a temperature dependent fast component (below 150*K*) of the order of 15*ns* and an intermediate lifetime of about 50*ns* (constant versus temperature). Conversely, our results are completely incompatible with a second lifetime of 50*ns* as shown in Figure 3 that reports simulation of theoretical curves with the lifetimes of reference [18]. A tentative best-fit with three lifetimes as free parameters does not substantially improve the fit of Figure 2.

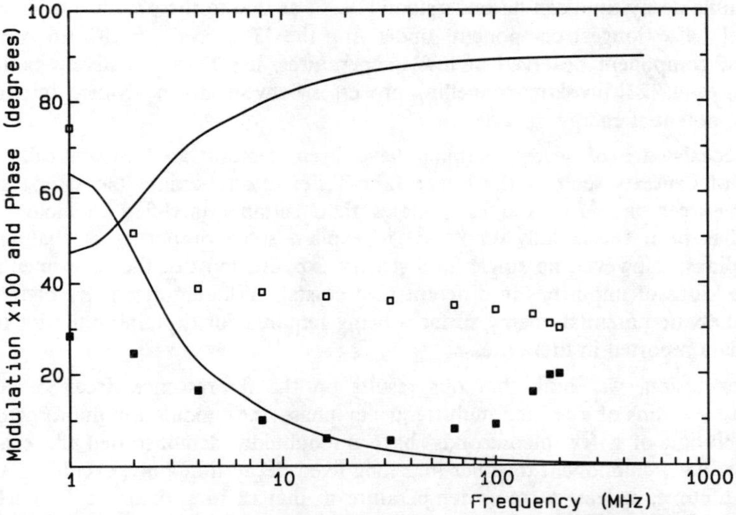

Figure 3. Frequency dependence of the phase (full squares) and modulation data (open squares) and simulation curves using two decay times $\tau_1 = 50ns$ and $\tau_2 = 200ns$ (see ref. [18]).

In fact the best-fit gives a partial fraction of the third lifetime near zero. Such results indicate, in our opinion, that including an additional component in the decay law is not justified by the experimental data. Furthermore, a best-fit with a single decay time is not capable to reproduce experimental data as demonstrated in Figure 4.

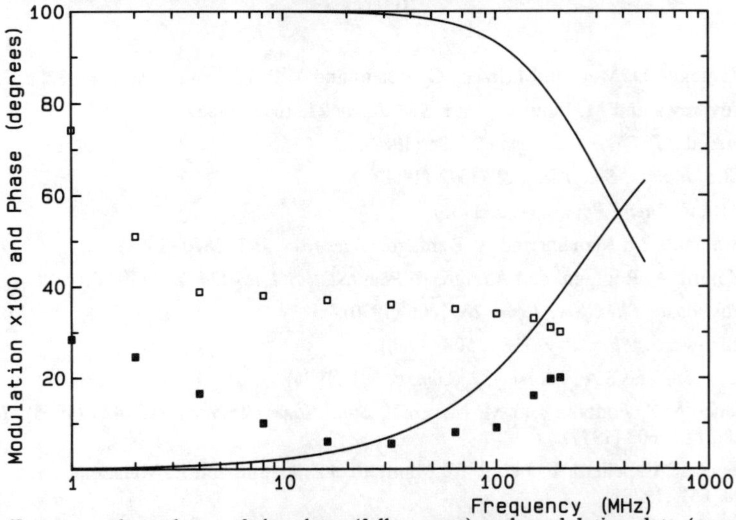

Figure 4. Frequency dependence of the phase (full squares) and modulation data (open squares) and best fit curves that gives a single decay time $\tau = 0.627ns$.

The fastest lifetime we found $\tau_1 = 0.288ns$ is, in our opinion, firstly observed because the present work is the first application of the phase and modulation method to the physics of the fluorescent emission of alkali-halides doped with thallium.

Furthermore, the experimental technique used in ref. [18] with a nominal resolution of the order of some nanoseconds could not allow the detection of the shortest component.

A double decay time can be understood [19,20] as due to the presence of a metastable trap level ${}^3A_{1u}$ (longest component) underlying the ${}^3T_{1u}$ level (short-lived component). The third component observed at low temperatures, has been tentatively explained by Ranfagni et al. [21] invoking tunnelling processes between not equivalent minima of the adiabatic potential energy surfaces.

The coexistence of several minima have been demonstrated theoretically [1] in a number of contexts such as the linear Jahn-Teller effect because the strong spin-orbit mixing between the ${}^3T_{1u}$ and ${}^1T_{1u}$ states that, suitably modified to include trigonal modes, has been successfully employed to explain some properties of thallium doped alkali halides. However, no single level scheme explains most of the experimental results for some kinds of impurities in different host crystals, different symmetry of the minima in the adiabatic potential energy surfaces being required for explaining the lot of experimental data reported in literature.

In conclusion, we think that our results on the fluorescence decay of $KBr:Tl^+$, obtained by means of a refined multifrequency phase and modulation fluorometer, with a time resolution of a few picoseconds, have undoubtedly demonstrated the existence of very short lived component together to a long lived decay time only, excluding any intermediate lifetime, at least at room temperature so that existing theoretical models should be revised in this new framework.

In particular, the fitting procedures of Ranfagni et al. [21] could be reconsidered to evaluate, by the *WKB* method, the emission intensities using a more plausible values of lifetimes.

References

1. A. Ranfagni, D. Mugnai, M. Bacci, G. Viliani and M.P. Fontana, *Adv. Phys.* **32**, 823, (1983).
2. Y. Toyozawa and M. Inoue, *J. Phys. Soc. Japan* **21**, 1663 (1966).
3. A. Fukuda, *J. Phys. Soc. Japan* **27**, 96 (1969).
4. K. Cho, *J. Phys. Soc. Japan* **25**, 1372 (1968).
5. F. Seitz, *J. Chem. Phys.* **6**, 150 (1938).
6. D. Bramanti, M. Mancini and A. Ranfagni, *Phys. Rev. B* **3**, 3670 (1971).
7. P. Moretti, A. Ranfagni and A. Agresti, *Phys. Stat. Sol. (b)* **124**, 297 (1984).
8. A. Fukuda, *J. Phys. Soc. Japan* **26**, 1006 (1970).
9. R. Illingworth, *Phys. Rev.* **136A**, 508 (1964).
10. P.W. Jacobs and S.A. Thorsley, *J. Lum.* **8**, 391 (1974).
11. S. Benci, M.P. Fontana and M. Manfredi, *Solid State Commun.* **18**, 1423 (1976); *Phys. Stat. Solid (b)* **81**, 603 (1977).
12. S. Benci, R. Capelletti, F. Fermi, M. Manfredi, J.Z. Damm and E. Mugenski, *Phys. Stat. Solid (b)* **90**, 657 (1978).
13. S. Aiello, S. Benci and M. Manfredi, *Solid State Commun.* **23**, 193 (1977).
14. D.M. Jamerson and E. Gratton, in: New Direction in Molecular Luminescence ASTM, Special Technical Publication, Phyladelphia (1982).
15. E. Gratton, D. Jamerson and R.D. Hall, *Ann. Rev. Biophys. Bioeng.* **13**, 105 (1984).
16. E. Gratton, D.M. Jamerson, N. Rosato and G. Weber, *Rev. Sci. Instrum.* **55**, 486 (1984).
17. E.F. Gudgin-Templeton and W.R. Ware, *J. Phys. Chem.* **88**, 4626 (1984).
18. S. Benci and M. Manfredi, *Phys. Stat. Sol. (b)* **99**, 531 (1980).

19. M. Bacci, A. Ranfagni, M.P. Fontana and G. Viliani, *Phys. Rev. B* **11**, 3052 (1975).
20. T. Shiraishi, *J. Phys. Soc. Japan* **55**, 4528 (1986).
21. A. Ranfagni, G. Viliani, M. Cetica and G. Molesini, *Phys. Rev. B* **16**, 890 (1977).

EXCITON INTERACTIONS

INTERACTION BETWEEN BOUND EXCITONS AND IMPURITIES AS WELL AS BETWEEN BOUND EXCITONS

W. CZAJA

Institut de Physique Appliquée
Ecole Polytechnique Fédérale de Lausanne
CH-1015 Lausanne
Switzerland

In a model system of isoelectronic impurities we review experiments on impurity-impurity interaction and give an interpretation of the experimental data. Two general results emerge: It seems to be possible under favourable conditions to propose a detailed analysis of the width of the bound exciton zero phonon recombination transition. Furthermore our data indicates that in systems of bound excitons with large Bohr radius the interaction between bound excitons is completely screened at least up to a concentration of $10^{18} cm^{-3}$ of isoelectronic impurities.

1. Introduction

Excitons bound to various impurities are well known and documented for a large variety of semiconductors [1]. Used as fingerprints for the presence of impurities they often play an important role in work devoted to materials analysis. Here a different aspect of the presence of bound excitons shall be emphasised: bound excitons will be used as model system to study the interaction between impurities, to obtain information on the intrinsic line width of exciton recombination transitions in solids and to study high excitation density effects. Since in the latter case unwanted side phenomena like the heating of the sample by strong optical excitation or the optical saturation of the impurity related transitions may mask the change of the spectral distribution due to the interaction, one is obliged to devote some attention to them.

1.1. THE MODEL SYSTEM *AgBr*:*I*

Although the phenomena one is going to be concerned with are of general significance most of them have been studied in connection with the 'model system' *AgrBr*:*I*. There are two reasons for this particular choice:

a) the spectra for this system are very conveniently located in the green spectral region.

b) much and detailed information is available on this particular substance and doping.

This latter information is important for a detailed analysis.

The luminescence and the absorption spectra for the isolated impurity iodine in *AgBr* are well understood and documented [2,3,4]. The luminscence spectra for an isolated iodine impurity due to the recombination of a bound exciton is displayed in Figure 1 for the purpose of reference. The intensity of the optical phonon replica which are superimposed on an unstructured background due to acoustical phonons has been shown to follow a Poisson distribution with rather good precision (see Ref. [2] and the literature quoted therein).

R. Girlanda et al. (eds.), Progress on Electron Properties of Solids, 161–173.

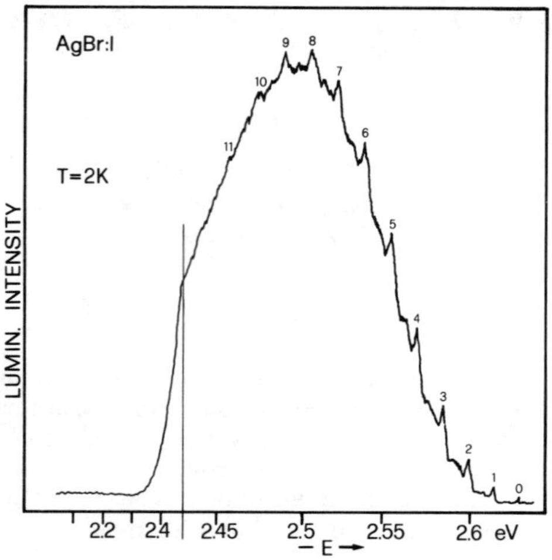

Figure 1. Luminescence spectrum of *AgBr:I*. Iodine concentration $4{,}3 \cdot 10^{15} cm^{-3}$. The numbered peaks are *LO* phonon replica.

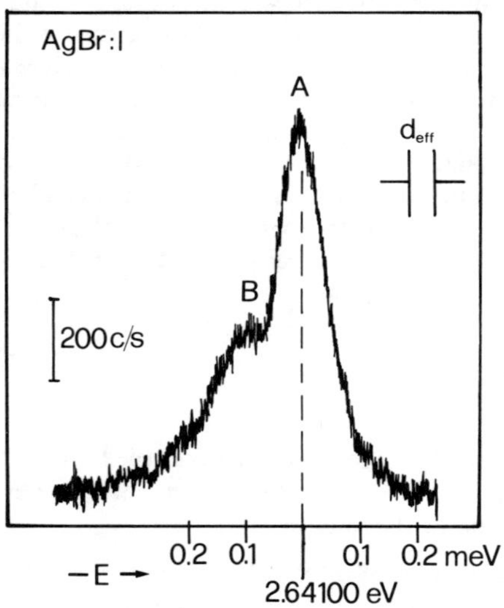

Figure 2. Zero phonon lines of the spectrum shown in Figure 1.

As almost all isoelectronic impurities, also iodine in *AgBr* has a zero phonon luminescence line split into two narrow components originating form spin 1/2 electrons and angular momentum 3/2 holes combined and localised to yield bound exciton states with total angular momentum $J = 2$ and $J = 1$, [5], as shown in Figure 2. The spectra of Figure 1 and Figure 2 will now be used for the detection of changes due to impurity-

impurity interaction.

2. Variable Iodine Concentration and Weak Excitation

2.1. Results and a model for the interaction between excited and non excited impurity atoms

The sharp $J = 1$ and $J = 2$ zero phonon transitions offer themselves as a rather convenient and sensitive detector for the influence of the interaction between the impurities. In Ref. [6] the results have been discussed in detail. In this context it is sometimes more convenient to use the 'tight binding' like notation of an excited impurity rather than that of an impurity bound exciton. In this sense one is considering here the case of one excited impurity surrounded by many other non excited impurity atoms since the data were taken at low excitation density. Figure 3 reproduces data from Ref. [6].

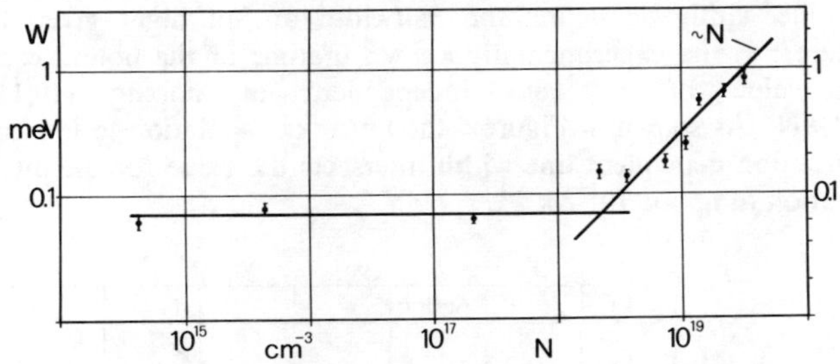

Figure 3a. Line width of the zero phonon A-transition as function of iodine concentration in $AgBr$ at $T = 2K$.

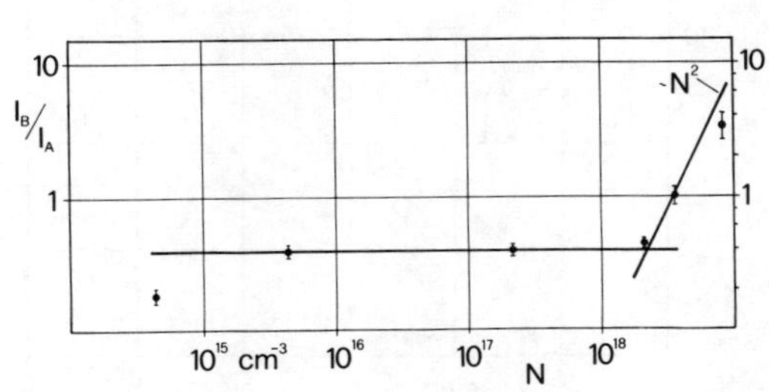

Figure 3b. Intensity ratio: dipol allowed B-transition divided by the dipol forbidden A-transition, as function of iodine concentration.

Plotted on a double logarithmic scale are line width W and the intensity ratio of the $J = 1$ and $J = 2$ transition I_B / I_A.

The interpretation of these data is based on the model of resonant energy transfer or self absorption known from atomic spectroscopy [6,7]. The excitation jumps from impurity atom to impurity atom until it recombines. The interaction between the atoms is a dipol - dipol interaction, as evidenced by the $1/r^3$ – dependence of the interaction potential [6]. This is equivalent to a linear dependence of the line width, and a quadratic of I_B / I_A, on the concentration N as visible in Figure 3. In KI single crystals the interaction between F - centers has been studied in Ref. [8] with the result that the best fit to the data is obtained assuming a dipol-dipol interaction, in perfect agreement with our result.

There is still a further effect on the spectrum of Figure 1 which is observed at low excitation with increased doping. The increase of the intensity of the phonon replica. Phenomenologically this increase is well described by a Huang-Rhys factor depending on the concentration N of iodine, $S = S(N)$, Ref. [9].

However, up to now, it has turned out to be very difficult to formulate a satisfactory microscopic description.

2.2. Some Consequences of the Resonant Energy Transfer: Line broadening and the intrinsic line width.

Given the linear dependence of the line width W on impurity concentration N above a doping of $10^{18} cm^{-3}$, as discussed in the preceding paragraph, one may ask the question whether there is any relationship of the measured line widths to the intrinsic line width. As intrinsic line width we define the halfwidth at full hight given by the relation $W \cdot \tau = \hbar$ where τ is the experimentally known lifetime of the bound exciton. With the experimental value of $\tau = 18 \mu s$ independent of doping [10,11], one finds $W \sim 3,5\ 10^{-5}\ meV$. As shown in Figure 4 the linear extrapolation to low concentrations N of the concentration dependent line width intersects the value for the intrinsic line width at a concentration of about $10^{15} cm^{-3}$.

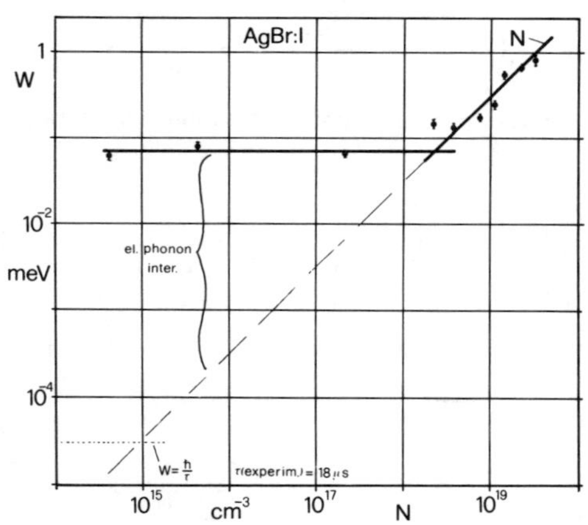

Figure 4. A-transition line width, extrapolation to low iodine concentrations in $AgBr$ at $T = 2K$.

The difference to the experimentally determined values ($\sim 10^{-1} meV$) in the low concentration range is presumed to be due to electron phonon coupling. The line widths observed in the low concentration regime display the typical dependence on temperature (see Appendix). An analysis not only must describe these dependencies but also the coupling strength, i.e. the increase in line width due to the phonon interaction. Such an analysis is at present under way. The intrinsic line width should be directly observable provided the bound exciton lifetime would be shorter than $2.10^{-9} sec$. Thus it seems that at least in an indirect way one may have access to the intrinsic linewidth of optical transition in solids, and that in the present case there are only two main contributions to the linewidth which actually can be studied separately. In a different context another study of intrinsic line widths has been published recently [12].

3. High Iodine Concentration and Variable Excitation

3.1. The Interpretation of Lineshapes

Whereas the interpretation of line shapes in the preceding sections in the limit of low excitation density proved to be rather straight forward some difficulties will be encountered when one moves to higher excitation densities. We limit ourselves to high doping concentrations.

Three contributions are generally to be expected to influence the line shape and luminescence intensity:

a) Optical saturation, i.e. the effect of a limited number of excitable states.

b) Increase of the sample temperature due to strong optical excitation in the intrinsic absorption region.

c) Change in the spectral intensity distribution due to the presence of an increasing number of excited impurities.

In order to identify clearly the contribution c), one has to correct for the two contributions a) and b).

The effect of optical saturation has been studied in Ref. [13]. Its main effect is to create a non linear dependence of the luminescence intensity on the excitation intensity. Postulating that each impurity can only be excited once, a suitable set of rate equations describes the experimental data well. The values of the fitting parameters agree well with data obtained in various different experiments. The optical saturation effect can therefore be eliminated.

Some more elaborate is the analysis of the increase of the sample temperature, ref. [14]. One uses the luminescence intensity as thermometer because its temperature dependence is the same for the whole spectrum and well understood [2]. If the sample excited via a quasi continuous pulsed excitation accumulates no heat from one pulse to the next, then one arrives at a stationary state where the optical energy input and the heat transfer from the sample to the surrounding bath of pumped liquid helium are equal. The rate of heat transfer to the bath (given by a Stefan-Boltzmann like law) fixes the sample temperature which can be determined [14]. Using this procedure the influence on the sample temperature for the case of $AgBr:I$ is eliminated in the following way: As a function of excitation power the zero phonon luminescence intensity shows a smooth dependence in agreement with the heat balance considerations [14], Figure 5a:

$$\frac{I}{P} = A \cdot \exp - \left\{ \gamma T_0^2 \left[1 + \frac{P}{a T_0^4} \right]^{1/2} \right\}. \tag{1}$$

The luminescence intensity from a different part of the spectrum divided by the zero phonon intensity must behave independent of the excitation power. Any deviations are then interpreted as being due to effects of the interaction between excited impurities.

3.2. Screening Effects

Luminescence spectra in a strongly doped sample with $N = 10^{19} cm^{-3}$ iodine taken as a function of excitation power reveal the result shown in Figure 5b. Apparently, the

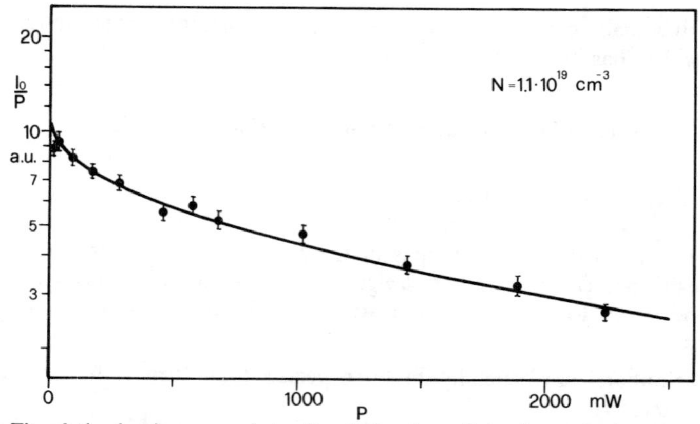

Figure 5a. Fit of the luminescence intensity I/P, where P is the excitation power, to Eq. (1). Closed and open circles are two different measurements. Sample doped with $1,1 \cdot 10^{19} cm^{-3}$ iodine.

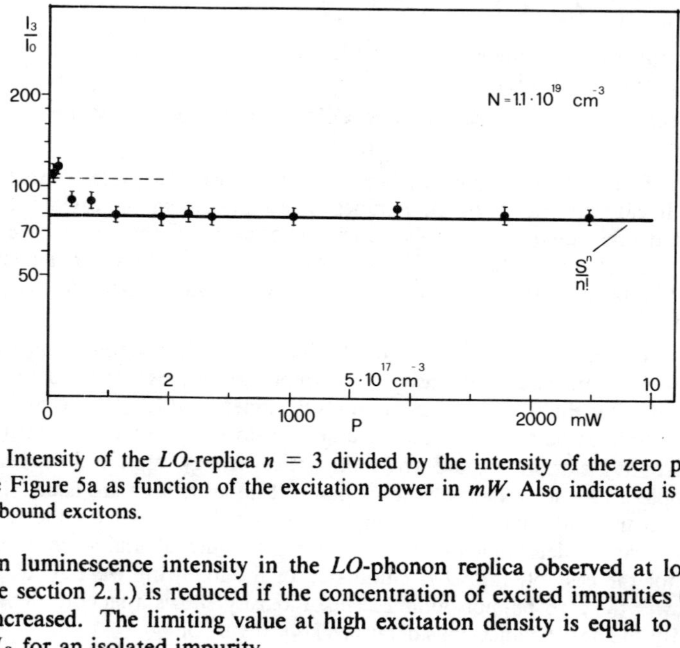

Figure 5b. Intensity of the *LO*-replica $n = 3$ divided by the intensity of the zero phonon line of the sample Figure 5a as function of the excitation power in mW. Also indicated is the estimated density of bound excitons.

increase in luminescence intensity in the *LO*-phonon replica observed at low excitation power (see section 2.1.) is reduced if the concentration of excited impurities (bound excitons) is increased. The limiting value at high excitation density is equal to the intensity ratio I_n / I_0 for an isolated impurity.

At first sight these observations are surprising. However, it has been shown [15] that this behaviour can be understood by a screening of the interaction between impurities due to the presence of an increased number of bound excitons. It must be remembered that the bound exciton Bohr radius as determined experimentally is as large as 20 Å [16]. Thus, these electrons may be treated in a first approximation as free. In agreement with this screening model is the observation that the line width of the zero phonon line which has increased due to the impurity-impurity interaction (comp. section 2.1.), is reduced with increasing excitation power as will be discussed in the next section.

3.3. Linewidth of the Zero Phonon Lines at High Iodine Concentration as Function of the Excitation Density

Figures 6a and 6b show zero phonon lines of the spectrum Figure 1 for two concentrations of iodine and as function of the excitation power.

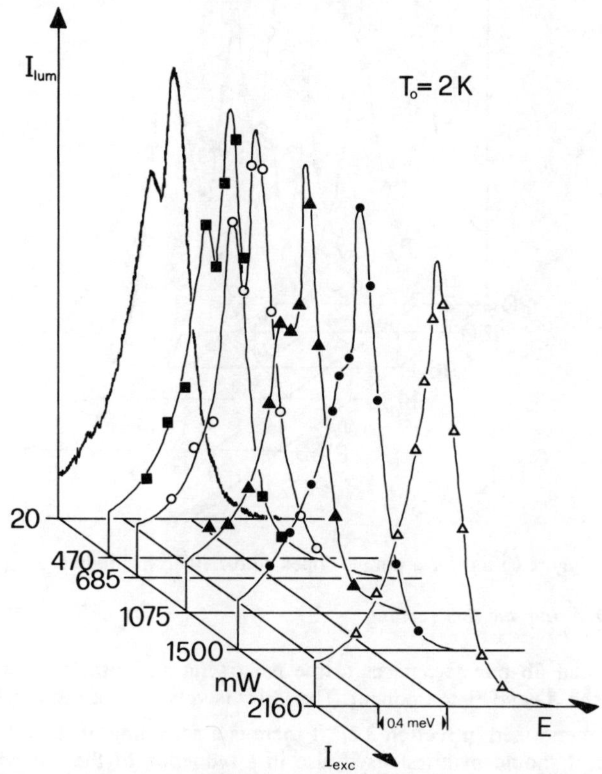

Figure 6a. Shapes of the zero phonon transition of a sample doped with $2 \cdot 10^{18} cm^{-3}$ iodine as a function of excitation power.

At the lowest excitation the spectra have been obtained by a continuous registration, whereas at higher excitations line shapes have been determined by a point by point measurement for the various energies under excitation conditions as described in Ref. [14]. All spectra have been normalised to the same hight at their maximal intensity. Three manifestations of strong excitation effects in the spectral intensity distribution can be distinguished starting from the line broadening observed initially in agreement with section 2.1:

1. Line narrowing due to screening.
2. Saturation of the *B*-transition.
3. Relative growth of the *A*-intensity on the expense of the *B*-intensity.

These three phenomena are important in different regions and will therefore be discussed separately.

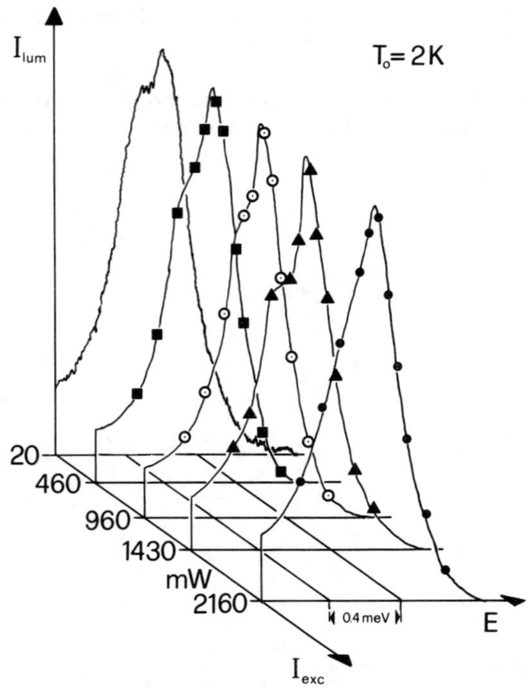

Figure 6b. As in Figure 6a but for a sample doped with $1 \cdot 10^{19} cm^{-3}$ iodine.

3.3.1. Line Narrowing due to screening

In Figures 6a and 6b one recognises a line narrowing and simultaneously an improved resolution for the A- and B-component. The latter is very pronounced in Figure 6a.

As already mentioned in section 3.2.: if there is a screening of the interaction between impurities then it should manifest itself also in a reduction of the line widths of the zero phonon lines for which one observes at low excitation an increase due to the impurity-impurity interaction. The observed line narrowing is then taken as a support for the screening model.

3.3.2. Saturation of the B-transition

This effect is not so much an effect of the interaction between the impurities but rather an effect of the strong excitation. By further inspecting Figure 6a a sudden reduction of the B-intensity relative to the A-intensity between 685 and 1074mW excitation power is visible. Around this excitation such a reduction is to be expected for samples doped with $2 \cdot 10^{18} cm^{-3}$ iodine impurities. In Figure 7 a schematic representation of the A- and B-levels and the relevant transitions is shown.

The following analysis is based on that given in Ref. [17]. Between the two levels one assumes thermal equilibrium. Therefore the populations are related by a Boltzmann term

$$\frac{N_A}{N_B} = e^{-\Delta E/kT}. \tag{2}$$

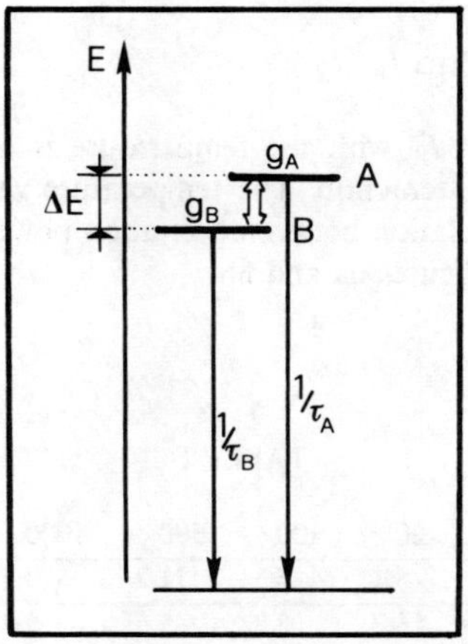

Figure 7. Schematic representation of the zero phonon transitions *A* and *B*.

The levels have degeneracies $g_A = 3$ and $g_B = 5$. The weighted average lifetime τ' is defined as

$$\frac{1}{\tau'} = \frac{\frac{1}{\tau_A} g_A N_A + \frac{1}{\tau_B} g_B N_B}{g_A N_A + g_B N_B}, \tag{3a}$$

from which one obtaines immediately

$$\frac{1}{\tau'} = \frac{1/\tau_B + 1/\tau_A \cdot g e^{-\Delta E/kT}}{1 + g e^{\Delta E/kT}}, \tag{3b}$$

with $g = g_A / g_B$. Note that τ' is the lifetime observed experimentally when the two components are not resolved. The experimental value is $\tau' = 18\mu s$, independent of doping (section 2.2). Moreover the experimentally known intensity ratio at low iodine concentration is $\tau_B^{-1} / \tau_A^{-1} \simeq 10$. Thus from

$$\frac{I_A}{I_B} = \frac{N_A \tau_B}{N_B \tau_A} \tag{4a}$$

the ratio of the individual lifetimes is also known

$$\frac{\tau_B}{\tau_A} = \frac{I_A}{I_B} e^{\Delta E/kT}. \tag{4b}$$

Finally, with the level splitting as obtained from the luminescence spectra $\Delta E = 0{,}097\,meV$ and Eq. (3b) and Eq. (4a) one has

$$\tau_A = 5\mu s \quad \tau_B = 100\mu s.$$

$2.2 \cdot 10^{22} e - h$ pairs are generated per sec and cm^{-3} at an excitation power of $10^3 mW$ and a light spot diameter of $2mm$. The absorption coefficient for the *Ar* laser *UV* wavelengths ($\lambda \sim 3660\text{Å} \cong 3.4\,eV$) is $\alpha = 325 cm^{-1}$. Thus in a sample doped with $2 \cdot 10^{18} cm^{-3}$ iodine impurities, each iodine trap is excited per second on the average 10^4 times. Therefore with a lifetime of 10^{-4} sec, the *B*-transition will show saturation. A similar effect is not visible in the spectra of the sample doped with $10^{19} cm^{-3}$ iodine impurities in agreement with the afore mentioned model.

3.3.3. Relative Intensity Ratio I_B / I_A

The variation of the ratio I_B / I_A with the temperature is again an effect of the strong excitation rather than of an interaction. The temperature variation is due to the optical pumping. Table I shows the relation between excitation power and temperature appropriate for the measurements of Figures 6a and 6b:

$$\frac{I_B}{I_A} = 20 \cdot e^{\Delta E / kT} \tag{5}$$

TABLE I

P	mW	0	20	100	500	1000	1500	2000
T	K	2.0	5.1	7.5	11.2	13.4	14.8	15.9
I_B / I_A	$x 10^2$	8.9	6.3	5.8	5.5	5.4	5.4	5.4

The last line of Table I also shows the ratio I_B / I_A according to Eq. (4a) which results form combining Eq. (2) with the values of I_A and I_B. On top of this temperature effect one would need to take into account also the reduction of the total intensity of the zero phonon transition with temperature according to the relation derived in Ref. [2]:

$$I(T) = I(0) e^{-\gamma T^2}. \tag{6}$$

Conclusions

As stated in the introduction the study of bound exciton transitions and in particular of excitons bound to isoelectronic traps has yielded a number of general insights not only into the nature of the impurity-impurity interaction but also into questions as to the important line broadening mechanism and the interaction between bound excitons. The results which have been discussed are believed to be of a much wider validity as just for the test system chosen.

The intrinsic line width for transitions between an excited and a ground state defined as the line width due to the intrinsic lifetime of the state in question will in pure systems often be hidden behind the line width produced by electron phonon interaction. For simple cases there may be a chance to completely analyse the observed line width and to obtain a reasonable estimate of the intrinsic line width.

The interaction between excited and nonexcited impurities turns out to be entirely equivalent to the case of resonant energy transfer known in atomic systems [6,7]. The energy transfer itself is due to a dipol-dipol interaction [6,8].

Finally, the interaction between bound excitons of large Bohr-radius (20Å in the present example) will be practically completely screened, even for relatively low bound exciton concentrations. From low concentrations on up to $10^{18} cm^{-3}$ the behaviour is that of isolated bound excitons. The phenomena observed in this range are consequences of an increased sample temperature and of the saturation of recombination channels and not due to the interaction between bound excitons. Whether at concentrations of higher than $10^{18} cm^{-3}$ bound excitons (corresponding to an average distance of ~60Å) effects due to the deformation of the wave functions become visible remains an open question.

Acknowledgement

This contribution is devoted to Franco Bassani on occasion of his sixtieth birthday.

During the work on the projects discussed in this contribution, the author has had the benefit of many stimulating discussions and suggestions by many colleagues. Among those were A. Baldereschi, Y. Burki, K. Maschke, A. Quattropani, P. Schwendimann, A. Testa, all from EPFL and R.K. Wehner from the University of Münster RFA. Part of this work has been supported by the Swiss National Science Foundation.

Appendix: Temperature dependence of the line width of the zero phonon line

In the range of iodine concentrations from $6 \cdot 10^{14} cm^{-3}$ to $2 \cdot 10^{18} cm^{-3}$ the temperature dependence of the line width of the A-transition is shown in Figure 8.

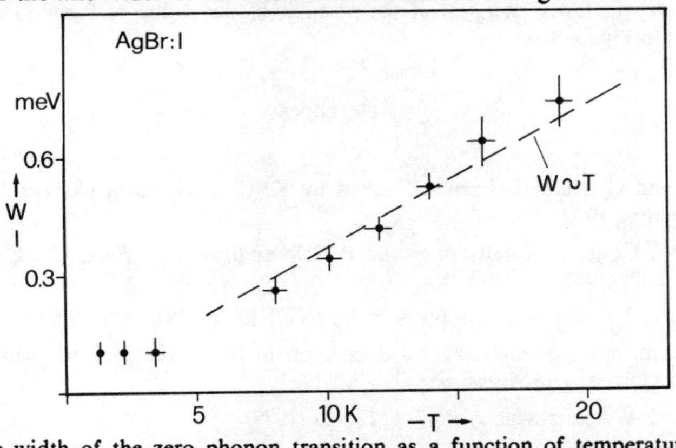

Figure 8. Line width of the zero phonon transition as a function of temperature for a sample doped with $4 \cdot 10^{15} cm^{-3}$ iodine.

As far as one can say the line is symmetric. Furthermore in Figure 9 the shift of the position of the A-line is presented as a function of temperature. Both dependencies are typical for processes based on the interaction with phonons. In ref. [18] some papers concerned with line broadening due to electron phonon interaction are collected.

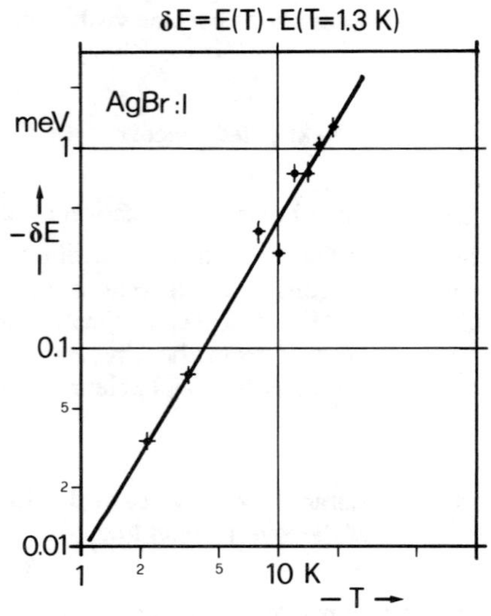

Figure 9. Shift δE (to lower energies) of the position of the A-line as function of temperature; same sample as in Figure 8.

References

1. P.J. Dean and C. Herbert, Excitons, edited by K. Cho, *Topics in Current Physics*, vol. 14, Springer, Berling 1979.
2. A. Testa, W. Czaja, A. Quattropani and P. Schwendimann, *J. Phys. C* **20**, 1253 (1987), *J. Phys. C* **21** 2189 (1988).
3. H. Kanzaki and S. Sakuragi, *J. Phys. Soc. Japan* **27**, 109 (1969).
4. W. v.d. Osten, Excitons and Exciton Relaxation in Silver Halides, International Advanced Study Inst., University of Antwerpen (U.I.A.) 1982.
5. W. Czaja and A. Baldereschi, *J. Phys. C* **12** 405 (1979).
6. W. Czaja, *J. Phys. C* **16**, 3197 (1983).
7. R.G. Breene, *Rev. Mod. Phys.* **29**, 94 (1957); I.I. Sobelmann, L.A. Vainshtein and E.A. Yukow, Excitation of atoms and broadening of spectral lines, *Springer series in Chemical Physics* Vol. 17, Springer Berlin, 1981.
8. D. Fröhlich and H. Mahr, *Phys. Rev.* **148**, 868 (1966).
9. A. Testa, W. Czaja, A. Quattropani and P. Schwendimann, Proc. 3rd International Conference on Shallow Impurities in Semiconductors, editor B. Monemar, Kinköping Sweden 1988, Inst. Phys. Conf. Ser. No. 95: Chapter 9.
10. F. Moser and S. Luy, *J. Luminescence* **3**, 447 (1971).
11. M. Tsukakoschi and H. Kanzaki, *J. Phys. Soc. Japan* **30**, 1423 (1971).
12. H. Navarro, E.E. Haller and F. Keilmann, *Phys. Rev. B***37**, 10822 (1988).
13. W. Czaja and R. Bertato, *Helv. Phys. Acta* **58**, 293 (1985).
14. W. Czaja and Y. Burki, *Helv. Phys. Acta* **60**, 809 (1987).
15. A. Testa, W. Czaja, A. Quattropani and P. Schwendimann, *Solid State Comm.* **69**, 1035 (1989).

16. A.P. Marchetti and D.S. Tinti, *Phys. Letters* **69**, 353 (1979).

17. J.D. Cuthbert and D.G. Thomas, *Phys. Rev.* **154**, 766 (1967).

18. M. Lax and E. Burstein, *Phys. Rev.* **100**, 592 (1955); D. Sampson and H. Margenau, *Phys. Rev.* **103**, 879 (1956); E.O. Kane, *Phys. Rev.* **119**, 40 (1960); K. Nishikawa and R. Barrie, *Can. J. Phys.* **44**, 1135 (1963); R. Barrie and K. Nishikawa, *Can. J. Phys.* **44**, 1823 (1963).

EXCITON INTERACTIONS
INTERACTIONS OF FREE EXCITONS

J.L. STAEHLI

Institut de Physique Appliquée
Ecole Polytechnique Fédérale
PH-Ecublens
CH-1015 Lausanne, Switzerland

We briefly survey the optical spectra of strongly excited semiconductor crystals. The luminescence spectra of *GaSe* are used to discuss exciton-exciton and exciton-free carrier recombination processes. Further, in the framework of a simple binary collision approximation, the effects of screening caused by the presence of a high electron - hole pair density are discussed.

1. Introduction

We consider an ensemble of interacting electrons (e) and holes (h) as a real gas. In the grand canonical distribution, i.e. for given temperature T and chemical potential $\mu = \mu_e + \mu_h$, we consider the virial expansion of the thermodynamic potential $\Omega = -pV$ (p is the pressure and V the crystal volume) [1]:

$$\Omega = k_B T \ln \sum_{N=0}^{\infty} \exp\frac{\mu N}{k_B T} \sum_n \exp\frac{-E_{nN}}{k_B T}. \tag{1}$$

Here $N \equiv N_e + N_h$ is the total number of particles and E_{nN} denotes the total energy of an N-particle system. For nondegenerate systems, the logarithm can be approximated by a power series:

$$\Omega = \sum_{\nu=1}^{\infty} \Omega_\nu \tag{2}$$

where Ω_ν is proportional to $n^\nu \propto \exp(\nu\mu/(k_B T))$ (n is the particle density). Ω_1 would be the thermodynamic potential of our electron-hole (eh) fluid if the interactions between the particles were absent. Ω_2 is a correction caused by the binary interactions, i.e. by the $e-e$, $h-h$ and eh interactions [2]. The most important contribution to Ω_2 is caused by the attractive eh interaction [3], which allows the well known formation of bound eh pairs, the excitons (X). This attractive interaction increases also the eh recombination probability. In literature this effect is called 'excitonic enhancement' [2].

Also three and four particle interactions can be observed through the optical spectra, as the X-free carrier and the $X-X$ interactions, respectively (the observation of processes in which only free carriers and/or dissociated eh pairs are involved is hard). An eh pair can recombine radiatively only if its centre of mass momentum $\hbar\mathbf{K}$ vanishes. In other words, before a luminescence photon escapes from the crystal, the excitonic polariton has to be scattered from an exciton like state to a photon like one [4]. Thus the above mentioned three and four particle interactions open recombination channels to eh pairs with $\mathbf{K} \neq 0$. Other important recombination channels are opened by the X-phonon [4] and the

R. Girlanda et al. (eds.), Progress on Electron Properties of Solids, 175–183.

X-impurity interactions. In this paper we will analyse the luminescence spectra of *GaSe*. For reasons which are not quite clear, neither the phonon assisted recombination of Γ excitons nor the formation of biexcitons (excitonic molecules) have been observed in this material. Therefore these two topics will not be discussed here.

For the discussion of the optical spectra of interacting X, one distinguishes three density regimes:

i) At low densities (and at low temperatures), after excitation the pairs relax rapidly into their bound states, and form an ideal X gas. However the interactions between X and crystal imperfections are always present and are most easily observed at low carrier densities. Some of these aspects are discussed in the paper by W. Czaja.

ii) At intermediate densities the interactions between the *eh* pairs become important. A broadening of the energy levels and higher order recombination processes, as those discussed above, are observed. $X-X$ and X-free carrier interactions are discussed in chapter 2.

iii) At the highest densities the distance between the pairs is comparable with the Bohr radius of the unperturbed X. The Coulomb interaction between the carriers is strongly screened and bound *eh* pairs do not exist. Thus, with increasing pair density, the *eh* fluid transits from an isolating X gas to a metallic *eh* plasma. This transition is called 'Mott transition' of the X [5]. As long as the system is not degenerate, it is still possible to think in terms of *eh* pairs. The effects of screening, which is a different aspect of the interactions, is discussed in static approximation in chapter 3.

2. Exciton-Exciton and Exciton-Free Carrier Recombination in *GaSe*

During such a process, one X with wavevector $\mathbf{K} \neq 0$ recombines while the momentum $\hbar\mathbf{K}$ and the energy required by the conservation law are transferred to another X or a single free carrier. Thus we have to consider the following processes [6]:

$$X^{\Gamma}(1,\mathbf{K}_1)+X^{\sigma}(1,\mathbf{K}_2)\rightarrow X^{\sigma}(n,\mathbf{K}_1+\mathbf{K}_2)+\hbar\omega_n^{\sigma} \tag{a}$$

$$X^{\Gamma}(1,\mathbf{K}_1)+X^{\sigma}(1,\mathbf{K}_2)\rightarrow e^{\sigma}h(\mathbf{K}_1+\mathbf{K}_2)+\hbar\omega_{eh}^{\sigma} \tag{b}$$

$$X^{\Gamma}(1,\mathbf{K}_X)+a^{\sigma}(\mathbf{k}_a)\rightarrow a^{\sigma}(\mathbf{K}_X+\mathbf{k}_a)+\hbar\omega_a^{\sigma}. \tag{c}$$

Here, $X^{\sigma}(n,\mathbf{K})$ represents a direct ($\sigma=\Gamma$) or an indirect ($\sigma=M$) exciton in a state with principal quantum number n and centre of mass wave vector $\mathbf{K}$, $e^{\sigma}h(\mathbf{K})$ is a dissociated *eh* pair with total momentum $\hbar\mathbf{K}$, $a^{\sigma}(\mathbf{k})$ is a single free carrier ($a = e$ or $a = h$), and $\hbar\omega^{\sigma}$ represents the emitted photon. We have to consider both conduction band edges Γ and M because in *GaSe* the M minimum is only 25 *meV* lower than the Γ one [7], therefore in our luminescence experiments both minima can be populated. At lower temperatures (i.e. neglecting kinetic energies), for $X-X$ recombination we can write [4]

$$\hbar\omega_n^{\sigma} = E_{X(1,0)}^{\Gamma}-E_R^{\sigma}(1-n^{-2}) \tag{3}$$

and

$$\hbar\omega_{eh}^{\sigma} \leqslant \hbar\omega_{\infty}^{\sigma}, \tag{4}$$

where $E_{X(1,0)}^{\Gamma}$ is the energy of the recombining X in its fundamental state and E_R^{σ} is the X Rydberg.

Experimental spectra are shown in Figure 1. It is difficult to discern the emission connected to each recombination channel because there are so many of them. At lower

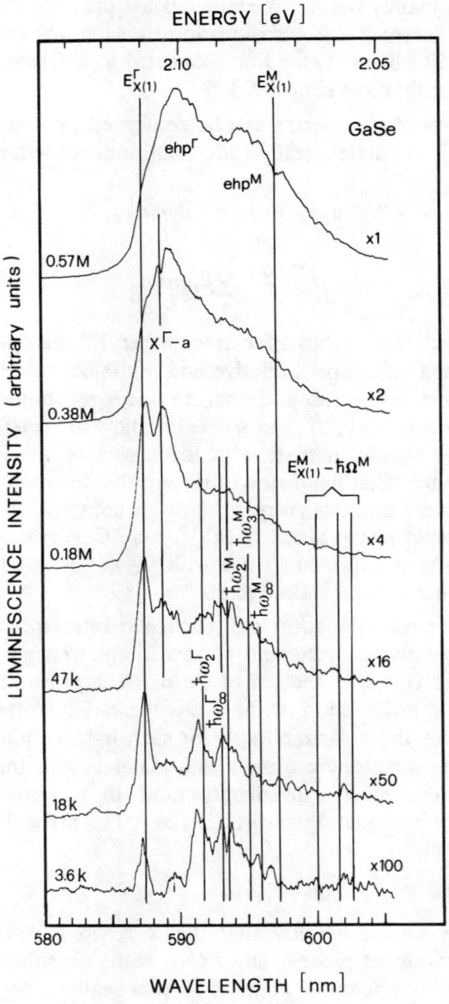

Figure 1. Series of luminescence spectra of a transport grown *GaSe* sample immersed in superfluid *He* [8,9] (sample thickness: of the order of $10\mu m$). On the left, the estimated excitation intensities in Wcm^{-2} are indicated for each spectrum (absolute accuracy: within a factor of about 2), the numbers on the right are relative scale factors for the luminescence intensity. The vertical line gives the spectral positions of the emissions connected to different *eh* recombination processes. The labels have the following meaning: $E^{\sigma}_{X(1)}$: fundamental Γ and M excitons; i: luminescence from a shallow impurity; $X^{\Gamma}-a$: exciton-free carrier recombination ($a=e^{\Gamma}, e^{M}$ and h); ehp^{σ}: *eh* plasma luminescence involving a Γ or an M electron; $\hbar\omega_n$: $X-X$ recombination processes as described in the text; $E^{M}_{X(1)}-\hbar\Omega^{M}$; indirect recombination of M excitons assisted by the emission of one of the three lowest M phonons of energy $\hbar\Omega^{M}$ [10]. Assuming an *eh* recombination lifetime of $0.1ns$ [11] and an absorption coefficient of $10^{3}cm^{-1}$ [12], the *eh* pair density of the top spectrum is about $1.5\cdot 10^{17}cm^{-3}$; the Mott transition is expected to occur at a density of dissociated pairs of roughly $0.8\cdot 10^{17}cm^{-3}$. For experimental details, see [8] and [9].

densities, when the number of dissociated *eh* pairs is still small, besides the 'simple' X recombination line, mainly the $X-X$ emissions are present. At higher densities, the lines broaden and the different $X-X$ emissions form a wide low energy tail. At pump intensities above roughly 50 $kWcm^{-2}$, the line connected to X-free carrier scattering appears. It is only slightly wider than the simple X line.

Our interpretation of the spectra can be confirmed by kinetic considerations. For the direct free X density n_X^Γ, under steady state conditions we write [6]

$$\frac{dn_x^\Gamma}{dt} = 0 = G - \alpha_r n_X^\Gamma - \alpha_D n_X^\Gamma - \beta_{\Gamma\Gamma}(n_X^\Gamma)^2 - \tag{5}$$

$$-\beta_{\Gamma M} n_X^\Gamma n_X^M - \sum_a \beta_{\Gamma a} n_X^\Gamma n_a^\Gamma$$

where G is the direct free X formation rate caused by the optical excitation, α_r and α_D are the rate constants for simple radiative and nonradiative recombination of the X, and the three coefficients β are proportional to the probabilities for radiative $X^\Gamma - X^\Gamma$, $X^\Gamma - X^M$ and $X^\Gamma - a$ ($a = e^\Gamma, e^M$ and h) scattering. The intensity of the direct-direct X recombination, lines should increase with the square of the density n_X^Γ, as long as the three first terms on the right hand side of (5) are the dominant ones. If one assumes that the ratio of the indirect to direct *eh* pair density is constant, then the direct-direct exciton recombination becomes proportional to $(n_X^\Gamma)^2$, too. Concerning the $X-a$ recombination, the emission intensity is expected to rise with $(n_X^\Gamma)^{3/2}$ [13]. These theoretically expected dependencies have been found indeed [6,8].

The X-free carrier recombination is of particular interest in *GaSe*. In contrast to other compounds [4] its relatively strong and narrow luminescence line is observed even at low lattice temperatures. This is connected to the particular symmetry of the *GaSe* structure: for light propagating in direction of the **c**-axis the radiative transitions are only partially allowed [14] and thus the polarizability γ for such light is quite small. Since *GaSe* has a layered structure the samples are usually thin platelets with the **c**-axis normal to the platelet plane. Therefore most conveniently (and thus most commonly) the observed luminescence is light propagating in direction of **c**. For the emitted luminescence intensity $I_{X-a}(\hbar\omega)$, we write [15]

$$I_{X-a}(\hbar\omega) = P_{X-a}(\hbar\omega) \cdot F(\hbar\omega) \tag{6}$$

where $P_{X-a}(\hbar\omega)$ is the probability that the X reach a state of momentum $\hbar\mathbf{K} = 0$ through an $X-a$ scattering process, and $F(\hbar\omega)$ is the probability that a polariton escapes from the crystal as a photon. $P_{X-a}(\hbar\omega)$ is practically independent of γ while $F(\hbar\omega)$ reflects the polariton dispersion in the direction of observation. The smallness of the polarizability γ of *GaSe* for light propagating in the **c** direction induces a narrow $X-a$ recombination line, as shown in Figure 2.

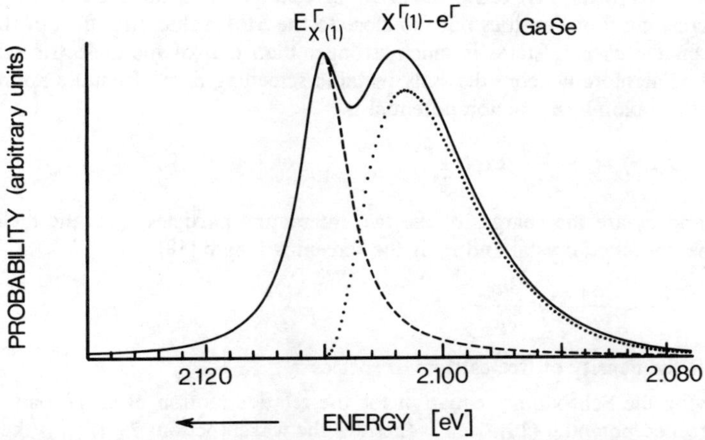

Figure 2. Calculated spectrum of the luminescence emitted by 'simple' recombination of the direct exciton $E^{\Gamma}_{X(1)}$ and by recombination assisted by direct X-direct e scattering ($X^{\Gamma}(1)-e^{\Gamma}$). The lineshape of the simple radiative X recombination is a Lorentzian of $5\,meV$ width, and the $X^{\Gamma}(1)-e^{\Gamma}$ line has been calculated following [15]. In order to obtain a spectrum which resembles the experimentally observed one, a temperature of $10K$ and a very small polarizability $4\pi\gamma/(\epsilon_0 E^{\Gamma}(1))$ of $6\cdot 10^{-5}$ has been used. The dielectric constant ϵ_0 and the effective masses have been taken from [12] and [16], respectively (courtesy of V. Capozzi and L. Pavesi).

3. Screening

We consider now our *eh* fluid in binary collision approximation, i.e. in the virial expansion of the thermodynamic potential Ω (Eq. (3)) we retain only the first two terms. For the correction Ω_2 caused by the binary interactions we found [2]

$$\Omega_2 = \frac{1}{2} k_B T V^2 \sum_a \sum_b n_a^0 n_b^0 \Lambda_{ab}^3 \int \Delta D_r^{ab}(E_r) \exp\frac{-E_r}{k_B T} dE \tag{7}$$

where n_a^0 would be (for given temperature and chemical potential) the density of the carriers of species a if the interactions were absent,

$$n_a^0 = \int D_a^0(E_a) \exp\frac{\mu_a - E_a}{k_B T} dE_a = 2\Lambda_a^{-3} \exp\frac{\mu_a}{k_B T}, \tag{8}$$

$\Lambda_a = 2\pi\hbar/(k_B T m_a)$ is the thermal wavelength of the particle a, $D_a^0(E_a)$ is the density of states (*DOS*) in a non-interacting gas, $\Delta D_r^{ab}(E_r)$ is the correction to the *DOS* caused by the interactions, and the subscript r indicates that the relative motion is concerned. It has been shown that the contributions of the repulsive interactions to Ω_2 are quite smaller than the attractive one [3], therefore equation 7 becomes

$$\Omega_2 = k_B T V^2 n_e^0 n_h^0 \Lambda_{eh}^3 \int \Delta D_r^{eh}(E_r) \exp\frac{-E_r}{k_B T} dE_r. \tag{9}$$

Thus, in our approximation, interpair interactions are neglected. However, these interpracticle interactions are usually considered indirectly: the presence of a high density of *eh* pairs screens the *eh* attraction in the pair. Screening is a many-body aspect of the interpair interactions.

At infinite temperatures, in general both bound and dissociated *eh* pairs are present.

The bound pairs (excitons) cause dielectric screening, while the dissociated ones cause metallic screening. For densities not too close to the Mott value, the effect of this metallic screening on the *eh* pair states is much stronger than that of the dielectric screening by the *X* [17]. Therefore we consider only metallic screening here. In static approximation, the screened Coulomb interaction potential is

$$U_{ab}(r) = -\frac{q_a q_b}{\epsilon_0}\exp(\frac{-r}{r_{sc}}) \tag{10}$$

where q_a and q_b are the charges of the two interacting particles, ϵ_0 is the dielectric constant of the unexcited crystal and r_{sc} is the screening length [18],

$$\frac{1}{r_{sc}^2} = \frac{4\pi}{\epsilon_0}\sum_a q_a^2 \frac{\partial n_{fa}}{\partial \mu_a}. \tag{11}$$

Here, n_{fa} is the density of free carriers of species *a*.

By solving the Schrödinger equation for the relative motion of an *eh* pair interacting via the screened potential (10), we get (besides the wavefunctions $F_{lm}(E_r,r)$) the *DOS*

$$D_r(E_r) = D_r^0(E_r)+\Delta D_r^{eh}(E_r). \tag{12}$$

Obviously, $\Delta D_r^{eh}(E_r<0)$ are the *X* states. With increasing carrier density, i.e. with decreasing screening length r_{sc}, the binding energy of the exciton states decreases. For $r_{sc} \leq 0.8399\, a_0$ (a_0 is the Bohr radius of the unperturbed *X*), no bound pair states exist [19], in other words the Mott transition takes place. Further, exchange and correlation effects of the metallic free carriers lead to a reduction of the band gap with E_g. The reductions of the band gap width and the *X* binding energy compensate each other almost completely [20], therefore the absolute energy of the fundamental *X* remains practically unshifted. This effect is clearly visible in the experimental luminescence and transmission spectra, as shown *eg* in figures 1 and 4.

We now discuss the optical interband density $\mathcal{D}^{eh}(\hbar\omega)$ (which is proportional to the absorption coefficient) of a semiconductor containing our dilute and nondegenerate *eh* fluid [2] (for a more general formulation valid also in the degenerate case, see e.g. [5]):

$$\mathcal{D}^{eh}(\hbar\omega) = |M_{cv}|^2 \int_{-\infty}^{\infty} dE_r D_{r,l=0}^{eh}(E_r)\Phi(E_r) \tag{13}$$

$$[F(-E_r,T)-F(+E_r,T)]A(E_g+E_r-\hbar\omega).$$

Here, M_{cv} is the interband momentum matrix element for optical transitions close to the minimum band gap width E_g, $\hbar\omega$ is the photon energy, $D_{r,l}^{eh}(E_r)$ is the density of *eh* pair states, while $F(\mp E_r,T)$ are the statistical probabilities to find the initial pair configurations appropriate for up and down radiative interband transitions. $A(E_g+E_r-\hbar\omega)$ is a spectral function which takes into account the finite width of the *eh* pair states. Finally, $\Phi(E_r)$ is the excitonic enhancement factor which is due to the *eh* attraction, in dilute systems

$$\Phi(E_r) = |F_{00}(E_r,r=0)|^2 \tag{14}$$

where $F_{lm}(E_r,r)$ is the envelope function of an *eh* pair having an angular momentum $\hbar l$, an interparticle distance *r* and an energy E_r.

For different screening lengths r_{sc}, we solved the above mentioned Schrödinger equation numerically and calculated the interband density $\mathcal{D}^{eh}(\hbar\omega)$, Eq. (13) [2]. We put the statistical factor in brackets equal to one, which is often appropriate for nondegenerate

***eh* systems at lower temperatures. For the spectral function *A* we used Lorentzians. The results are shown in figure 3.**

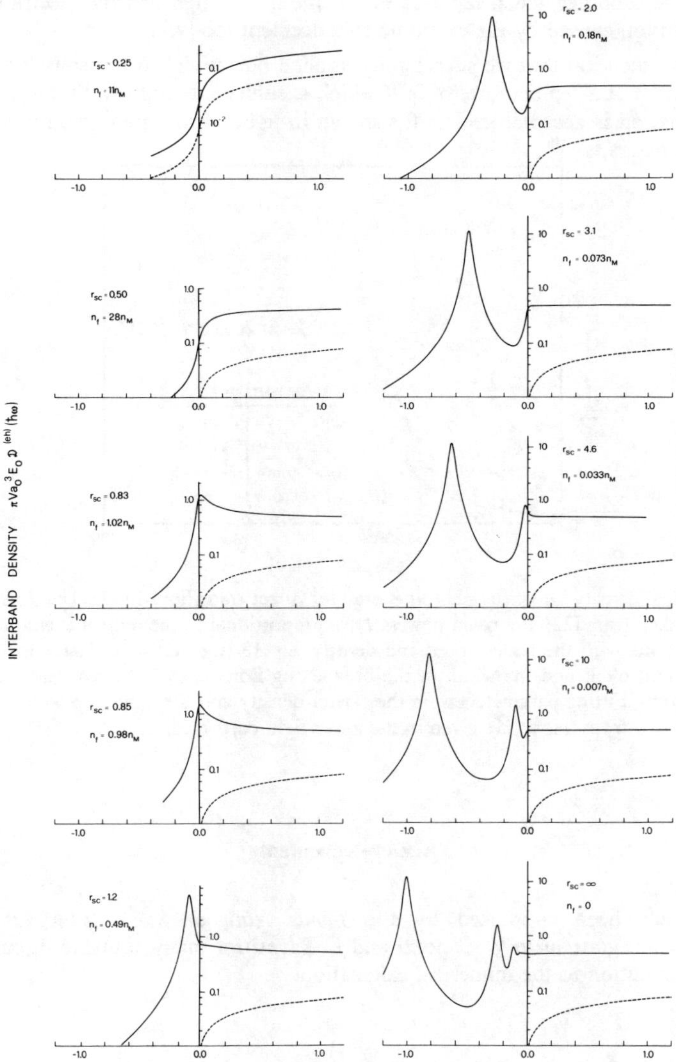

Figure 3. Optical interband density $\mathcal{D}^{(eh)}$ for direct transitions versus photon energy $\hbar\omega$, shown for different screening lengths r_{sc}. E_0 and a_0 are the binding energy and the Bohr radius of the unscreened *X*, E_g is the width of the energy gap, *V* is the crystal volume, r_{sc} is given in units of a_0 and the free carrier density n_f is given in units of the Mott density n_M. Full curves are for *eh* pairs interacting via a statically screened Coulomb potential (eq. 10), the dashed ones for non-interacting particles.

Looking at this figure it becomes clear that the Mott transition is continuous. The influence of the *eh* interactions on the spectrum of the continuum states is obviously important; it is mainly due to the excitonic enhancement, the corrections $\Delta D^{eh}_{r,l}(E_r > 0)$ being small. In particular, the X peak in the interband density is already visible at the transition. The shoulder which replaces the X line in the high density spectra of figure 1 is, to our opinion, caused by the excitonic enhancement too [9,21].

Finally, we mention that we successfully applied our model to transmission spectra of *Ge* [2], measured at a temperature of $30K$ which is sufficiently high so that our semiclassical approximation is acceptable. The fits shown in figure 4 are quite good and yield reasonable parameters.

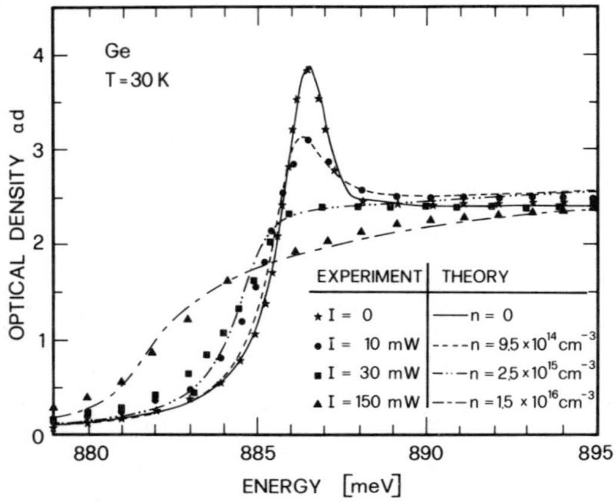

Figure 4. Optical density αd versus photon energy for direct transitions in *Ge*. The dots are experimental data taken from [22], the given powers I are proportional to the optical excitation intensity. The curves are fits with the scaled interband density, eq. 13 (the statistical factor in the brackets has been replaced by 1, and the width of the broadening Lorentzian has been read from the zero density spectrum). Fitting parameters were the carrier density and the band gap width. The *eh* pair densities extracted from this fit are given in the lower right corner [2].

Acknowledgement

This work has been supported by the *Fonds National Suisse pour la Recherche Scientifique*. I am grateful to V. Capozzi and L. Pavesi for many valuable discussions and for their contribution to the numerical calculations.

References

1. E.M. Lifshitz and L.P. Pitaevskii, Statistical Physics 3rd edition part 1, (Pergamon, Oxford, 1980) Chap. 75 and Chap. 77.
2. L. Pavesi et al., *J. Phys. C Solid State Phys.* **21**, 1485 (1988).
3. H. Stolz and R. Zimmermann, *phys. status solidi (b)* **94**, 135 (1979).
4. See e.g. C. Klingshirn and H. Haug, *Phys. Reports* **70**, 315 (1981).
5. H. Haug and S. Schmitt-Rink, *Prog. Quant. Electr.* **9**, 5 (1984).
6. V. Capozzi and J.L. Staehli, *Phys. Rev.* **B28**, 4461 (1983).
7. Le C.T. and C. Depeursinge, *Solid State Com.* **21**, 317 (1977); A. Mercier et al., *Phys. Rev.*

B12, 4307 (1975).

8. L. Pavesi, *Tesi di Laurea*, Univ. di Trento (1985).
9. L. Pavesi et al., *Phys. Rev.* **B39** (1989), 10982.
10. S. Jandl et al., *Phys. Rev.* **B13**, 686 (1976).
11. T. Kushida et al., *Nuovo Cimento* **39B**, 650 (1977); J. Collet, private communication.
12. R. LeToullec et al., *Phys. Rev.* **B22**, 6162 (1980); *Nuovo Cimento* **38B**, 159 (1977).
13. C. Benoit à la Guillaume et al., *Phys. Rev.* **177**, 567 (1969).
14. E. Mooser and M. Schlüter, *Nuovo Cimento* **18B**, 164 (1973).
15. B. Hönerlage et al., *phys. stat. solidi (b)* **78**, 599 (1976).
16. G. Ottaviani et la., *Solid State Com.* **14**, 933 (1974).
17. G.W. Fehrenbach et al., Proc. 17th Int. Conf. Phys. Semiconductors (San Francisco, 1984), 1251.
18. W. Ebeling, W.D. Kraeft and D. Kremp, Theory of Bound States and Ionization Equilibrium in Plasmas and Solids (Academie, Berlin, 1976) Chap. 2.
19. F.J. Rogers et al., *Phys. Rev.* **A1**, 1577 (1970).
20. R. Zimmermann et al., *phys. status solidi (b)* **90**, 175 (1978).
21. L. Pavesi et al., *Physica scripta* **38**, 627 (1988) and *Solid State Com.* **61**, 321 (1987).
22. H. Schweizer et al., *Phys. Rev. Let.* **51**, 698 (1983).

THE SPECTRAL MANIFESTATION OF WEAKLY COUPLED MIXED STATES OF EXCITONS AND LATTICE DEFORMATION

M.S. BRODIN, I.V. BLONSKIJ AND S.V. MARISOVA

Institute of Physics, Ukr. Academy of Sciences
Prospekt Nauki 46,
Kiev, 252028 USSR

The secondary emission spectra (*SES*) were measured as functions of λ_{ex} and T, together with the excitation spectra (*ES*) of exciton bands in HgI_2 and PbI_2 semiconductors specified by the intermediate exciton-phonon coupling. In the low-temperature *SES* ($T < 50K$) a two series structure of bands is discovered. Apart from the *A*-series of bands of polariton origin, we observed another series (*B*) of bands shifted with respect to the *A*-series by $7\,meV$ in HgI_2 and $2 \div 3\,meV$ in PbI_2. The *B*-series arises at a later stage of polariton relaxation and is due to specific states of excitons weakly bound with lattice deformation. Pertinent theoretical models are reviewed. Some features which have been revealed in the low temperature spectra of transmission, reflection and emission of anthracene crystals are also attributed to existence of the bound exciton polaritons with elastic lattice deformation.

1. Introduction

The main features of exciton-phonon interaction have been studied in great detail for two limiting cases of weak ($\Lambda \ll 1$) and strong ($\Lambda \gg 1$) coupling, the dimensionless coupling constant Λ being usually defined as the ratio of the lattice-deformation energy E_D to the halfwidth of the exciton band L, i.e. $\Lambda = E_D / L$. Under the weak coupling, the quasi-elastic scattering processes lead to establishing of the statistical equilibrium between free (*F*) excitons and phonons, whereas their dynamical effects amount to an insignificant, at least at small values of quasi-momentum, renormalization of the energy spectrum of *F*-excitons. In the opposite case ($\Lambda \gg 1$), the exciton photoexcitation itself is accompanied by the lattice distortion that facilitates appearance of self-trapped (*S*) states; the latter may co-exist with the *F*-excitons, provided there is an autolocalization barrier, and may be stable or metastable depending on the magnitudes of E_D, Frank-Condon energy $E_{F\text{-}C}$ and the barrier height W (see the review by Rashba and Lushtchik in [1]).

The case of intermediate exciton-phonon coupling ($\Lambda \cong 1$), corresponding to the transition from F to S state, has been studied less of all. This holds especially for the crystals with high oscillator strength of the exciton transition. Some of these crystals exhibit also strong light-exciton mixing. One may suppose that in such materials the lattice deformation accompanying the exciton generation leads to localization of the excitations, whereas the strong light-exciton coupling or the high kinetic energy of the excitons hinder their localization. At present the result of these two counteracting effects cannot be estimated in terms of the available theory. This is connected with the well known difficulty to allow for nonadiabatic corrections in the problem of photon-exciton-phonon interaction. Thus the results of experimental studies are of great interest.

This paper deals with the problem of the observation and the study of spectral peculiarities of the crystals with intermediate strength of exciton-phonon coupling and high light-exciton interaction. Two types of materials were chosen for our studies: layered ionic semiconductors HgI_2 and PbI_2, and the molecular crystal of anthracene. HgI_2 and

R. Girlanda et al. (eds.), Progress on Electron Properties of Solids, 185–195.

PbI_2 are characterized by almost equal value of light-exciton and exciton-phonon interaction strength and large exciton band width ($L \cong 1\, eV$) [2].

The crystal of anthracene is characterized by intermediate strength of the exciton-phonon coupling and high transverse-longitudinal splitting E_{LT} ($E_{LT} \cong L \cong 10^{-1}\, eV$) which indicates strong exciton-photon interaction [3].

2. Experimental

We studied the emission spectra of the vapour-grown samples of HgI_2 (the red modification) and of the Bridgman-grown samples of the $2H$-polytype of PbI_2. Secondary emission spectra (*SES*) were measured as functions of T and λ_{ex}, in the photon-counting operation mode of the automatic spectral setup with the multipurpose spectrometer *SDL*-2 used as a basic unit. The exciting monochromatic light came from the *DKSH*-250 lamp through the monochromatic illuminator *MDR*-2 with the relative aperture equal to 1:2. The spectral width of the source slit did not exceed 3Å, whereas the detector slit was 2Å-wide. The accuracy of temperature stabilization and measurements was about $0.1K$.

3. Two-Series Structure of *SES* in HgI_2 and PbI_2

a) Experimental results

The layered ionic crystals HgI_2 and PbI_2 belong to the direct-transition semiconductors characterized by great magnitudes of the oscillator strength of the exciton transitions. Earlier investigations of the secondary emission of these compounds aimed to studying the basic properties of the exciton-photon interaction, with emphasis given to the interaction anisotropy, as well as to the energy and phase relaxation of polaritons. The obtained results have been exposed in the monograph [2]. Below we report on another peculiarity of the secondary emission observed at $T < 50K$, i.e. on the two-series structure of the spectra.

Figure 1 (1) shows the *SES* of the layer plane of HgI_2 measured at low levels of the band-to-band excitation for $E \perp C$ and $T = 4.5K$. Such a shape of the spectrum was also reported by other authors [4,5] who were the first to suggest an interpretation of the main emission bands which has become generally accepted. The assumption underlying their interpretation is that in these semiconductors there exist only the *F*-excitons strongly coupled with light. Consequently, the bands A_U ($529.8nm$) and A_L($531.1nm$) were explained as the emission from the upper (*UPB*) and lower (*LPB*) polariton branches, respectively. The band $532.1nm$ was attributed to exciton-impurity complexes (*EIC*), and the bands $A_L - 2E'$($532.9nm$), $A_L - A_{1g}$($533.9nm$), $A_L - PC_1$($535.1nm$), $A_L - 2A_{1g}$($536.7nm$), $A_L - PC_2$($537.9nm$) were interpreted as the radiation from *UPB* accompanied by simultaneous emission of one or several phonons of different dispersion branches. The conclusion about the character of the bands $A_L - 2E$, $A_L - A_{1g}$, $A_L - PC_1$, $A_L - 2A_{g1}$ and $A_L - 2PC_2$ was drawn by Goto and Nishina [5] from their analysis of the typical asymmetry of the respective band-shapes where a flatter shortwave wing, sensitive to λ_{ex} and T, is observed (see [5] and also the clip in Figure 1). In fact, Goto and Nishina followed the earlier study due to Gross et al. [6] who had proved such asymmetry of the band-shape to be indicative of the temperature distribution of exciton-like polaritons in the quasi-momentum space and, as a result, introduced the well-known method of the

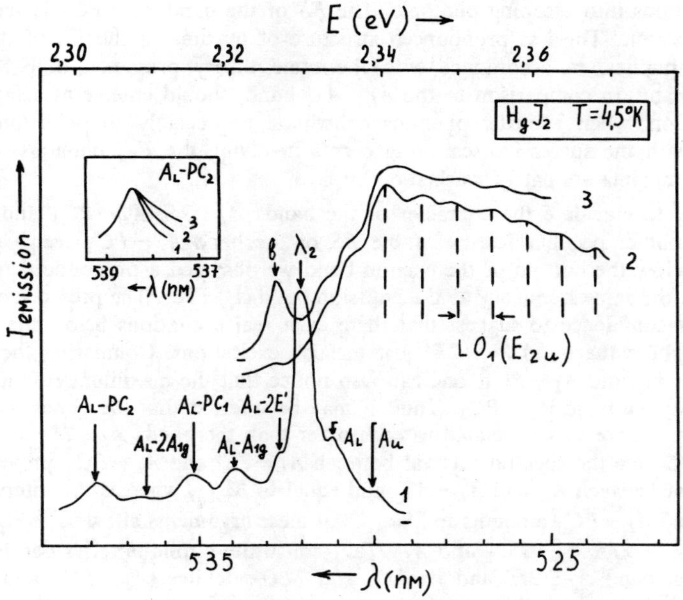

Figure 1. (1) Low-temperature *SES* of HgI_2 recorded for the band-to-band excitation at $T = 4.5K$, (2) *ES* of the $A_L - A_{1g}$ band, (3) *ES* of the $A_L - PC_1$ bad. Temperature-induced variation of the band-shape for $A_L - PC_1$ is illustrated in the clip where the band-shapes 1,2,3 correspond to $T = 4.5$, 25 and 40*K*, respectively.

thermometry of exciton gas. However, further comparison of the spectral intervals between the bands belonging to this group and the A_L-band with the phonon energies reported for HgI_2 (see [2]) showed that only the $A_L - A_{1g}$ and $A_L - 2A_{1g}$ bands fit the interpretation scheme proposed in [5]: they correspond to the *LPB* emission together with that of one or two A_{1g}-phonons of energy $14\,meV$. In order to account for the spectral position of the remaining bands, they were assumed to result from the *LPB* emission accompanied by the simultaneous birth of an unknown combination of different *LO*-phonons.

Below we present the results of our studies which suggest a new interpretation of the bands $A_L - 2E'$, $A_L - PC_1$, $A_L - PC_2$. In this connection, it seems appropriate to consider the low-temperature *ES* of a couple of the neighboring emission bands $A_L - A_{1g}$ and $A_L - PC_1$ (see Figure 1 (2,3)). Comparison of the *ES* of these bands reveals two main differences between them, i.e. the discrete structure of the maxima is more pronounced in the *ES* of the band $A_L - A_{1g}$, and in the *ES* of the band $A_L - PC_1$ there is a strong maximum *B* shifted by $7\,meV$ below the energy E_T of the transversal exciton and coincident with the emission band $A_L - 2E'$.

As regards the discrete structure of maxima in the *ES* of the band $A_L - A_{1g}$, it should be noted that it was also described previously [2,5]. It was inferred, in the frame of the theory of polariton kinetics [7,8], that the emission under consideration arises at an early stage of polariton relaxation, that is, after several successive scattering events involving the dipole active LO-phonons of the E_{2u} symmetry (with energy $\hbar\omega_{E_{2u}} = 3.5\,meV$, as seen from the frequency spacing between the neighboring maxima of ES in the plateau region), and finally appears as result of the A_{1g}-phonon-assisted conversion of the scattered polaritons into escaping photons. The ES of the band A_L-PC_1 is reported here for the first time. The less pronounced structure of maxima in the ES of this band at $T = 4.5K$ implies, in accordance with the same theory presented in [7,8], that the A_L-PC_1 band, in comparison to the A_L-A_{1g} band, should emerge at a later stage of the relaxation, when acoustic phonons contribute appreciably to polariton scattering alongside with the successive scattering events involving the E_{2u} phonons. Due to the former the maxima are partly 'washed out'.

In order to elucidate the character of the bands A_L-2E, A_L-PC_1 and A_L-PC_2, however, another peculiar feature of the ES of the band A_L-PC_1 seems much more effective. Below the bottom of the exciton band we observed a pronounced maximum B occuring at the same frequency as the emission band A_L-2E'. The presence of this maximum gives confidence to suggest that there exist real excitations below the energy E_T, and to attribute the band A_L-2E' just to such excitations. Comparing the ES of the bands A_L-A_{1g} and A_L-PC_1, one can also notice that the maximum B is more typical in the ES of the band A_L-PC_1. Thus, it may be inferred that the bands A_L-2E' and A_L-PC_1 are more closely related to each other than the bands A_L-2E' and A_L-A_{1g}. At the same time the spectral interval between A_L-2E' and A_L-PC_1 proves to be the same as that between A_L and A_L-A_{1g} and equal to $\hbar\omega_{A_{1g}}$, whereas the interval between A_L-2E' and A_L-PC_2 amounts to $2\hbar\omega_{A_{1g}}$. All these arguments allow us to conclude that the bands A_L-2E', A_L-PC_1 and A_L-PC_2 constitute a unique series consisting of the zero-phonon band A_L-2E' and its LO- and $2LO$-satellites (A_L-PC_1 and A_L-PC_2, respectively). Henceforth this group of bands will be referred to as the B-series, distinct from the A-series that comprises the bands of polariton origin (A_U,A_L,A_L-A_{1g}, A_L-2A_{1g}). A and B series are separated by an energy interval of about $7\,meV$ and correspond to two different stages of electron excitation relaxation.

Another layered semiconductor, $2H$-PbI_2, also exhibits a two-series character of the SES. This has been pointed out for the first time in [9]. However, the energy interval between A and B series in $2H$-PbI_2 is only $2 \div 3\,meV$, what hinders their resolution. One can manage to resolve them by the different behaviour of the A and B series with λ_{ex} and T. This clearly results from Figure 2. Two remarkably different properties of the A and B series should be stressed. These are the temperature stability of the A series in the interval $1.5 \leqslant T \leqslant 150K$ [2] whereas the B series disappears from the spectrum as temperature grows up to $20K$; and the nonthermalized type of luminescence corresponding to the A series, while the thermalized one is related to the B-series [9]. The latter conclusion results from the analysis of the excitation spectra of the A and B series in terms of known ideas about spectral evidences of the exciton lifetime [6,8].

b) Discussion

The low-temperature SES of the HgI_2 and PbI_2 crystals have been shown above to exhibit a two-series structure of the bands arising at different stages of relaxation of excitations; the A-series results from an earlier stage and the B-series emerges later. Naturally, this fact implies that there should be two different mechanisms which give rise to the outgoing emission. The first mechanism, governing the A-series, is typical for most of direct-transition semiconductors with strong exciton-photon coupling. The secondary emission of this type is generated at the crystal boundary due to a dissociation of the polaritons coming to the surface at such angles that their total internal reflection does not take place. The spectral distribution of intensity of this emission is given by [7]

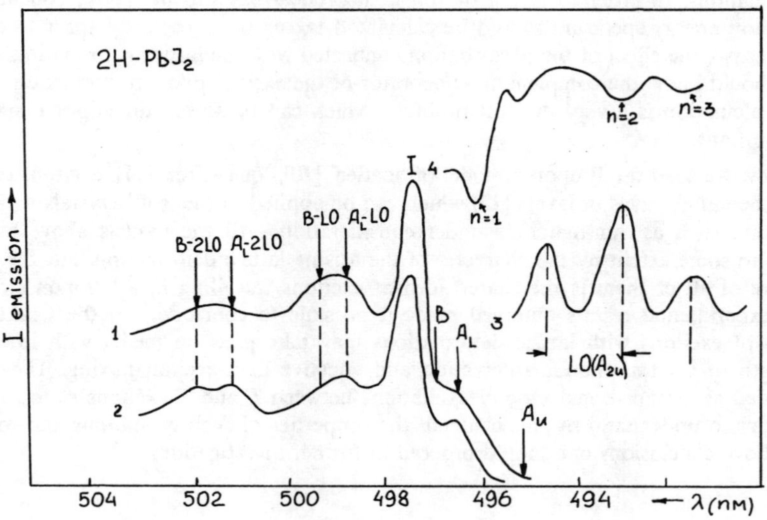

Figure 2. *SES* of the $2H$-PbI_2 recorded at $T = 4.5K$ (1) and $T = 10K$ (2) and *ES* of the bands $A_L - A_{1g}$ (3) and $B - A_{1g}$ (4).

$$I(E)d\Omega' = \left\{1 - \left[\frac{n(E)-1}{n(E)+1}\right]^2\right\}\rho(E)g(E)v(E)\frac{d\Omega}{4\pi}.$$

Here $d\Omega$ is the solid angle of the emission detection, $n(E)$ is the dispersion of the refractive index, $\rho(E)$, $v(E)$ and $g(E)$ denote distribution function, group velocity and state density of polaritons, respectively. For low T such emission is referred to as the unthermalized luminescence.

The evolution of polaritons totally reflected by the crystal surface proves to be qualitatively different. Being 'captured' inside crystal bulk, they take part in the multiple scattering involving acoustic phonons and only at a later stage of their relaxation a part of them is converted into secondary emission. Therefore, in many semiconductors such as *CdS* [8], the low-temperature *SES* results from two emission components generated at the earlier and at the later relaxation stages, respectively, and - which is not less important - the spectral position of the bands of the same origin related to the two emission components is almost identical.

In contrast to *CdS*, the distinguishing feature of the *SES* recorded in HgI_2 and PbI_2 is that the emission component due to the delayed relaxation is spectrally shifted with respect to the *A*-series by 7 and $2 \div 3\,meV$, respectively.

The reason of the observed Stokes shift of the emission component arising at later stage of the exciton relaxation can be accounted for in the case when - apart from the traditionally considered processes of elastic polariton-phonon interaction determining energy relaxation of the polaritons - one also takes into account the many-phonon processes causing deformations in the medium. In the crystals in question two favourable circumstances give rise to many-phonon-polariton relaxation: the strength of exciton-phonon coupling is higher than in usual semiconductors; the mechanical strength is low so that the deformation can easily arise in these 'soft' crystals even under weak perturbation (the electron subsystem excitation). As a result, we are led to the following interpretation of the *B*-series: it is due to weakly bound states of excitons with lattice deformations. In order to give a thorough theoretical basis to the above conclusion, the polariton energy spectrum should be calculated taking into account - apart from polariton decay - the effect of the perturbation connected with the lattice deformation. For this, one should know the complete mass operator of the exciton-phonon interaction. Its accurate calculation is a very difficult problem, which can be solved under some simplifying assumptions.

Now we shall dwell upon a model of 'loaded' [10], 'quasi-free' [11], excitons connected with 'bending' waves in layers [12] which can be applied to the 'soft' crystals with layered structure such as our materials under consideration. All the models above mentioned differ to some extent by the character of the arising lattice deformation, but the common feature of all of them is the shared idea of excitons travelling in a 'phonon cloud' [13]. The experimental results obtained made it possible to conclude that the weakly bound states of excitons with lattice deformations may take place in media with intermediate strength of exciton-phonon interaction and effective light-exciton mixing. They may be regarded as a transitional type of excitations between F and S excitons states. However, in order to understand in greater detail the properties of such excitations and to confirm the above conclusions one should proceed to further investigations.

4. Peculiarities of the Polariton Spectra in Anthracene

The penetration of an electromagnetic wave through a crystal manifests itself in various optical effects, such as reflection, transmission, luminescence etc. Their properties are governed by the elementary excitations interacting with photons and the subsequent relaxation processes. Below we consider in detail the optical properties of anthracene single crystals in the vicinity of a strong exciton resonance, i.e. the lowest component of Davydov's exciton doublet (the resonance frequency E_0 is about 25095 cm^{-1}). In view of the very great oscillator strength of the optical transition in anthracene, this case proves to be pretty instructive in studies of exciton-photon interactions. Here, in considering the relaxation of molecular excitons due to their scattering by phonons, one should first take into account the intensive exciton-photon mixing (the polariton character of the excitations induced in crystal by the incident electromagnetic wave) [3,14]. There exists a substantial wealth of experimental evidence [3,15-17] that the polariton theory describes fairly well the main features of secondary emission spectra in anthracene at low temperature. According to polariton models, the exciton luminescence band-shape is determined by the spectral distribution function along the lower polariton branch and by the boundary conditions imposed at crystal surface. The nonequilibrium excitations, generated by external irradiation at some point of the band, are accumulated at the band bottom E_0 during their thermalization and, with acoustic phonons vigorously involved in the process, eventually they form a narrow peak ($\sim kT$ wide) of the polariton spectral distribution. However, in contrast to the case of HgI_2 and PbI_2 crystals discussed above, in anthracene this spectral density maximum does not result in a well-pronounced 'zero-phonon' line in the luminescence spectrum. There are several channels of polariton escape from this density maximum into the energy region below E_0 where the polariton free path becomes comparable (or even greater) with crystal dimensions. Scattering of E_0-polaritons by high-frequency phonons, such as intramolecular vibrations or lattice optical phonons, followed by the conversion of polaritons into escaping photons gives rise to a series of the narrow luminescence bands which may be regarded as phonon repetitions of the main maximum of polariton spectral density, that is phonon satellites of the zero-phonon line. The transitions of polaritons from the primary distribution maximum $\sim E_0$ below the band bottom which are induced by acoustic modes create the secondary maximum E_1 in the polariton distribution. The latter manifests itself in the emission spectrum as a wide and intense band identified by the shortest wave-lengths of the outgoing luminescence. It is this band that opens the series of wide emission bands. Thus, the low-temperature luminescence spectrum of perfect single crystals has the form of two series of bands different in width, shape, relative lifetime and other characteristics, although all these bands come forth from one and the same fundamental type of elementary excitations and result from conversion of polaritons in the course of their relaxation.

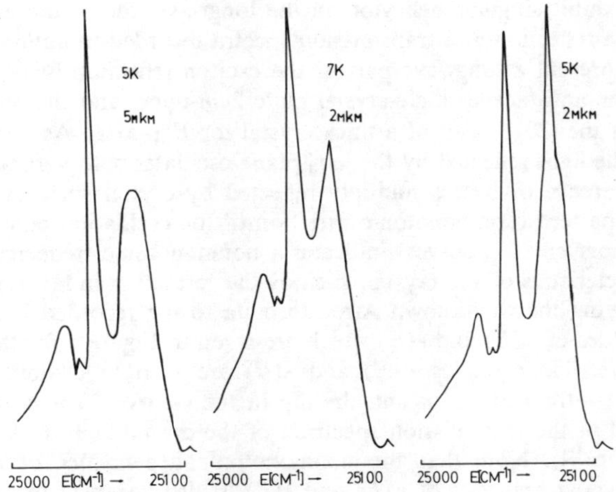

Figure 3. The initial part of the exciton luminescence of two different anthracene crystals.

As an example, consider the initial part of the luminescence spectrum recorded in anthracene (Figure 3). It is worth being noted the sensitivity of shape and intensity of the first ('acoustic') emission band to variations of temperature and crystal thickness. In a wedge-shaped crystal with thickness $d = 2\mu m$ a pronounced spike of the emission intensity is observed at $T = 5K$, at higher temperatures the intensity decreases rapidly at the band maximum. The luminescence spectrum of a plane-parallel crystal plate, also $\sim 2\mu m$ thick, is almost identical with that shown in Figure 3 though it is modulated due to interference of multiply reflected waves inside the crystal volume. In the ultra-thin crystals ($d < 1\mu m$) the mode structure of the polariton spectrum is clearly seen, with the emission intensity being redistributed nonmonotoneously among the modes in crystals of different thickness. A singularity of the excitation spectrum, first reported by Ferguson [18], is observed at the maximum of the 'acoustic' luminescence band; this singularity is not present in all crystals and its interpretation requires further studies.

It should be noted that the standard theory of polariton emission, which accounts for exciton-phonon scattering in the framework of the one-phonon approximation, fails to provide a proper description of the initial 'acoustic' band, in particular of its great width and high intensity; with no selection of the parameters one can fit the experimental data. Another theoretical challenge is the longer luminescence lifetime in the frequency region of the 'acoustic' band; here, it exceeds the lifetimes of all narrow-emission-bands-one-phonon repetitions of the maximum of the polariton spectral density [15,16]. The estimations of mean energy of the exciton-acoustic-phonon interaction presented in [3,15,19] show that such interaction cannot be assumed weak and, therefore, the multiphonon transitions may play an essential role in relaxation processes. Allowing for the specific dispersion of exciton polaritons and acoustic phonons in anthracene, we have analyzed the behavior of the functions $\Gamma^{\mp}(k)$ accounting for scattering of polaritons by acoustic modes [3,20]. It has been found that those polaritons which evolve into the states with the group velocity equal to that of sound intensively absorb acoustic phonons. Mathematically, this effect is revealed in a singular behavior of the imaginary part $\mathrm{Im}\, M(\omega)$ of the self-energy operator of exciton-phonon interaction. Presence of such a singularity together with intermediate values of the exciton-phonon coupling constant are indicative of the importance of multiphonon-polariton scattering in anthracene.

If all our arguments are adequate, then also other optical characteristics of anthracene crystals should exhibit singular behavior on the longwave-side of the exciton band bottom. Consider the reflection and transmission spectra recorded in anthracene at $T \leqslant 10K$. Figures 4 and 5 present a longwave part of the exciton reflection band recorded for the $[ab]$ - plane of the anthracene single-crystal plate $2\mu m$-thick, and the reflection spectrum $R(E)$ recorded at the $[20\bar{1}]$ - end of a thick crystal for $E \| b$-axis. As seen from Figure 4, the intensity of the light reflected by the $[ab]$-plane oscillates with variations of frequency due to the interference of beams multiply reflected by crystal surfaces inside the bulk. The interference pattern (nonmonotonic variation of the oscillation period and shapes of the lower and upper envelop curves) indicates a nonmonotonic frequency dependence of the optical characteristics of the crystal, such as the refraction index n and the damping constant κ. Applying the well-known Airey formula to the recorded spectra $R(E)^{*)}$, we calculated the values of $n(E)$ and $\kappa(E)$ which are given in Figure 4 for the case of a crystal with $d = 2\mu m$. These curves $n(E)$ and $\kappa(E)$ are seen to exhibit a singularity at $E_1 = 25077 cm^{-1}$ - the spike of n and the dip in the κ-curve. The corresponding maximum is observed in the transmission spectrum of the crystal $2\mu m$-thick at the same frequency (see Figure 4). Note that the nonmonotonic dependence of $R(E)$ and $T(E)$ caused by the singular behavior of $n(E)$ and $\kappa(E)$ is also observed in the spectra of the wedge-shaped crystal with $d = 2\mu m$, though the interference pattern became blurred here. At the same time we did not observe any singularity in the reflection spectrum $R(E)$ recorded at the end plane of a thick single crystal ($d = 100\mu m$) for the same polarization $E \| b$ (Figure 5). With increasing temperature, the singularity of $R(E)$ and $T(E)$ at $E_1 = 25077 cm^{-1}$ quickly disappears and becomes almost indiscernible at $10K$ (Figure 5). At the same time the emission intensity at the maximum of the 'acoustic' band drops markedly, though at such a temperature the high-energy tail of exciton luminescence is not affected by reabsorption of light.

Having analysed the whole wealth of the obtained data we inferred that, apparently, some deformation of the crystal lattice in such molecular crystals as anthracene should accompany the Frank-Condon relaxation of excitations and, thus, some bound states - the deforming exciton polaritons - might arise. Until now, to our knowledge, no theoretical model has been developed which could rigorously account for multiphonon scattering of polaritons and its effect on optical dispersion and emission spectra. Yet, we are inclined to believe that the self-trapping effects, though hindered by intense exciton-photon mixing, may be of importance in interpretation of our data. Such being the case, in crystal there should exist the exciton polaritons accompanied by lattice distortions rather than free exciton polaritons. The size of the distorted crystal area may be sufficiently great due to large values of the resonance interaction. We have roughly

*) It is known [3] that in anthracene, due to the growth of the frequencies of the local electronic transition at crystal boundary, there exist the subsurface exciton states with the energies $E_s > E_0$; they manifest themselves in an additional structure of $R(E)$ observed above E_0 (deep reflection dips arise in the reflection spectra, one of them is shown in Figure 5). In the frequency region below E_0 presence of such a subsurface transition layers leads to a monotonic decrease of R and n (see the dashed line in Figure 4).

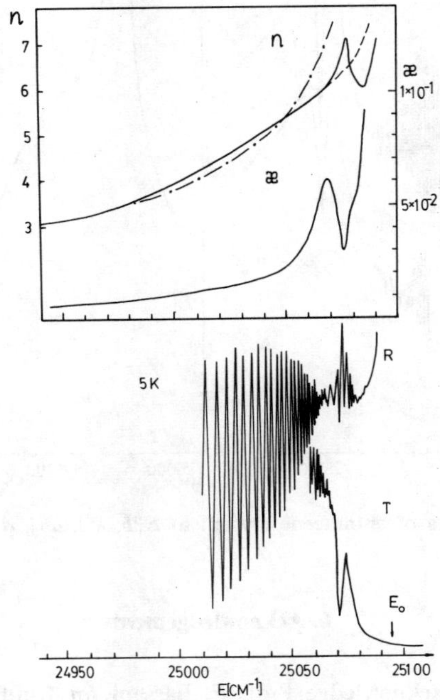

Figure 4. Reflection and transmission spectra of anthracene and the refractive index $n(E)$ and damping constant $\kappa(E)$ calculated according to the data [the dotted-dashed line presents the $n(E)$ dependence given by the dispersion formula for $\epsilon(E)$].

estimated this area ($\leqslant 2\mu m$) using the data on the sound velocity and the rise-time of the 'acoustic' luminescence band in anthracene. The corresponding estimation of the energy of phonons bound with such polaritons amounts to several fractions of $1 cm^{-1}$.

5. Conclusions

It has been shown for the first time that in the layered crystals PbI_2, HgI_2 and the molecular crystal of anthracene with the intermediate exciton-phonon coupling and short exciton lifetime, the secondary emission spectra exhibit a two-series structure. This peculiarity is accounted for on the basis of the hypothesis that there are two types of excitations in the 'soft' media, that is mixed light-exciton states responsible for the short-wave series of bands, and weakly bound states of exciton polaritons with lattice deformations giving rise to one more series of bands. Some features which have been revealed in low temperature spectra of transmission, reflection and emission of the anthracene crystals are also attributed to the existence of the exciton polaritons band with elastic deformation. The data obtained may be of importance to understand features of the spectra in the crystals with intermediate strength of exciton-phonon coupling and transitions from F and S states.

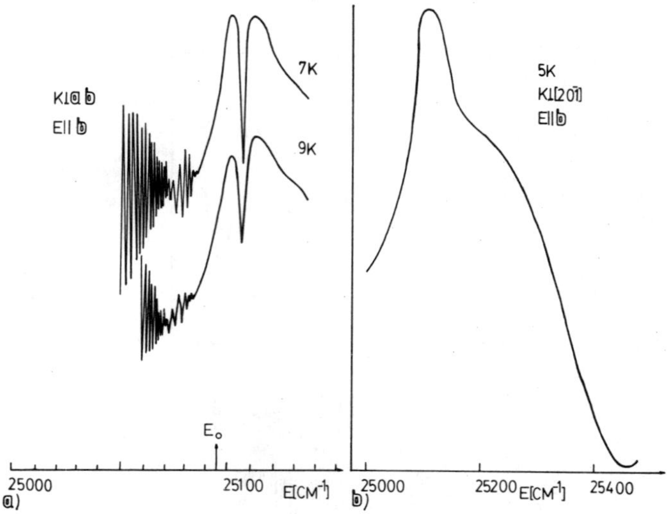

Figure 5. Reflection spectra of anthracene crystal: a) $E \| b$, $k \perp [ab]$, $d = 2\mu m$; b) $E \| b$, $k \perp [20\bar{1}]$, $d \cong 100\mu m$, $T \cong 5K$.

6. Acknowledgements

The authors gratefully acknowledge Prof. F. Bassani for fruitful discussions on topics dealt with in the paper while being in Kiev.

References

1. *Excitons*, ed. by E. Rashba and M. Sterge, North-Holland Publ. Co., Amsterdam, 1982.
2. M.S. Brodin, I.V. Blonskii, *Exciton Processes in Layered Crystals*, Naukova Dumka, Kiev, 1986.
3. M.S. Brodin, E.N. Myasnikov and S.V. Marisova, *Polaritoni v kristallooptike*, Naukova Dumka, Kiev, 1984.
4. I.S. Akopyan, B.M. Novicov and S.A. Permogorov, *Phys. Stat. Sol. (b)* **70**, 353 (1975).
5. T. Goto and Y. Nishina, *Phys. Rev. B* **17**, 4565 (1978).
6. E.F. Gross, S.A. Permogorov and B.S. Razbirin, *Uspekhi Fiz. Nauk* **103**, 431 (1971).
7. H. Sumi, *J. Phys. Soc. Japan* **41**, 526 (1976).
8. S.A. Permogorov, *Phys. Stat. Sol. (b)* **68**, 9 (1975).
9. I.V. Blonskii, *Ukr. Fiz. Zh.* **32**, 1461 (1987).
10. A. Sumi and H. Sumi, *Technical Report of ISSP, ser. A*, 1984, N 1478, pp. 1-36.
11. J. Singh and A. Matsui, *Phys. Rev. B* **36**, 6094 (1987).
12. V.M. Agranovich, B.P. Antonyuk and E.P. Ivanova, *JETP* **72**, 614 (1977).
13. A.S. Davydov and G.M. Pestryakov, Preprint ITP-81-127P.
14. V.M. Agranovich, *Teoriya eksitonov*, Izd. Nauka, Moskwa, 1968.
15. M.D. Galanin, E.N. Myasnikov and Sh.D. Khan-Magometova, *Isvestiya AN SSSR, ser. fiz.* **44**, 730 (1980).

16. M.D. Galanin and Sh.D. Khan-Magometova, *Solid State Commun.* **49**, 655 (1984).
17. N.A. Vidmont, A.A. Maksimov and I.I. Tartakovskii, *Fiz. Tverd. Tela* **24**, 1384 (1982).
18. J. Ferguson, *Molec. Cryst. Liq. Cryst.* **58**, 55 (1980).
19. S.V. Marisova, E.N. Myasnikov and A.N. Lipovchenko, *Phys. Stat. Sol. (b)* **115**, 649 (1983).
20. E.N. Myasnikov and G.V. Fomin, *Fizika nizkikh temperatur* **7**, 1053 (1981).

TWO-PARTICLE COULOMB SYSTEMS IN A MAGNETIC FIELD

A. BALDERESCHI* AND N. BINGGELI

Institut de Physique Appliquée
EPF-Lausanne, Switzerland

Dipartimento di Fisica Teorica e GNSM-CNR
Università di Trieste, Italy

The solution of two- and three-dimensional two-body Coulomb systems in a homogeneous magnetic field are investigated in the low- and intermediate-field regimes. In the two-dimensional case we give exact particular solutions for hydrogenic and two-electron systems. We also present results of variational calculations performed with a basis set which reproduces exactly the low field particular solutions as well as the asymptotic high-field behavior. For the three-dimensional case, we restrict ourselves to the problem of magneto-excitons and discuss particular features of the magneto-optical spectra obtained by means of variational calculations.

1. Introduction

The present article is dedicated to Professor Franco Bassani on the occasion of his sixtieth birthday. The topic has been chosen in order to stress the relevance of his contributions to the study of electron states and optical properties of solids. In addition, the investigation of two-particle Coulomb systems in an external magnetic field represents for both authors the first research experience in theoretical solid-state physics. The study of magneto-optical properties of free and bound Coulomb systems in solids was selected by Professor Bassani as the thesis subject for one of the present authors.

The topic we will discuss has been and still is of relevance in at least three fields: atomic physics, solid-state physics, and astrophysics. The spectroscopy of hydrogenic systems in magnetic fields has played an important role in the development of atomic physics since the end of last century. Very recently, renewed interest arose in connection with the problem of quasi-Landau levels in atomic spectra and the search for a quantum analog to classical chaos [1]. In solid state physics as well as astrophysics, the problem of hydrogenic systems in magnetic fields has received considerable attention since at least two decades [2,3]. In the context of astrophysics the Hydrogen atom has been studied especially in the Landau limit, beyond the first and second order Zeeman effect, to understand atomic processes near or on the surface of neutron stars (characteristic magnetic field $B \sim 10^7 - 10^9 T$) or white dwarfs ($B \sim 10^2 - 10^4 T$). In condensed matter physics, the analogous problem of an exciton in an external magnetic field is a subject of strong interest, since magnetospectroscopy of excitons in semiconductors, for example, allows one to determine accurately the fundamental band structure parameters [3]. The case of particles with identical charge is also of interest in condensed matter physics. In the last few years, in connection with the discovery of the fractional quantum hall effect [4], where correlation effects are important, particular attention has been given to the problem of two-dimensional systems consisting of a few electrons in a magnetic field [5].

Coulomb systems of great interest in today's condensed-matter physics correspond to

R. Girlanda et al. (eds.), Progress on Electron Properties of Solids, 197–209.

intermediate situations between the two-dimensional and the three-dimensional cases. This is especially true of excitons and substitutional impurities in strongly anisotropic materials, such as superlattices and quantum well systems. In the following pages, we consider some general properties of two-body Coulomb systems in a homogeneous magnetic field, and we present exact and variational solutions which include new results obtained for the hydrogenic systems and for the two-dimensional electronic system. In view of the extensive literature available on the subject, the present paper is not intended to be an exhaustive review of the state of the field.

2. Formulation of the Problem

The Hamiltonians of a hydrogenic system and a two-electron system in a homogeneous magnetic field **H** are given by the expression:

$$H = \frac{1}{2m_1}(\mathbf{p}_1 + \frac{|e|}{c}\mathbf{A}(r_1))^2 + \frac{1}{2m_2}(\mathbf{p}_2 \mp \frac{|e|}{c}\mathbf{A}(\mathbf{r}_2))^2 \mp \frac{e^2}{|\mathbf{r}_1 - \mathbf{r}_2|}, \tag{1}$$

where the upper signs pertain to the hydrogenic situation, the lower signs to the system of like charges, and m_1, $-|e|$, $\mathbf{r}_1$,$\mathbf{p}_1$ and m_2, $\pm|e|\mathbf{r}_2$,$\mathbf{p}_2$ are the mass, charge, position and momentum of the electron and ion or of the two electrons. $\mathbf{A}(\mathbf{r})$ is the vector potential and we choose the symmetric gauge:

$$\mathbf{A}(\mathbf{r}) = \frac{1}{2}\mathbf{H}\wedge\mathbf{r}. \tag{2}$$

Spin dependent terms are not included in (1). The well known Zeeman contribution due to the Pauli term may be simply added to the eigenvalues of (1). For excitons the electron-hole exchange term should be added to eq. (1) in those cases where the singlet-triplet spitting is comparable to or larger than the magnetic energies involved.

We change to center of gravity **R**,**P** and relative **r**,**p** coordinates:

$$\begin{cases} \mathbf{R} = \dfrac{m_1\mathbf{r}_1 + m_2\mathbf{r}_2}{M} \\ \mathbf{r} = \mathbf{r}_1 - \mathbf{r}_2, \end{cases}$$

and introduce the total and reduced masses of the system $M = m_1 + m_2$ and $\mu = \dfrac{m_1 \cdot m_2}{M}$.

Considering first the case of particles of opposite charges, we write the solutions of (1) as:

$$\Phi(\mathbf{r},\mathbf{R}) = e^{i(\mathbf{K} + \frac{|e|}{2\hbar c}\mathbf{H}\wedge\mathbf{r})\mathbf{R}}\,\Psi(\mathbf{r}), \tag{3}$$

since (3) is the general eigenfunction of the constant of motion [6]:

$$\mathbf{P}_0 = \mathbf{P} - \frac{|e|}{2c}\mathbf{H}\wedge\mathbf{r}. \tag{4}$$

Inserting the general solution (3) in the eigenvalue problem associated with the Hamiltonian (1), one finds that the solutions $\Psi(\mathbf{r})$ for the relative motion of the two particles are given by:

$$\{\frac{\mathbf{p}^2}{2\mu} + \frac{|e|}{2c}\left[\frac{1}{m_e} - \frac{1}{m_i}\right]\mathbf{L}\cdot\mathbf{H} + \frac{e^2}{8\mu c^2}(\mathbf{H}\wedge\mathbf{r})^2 - \frac{e^2}{r} \tag{5}$$

$$+\frac{\hbar^2\mathbf{K}^2}{2M}+\frac{|e|\hbar}{Mc}(\mathbf{K}\wedge\mathbf{H})\cdot\mathbf{r}\}\Psi(\mathbf{r}) = E\Psi(\mathbf{r}).$$

where subscripts e and i characterize electron and ion parameters, respectively. The first four terms on the left side of (5) can be interpreted as, respectively, the kinetic energy operator for the relative motion, the Zeeman and diamagnetic operators and the electron-ion interaction. The last two terms depend on the translational motion of the center of mass, and correspond, respectively, to the kinetic energy of the overall atomic system and a coupling term equivalent to the effect of an external electric field $\mathbf{E} = \frac{\hbar}{Mc}\mathbf{K}\wedge\mathbf{H}$ on the two-particle system. For excitonic levels observed in optical absorption, experimental evidence of the translational motion of the excitonic system with the photon wave-vector $\mathbf{K}$ has been given by Thomas and Hopfield who balanced the effect of the coupling term with the Stark effect induced by an applied external electric field [7]. The coupling term vanishes in the Faraday geometry and is generally negligible relative to the Zeeman and diamagnetic terms for an exciton system due to the small value of the momentum K. In what follows, we will not consider this term, and we will assume $K = 0$. Without the coupling term, Hamiltonian (5) is invariant under rotations about the magnetic field, and its solutions can be written in cylindrical coordinates ($\mathbf{H}_{/\!/}\epsilon_z$) as:

$$\Psi(r) = R(\rho,z)e^{im\varphi}. \tag{6}$$

The spectra corresponding to the internal Hamiltonian (5) for a two- and three-dimensional hydrogenic system are analyzed for $K = 0$ in sections 3 and 4, respectively.

Contrary to the case of Hydrogen-like centers, the center of mass and internal motions are independent for the system of identical particles:

$$H = H_{CM}(\mathbf{R},\mathbf{P})+H_{Int}(\mathbf{r},\mathbf{p}), \tag{7}$$

where the Hamiltonians for the center of mass and relative motions of the two-electron system are given by, respectively:

$$H_{CM} = \frac{1}{4m_e}(\mathbf{P}+\frac{2|e|}{c}\mathbf{A}(\mathbf{R}))^2 \tag{8}$$

$$H_{Int} = \frac{1}{m_e}(\mathbf{p}+\frac{|e|}{2c}\mathbf{A}(\mathbf{r}))^2+\frac{e^2}{r}. \tag{9}$$

The Hamiltonian (8) describes a free particle of mass $2m_e$ and charge $-2|e|$ in the homogeneous magnetic field H. The solutions of (8) for the translational motion of the system are Landau states corresponding to the classical electronic cyclotron frequency $\omega = \frac{|e|H}{m_e c}$ and the magnetic length $a_0^{CM} = \frac{1}{\sqrt{2}}a_0$, where $a_0 = \sqrt{\frac{\hbar c}{|e|H}}$ is the single electron magnetic length. The total energy is given by $E^{tot} = (N+\frac{1}{2})\hbar\omega+E$, where N is the Landau quantum number, and E is the energy associated with the internal degrees of freedom. The Hamiltonian (9) for the internal motion is identical to that of a free particle of mass $m_e/2$ and charge $-|e|/2$ in a magnetic field H and a repulsive Coulomb potential $\frac{e^2}{r}$. The magnetic field produces thus a qualitative change in the internal motion of the Coulomb system, since the field confines the particles, which would otherwise repel each other, around the center of gravity. The eigenstates of (9) are also eigenstates of the operator L_z, and the general form (6) can be used for the wavefunctions of the relative motion. The spectrum associated with Hamiltonian (9) for a two-dimensional system, is discussed in the following section.

3. Two-Dimensional Systems

In this section, we consider the case of an extremely anisotropic material where the two particles are confined in a plane perpendicular to the magnetic field. We study the internal motion of the two-dimensional Hydrogen atom (see eq. (5)) with $K = 0$ as well as of the two-dimensional electron-pair system (see eq. (9)) in a magnetic field of arbitrary strength. We note that pure two-dimensional Coulomb systems contain a logarithmic potential whereas we use a $1/\rho$ potential. As mentioned before, the physical situation is generally intermediate between the three- and the two-dimensional cases, and the system we are considering here is a useful limiting case.

The two-dimensional limit was considered previously by Akimoto and Hasegawa using the semi-classical *WKB* approach [8], by Shinada and Tanaka who integrated numerically the radial Schrödinger equation [9], and by MacDonald and Ritchie who used a two-dimensional Padé approximant to interpolate between the low and large magnetic field perturbative expansions of the eigenvalues [10].

We first look for exact special solutions to the problem of the radial motion. Our approach is then variational. We choose a basis set which is flexible enough to reproduce the pure Landau states, the exact special solutions at intermediate fields, and yields the correct behavior of the eigenstates at small radii, where the Coulomb potential dominates.

For the hydrogenic system, we choose units of length and energy given by the effective Bohr radius $a_B' = \frac{\hbar^2}{\mu e^2}$ and the effective Rydberg $R_y' = \hbar^2/2\mu a_B'^2$. In these units, and with the form (6) for the wave-functions, the equation for the radial motion is:

$$\left\{-\frac{d^2}{d\rho^2}-\frac{1}{\rho}\frac{d}{d\rho}+\frac{m^2}{\rho^2}+\alpha\gamma m+\frac{\gamma^2}{4}\rho^2-\frac{2}{\rho}\right\}R(\rho) = ER(\rho) \tag{10}$$

where we have introduced $\alpha = \frac{m_i - m_e}{M}$, and the parameter $\gamma = \frac{1}{2}\hbar\frac{|e|H}{\mu c}/R_y' = \frac{\hbar^3 H}{\mu^2|e|^3 c}$, proportional to the field and corresponding to the ratio between cyclotron and Rydberg energies.

For the internal Hamiltonian of the two-electron system, the relevant units are the magnetic units of length $a_0' = \sqrt{2}a_0$ and energy $E_H = \frac{1}{2}\hbar\omega$ for a free particle of charge $-\frac{|e|}{2}$ and mass $\frac{m_e}{2}$. Using the wave-function (6), the radial motion is described by:

$$\left\{-\frac{d^2}{d\rho^2}-\frac{1}{\rho}\frac{d}{d\rho}+\frac{m^2}{\rho^2}+m+\frac{\rho^2}{4}+\sqrt{\frac{2}{\gamma}}\frac{1}{\rho}\right\}R(\rho) = ER(\rho), \tag{11}$$

where $\gamma = \frac{\hbar^3 H}{m_e^2|e|^3 c}$. In order to have an equation similar to (10), and to solve in parallel the problem of internal motion of the hydrogenic and electronic systems, we choose new units $a_0^* = \sqrt{\gamma}a_0'$ and $E_0^* = E_H/\gamma = Ry$. In these units eq. (11) becomes:

$$\left\{-\frac{d^2}{d\rho^2}-\frac{1}{\rho}\frac{d}{d\rho}+\frac{m^2}{\rho^2}+\gamma m+\frac{\gamma^2}{4}\rho^2+\frac{\sqrt{2}}{\rho}\right\}R(\rho) = ER(\rho). \tag{12}$$

For the determination of analytical solutions to (10) and (12), we propose to study the class of trial functions:

$$R(\rho) = \sum_{i=0}^{I} C_i \rho^{|m|+i} e^{-\frac{\gamma}{4}\rho^2}, \qquad (13)$$

which at long range correctly behave as states confined in a magnetic field. Inserting (13) in equation (10) or (12), we are left with the following equation for the coefficients C_i:

$$\sum_{i=0}^{I} C_i\{i(i+2|m|)\rho^{i-2}+\beta\rho^{i-1}+(E-\gamma(i+|m|+\alpha m+1))\rho^i\} = 0 \qquad (14)$$

where $\beta = 2$ and $0 \leqslant \alpha \leqslant 1$ for a hydrogenic system, while $\beta = -\sqrt{2}$ and $\alpha = 1$ for the two-electron system. The solutions are given by the recursion relations:

$$\begin{cases} C_1 = -\dfrac{\beta}{2|m|+1} C_0 \\ C_{i+2} = -\dfrac{\beta}{(i+2)(i+2|m|+2)} C_{i+1} + \\ \qquad \dfrac{\gamma(i+|m|+\alpha m+1)-E}{(i+2)(i+2|m|+2)} C_i, \qquad i \geqslant 0. \end{cases} \qquad (15)$$

Imposing that the summation (13) reduces to a finite number of terms, i.e. $C_{I+1} = C_{I+2} = 0$, we obtain the conditions:

$$\begin{cases} C_I = -\dfrac{\gamma}{\beta} C_{I-1} \\ E = (I+|m|+\alpha m+1)\gamma \end{cases} \qquad (16)$$

which define the exact special solutions $E(\gamma_j)$, $R_{\gamma j}(\rho)$ for (10) and (12). The number of solutions for each I is $I/2$ for even I, and $(I+1)/2$ for odd I.

We consider here, as an example, exciton states with $m = 0$, which are observable in one-photon absorption spectra since the transition probability, proportional to $|\Psi(0)|^2$, does not vanish for these states. The special solutions with lowest I are:

$$\begin{aligned} &I = 1:\ \gamma = 4 && E = 8 && R(\rho) = C_0(1-2\rho)e^{-\rho^2} \\ &I = 2:\ \gamma = \frac{2}{3} && E = 2 && R(\rho) = C_0\left(1-2\rho+\frac{2}{3}\rho^2\right)e^{-\frac{1}{6}\rho^2} \\ &I = 3:\ \gamma = \frac{4(10\pm\sqrt{73})}{27} && E = \frac{16(10\pm\sqrt{73})}{27} && R(\rho) = C_0\left(1-2\rho-\frac{1\pm\sqrt{73}}{9}\rho^2+\right. \\ &&&&& \left.\frac{2(83\pm11\sqrt{73})}{243}\rho^3\right)e^{-\frac{10\pm\sqrt{73}}{27}\rho^2} \\ &I = 4:\ \gamma = \frac{25\pm3\sqrt{33}}{82} && E = \frac{5(25\pm3\sqrt{33})}{82} && R(\rho) = C_0\left(1-2\rho+\frac{57\mp3\sqrt{33}}{82}\rho^2+\right. \\ &&&&& \left.\frac{2(3\pm2\sqrt{33})}{123}\rho^3-\frac{273\pm59\sqrt{33}}{10086}\rho^4\right)e^{-\frac{25\pm3\sqrt{33}}{328}\rho^2} \end{aligned} \qquad (17)$$

States with higher I can be found by determining numerically the special values γ_j which fulfill the condition $C_{I+1} = 0$, given the recursion relations (15) and the condition on the energy (16).

The lowest exact energy levels with $m = 0$ for the exciton system, and with

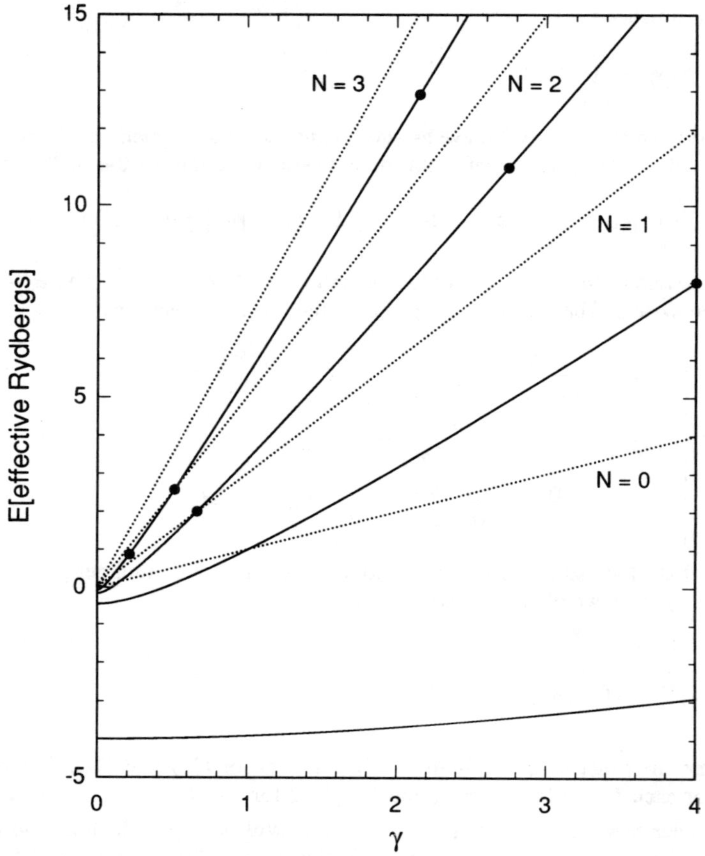

Figure 1.Lowest four energy levels with $m = 0$ of the two-dimensional exciton as a function of the external magnetic field. The solid lines correspond to the results of the variational calculations, and the dots to the exact solutions discussed in the text. For convenience, the corresponding free Landau levels are also given (dotted lines).

$m = 0, \pm 1$ for the two-electron system are shown by dots in Figures 1 and 2, respectively. The spins contribute an additional Zeeman term $(g_e S_{ez} + g_{i(e)} S_{i(e)z})\mu_B H$ to the magneto exciton (two-electron) energy levels, and in the case of the two-electron system, due to the Pauli principle, the levels with even (odd) m have to be associated with the singlet (triplet) total spin state, respectively. The solid lines in Figure 1, and solid, dashed and dashed-dotted lines in Figure 2 are the result of variational calculations that will be discussed in detail in the following part of this section. Also shown in Figures 1 and 2 are the first Landau levels (dotted lines) $E = (2N+1)\gamma Ry^{(*)}$, $N = 0,1,2$ (Figure 2) and $N = 0$ to 3 (Figure 1) corresponding to the limit of vanishing Coulomb potential. We note that since the special solutions are found on the straight lines corresponding to $E = (I + |m| + m + 1)\gamma Ry^{(*)}$, some of them occur at the intersection of the curves representing the Landau levels and the true levels of the system.

The variational calculations have been performed with the basis functions $\rho^{|m|+i} e^{-\frac{\gamma}{4}\rho^2}$ with $i = 0$ to 15 for $0 < \gamma < 100$. Inclusion of the odd parity values of i in

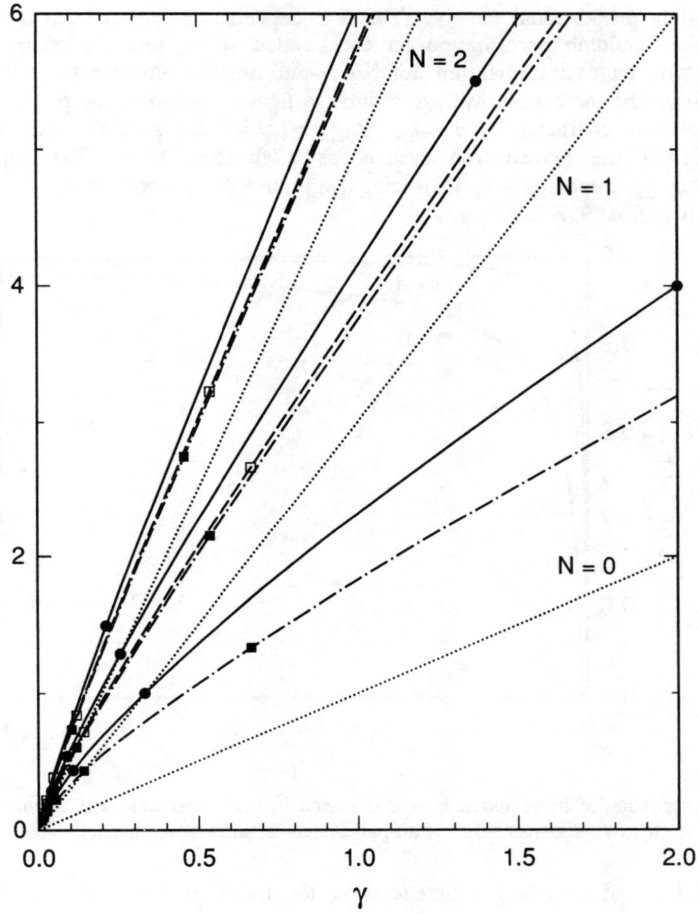

Figure 2. Energy levels with $m = 0, \pm 1$ corresponding to the adiabatic quantum number $N = 0, 1, 2$ for the two-dimensional electron-pair system in an external magnetic field. The solid ($m = 0$), dashed ($m = 1$) and dashed-dotted ($m = -1$) lines are the results of the variational calculations, while the dots ($m = 0$), empty squares ($m = 1$) and filled squares ($m = -1$) are the exact solutions discussed in the text. The states associated with a given adiabatic quantum number N have $m \leqslant N$, and the corresponding energy levels converge for $m \to -\infty$ to the free Landau level (dotted line).

the magnetic basis set has a strong impact on convergence, especially in the low field regime. For the two-electron system, convergence tests for $\gamma = 1$ show that 0.1% accuracy on the ground state is obtained with a basis only formed by the two functions with $i = 0$ and 1, while seventeen basis functions with lowest even powers are required to obtain the same accuracy. In the case of the exciton system for $\gamma = 1$, an accuracy of 0.1% is reached with the lowest 7 functions with even or odd powers while 28 functions with only even powers do not even insure an accuracy of 10%.

Figures 1 and 2 illustrate the transition between the low and high field regimes. In the high field regime, due to the Coulomb term, the energy levels are below (hydrogenic systems) or above (two-electron system) the corresponding Landau level by a quantity approximately proportional to $\sqrt{\gamma}$. The $\sqrt{\gamma}$ dependence derives from the first order term in the Coulomb perturbation on the Landau states, and the related correction decreases with increasing quantum number N and angular momentum $|m|$ (Figures 1 and 2) due to the increase in average radius. In figure 3 we show the proportion of free Landau $N = 0$ character $|<\psi_{N=0,m}|\Psi_{GS,m}>|^2$ in the ground state $\Psi_{GS,m}$ with $m = 0, -1$ of the two-electron system, as a function of γ. The approximation $\Psi_{GS,m} \approx \Psi_{N=0,m}$ improves with increasing $|m|$, and the proportion of $N = 0$ Landau character is at least 95% for $\gamma > 4$.

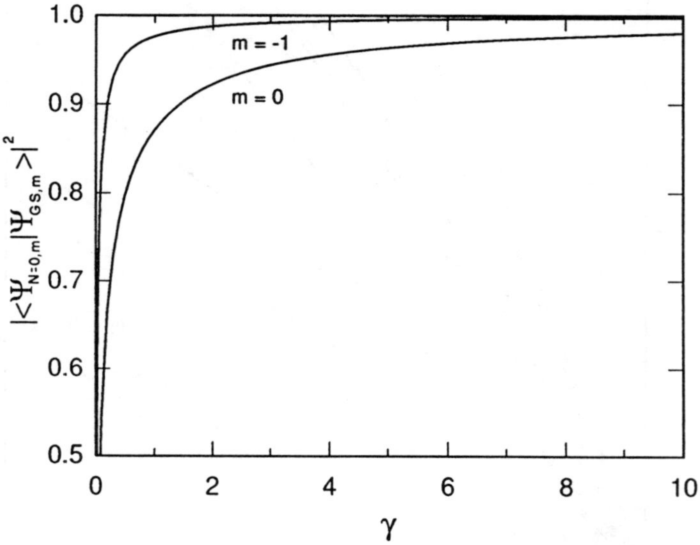

Figure 3. Proportion of free Landau $N = 0$ character in the lowest state with angular momentum $m = 0$, -1 for the two-dimensional electron-pair system in an external magnetic field.

In the limit of vanishing magnetic field, the levels in Figure 1 tend to the two-dimensional Hydrogen limit $E_n = 1/(n+\frac{1}{2})^2$. When the magnetic field is increased, the eigenfunctions contract by amounts that depend on their average radius. The ground state $m = 0$ is almost unaffected in the range $0<\gamma<1$. Only for higher magnetic fields, when the magnetic length is comparable to or smaller than the Bohr radius, the ground state level changes to follow asymptotically the first Landau level. The shrinking of the hydrogenic orbitals is also illustrated in Figure 4, where we show the relative oscillator strength of the first four optically allowed exciton states as a function of γ. The magnetic contraction of the extended hydrogenic states with $n>0$ occurs for $\gamma \ll 1$, and the oscillator strength of such states is already enhanced by at least one order of magnitude at $\gamma = 1$.

Concerning the use of semiclassical *WKB* approximation for the two-dimensional magneto-exciton, we note that the *WKB* $m = 0$ spectrum obtained by Akimoto and Hasegawa [8] is in good agreement with our results of Figure 1, except for the ground state which, according to our results, should shift to higher energy more rapidly as a function of γ. We find, in particular, that the ground state energy vanishes at $\gamma \approx 9.4$, while Akimoto and Hasegawa predict that the crossing with the $E = 0$ line occurs at $\gamma \approx 12.5$.

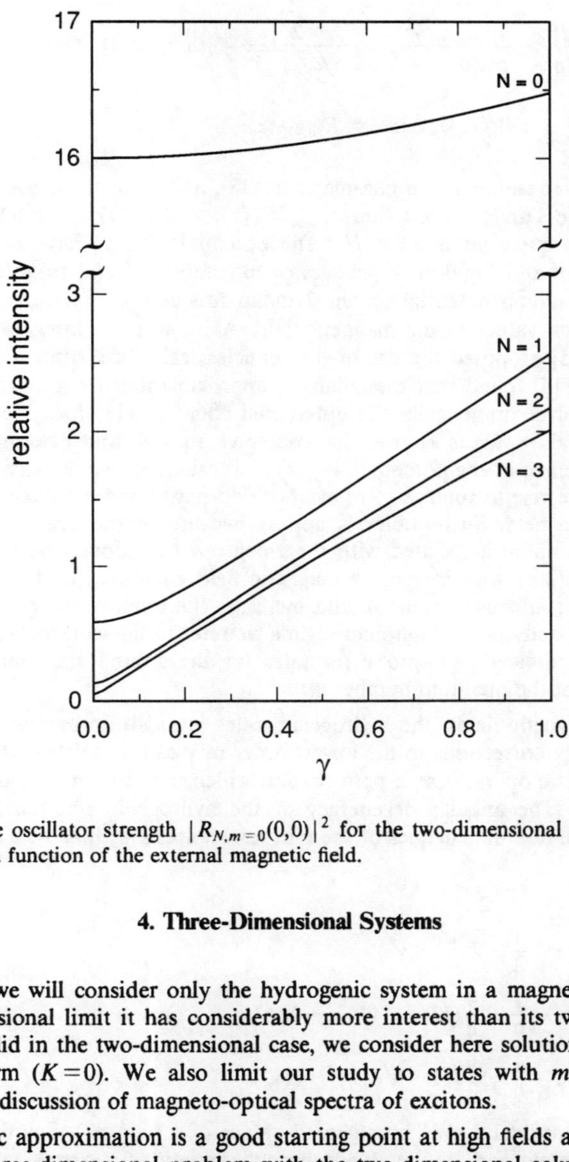

Figure 4. Relative oscillator strength $|R_{N,m=0}(0,0)|^2$ for the two-dimensional exciton states with $N = 0,1,2,3$, as a function of the external magnetic field.

4. Three-Dimensional Systems

In this section we will consider only the hydrogenic system in a magnetic field, since in the three-dimensional limit it has considerably more interest than its two-electron counterpart. As we did in the two-dimensional case, we consider here solutions of eq. (5) with no coupling term ($K=0$). We also limit our study to states with $m = 0$, which are relevant for the discussion of magneto-optical spectra of excitons.

The adiabatic approximation is a good starting point at high fields and allows one to approach the three-dimensional problem with the two-dimensional solutions in a model potential obtained by correcting the Coulomb potential for the third dimension. This approach goes back to the early work of Schiff and Snyder [11], and was used later on by Elliott and Loudon [12,13]. It allows one to obtain accurate solutions in the large to intermediate field regime ($\gamma n^3 > 1$), where the high frequency motion in the plane perpendicular to the field is well separated from the motion parallel to the field [14]. In this regime, writing the wave-function as $\Psi(r) = f(\rho,z)g(z)$, inserting it into (5) and using the adiabatic approximations $\frac{d}{dz}f(\rho,z) = \frac{d^2}{dz^2}f(\rho,z) = 0$, yields the two coupled equations:

$$\left\{-\frac{d^2}{d\rho^2}-\frac{1}{\rho}\frac{d}{d\rho}+\frac{\gamma^2}{4}\rho^2-\frac{2}{\sqrt{\rho^2+z^2}}\right\}f_N(\rho,z) = W_N(z)f_N(\rho,z) \tag{18}$$

$$\left\{-\frac{d^2}{dz^2}+W_N(z)\right\}g_{Ni}(z) = E_{Ni}g_{Ni}(z). \tag{19}$$

The z-position is considered as a parameter in (18), and since the eigenvalues for large z correspond to the Landau levels ($\lim_{|z|\to\infty} W_N(z) = (2N+1)\gamma$), we label the solutions with the Landau quantum number N. The potentials $W_N(z)$ have been computed in many ways. Elliott and Loudon [13] suggested to compute $W_N(z)$ from first order perturbation in the Coulomb potential on the Landau functions. The treatment, however, is valid only for high values of the magnetic field. Akimoto and Hasegawa [8] and Zhilich and Monozon [15] proposed the use of the semiclassical *WKB* approximation. Baldereschi and Bassani [14] found that the adiabatic approximation has a much larger range of validity if one solves numerically the differential equation (18) for different values of z. Once the potential $W_N(z)$ is known, one can solve eq. (19) and determine the magnetic levels E_{Ni}. The effect of the potentials $W_N(z)$ is to raise the two-dimensional spectrum of Figure 1, and give rise to sublevels i of even or odd parity under reflexion in a plane perpendicular to the field. Such sublevels appear because of the quantization of the one dimensional continuum associated with the z-degree of freedom. The localization of the wave-function $f_N(\rho,z)$ with increasing magnetic field enhances the Coulomb term. This produces a deeper effective potential, and increases the binding energies $(2N+1)\gamma-E_{Ni}$. For an extended study of the high field regime we refer to the work of Avron, Herbst and Simon [16], who derived asymptotic formulas for the ground state binding energy for arbitrary values of the quantum number m.

For small magnetic fields, the hydrogenic states are split and shifted by the diamagnetic term. Energy corrections to the lowest order in γ can be calculated [11,17] by treating the diamagnetic operator as a perturbation which is not strong enough to mix states with different n. The angular degeneracy of the hydrogenic states is removed by the diamagnetic term, and the energies of the four lowest states with $m = 0$ are given by:

$$E_1 = -1+\frac{1}{2}\gamma^2+0(\gamma^4)$$

$$E_2 = -\frac{1}{4}+7\gamma^2+0(\gamma^4)$$

$$E_3 = -\frac{1}{9}+\frac{9}{4}(11-\sqrt{41})\gamma^2+0(\gamma^4)$$

$$E_4 = -\frac{1}{9}+\frac{9}{4}(11+\sqrt{41})\gamma^2+0(\gamma^4).$$

Such expansions are valid only up to small values of γ ($\gamma n^3 \approx 1$). Solutions in the intermediate field regime can be obtained through variational, expansion-perturbation or modified perturbation methods. Results for the fundamental state have been obtained by a number of authors [18] using various trial functions. Variational solutions and solutions obtained by numerical integrations are also available for a large number of excited states and for any given strength of the magnetic field [19]. Analytical approximations to the ground state for arbitrary momentum m have also been obtained recently through a modified perturbation method including variational ideas [20]. In Figure 5 we present the relative oscillator strength (a) and energy levels (b) of the first four optically active exciton states calculated with the Gaussian basis set given in ref. [21]. As opposed to the two-dimensional situation, in the three-dimensional case the oscillator strength of the excited states does not increase monotonically as a function of γ. This particular behavior of the oscillator strength is illustrated in Figure 5a and is due to the anticrossings of the energy levels.

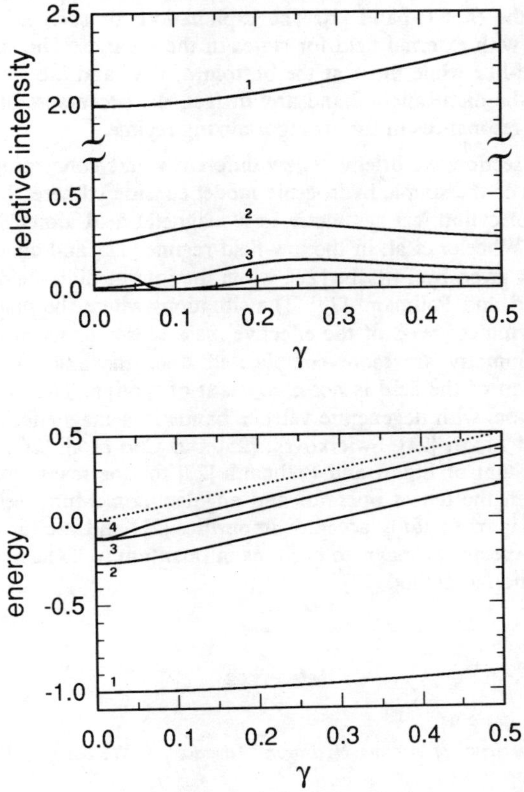

Figure 5. (a) Relative oscillator strength, and (b) energy levels for the four lowest even-parity $m = 0$ exciton states, as a function of the external magnetic field. The dotted line in (b) corresponds to the Landau level $N = 0$. Energies are given in units of the effective Rydberg.

For the ground state, which is well separated in energy from the excited states, no such effect appears and the oscillator strength is continuously enhanced by the magnetic field.

A problem of practical interest is the connection between the hydrogenic levels at low fields and the adiabatic levels for high magnetic fields. Elliott and Loudon [13] suggested to connect the levels according to the nodal surfaces of the wave-function. However, no solid argument was presented to support this conjecture according to which states of the same symmetry would have to cross for some values of the magnetic field. All bound exciton levels calculated till now [14,19] join the sublevels of the lowest Landau state ($N=0$) without any crossing. According to the non-crossing rule, the quantum number correspondence for the lowest four exciton states is:

nl	$\rightarrow N$	i
$1s$	0	0
$2s$	0	2
$[(3s),(3d)]_1$	0	4
$[(3s),(3d)]_2$	0	6

Recent numerical studies of highly excited states show the presence of exponentially small anticrossings, which scale as e^{-2n}, between levels of the top of the exciton n-band and the bottom of the $(n+1)$-band [1]. The explanation for such a behavior is in the change of symmetry with external field for states in the n-bands. The states at the top of a band are rotational-like while those at the bottom of the band are vibrational-like. The states of the top of the diamagnetic band are, in fact, the precursors of those giving rise to the quasi-Landau resonances in the strong n-mixing regime.

To conclude this section, we briefly review different works concerning exciton systems which are extensions of the simple hydrogenic model considered here. The exciton system with a mass tensor of cylindrical symmetry in a magnetic field along the symmetry axis has been studied by Wheeler et al. in the low field regime [17], and adiabatic solutions at high fields have been given by Fritsche [22], while the intermediate field regime has been investigated by Lipari and Pollmann [23]. The situations where the magnetic field is not oriented along the symmetry axis of the effective mass tensor or when the effective mass tensor has lower symmetry are more complicated since the angular momentum component in the direction of the field is not a constant of motion. The problem of excitons in cubic semiconductors with degenerate valence bands in a magnetic field has been studied by Altarelli and Lipari [24], Swierkovski [25] and Cho et al. [26]. Their variational approach, as well as that of Lipari and Pollmann [23] for the mass tensor of cylindrical symmetry, is based on the tensor operator and angular momentum techniques proposed by Baldereschi and Lipari to study acceptor impurities [27] and excitons [28] in semiconductors. Finally, the spectra of magneto excitons in quantum wells have also been studied recently by the variational method [29].

References

1. J.C. Gray, *The Spectrum of Atomic Hydrogen, Advances*, G.W. Series, ed. (World Scientific, Singapore, 1988), pp. 369.
2. For reviews on excitons in external magnetic fields see: H. Hasegawa in *Physics of Solids in Intense Magnetic Fields*, E.D. Haidemanakis, ed. (Plenum, New York, 1969), pp. 246; F. Bassani and A. Baldereschi, *Surface Science* **37**, 304 (1973); H.J. Zeiger and G.W. Pratt, *Magnetic Interactions in Solids* (Clarendon Press, Oxford, 1973) Chap. 7.
In the context of astrophysics see: R.H. Garstang, *Rep. Prog. Phys.* **40**, 105 (1977); G. Wunner, H. Herold and H. Ruder, *J. Phys. B* **16**, 2937 (1983), and references therein.
3. D. Bimberg, *Festkörperprobleme XVII, Adv. in Solid State Physics*, J. Treusch, ed. (Pergamon-Vieweg, Braunschweig, 1977) pp. 195.
4. D.C. Tsui, H.L. Störmer and A.C. Gossard, *Phys. Rev. Lett.* **48**, 1559 (1982).
5. R.B. Laughlin, *Phys. Rev. B* **27**, 3383 (1983).
6. G. Wunner and H. Ruder, *J. Physique Coll. C2 Suppl. 11* **43**, 137 (1980).
7. D.G. Thomas and J.J. Hopfield, *Phys. Rev.* **124**, 657 (1961).
8. O. Akimoto and H. Hasegawa, *J. Phys. Soc. Japan* **22**, 181 (1967).
9. M. Shinada and K. Tanaka, *J. Phys. Soc. Japan* **29**, 1258 (1970).
10. A.H. MacDonald and D.S. Ritchie, *Phys. Rev. B* **33**, 8336 (1986).
11. L.I. Schiff and H. Snyder, *Phys. Rev.* **55**, 59 (1939).
12. R.J. Elliott and R. Loudon, *J. Phys. Chem. Solids* **8**, 382 (1959).
13. R.J. Elliott and R. Loudon, *J. Phys. Chem. Solids* **15**, 196 (1960).
14. A. Baldereschi and F. Bassani, Proc. X Int. Conf. Phys. Semiconductors, (Cambridge

Massachusetts, 1970), pp. 191.

15. A.G. Zhilich and B.S. Monozon, *Fiz. Tverd. Tela* **8**, 3559 (1966) (English translation *Sov. Phys. Solid State* **8**, 2846 (1967).)
16. J.E. Avron, I.W. Herbst and B. Simon, *Phys. Rev. A* **20**, 2287 (1979).
17. R.G. Wheeler and J.O. Dimmock, *Phys. Rev.* **125**, 1805 (1962); J.C. Miklosz and R.G. Wheeler, *Phys. Rev.* **153**, 913 (1967).
18. Y. Yafet, R.W. Keyes and E.N. Adams, *J. Phys. Chem. Solids* **1**, 137 (1956); E.P. Pokatilov and M.M. Rusanov, *Fiz. Tverd. Tela* **10**, 3117 (1968) (English translation *Sov. Phys. Solid State* **10**, 2458 (1969)); D.M. Larsen, *J. Phys. Chem. Solids* **29**, 271 (1968); D. Cabib, E. Fabri and G. Fiorio, *Solid State Commun.* **9**, 1517 (1971); W. Ekardt, *Solid State Commun.* **16**, 233 (1975); M. Cohen and S. Kais, *J. Phys. B* **17**, 3049 (1984); J.A.C. Gallas, *J. Phys. B* **18**, 2199 (1985); P.C. Rech, M.R. Gallas and J.A.C. Gallas, *J. Phys. B* **19**, L215 (1986).
19. D. Cabib, E. Fabri and G. Fiorio, *Nuovo Cimento* **10B**, 185 (1972); N. Lee, D.M. Larsen and B. Lax, *J. Phys. Chem. Solids* **34**, 1059 (1973); Y. Chen, B. Gil and H. Mathieu, *Phys. Rev. B* **34**, 6912 (1986).
20. B. Gerlach, D. Richter and J. Pollmann, *Z. Phys. B* **66**, 419 (1987).
21. A. Baldereschi, *Theoretical Aspects on New Developments in Magneto-Optics*, J.T. Devreese, ed. (Plenum, New York, 1980), pp. 291.
22. L. Fritsche, *Phys. Stat. Sol.* **34**, 195 (1969); L. Fritsche and F.D. Heidt, *Phys. Stat. Sol.* **35**, 987 (1969).
23. N.O. Lipari and J. Pollmann, *Inst. Phys. Conf. Ser.* **43**, 1097 (1979).
24. M. Altarelli and N.O. Lipari, *Phys. Rev. B* **37**, 3798 (1973); *Phys. Rev. B* **9**, 1733 (1974); N.O. Lipari and M. Altarelli, *Solid State Commun.* **33**, 47 (1980).
25. L. Swierkovski, *Phys. Rev. B* **10**, 3311 (1974); *Nuovo Cimento B* **29**, 340 (1975).
26. K. Cho, S. Suga, W. Dreybrodt and F. Willmann, *Phys. Rev. B* **1**, 1512 (1975); K. Cho, W. Dreybrodt, P. Hiesinger, S. Suga and F. Willmann, Proc. XII Int. Conf. Phys. Semiconductors, Stuttgart 1974, pp. 945.
27. A. Baldereschi and N.O. Lipari, *Phys. Rev. B* **3**, 439 (1971); N.O. Lipari and A. Baldereschi, *Phys. Rev. B* **5**, 2497 (1971); *Phys. Rev. B* **6**, 3764 (1972).
28. A. Baldereschi and N.O. Lipari, *Phys. Rev. B* **8**, 2697 (1973); *Phys. Rev. B* **9**, 1525 (1974); N.O. Lipari, *Solid State Commun.* **25**, 665 (1978).
29. See for example, G.E.W. Bauer and T. Ando, *Phys. Rev. B* **37**, 3130 (1988).

A SIMPLE APPROACH FOR ESTIMATING AUGER DECAY RATES AND RESONANCE ENERGIES IN MOLECULES

R. COLLE AND S. SIMONUCCI

Scuola Normale Superiore
56100 Pisa
Italy

A new and simpler version of the method for calculating Auger decay rates, recently proposed by the authors in [1-2], is presented here. Decay rates and resonance energies are calculated for the Auger process involving the *LiF* molecule ionized in its deepest shell.

1. Introduction

Recently there has been a remarkable revival of interest in the Auger effect and nonradiative decay processes, because of the relevant technical developments, which allow both the selective preparation of the initial - ionized or highly excited - state of the system and make also possible accurate measurements of fluorescence and electronic yields [3-7]. The Auger spectra are increasingly used as a tool in the investigation of solid surfaces, in the analysis of surface impurities and the Auger transitions are also important for the understanding of mesic atoms, nuclear internal conversion and orbital capture of electrons, ion-atom collisions etc. Furthermore the Auger effect, involving as it does both discrete and continuous states, can be used as a good testing ground for various theoretical models, for the simplifications resulting from them and for the corresponding wave functions.

From the standpoint of theory, the Auger effect is an example of a resonant scattering process, the analysis of which has usually been developed within the framework of the following two approximations [8]: 1) neglect of interactions between the primary electron and the Auger electron; 2) complete independence of the phenomena of electron emission and photon emission. This means that the Auger effect is interpreted as a two-step process, where the decay is treated separately from the initial ionization and the resonance originates from the interaction between the quasi-bound state of the initial ion and the continua of the doubly ionized target, assumptions that are in general appropriate for the usual systems where the resonances are widely spaced and not too close to the double ionization threshold. For a more general approach, that takes into account also the interaction between the initial and final ionization continua (i.e. post-collisional effects) or includes other scattering particles like photons, we refer to the paper of Åberg and Howat [9], whose formalism will also be used in the next section.

From the standpoint of computation one notes the existence of an ample body of methods (often accompanied by fully developed applications) for the study of the Auger effect in atomic systems [9-11], whereas for molecules the number is much more limited, because of computational difficulties due to the reduced symmetry. In particular, while there are a number of calculations of the *energies* of the emitted Auger electrons [5,6,12] (which can be carried out accurately by the standard *ab initio* methods of quantum chemistry for bound states), there are few calculations of emission *probabilities* for molecular

R. Girlanda et al. (eds.), Progress on Electron Properties of Solids, 211–220.

systems [13-17]. On the other hand for such systems the Auger spectra now available are of such an accuracy that an effort to obtain a corresponding accurate prediction of energy shifts, intensities and lineshapes from the first principle calculations seems desirable.

In this paper we present a new technique for calculating Auger decay rates and resonance energies, that is a simpler and more efficient version of the method recently proposed by the authors in [1,2] and successfully applied to atoms and molecules. The main advantage of this new approach with respect to the previous one is that it allows to bypass the construction of the continuum orbital for the Auger electron and to obtain directly the observable quantities of interest.

The outline of this paper is the following: in Section 2, using the approach developed by Fano for treating autoionization problems [18], we derive the analytic expressions of the quantities we want to compute. In Section 3 we illustrate our method showing the advantages of the new approach. In Section 4 we calculate decay rates and energy shifts for the Auger spectrum of the *LiF* molecule ionized in its deepest shell.

2. The Theory

Let us consider first the internal structure of the wave function $\Psi^-_{\alpha,\mathcal{E}}$, that describes the $(N-1)$-electron system produced by the initial, single ionization process. We impose to $\Psi^-_{\alpha,\mathcal{E}}$ the ingoing wave boundary condition, since we are interested in the probability distribution of the various entrance channels for an Auger electron escaping into a given channel α, with a kinetic energy $\mathcal{E}$ measured from the threshold energy E_α. Since we are interested in a problem where an isolated resonance state $|\Phi>$ interacts with N_c continua (N_c may be infinite), following Fano [18] we represent $\Psi^-_{\alpha,\mathcal{E}}$ using a configuration interaction expansion of the type:

$$\Psi^-_{\alpha,\mathcal{E}} = a^-_\alpha(\mathcal{E})\Phi + \sum_{\beta=1}^{N_c}\int_0^\infty \chi^-_{\beta,\tau} C^-_{\beta,\alpha}(\tau,\mathcal{E})d\tau \tag{1}$$

with the normalization condition

$$<\Psi^-_{\alpha,\mathcal{E}}|\Psi^-_{\beta,\mathcal{E}'}> = \delta_{\alpha\beta}\delta(E_\alpha+\mathcal{E}-E_\beta-\mathcal{E}'). \tag{2}$$

In (1) $\chi^-_{\beta,\tau}$ is a continuum function representing a direct scattering channel, whose asymptotic behaviour is given by

$$\lim\chi^-_{\beta,\tau} = \sum_{\gamma=1}^{N_c}\frac{\Omega_\gamma}{2ir}(e^{i\theta_\gamma(\tau,r)}\delta_{\gamma\beta} - e^{-i\theta_\gamma(\tau,r)}S'^{+}_{\gamma\beta}). \tag{3}$$

The symmetry adapted wave functions $\{\Omega_\gamma\}$ describe possible states of the doubly ionized target and contain also the angular coordinates of the outgoing electron, whose radial phase $\theta_\gamma(\tau,r)$ in the channel γ depends on the nature of the long-range interaction.

The problem is to determine the expansion coefficients in (1) and without any loss of generality one can assume that the channel functions $\{\chi^-_{\beta,\tau}\}$ do not interact among themselves, that means that they have been obtained from a given set of interacting continuum functions $\{\chi_{\beta,\epsilon}\}$ through the diagonalization of the corresponding Hamiltonian matrix:

$$<\chi_{\beta,\epsilon}|H-E|\chi_{\gamma,\epsilon'}> = (E_\beta+\epsilon-E)\delta(E_\beta+\epsilon-E_\gamma-\epsilon')+V_{\beta\gamma}(\epsilon,\epsilon';E). \tag{4}$$

Note that from the solution of this set of equations, one derives also the expression of the scattering matrix S', whose adjunct is used in (3). The expansion coefficients in (1) can be determined by requiring that

$$<\Phi|H-E|\Psi^-_{\alpha,\mathcal{E}}> = <\chi^-_{\beta,\epsilon}|H-E|\Psi^-_{\alpha,\mathcal{E}}> = 0 \tag{5}$$

and solving the following system of equations under the assumption of non-interacting channels:

$$(E_\Phi - E)a^-_\alpha(\mathcal{E}) + \sum_{\beta=1}^{N_c}\int_0^\infty M^-_\beta(\tau,\mathcal{E})C^-_{\beta\alpha}(\tau,\mathcal{E})d\tau = 0 \tag{6a}$$

$$a^-_\alpha(\mathcal{E})M^-_\beta(\epsilon,\mathcal{E})^* + (\epsilon + E_\beta - E)C^-_{\beta\alpha}(\epsilon,\mathcal{E}) = 0 \tag{6b}$$

with

$$M^-_\beta(\epsilon,\mathcal{E}) = <\Phi|H-E|\chi^-_{\beta,\epsilon}>;\ (E_\Phi - E) = <\Phi|E-H|\Phi>. \tag{7}$$

The solution of this set of equations can be carried out exactly using the technique proposed by Fano [18] to remove the singularity at $E = E_\beta + \epsilon$. This leads to the following expression for $\Psi^-_{\alpha,\mathcal{E}}$ in (1):

$$\Psi^-_{\alpha,\mathcal{E}} = \chi^-_{\alpha,\mathcal{E}} + \frac{M^-_\alpha(\mathcal{E},\mathcal{E})}{E-E_r-i\frac{\Gamma}{2}}[\Phi + \lim_{\nu\to 0}\sum_{\beta=1}^{N_c}\int\frac{\chi^-_{\beta,\tau}M^-_\beta(\tau,\mathcal{E})^*}{E-E_\beta-\tau-i\nu}d\tau] \tag{8}$$

with the quantities Γ and E_r defined as follows

$$\Gamma = \sum_{\beta=1}^{N_c}\Gamma^-_\beta = 2\pi\sum_{\beta=1}^{N_c}|M^-_\beta(\epsilon,\mathcal{E})|^2 \tag{9}$$

$$E_r = E_\Phi + \Delta;\ \Delta = \sum_{\beta=1}^{N_c}\mathcal{P}\int_0^\infty\frac{|M^-_\beta(\tau,\mathcal{E})|^2}{E-E_\beta-\tau}d\tau. \tag{10}$$

Eq. (10) shows that the resonance energy E_r differs from that of the quasi-bound state $|\Phi>$ because of the presence of the energy shift Δ, whose expression is given in terms of the principal part $\mathcal{P}$ of the integrals in (10). Furthermore, the quantity Γ defined in (9) is related to the total probability of nonradiative transitions from the resonance state $|\Phi>$, and consistently $\Gamma^-_\beta = 2\pi|M^-_\beta(\epsilon,\mathcal{E})|^2$ to the partial, nonradiative decay probability into channel β.

In addition one can derive from Eqs. (3) and (8) the following expression for the scattering matrix S, whose adjoint characterizes the asymptotic behaviour of $\Psi^-_{\alpha,\mathcal{E}}$:

$$S_{\beta\alpha}(\mathcal{E}) = S'_{\beta\alpha}(\mathcal{E}) - 2\pi i\frac{M^+_\alpha(\mathcal{E},\mathcal{E})\cdot M^-_\beta(\epsilon,\mathcal{E})^*}{E-E_r+i\frac{\Gamma}{2}}. \tag{11}$$

The matrix elements in (11) describe the influence of a single resonance on the scattering matrix in presence of direct scattering, whose amplitude ($S'_{\alpha\beta}$) characterizes the asymptotic behaviour of the channel function $\chi^-_{\beta,\tau}$ according to Eq. (3).

3. A New Computational Approach for Molecular Problems

In the previous section we have obtained quite general, analytical expressions that can be used to interpret a typical Auger spectrum in terms of resonance energies and decay probabilities. However in the effective application of the theory to a specific molecular problem a relevant difficulty is represented by the construction of the continuum functions $\{\chi_{\beta,\tau}^{-}\}$ necessary to calculate the quantities of interest. The main obstacles are constituted by the non central-symmetric nature of the effective potential to which the outgoing electron is subjected in a molecule, and by the fact that a good representation of the $(N-1)$-electron system is required also in the inner molecular region for obtaining reasonably accurate matrix elements in (9) and (10).

As a matter of fact, while a large number of theoretical predictions of Auger decay probabilities is available for atomic systems [9-11,19], only very few calculations have been performed on molecules and these ones essentially based on semiquantitative approaches [12,14,16] or restricted to those types of molecules, for which a monocentric expansion of the wave function and of the potential can be reasonably performed [13,15,17].

In what follows we present a simple technique for evaluating Auger decay probabilities, that can be applied to any type of polyatomic system without any specific difficulty. In this article we describe our method in the limit of non-interacting decay channels, however its generalization to include the correlation effects due to the interchannel coupling is straightforward.

Let us consider first a few simplifications that are important for a method that can be efficiently applied to the calculation of decay probabilities and energy shifts in molecules. The first simplification we introduce is the independent particle representation of the states, given as antisymmetrized products of spin-orbitals with the correct spin symmetry. This means that we disregard the interchannel coupling that characterizes $\chi_{\beta,\tau}^{-}$ in (3) and represents each decay channel β in the following way:

$$\chi_{\beta,\tau}^{-}(1,2,\cdots,N-1) = \Psi_{\beta,\mathbf{k}}(1,2,\cdots,N-1) = = M\hat{A}\{\eta_{\beta,\mathbf{k}}(1)\Theta_{\beta}(2,\cdots,N-1)\} \tag{12}$$

with the normalization condition:

$$<\Psi_{\beta,\mathbf{k}}|\Psi_{\beta',\mathbf{k}'}> = \delta_{\beta\beta'}(2\pi)^3\delta(\mathbf{k}-\mathbf{k}'). \tag{13}$$

In (12) $\hat{A}$ represents the antisymmetrizer with the proper normalization constant M, Θ_β the HF wave function for the doubly charged ion in the state β, whose energy is E_β and $\eta_{\beta,\mathbf{k}}(1) = \varphi_{\beta,\mathbf{k}}(\mathbf{r}_1)\sigma_\beta(s_1)$ the continuum spin-orbital asymptotically defined by the momentum $\mathbf{k}$ and by the energy $\epsilon_k = \dfrac{\hbar^2k^2}{2m} = E_r - E_\beta$ of the outgoing electron. Furthermore, since the Auger spectra we want to interpret identify each decay channel only in terms of the kinetic energy of the outgoing electron, the quantity we compare with the measured Auger rate for channel β is the following:

$$W_\beta = \frac{2\pi}{\hbar}\int|<\Phi|H-E_r|\Psi_{\beta,\mathbf{k}}>|^2\delta(E_r-E_\beta-\frac{\hbar^2k^2}{2m})\frac{d\mathbf{k}}{(2\pi)^3}. \tag{14}$$

This expression can be further simplified by approximating the resonant energy with $E_r \simeq E_\Phi$, i.e. by neglecting the energy shift Δ defined in (10) and thus the final expression for the partial decay rate into each channel β becomes:

$$W_\beta = \frac{mk}{(2\pi)^2\hbar^3}\int|<\Phi|H-E_\Phi|\Psi_{\beta,k,\hat{k}}>|^2d\hat{k} \tag{15}$$

with $k = \dfrac{1}{\hbar}\sqrt{2m(E_\Phi - E_\beta)}$.

For calculating W_β in a specific molecular problem we propose two alternative approaches: the first one, that requires the construction of the continuum orbital $\varphi_{\beta,\mathbf{k}}$, the second one instead which gives directly the quantities W_β and Δ without introducing any specific representation of the continuum orbital.

a) The construction of the continuum orbital

Since in our approach both the initial quasi-bound state $|\Phi>$ and the final decay states $|\Theta_\beta>$ of the doubly ionized target are represented by *HF* wave functions derived from independent *SCF* processes, the natural way for obtaining the continuum orbital $\varphi_{\beta,\mathbf{k}}$ is via solution of an *HF* equation of the type

$$\hat{F}_\beta(\mathbf{r})\varphi_{\beta,\mathbf{k}}(\mathbf{r}) = \epsilon_k\varphi_{\beta,\mathbf{k}}(\mathbf{r}) \tag{16}$$

with $\hat{F}_\beta$ defined in terms of the nonlocal static-exchange potential $\hat{V}_\beta$, due to the bound orbitals in Θ_β:

$$\hat{F}_\beta(\mathbf{r}) = -\frac{1}{2}\hat{\nabla}_\mathbf{r}^2 - \sum_\alpha \frac{Z_\alpha}{|\mathbf{r}-\mathbf{R}_\alpha|} + \sum_{j=1}^{N} [a_j^{(\beta)}\hat{J}_j^{(\beta)}(\mathbf{r}) - b_j^{(\beta)}\hat{K}_j^{(\beta)}(\mathbf{r})] \tag{17}$$

$$= -\frac{1}{2}\hat{\nabla}_\mathbf{r}^2 + \hat{V}_\beta(\mathbf{r},\{\mathbf{R}\}).$$

In (17) Z_α, $\mathbf{R}_\alpha$ are respectively the charge and position of the αth nucleus in the molecule, while $\hat{J}_j^{(\beta)}$ and $\hat{K}_j^{(\beta)}$ are the usual Coulomb and exchange operators weighted by coefficients that are related to the occupation number of the jth orbital in Θ_β.

A possible way for solving Eq. (16) is to transform it into a Lippmann-Schwinger (*LS*) integral equation by partitioning $\hat{F}_\beta$ into a zero order Hamiltonian, whose eigenfunctions are known, and a perturbing potential $\hat{V}'_\beta$. Using such an approach the solutions of Eq. (16) will be simply given by

$$\varphi_{\beta,\mathbf{k}}(\mathbf{r}) = \varphi_{\beta,\mathbf{k}}^{(0)}(\mathbf{r}) + \hat{G}_\beta^{(0)}(\epsilon_k)\hat{T}_\beta(\epsilon_k)\varphi_{\beta,\mathbf{k}}^{(0)}(\mathbf{r}) \tag{18}$$

where $\varphi_{\beta,\mathbf{k}}^{(0)}(\mathbf{r})$ is an eigenfunction of the zero order Hamiltonian, $\hat{G}_\beta^{(0)}(\epsilon_k)$ the corresponding Green function and $\hat{T}_\beta(\epsilon_k)$ the transition operator. However, we observe that in the Auger problem $\hat{V}_\beta(\mathbf{r},\{\mathbf{R}\})$ is a long range potential:

$$\hat{V}_\beta(\mathbf{r},\{\mathbf{R}\})_{r\rightarrow\infty} - \frac{2}{r} + \mathcal{O}(\frac{1}{r^n}) \tag{19}$$

with a tail represented by a Coulomb potential plus higher order corrections that depend on the type of molecule considered. Because of this fact, one should include in the zero order Hamiltonian at least the Coulomb and static dipole components of the potential to obtain a solution of the type given in (18), but the technical implementation of this requirement would present serious difficulties in a molecular context. The alternative approach we suggest consists in the use of the kinetic energy operator, as zero order Hamiltonian, and in the approximate representation of the scattering potential $\hat{V}_\beta$ by means of its expansion $\hat{V}_\beta^t$ in a finite discrete basis set of L^2-functions $\{<\mathbf{r}|\lambda>\}$, chosen in such a way as to minimize the difference $(\hat{V}_\beta - \hat{V}_\beta^t)|\varphi_{\beta,\mathbf{k}}>$ inside the region of interest:

$$\hat{V}_\beta(\mathbf{r},\{\mathbf{R}\}) \sim \hat{V}_\beta^t(\mathbf{r},\{\mathbf{R}\}) = \sum_{\lambda\mu\nu\tau} |\lambda>S_{\lambda\mu}^{-1}<\mu|\hat{V}_\beta|\nu>S_{\nu\tau}^{-1}<\tau|; \tag{20}$$

$$S_{\lambda\mu} = <\lambda|\mu>.$$

The replacement $(\hat{V}_\beta^t \rightarrow \hat{V}_\beta)$ allows one to solve the *LS* equation for the transition operator $\hat{T}_\beta^t$ at ϵ_k:

$$\hat{T}_\beta^t = \hat{V}_\beta^t + \hat{V}_\beta^t\hat{G}_0^-(\epsilon_k)\hat{T}_\beta^t \tag{21}$$

where $G_0^-(\epsilon_k)$ is now the free-particle Green function for the ingoing wave boundary conditions, and to obtain the following matrix representation of $\hat{T}^t$:

$$\mathbf{T}^t = \mathbf{V}^t_\beta[\mathbf{V}^t_\beta - \mathbf{V}^t_\beta \mathbf{G}^-_0(\epsilon_k)\mathbf{V}^t_\beta]^{-1}\mathbf{V}^t_\beta. \tag{22}$$

The corresponding *LS* wave function, that satisfies the ingoing wave boundary condition and represents an eigenvector of $\hat{F}^t_\beta$, i.e. of the effective Hamiltonian defined in (17) but with $\hat{V}_\beta$ replaced by $\hat{V}^t_\beta$, is given by

$$\varphi_{\beta,\mathbf{k}} = e^{i\mathbf{k}\cdot\mathbf{r}} + \hat{G}^-_0(\epsilon_k)\hat{T}^t_\beta e^{i\mathbf{k}\cdot\mathbf{r}}. \tag{23}$$

Obviously the crucial point in this type of approach is represented by the replacement $(\hat{V}^t_\beta \rightarrow \hat{V}_\beta)$, which seems particularly delicate, involving as it does truncation of the long range part of the *HF* potential, even if it is clear that the Auger matrix element is really sensitive to the form of the continuum orbital only inside the molecular region.

The way we suggest to properly perform such a replacement is based on the idea that it should be sufficient to have a projected potential $\hat{V}^t_\beta$, which applied to $\varphi_{\beta,\mathbf{k}}$ correctly reproduces the effect of the true potential $\hat{V}_\beta$ on $\varphi_{\beta,\mathbf{k}}$ continuing into at least a part of the asymptotic region. In that case the *LS* wave function, within this intermediate asymptotic region, will be a linear combination of the exact long range eigenfunctions of $\hat{F}_\beta$ corresponding to the Auger energy, and since these eigenfunctions for the usual problems are known analitically or in any case can be easily obtained by using standard numerical techniques, one can set up procedures for matching them to the *LS* wave function, thus obtaining the correct asymptotic behavior.

The implementation of this method requires establishing a criterion for the choice of a basis set able to properly represent the scattering potential $\hat{V}_\beta$. The authors have shown in [1] that the more convenient one is represented by the simulateneous request of: a) a stable matching between the *LS* orbital and the analytic solutions in the asymptotic region; b) the variational stability of the Auger matrix element in (15) with respect to an uniform scaling of the orbital exponents of the basis functions.

Having solved the problem of the construction of the continuum orbital, one simply has to evaluate the matrix element in (15), by taking into account the fact that the orbitals of the initial and final states are mutually non-orthogonal because of the different *SCF* processes used for their derivation. The calculation must be repeated for various direction of $\hat{k}$ in (15) and the results interpolated with an analytic function of $\hat{k}$ to perform the final integration. This point represents the main drawback of the present approach, which can be eliminated by using the technique described in the next section, however the advantage of constructing explicitly the continuum orbital is represented by the fact that through the matching procedure, suggested by the authors in [1,2], one can correct the wrong asymptotic behaviour of the *LS* orbital, obtaining the exact phase shift and amplitude.

b) The direct evaluation of W_β and Δ.

Let us consider the expression of the Auger decay amplitude $A_{\beta,\mathbf{k}}$ for channel β

$$A_{\beta,\mathbf{k}} = <\Phi | H - E_\Phi | \Psi_{\beta,\mathbf{k}}>. \tag{24}$$

Because of the form of the $\Psi_{\beta,\mathbf{k}}$ defined in (12) and for a given choice of $|\Phi>$ and $|\Theta_\beta>$, $A_{\beta,\mathbf{k}}$ is a linear functional of $\varphi_{\beta,\mathbf{k}}$ and therefore one can define a vector $|\mu_\beta>$ in the 1-particle space, such that

$$<\mu_\beta | \varphi_{\beta,\mathbf{k}}> = <\Phi | H - E_\Phi | \Psi_{\beta,\mathbf{k}}> = A_{\beta,\mathbf{k}}. \tag{25}$$

The $\mathbf{r}$-representation of $|\mu_\beta>$ is simply given by

$$\mu_\beta(\mathbf{r}) = <\mathbf{r} | \mu_\beta> = \tag{26}$$

$$= <M\hat{A}\{\Theta_\beta(1,2,\cdots,N-2)\delta(\mathbf{r}-\mathbf{r}_{N-1})\}|H-E_\Phi|\Phi>$$

while its expansion coefficients over a set of orthonormal functions $\{|e_n>\}$ by

$$<e_n|\mu_\beta> = <M\hat{A}\{\Theta_\beta(1,2,\cdots,N-2)e_n(N-1)\}|H-E_\Phi|\Phi>. \quad (27)$$

Using definitions (25) and (26) in Eq. (14) with E_r substituted by E_Φ, one reduces the $(N-1)$-electron problem to the calculation of the following simple expression that contains only 1-electron functions:

$$W_\beta = \frac{2\pi}{\hbar}\int <\mu_\beta|\varphi_{\beta,\mathbf{k}}>\delta(E_\Phi-E_\Phi-\frac{\hbar^2k^2}{2m})<\varphi_{\beta,\mathbf{k}}|\mu_\beta>\frac{d\mathbf{k}}{(2\pi)^3}. \quad (28)$$

We observe that since $\varphi_{\beta,\mathbf{k}}$ is an eigenfunction, at the positive energy ϵ_k, of the HF operator $\hat{F}^t_\beta$, previously defined, it can be easily proved that

$$\int|\varphi_{\beta,\mathbf{k}}>\delta(E_\Phi-E_\beta-\frac{\hbar^2k^2}{2m})<\varphi_{\beta,\mathbf{k}}|\frac{d\mathbf{k}}{(2\pi)^3} = \delta(E_\Phi-E_\beta-\hat{F}^t_\beta). \quad (29)$$

This equality follows from the definition of the Green functions $\hat{G}^\pm_\beta(E_\Phi)$ for $\hat{F}^t_\beta$, that can be given in terms of the complete set of orthonormal, continuum $\{\varphi_{\beta,\mathbf{k}}\}$ and discrete $\{\tau_{\beta,n}\}$ eigenfunctions of $\hat{F}^t_\beta$:

$$\hat{G}^\pm_\beta(E_\Phi) = \int|\varphi_{\beta,\mathbf{k}}>\frac{1}{(E_\Phi-E_\beta-\dfrac{\hbar^2k^2}{2m}\pm i\epsilon)}<\varphi_{\beta,\mathbf{k}}|\frac{d\mathbf{k}}{(2\pi)^3}+ \quad (30)$$

$$\sum_n|\tau_{\beta,n}>\frac{1}{E_\Phi-E_\beta-E_n}<\tau_{\beta,n}|$$

or alternatively using the identity

$$\hat{G}^\pm_\beta(E_\Phi) = \frac{1}{(E_\Phi-E_\beta-\hat{F}^t_\beta\pm i\epsilon)} = \quad (31)$$

$$= \mathcal{P}\frac{1}{(E_\Phi-E_\beta-\hat{F}^t_\beta)}\mp i\pi\delta(E_\Phi-E_\beta-\hat{F}^t_\beta).$$

Equating separately the real and imaginary part of Eqs. (30) and (31), one obtains Eq. (29) and then, using Eq. (29) in (28), the following simple expression for the decay rate into channel β:

$$W_\beta = \frac{2\pi}{\hbar}<\mu_\beta|\delta(E_\Phi-E_\beta-\hat{F}_\beta)|\mu_\beta> = \quad (32)$$

$$= \frac{i}{\hbar}<\mu_\beta|\hat{G}^+_\beta(E_\Phi)-\hat{G}^-_\beta(E_\Phi)|\mu_\beta>.$$

Analogously, applying to Eq. (10) the same procedure used to derive Eq. (32), one obtains the following expression for the energy shift:

$$\Delta = \sum_{\beta=1}^{N_c}\Delta_\beta = \frac{1}{2}\sum_{\beta=1}^{N_c}<\mu_\beta|\hat{G}^+_\beta(E_\Phi)+\hat{G}^-_\beta(E_\Phi)|\mu_\beta>. \quad (33)$$

The practical utility of Eqs. (32) and (33) rests on the fact that $<\mathbf{r}|\mu_\beta>$ is entirely contained in the space of the L^2-basis functions used for the expansion of the bound orbitals in $|\Phi>$ and $|\Theta_\beta>$, and therefore only the calculation of the matrix elements of $\hat{G}^\pm_\beta(E_\Phi)$ over such basis functions is required. These elements, because of the form of $\hat{V}^t_\beta$, can be easily expressed in terms of the analitically known matrix elements of $\hat{G}^\pm_0$ and $\hat{V}^t_\beta$ over such basis functions:

$$<\alpha|\hat{G}^{\mp}_{\beta}(E_\Phi)|\beta> = <\alpha|(1-\hat{G}^{\mp}_{0}(E_\Phi)\hat{V}^{t}_{\beta})^{-1}\hat{G}^{\mp}_{0}(E_\Phi)|\beta>. \quad (34)$$

Therefore this method allows one to obtain W_β and Δ without constructing explicitly the continuum orbital $\varphi_{\beta,\mathbf{k}}$ and thus avoiding the integration over $\mathbf{k}$. Obviously using such a technique one neglects the corrections due to the fact that $\hat{V}^{t}_{\beta}$ reproduces correctly the true potential $\hat{V}_\beta$ at most in a limited region of space and therefore the eigenfunctions of $\hat{F}^{t}_{\beta}$ are characterized by an incorrect asymptotic behaviour, however one can hope that such corrections are not so important as long as the Auger amplitude is involved, which is appreciably different from zero only inside the molecular region.

4. Calculation of Auger Decay Rates and Resonance Energies for the *LiF* Molecule

As an example of application of our method to a specific molecular problem, we have considered the Auger spectrum of the *LiF* molecule, ionized in its deepest shell, for which sufficiently accurate measurements are available [6] and furthermore semiquantitative predictions of the Auger decay rates have been recently proposed by the authors [20]. Here we present the new results obtained from the application of our method, as explained in Section 3b, and compare them with the experimental values given in [6].

The *HF* calculations for the initial quasi-bound state $|\Phi>$ and for the final states $\{|\Theta_\beta>\}$ of the doubly ionized molecule have been performed using a basis set of 62 modified Gaussian functions, given in Ref. [20]. The calculated transition energies, that are shown in Table I, compare very well with the experimental values because of the cancellation of the correlation errors between the initial and final states.

TABLE I

HF energies for the *LiF* ion calculated, at internuclear distance $R = 2.955$ a.u., by means of independent *SCF* processes for each state. Δ_β indicates the contribution to the total energy shift (Δ) due to channel β, ΔE and ΔE_{exp} are respectively the transition energy calculated with respect to the resonance energy ($E_{HF}(^2\Sigma^+)+\Delta$) and the corresponding experimental quantity Ref. [6].

Transition	Hole Configuration State	E_{HF}(a.u.)	$\Delta E(eV)$	$\Delta_\beta(meV)$	$\Delta E_{exp}(eV)$
	$1\sigma{:}^2\Sigma^+$	-81.741500		$\Delta = 106.30$	
1)	$\pi_-\pi_-{:}^3\Sigma^-$	-105.734588	652.99	0.0	652.2
2)	$4\sigma\pi_+{:}^3\Pi$	-105.714699	652.45	0.2	650.3
3)	$\pi_+\pi_+{:}^1\Delta$	-105.629468	650.13	21.8	648.5
4)	$4\sigma\pi_+{:}^1\Pi$	-105.609632	649.59	21.5	647.7
5)	$\pi_+\pi_-{:}^1\Sigma^+$	-105.526636	647.33	10.1	646.6
6)	$4\sigma4\sigma{:}^1\Sigma^+$	-105.539617	647.68	8.1	644.8
7)	$3\sigma\pi_+{:}^3\Pi$	-104.923948	630.93	25.4	630.5
8)	$3\sigma4\sigma{:}^3\Sigma^+$	-104.908740	630.51	12.8	629.6
9)	$3\sigma\pi_+{:}^1\Pi$	-104.548457	620.71	4.3	621.8
10)	$3\sigma4\sigma{:}^1\Sigma^+$	-104.540838	620.50	1.7	620.3
11)	$3\sigma3\sigma{:}^1\Sigma^+$	-103.810355	600.63	0.5	602.6

In Table I we give also the partial energy shifts for the various transitions and the total energy shift that has to be added to the energy of the initial quasi-bound state for obtaining the corrected transition energies. As one can see from these results, the contribution of the total energy shift is quite negligible, being of the order of fractions of *eV*, and therefore the approximation $E_r \simeq E_\Phi$, used in Section 3, is largely justified.

Finally in Table II we compare the partial Auger decay rates, calculated using our method, with the experimental quantities. Note that in Table II we have reported only the relative experimental rates and not the absolute quantities since the total Auger rate has not been clearly measured.

TABLE II

Values of the absolute (W_{abs}) and relative (W_{rel}) transition rates for various decay channels of the *LiF* ion, calculated with our method and compared with the experimental relative quantities (W_{exp}).

Transition	Hole Configuration State	$W_{abs}(10^{-3}$a.u.)	W_{rel}	W_{exp}
1)	$\pi_+\pi_-:{}^3\Sigma^-$	0.0000	0.0000	0.01
2)	$4\sigma\pi_+:{}^3\Pi$	0.0031	0.0017	0.02
3)	$\pi_+\pi_+:{}^1\Delta$	1.8402	1.0000	1.00
4)	$4\sigma\pi_+:{}^1\Pi$	1.8184	0.9882	0.91
5)	$\pi_+\pi_-:{}^1\Sigma^+$	0.6231	0.3386	0.27
6)	$4\sigma4\sigma:{}^1\Sigma^+$	0.7317	0.3976	0.32
7)	$3\sigma\pi_+:{}^3\Pi$	0.4934	0.2681	0.20
8)	$3\sigma4\sigma:{}^3\Sigma^+$	0.2363	0.1284	0.07
9)	$3\sigma\pi_+:{}^1\Pi$	1.1819	0.6423	0.42
10)	$3\sigma4\sigma:{}^1\Sigma^+$	0.5600	0.3043	0.17
11)	$3\sigma3\sigma:{}^1\Sigma^+$	0.8909	0.4841	0.21
	$W_{tot}(10^{-3}$a.u.)	8.3790	-	?

For the calculations of decay rates and energy shifts we have enlarged the *SCF* basis set to include other basis functions necessary for an accurate representation of the scattering potential $\hat{V}_\beta$ as shown in Eq. (20). Such functions have been chosen using the criteria discussed in Section 3 and proposed by the authors in [1].

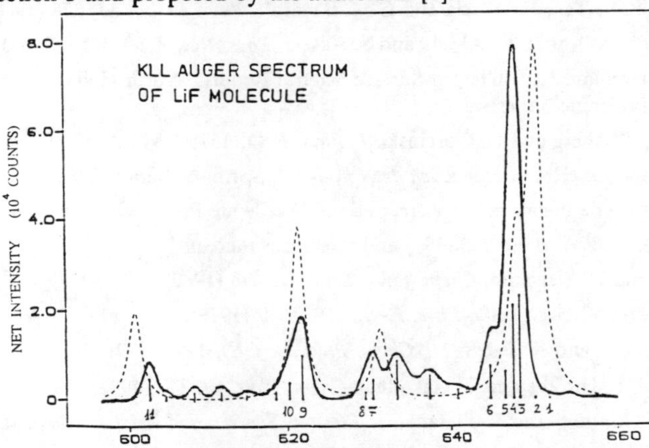

Figure 1. Experimental Auger spectrum (—), of gaseous *LiF* (Ref. [6]), compared with that obtained by means of superposition of Lorentzians (---), whose parameters have been chosen as explained in the text.

The results obtained are quite satisfactory, if one consider that neither correlation contributions nor interchannel coupling have been taken into account and furthermore the effects of the nuclear motion have been completely neglected. The results of Table II are visualized in Figure 1, where we compare the experimental spectrum with a 'model' one obtained from the superposition of Lorentzians of the type: $\Gamma_\beta/[(E-E_\beta)^2+\Gamma^2/4]$, with Γ_β taken from Table II and Γ from Ref. [6]. We observe that the presence of other transitions in the energy range between 608-612 eV and $633-640\,eV$ of the experimental spectrum has been attributed to KLL shake-up and $KL-KLL$ shake-off processes, that have not been included in our calculations.

To conclude we observe that an accurate prediction of the experimental Auger rates requires the inclusion of the correlation effects and also, in several cases, of the coupling between electronic and nuclear motion. However, as a preliminary step in the interpretation of the experimental results, it is useful to have an accurate estimate of the best values of the transition rates and phase shifts obtainable using an independent particle approach. To this end we have proposed a new and simple approach, that satisfies this need and has allowed us to obtain a very satisfactory reproduction of the experimental Auger spectrum of the LiF molecule.

*****

This work is dedicated to Prof. F. Bassani, to whom we are deeply indebted for scientific guidance and human qualities.

References

1. R. Colle and S. Simonucci, *Phys. Rev. A*, in press (1989).
2. R. Colle and S. Simonucci, *Nuovo Cimento*, in press (1989).
3. V. Drchal and J. Kudrnovsky, *Phys. Stat. Sol.* **146**, 597 (1988).
4. D.D. Laks and G.F. Neumark, *Phys. Rev. Lett.* **61**, 1229 (1988).
5. S. Aksela, K.H. Tan, H. Aksela and G.M. Bancroft, *Phys. Rev. A* **33**, 258 (1986).
6. M. Hotokka, A. Ågren, H. Aksela and S. Aksela, *Phys. Rev. A* **30**, 1855 (1984).
7. R. Weissmann and K. Müller, in Surface Science Reports, North-Holland Publishing **105**, 251 (1981) and references therein.
8. G. Howat, T. Åberg and O. Goscinski, *J. Phys. B* **11**, 1575 (1978).
9. T. Åberg and G. Howat, *Handbuch der Physik* **31**, Springer, Berlin (1982).
10. D. Chattarji, The theory of Auger transitions (Academic Press, New York, 1976).
11. J. Bruneau, *J. Phys. B* **20**, 713 (197) and references therein.
12. H. Ågren and H. Siegbahn, *Chem. Phys. Lett.* **72**, 498 (1980).
13. K. Faegri Jr. and H.P. Kelly, *Phys. Rev. A* **19**, 1649 (1979).
14. O.M. Kvalheim and K. Faegri Jr., *Chem. Phys. Lett.* **67**, 127 (1979).
15. M. Higashi, E. Hiroike and T. Nakajima, *Chem. Phys.* **68**, 377 (1982).
16. M. Cini, F. Maracci and R. Platania, *J. Electron Spectr. and Relat. Phenom.* **41**, 37 (1986).
17. V. Carravetta nd H. Ågren, *Phys. Rev. A* **35**, 1022 (1987).
18. U. Fano, *Phys. Rev.* **124**, 1866 (1961).
19. H.P. Kelly, *Phys. Rev. A* **11**, 556 (1975).
20. R. Colle, S. Simonucci and T.O. Woodruff, *Phys. Rev. A* **38**, 694 (1988).

ORDERING TRANSITIONS INDUCED BY COULOMB INTERACTIONS

M. ROVERE, G. SENATORE* AND M.P. TOSI**

Department of Theoretical Physics
University of Trieste, Italy

**Department of Physics*
University of Pavia, Italy

***International Centre for Theoretical Physics*
Trieste, Italy

We briefly review recent progress in treating phase transitions to ordered states driven by Coulomb interactions. Wigner crystallization of the one-component plasma, both in the degenerate Fermi limit and in the classical limit, is the foremost example and developments in its theory are discussed in some detail. Attention is also given to quasi-two-dimensional realizations of the plasma model in the laboratory. The usefulness of these ideas in relation to freezing and ordering transitions is illustrated with reference to alkali metals, elemental and polar semiconductors, and various types of ionic systems (molten salts, colloidal suspensions and astrophysical plasmas).

1. Introduction

It is now more than fifty years since Wigner [1], in dealing with the exchange and correlation energy of conduction electrons in simple metals, introduced the idea that the degenerate electron fluid on a uniform neutralizing background can crystallize into a bcc lattice under the effect of the electron-electron Coulomb repulsions. A very wide variety of theoretical estimates has been produced over the years for the coupling strength at the phase transition [2]. The electron-electron coupling strength is in general defined as the ratio of average potential to average kinetic energy and is conventionally measured in the degenerate electron fluid by the parameter r_s, equal to the radius of the electron-volume sphere ($r_s = (4\pi n/3)^{-1/3}$ where n is the electron density). The best current assessment of the location of Wigner crystallization comes from quantal simulation work of Ceperley and Alder [3], which yields $r_s = 100\pm20$ Bohr radii at the phase transition from a fully spin-polarized fluid state to a bcc crystal.

Wigner crystallization in the fully classical equivalent of the electron fluid (the one-component classical plasma, or OCP) appears to have been first noticed in Monte Carlo work by Brush, Sahlin and Teller [4]. The coupling strength is measured in the OCP by the plasma parameter $\Gamma = e^2/(r_s k_B T)$, since the classical thermal energy $k_B T$ replaces the Fermi kinetic energy (which is proportional to r_s^{-2}). The latest assessment of crystallization in the OCP from computer simulation work [5] puts it at $\Gamma = 180\pm1$.

A clear identification of instances of Wigner crystallization in real laboratory systems has been elusive for a long time [2]. The situation has changed to a considerable extent in recent years. Firstly, quasi-two-dimensional physical systems closely approaching the ideal plasma model can now be prepared and their fluid-to-solid transition can be experimentally revealed. Of course, solidification in reduced dimensionality has a profound physical interest of its own. Secondly, Wigner's idea has been extremely useful in

R. Girlanda et al. (eds.), Progress on Electron Properties of Solids, 221–238.

providing simple yet deep insight into freezing and ordering transitions in several types of condensed-matter systems. At the same time, the availability of computer simulation data on crystallization in the ideal plasma model has motivated progress in the microscopic theory of this phase transition. The purpose of this article is to review these recent developments [6].

2. Freezing under Pure Coulomb Repulsions

In general, the state of a many particle system can be described through a suitable thermodynamic potential, which is a functional of the one-body density and attains its minimum in correspondence to the stable density distribution [7-9]. Crystallization in equilibrium conditions is microscopically characterized by the change of the one-body density from homogeneous to periodic across the transition from liquid to crystal. For the sake of simplicity we shall restrict ourselves to systems with only one type of particle, whose density profile we denote by $n(\mathbf{r})$, the generalization to multicomponent systems being straightforward. In particular, in a periodic solid the density is represented by the Fourier series

$$n(\mathbf{r}) = n + \sum_{\mathbf{G}\neq 0} n_{\mathbf{G}} \exp(i\mathbf{G}.\mathbf{r}) \tag{1}$$

where the $\mathbf{G}$'s are the reciprocal lattice vectors (RLV).

The relevant functional is the Helmholtz free energy $F[n(\mathbf{r})]$, reducing in the limit of vanishing temperature - which is meaningful only for a quantum system - to the ground state energy functional $E_g[n(\mathbf{r})]$. To be more precise, in the presence of an external potential $v(\mathbf{r})$ it is the internal free energy F_I,

$$F_I[n(\mathbf{r})] = F[n(\mathbf{r})] - \int d\mathbf{r} n(\mathbf{r}) v(\mathbf{r}), \tag{2}$$

which is a functional solely of the density $n(\mathbf{r})$. Possible constraints on the many body system can be taken into account by introducing suitable Lagrange multipliers. Thus, in order to fix the number of particles one introduces the chemical potential μ and minimizes the grand potential $\Omega = F - N\mu$. A *mechanically* stable density must therefore satisfy the equilibrium condition

$$\frac{\delta\Omega}{\delta n(\mathbf{r})} = \frac{\delta F_I}{\delta n(\mathbf{r})} - \mu + v(\mathbf{r}) = 0 \tag{3}$$

whereas thermodynamic stability requires the stronger condition that $\Omega[n(\mathbf{r})]$ be at its absolute minimum.

It should be noticed that some subtleties arise in the definition of thermodynamic functions when dealing with systems with long range forces. In particular, the chemical potential of a single-phase plasma on a uniform background (jellium model) is discussed in detail in the Appendix, showing that in the thermodynamic limit it is necessarily referred to the spatial average V_M of the Madelung potential. Indeed, owing to the long range of Coulomb interactions, the value of V_M depends on the boundary conditions assumed in taking the limit. However, the distinction between the full electrochemical potential μ and the quantity $\mu^* = \mu - V_M$ for a single-phase plasma becomes crucial in phase equilibria [10]. Referring specifically to the freezing transition of jellium as investigated in numerical experiments [3, 5], namely at fixed background density, coexistence between the solid (s) and liquid (l) phase at constant temperature evidently requires $F_s = F_l$ and $\mu_s = \mu_l$. Hence, equilibrium between the two phases is maintained by an interfacial potential drop equal to $\Delta V_M \equiv V^s_M - V^l_M = -(\mu^*_s - \mu^*_l) \equiv -\Delta\mu^*$. This balances an interfacial pressure drop $P_s - P_l = n\Delta\mu^*$.

Returning to microscopic aspects, the equilibrium condition in eqn. (3) can also be put in a differential form by taking advantage of the one-to-one correspondence between the density $n(\mathbf{r})$ and the external potential $v(\mathbf{r})$. One immediately obtains, by considering an infinitesimal rigid translation of the whole system [11], that

$$\nabla(\mu - v(\mathbf{r})) = -\int d\mathbf{r}' \chi^{-1}(\mathbf{r},\mathbf{r}') \nabla n(\mathbf{r}') \tag{4}$$

reducing in the classical limit to [12]

$$\nabla(\mu - v(\mathbf{r})) = \nabla n(\mathbf{r})/n(\mathbf{r}) - \int d\mathbf{r}' c(\mathbf{r},\mathbf{r}') \nabla n(\mathbf{r}'). \tag{5}$$

This form of the equilibrium condition emphasizes the role of the linear density response function $\chi(\mathbf{r},\mathbf{r}')$ of the system at density $n(\mathbf{r})$, or equivalently of the two-body direct correlation function $c(\mathbf{r},\mathbf{r}')$ in the classical case, in determining the density profile. In relation to freezing, $v(\mathbf{r})$ can be looked upon as an external potential of the desired lattice periodicity, which is applied to the liquid approaching crystallization and modulates it with density waves. At liquid-crystal coexistence the modulating potential can be allowed to vanish, while the amplitudes of the density waves (the quantities $n_\mathbf{G}$ in eqn. (1)) retain finite values as the order parameters of the new phase.

A further general result follows by separating out from F_l an ideal term F_0, which is the Helmholtz free energy for noninteracting particles at density $n(\mathbf{r})$. Specializing to the jellium model one writes

$$F[n(\mathbf{r})] = F_0[n(\mathbf{r})] + \int d\mathbf{r}\, n(\mathbf{r}) v(\mathbf{r}) + \frac{e^2}{2}\int\int d\mathbf{r} d\mathbf{r}' \frac{n_Q(\mathbf{r}) n_Q(\mathbf{r}')}{|\mathbf{r}-\mathbf{r}'|} + F_c[n(\mathbf{r})] \tag{6}$$

where $n_Q(\mathbf{r}) = n(\mathbf{r}) - n$ with n the background density, and F_c contains all effects due to correlations and - in the quantum case - to exchange. Equation (3) takes the form

$$\frac{\delta F_0}{\delta n(\mathbf{r})} + v(\mathbf{r}) + e^2 \int d\mathbf{r}' \frac{n_Q(\mathbf{r}')}{|\mathbf{r}-\mathbf{r}'|} + \frac{\delta F_c}{\delta n(\mathbf{r})} \equiv \frac{\delta F_0}{\delta n(\mathbf{r})} + v_{\text{eff}}(\mathbf{r}) = \mu \tag{7}$$

showing that the problem of interacting particles is equivalent to that of noninteracting particles in an *effective selfconsistent potential* $v_{\text{eff}}(\mathbf{r})$ defined by eqn. (7) [13].

The above considerations are formally exact. However, practical implementation of a Density Functional Approach (DFA) to freezing requires suitable approximations on the free energy functional. One can try to approximate either $F[n(\mathbf{r})]$ or directly $\Delta F = F[n(\mathbf{r})] - F[n]$. We shall present below the consequences of the latter approach in studying freezing of the jellium model, in both its quantal and classical versions. Here we may stress that, if the properties of the liquid are known, the task of directly approximating ΔF may be expected to be easier, since near the transition the free energy difference is exceedingly small relative to the total free energy of either phase.

2.1. The One-Component Quantum Plasma (OQP)

Ceperley and Alder [3] have studied the freezing of the OQP in the full degeneracy limit ($T=0$) by Diffusion Monte Carlo. In their phase diagram cyrstallization is seen to occur from a spin-polarized liquid. One has thus to find inhomogeneous densities (if any) which make the ground state energy functional $E_g[n(\mathbf{r})]$ stationary and compare the energies relative to such densities with that of the homogeneous state. In practice one chooses to investigate the stability of given crystalline structures with respect to the liquid. This requires the calculation of $\Delta E_g = E^s_g - E^l_g$.

A convenient approach has recently been proposed [14], resorting to a functional expansion of the exchange and correlation part of ΔE_g with respect to the density difference $n_Q(\mathbf{r}) = n(\mathbf{r}) - n$ between solid and liquid. This is the quantal analogue of the density wave theory of freezing for classical liquids proposed by Ramakrishnan and Yussouff [15] (see also Haymet and Oxtoby [16]). In the present quantal case one obtains

$$\Delta E_g = T_0[n(\mathbf{r})] - \frac{3}{5}N\epsilon_F + \frac{1}{2}\int\int d\mathbf{r}d\mathbf{r}' n_Q(\mathbf{r})n_Q(\mathbf{r}')[v(\mathbf{r}-\mathbf{r}') - K_{xc}(\mathbf{r}-\mathbf{r}')] + ... \quad (8)$$

Here, T_0 denotes the kinetic energy of noninteracting electrons at density $n(\mathbf{r})$, ϵ_F is the Fermi energy of the liquid phase and $v(\mathbf{r})$ the bare Coulomb coupling. Also, $-K_{xc}$ indicates the second functional derivatives of the exchange-correlation energy functional E_{xc} with respect to $n(\mathbf{r})$ evaluated in the liquid, while the dots indicate all higher order terms in $n_Q(\mathbf{r})$ originated by the expansion of E_{xc}.

Let us consider only the terms of the expansion that are explicitly shown in eqn. (8) and assume for the moment that K_{xc} is known. Imposing the extremum condition on the approximate functional for ΔE_g is equivalent to solving a problem of independent electrons in an effective periodic selfconsistent potential, whose Fourier components are explicitly given by

$$v_{\text{eff}}(\mathbf{G}) = n_{\mathbf{G}} v(\mathbf{G})[1 - G(\mathbf{G})] \quad (9)$$

with $v(\mathbf{G})$ the Fourier component of the bare Coulomb potential and $G(\mathbf{G})$ the so-called static local field factor in the dielectric function of the homogeneous electron gas [17]. Hence, one has to selfconsistently solve the Kohn-Sham equations [13]

$$\left[-\frac{\hbar^2}{2m}\nabla^2 + v_{\text{eff}}(\mathbf{r}) \right] \psi_i(\mathbf{r}) = \epsilon_i \psi_i(\mathbf{r}) \quad (10)$$

for the Bloch orbitals ψ_i, the density being given by $n(\mathbf{r}) = \Sigma_i |\psi_i(\mathbf{r})|^2$. In terms of the Bloch eigenvalues ϵ_i and of the components $n_{\mathbf{G}}$ of the selfconsistent density, the energy difference between the crystalline and the liquid phase is given by

$$\Delta E_g = \sum_i \epsilon_i - \frac{3}{5}\epsilon_F - \frac{V}{2}\sum_{\mathbf{G}\neq 0} v(\mathbf{G})[1 - G(\mathbf{G})]|n_{\mathbf{G}}|^2. \quad (11)$$

The coexistence point, characterized by the vanishing of ΔE_g locates the critical r_s for crystallization. At larger r_s one should find $\Delta E_g < 0$, since the crystal should be stable, while $\Delta E_g > 0$ at smaller r_s, the liquid being stable.

At variance from the case of classical plasmas, for which the direct correlation function is available from simulation, direct tests of approximate forms for the local field factor in the OQP are not possible. The so-called STLS scheme [18] has been employed, in which G is evaluated from the static structure factor of the electron liquid. The latter is known from Quantum Monte Carlo [19] at a few values of r_s ranging from 2 to 100. The static local field factor at the same values of r_s can readily be calculated and then extrapolated to all relevant values of the electron density. An illustration of the local field factor that one obtains for spin-polarized electrons is given in Fig. 1. The calculation of the Bloch orbitals ψ_i requires the solution of Kohn-Sham equations for the crystal, i.e. a band structure calculation. However, on the grounds that the electrons should be well localized in the crystalline phase even at coexistence [20], one may first try to variationally calculate ΔE_g by the Ansatz that the ψ_i's can be constructed from a single Gaussian orbital $\phi(r) = \exp(-Cr^2)$ per site, with C a variational parameter. Bloch sums are constructed and overlap is exactly handled [14], even though expected to be small. This calculation yields a critical r_s of 107 for the freezing of the spin-polarized electron liquid into a bcc crystal. It is also found that the optimal value of C at coexistence is indeed

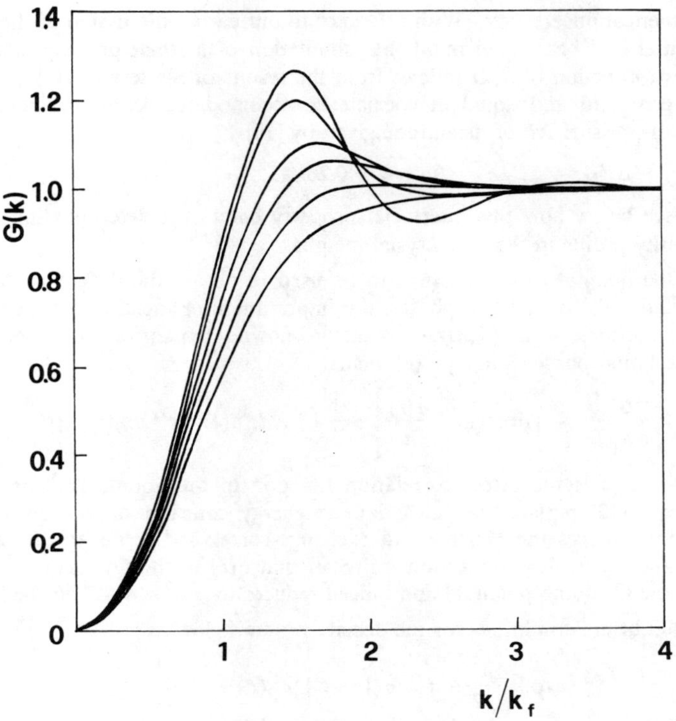

Figure 1. Local field factor $G(k)$ for the spin-polarized electron fluid at increasing values of the coupling strength parameter $\mathbf{r}_s$. In order of increasing peak height the curves refer to $\mathbf{r}_s = 2, 5, 10, 20, 50$ and 100. From results of G. Pastore and G. Senatore [14].

giving pronounced localization of the electrons on lattice sites, yielding a Lindemann ratio of 0.30 which is in nice agreement with the value found by Ceperley [20] in his work on crystallization of quantum systems. Morover, the results appear to be practically insensitive to overlap and to the number of RLV stars included in the calculation.

State-of-art band structure calculations have also been performed [14] to transcend the Gaussian Ansatz. They are based on plane wave expansions involving up to about 500 plane waves for each of the special points [21] in the irreducible wedge of the Brillouin zone, up to 40 such points being used. The calculation shows that solution of the Kohn-Sham equations has only moderate quantitative effects in comparison with the simpler calculation. The electrons are found to be very well localized at the melting point. This is now found to lie at $r_s = 129$, in satisfactory agreement with simulation.

2.2. The One-Component Classical Plasma (OCP)

Soon after the problem of phase transitions in the OCP at large coupling strength Γ was posed by the Monte Carlo runs of Brush et al. [14], van Horn [22] predicted from Lindemann's criterion of melting that the model would crystallize at $\Gamma \simeq 170$. Progressively more accurate simulation studies of bulk phases [5] have led to the current value $\Gamma = 180 \pm 1$ for crystallization into a bcc lattice at constant mean density. Of course, this result fixes completely the melting curve $T_m \propto n^{1/3}$ for the classical bcc solid, within the stated numerical uncertainty. With reference to our earlier discussion of the electrochemical potential and bearing in mind that simulation of a single-phase plasma determines μ^*/k_BT as a function of Γ, it follows from the results of Slattery et al. [5] that the interface between solid and liquid at coexistence accomodates drops of electrical potential ΔV_M and of pressure ΔP of magnitude given by [10]

$$-\Delta V_M = \Delta P/n = \Delta\mu^* \approx -0.26k_BT. \tag{12}$$

We shall see below how this interfacial property enters the determination of the equilibrium density profile in the bulk crystalline phase.

As in the quantal case, eqn. (6) can be used to obtain the difference ΔF of the free energy functionals for the two phases, the important simplification for a classical system being that the ideal term $F_0[n(\mathbf{r})]$ is exactly known. Expansion of the non-ideal terms around the liquid phase as in eqn. (8) yields

$$\frac{\Delta F}{Nk_BT} = \int d\mathbf{r}n(\mathbf{r})\ln\left[\frac{n(\mathbf{r})}{n}\right] - \frac{1}{2}\int\int d\mathbf{r}d\mathbf{r}'c(|\mathbf{r}-\mathbf{r}'|)n_Q(\mathbf{r})n_Q(\mathbf{r}')+... \tag{13}$$

where $c(|\mathbf{r}-\mathbf{r}'|)$ is the direct correlation function in the liquid. The first term on the r.h.s. of eqn. (13) replaces the ideal kinetic energy contribution to eqn. (8), while the second term replaces the Hartree and exchange-correlation terms to the same order in $n_Q(\mathbf{r}) = n(\mathbf{r})-n$. In this connection we recall that $c(r)$ in the OCP contains a contribution from the Coulomb potential and indeed reduces to $-e^2/(rk_BT)$ in the limit $r \to \infty$.

The equilibrium conditions for the density profile follow as [10]

$$\frac{n(\mathbf{r})}{n} = \exp\left[\frac{\Delta\mu^*}{k_BT} + \int d\mathbf{r}'c(|\mathbf{r}-\mathbf{r}'|)n_Q(\mathbf{r}') + \cdots\right]. \tag{14}$$

The contribution from the interfacial potential drop has been explicitly included and the integral in eqn. (14) is to be evaluated under the condition $V_M = 0$. By taking Fourier transforms and using eqns. (1) and (14), the coexistence condition becomes

$$\frac{\Delta F}{Nk_BT} = \frac{\Delta\mu^*}{k_BT} + \frac{V}{2}\sum_{\mathbf{G}\neq 0} c(\mathbf{G})|n_\mathbf{G}|^2 + \cdots = 0 \tag{15}$$

which is to be compared with eqn. (11). In eqn. (15) $c(\mathbf{G})$ denotes the Fourier transform $c(k)$ of the direct correlation function of the liquid, evaluated at the RLV's. Finally, Fourier transform of eqn. (14) yields a set of coupled equations for the order parameters $n_\mathbf{G}$ as well as an equation for $\Delta\mu^*$ from the condition that the mean density be the same in the two phases.

The coefficients $c(\mathbf{G})$ as functions of Γ are the input of the calculation. Accurate theoretical approaches are available for the evaluation of the function $c(k)$ in the liquid, which are known as the generalised mean spherical approximation [23] (GMSA) and the modified hypernetted chain apprximation [24] (MHNC). The high degree of agreeement with simulation data is illustrated in Fig. 2 at $\Gamma = 160$. In relation to freezing, the arrows on the horizontal axis in Fig. 2 mark the positions of the first four stars of RLV's in the bcc lattice. High (low) values of $c(k)$ imply high (low) deformability of the liquid under modulation of its density by density waves. One sees immediately from the Figure that the liquid phase of the OCP is soft against modulation with a periodicity given by any **G** vector of the bcc lattice except those in the (200) star. The rigidity is sufficiently strong to prevent crystallization in the bcc lattice [10], even if one extends the linearized theory that we have discussed so far to include higher-order terms arising from the Γ dependence of $c(k)$. However, one expects that the corresponding order parameter is small in the crystal near melting and one can therefore look for an approximate solution

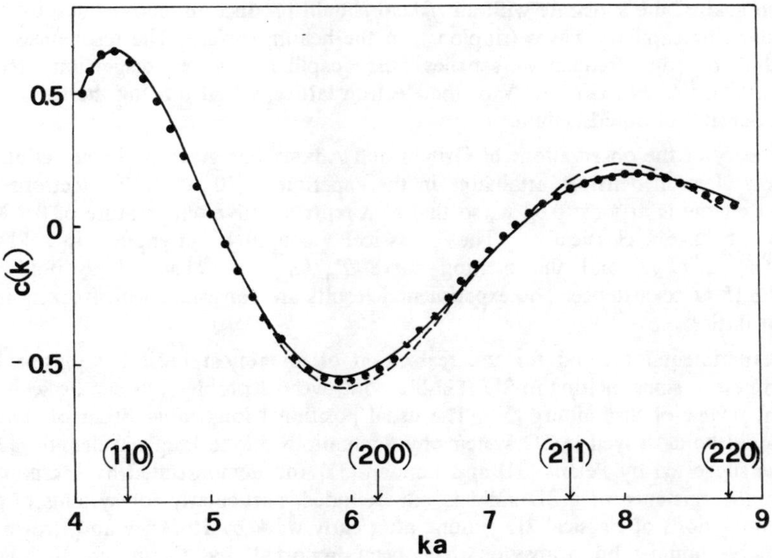

Figure 2. Direct correlation function $c(k)$ for the OCP in 3D at $\Gamma = 160$, from the GMSA theory (full curve) and the MHNC theory (dashes) compared with computer simulation results (dots). The arrows mark the location of the first four stars of RLV's for the bcc lattice. From M. Rovere and M.P. Tosi [10].

of the problem under the assumption $n_{(200)} = 0$. With this approximation the liquid-solid transition is found at $\Gamma = 158$ in the GMSA and at $\Gamma = 142$ in the MHNC. The calculation also yields an entropy change $\Delta S / k_B \approx 1.\div 0.8$ on melting, against $\Delta S / k_B \approx 0.80$ from simulation [5], and an interfacial potential drop $\Delta V_M \approx 0.6 k_B T$, of the same sign and magnitude as the simulation value quoted in eqn. (12).

The above calculation highlights the role of microscopic couplings between order parameters in stabilizing the bcc crystal phase. These are described in lowest order by the three-body direct correlation function of the liquid, whose evaluation is a difficult task. An approximate calculation of three-body terms for the OCP by Barrat et al. [25] shows that their inclusion yields a contribution stabilizing the bcc solid, even though it does not lead to a completely accurate determination of the phase transition. Similar results have also been obtained by Iyetomi and Ichimaru [26], who analyze the problem with the help of a Gaussian representation of the density profile.

2.3. Crystallization of the Twodimensional Electron Liquid

Grimes and Adams [27] where the first to realize a quasi-2D electron system closely approaching the ideal jellium model and to observe its transition from a liquid to a solid state. The system is a monolayer of electrons deposited on the surface of liquid helium, which acts as a very smooth and essentially impenetrable substrate binding electrons through classical image forces and constraining them to quasi-2D motion parallel to the surface. An electric field is applied perpendicularly to the monolayer by means of a capacitor, with the positive plate submerged in helium to provide a 'neutralizing background'. The phase transition is observed by a variant of an experiment proposed by Shikin [28]. He suggested that when the electrons have formed a crystal then driving the crystal up and down against the substrate with an *rf* field should produce resonances due to excitation of standing capillary waves (ripplons) on the helium surface. The resonances occur when the driving frequency satisfies the capillary wave dispersion relation $\omega = (\gamma/\rho)^{1/2}G^{3/2}$ with G an RLV of the electron lattice, γ and ρ being the surface tension and density of liquid helium.

The theory of the observations of Grimes and Adams was given by Fisher et al. [29]. At the low electron densities attainable in the experiment ($10^5 \leqslant n \leqslant 10^9$ electrons/cm^2) the Fermi energy is at most 0.03 K, so that at a representative temperature of 0.5 K the electrons behave classically. The classical coupling strength in 2D is $\Gamma = \pi^{1/2}n^{1/2}e^2/k_BT$ and the melting curve $T_m \propto n^{1/2}$ is observed to hold, with $\Gamma = 137 \pm 15$ at coexistence. The experimental results are consistent with freezing into a triangular lattice.

The experimental method for the realization of an almost ideal 2D plasma is of relevant interest, since melting in 2D is still a controversial problem, as can be seen from the recent review of Strandburg [30]. The usual positional long range order of the solid phase should be destroyed in 2D systems by fluctuations due to long wavelength phonon modes, as suggested by Peierls [31] and Landau [32] and demonstrated by Mermin [33]. However, the existence of a 2D solid is not excluded, particularly for systems of finite size. In simulations of classical 2D jellium, after early work by Hockney and Brown [34], an ordinary liquid-solid transition has been reported by Gann et al. [35] at $\Gamma = 110 \div 140$ and by Kalia et al. [36] at $\Gamma = 118 \div 130$. The latter authors report hysteretic behaviour of both the solid and the liquid phase and an entropy change of $0.3k_B$ across the transition.

The density wave theory described in section 2.2 yields first-order freezing into a stable lattice [37], as expected from a mean field theory which does not take fluctuations into account. The liquid is indeed structurally predisposed to crystallization into a triangular lattice, as can be seen from Fig. 3 showing that the first three stars of RLV's for this lattice are located in correspondence with the main two peaks of the direct correlation function. First-order freezing is found in the theory to occur at $\Gamma = 149$ with an entropy change of $1.2k_B$. Thus, even if omission of higher order correlations prevents safe conclusions, it appears that an ordinary freezing/melting transition would occur at a lower temperature and with a stronger first-order character than are compatible with the simulation data.

In fact, Morf [38] has noted possible doubts left by the aforementioned simulation work on the 2D plasma, in relation to the attainment of equilibrium near the transition, and has devised simulation tests of the transition mechanism proposed by Halperin and Nelson [39] and by Young [40] for a 2D system on a smooth substrate, starting from ideas of Kosterlitz and Thouless [41] (KTHNY theory). The 2D solid would be characterized at low temperature by the presence of topological defects, which are dislocations bound in pairs. On increasing temperature, as the entropic terms increase, the dislocation pairs start to dissociate, driving a continuous transition to a so-called hexatic phase, in which translational order is lost but orientational order persists until a true liquid is realized at higher temperatures. The melting temperature T_m for the loss of translational order is related to the Lamè elastic constants λ and μ by the criterion

$$\frac{4\pi k_B T_m}{a_L^2} = \left[\frac{\mu(T)(\mu(T)+\lambda(T))}{2\mu(T)+\lambda(T)}\right]_{T\to T_m} \tag{16}$$

where a_L is the lattice constant. For 2D jellium Thouless [42] estimated $\Gamma = 78$ at melting from eqn. (16).

Morf [38] has studied the temperature dependence of the shear modulus μ, which

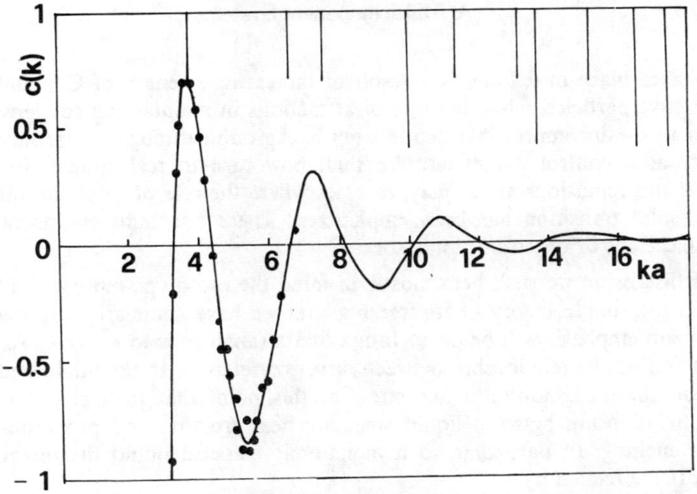

Figure 3. Direct correlation function $c(k)$ for the OCP in 2D at $\Gamma = 90$, from theory (curve) and computer simulation (dots). The vertical bars mark the location of the stars of RLV's for the triangular lattice. From P. Ballone et al. [37].

enters eqn. (16). In the solid phase μ is found to decrease linearly with increasing temperature. A drop in the shear wave frequency and onset of particle diffusion occur for $120 < \Gamma < 140$. From the behaviour of $\mu(T)$ and with an estimate of the core energy of dislocations, Morf obtains $\Gamma = 128$ at melting from eqn. (16). He also reports indications for the existence of an hexatic phase.

Gallet et al. [43] have been able to measure the shear modulus as a function of temperature in the quasi-2D system of electrons on liquid helium. Application of a magnetic field makes it possible to separate the transverse and longitudinal branches at long wavelengths. Resonances of the transverse modes in interaction with an oscillating electromagnetic field are detected in the solid phase and disappear at melting. A rapid decrease of shear modulus is observed near melting, in agreement with the KTHNY theory and with Morf's simulation. Further relevant experiments have been reproted by Mehrotra et al. [44], who measure the electron mobility and by Guo et al. [45], who study the scattering of electrons against ripplons.

The region of thermodynamic parameters where quantal effects would arise and melting would take place at increasing density is not easily accessible experimentally for the quasi-2D electron system. A crude estimate of the full phase diagram can be given [46] on the assumption that the critical coupling strength (ratio of potential to kinetic energy) for melting does not change much in going from the classical to the quantum limit. Indeed, Ceperley [20] estimates by simulation a value $r_s = 33$ for the Wigner transition in 2D quantal jellium, which corresponds to about half the value of the coupling strength for the transition in the classical limit. A similar approximate relation holds also in 3D. The phase diagram would be appreciably modified [47] by depositing the electron monolayer on a thin helium film overlying a dielectric substrate, owing to image-charge modifications of the electron-electron interaction. In particular, in such a configuration the electron system would be maintained in a quantal fluid state at low density and temperature.

3. Multicomponent Systems

Freezing takes place in jellium as a result of increasing strength of Coulomb repulsions between active particles, while the role of attractions in maintaining condensed phases is attributed to the incompressible neutralizing background through the artifice of treating its density as a control parameter. We shall now turn to real systems in which both attractions and repulsions are at play, in cases where the role of Coulomb interactions in the liquid-solid transition has been emphasized. These are multicomponent systems of ions and electrons or of various ionic species.

In our discussion we shall bear mostly in mind the prototype represented by the OCP in 3D and the simple theory of its freezing that we have summarily exposed in section 2.2. Our main emphasis will be on melting criteria (analogous to $e^2/(r_s k_B T_m) \approx 180$ for the OCP) and on the relationship between direct correlations in the liquid and crystallization in equilibrium conditions. We stress at this point that in a classical system this implies a relationship between liquid structure near freezing and properties of the hot solid near melting. In particular, in a monatomic classical liquid the direct correlation function $c(k)$ is related by

$$nc(k) = 1 - \frac{1}{S(k)} \tag{17}$$

to the liquid structure factor $S(k)$, which is known in many systems from diffraction experiments. A similar relation exists for an ionic classical liquid between the direct correlation functions for the various pairs of ionic components and the partial structure factors, which can be resolved from the total diffraction pattern of the liquid by the technique of isotopic enrichment in neutron diffraction [6].

The import of the relation that we have just recalled is that simple matching of the main structural features of the liquid from diffraction data with the location of reciprocal lattice points for the crystal already allows qualitative inferences on crystallization and hot-solid properties. Of course, the theory can be developed up to a full evaluation of the phase transition from given interatomic pair potentials through a calculation of bulk liquid structure and of the change in the free energy functional between the two phases. Further contact between theory and experiment could then be made through the order parameters n_G of the solid phase, which yield directly the values of the Debye-Waller factor at the Bragg reflections. We shall not dwell on such calculations, as our discussion of jellium in section 2 should suffice for an illustration of the approach and of its quantitative limitations.

3.1. Alkali Metals

The alkali metals, with perhaps the exception of Li, are ideal candidates for parallelism with the classical jellium model, in view of the relative smallness of their ionic cores and weakness of their electron-ion interactions. Standard-pressure melting occurs from a bcc crystal at values of the *ion-ion* coupling strength $e^2/(r_s k_B T_m)$ in the range from 210 for Li and Na to 180 for Cs. The entropy change on melting ($\Delta S/k_B = 0.80 \div 0.85$) is also close to the OCP value. However, a distinctly small but finite volume change on melting ($\Delta V/V = 0.016 \div 0.026$) reflects the finite compressibility of the 'background' of conduction electrons. Measurements of the melting curve of Na under pressure indicate [48] that the coupling strength at melting tends to a value of order 150 as the volume change on melting tends to zero with increasing pressure.

From the structural viewpoint, the OCP is preferable to the classical hard-spheres model as a reference liquid for variational calculations of thermodynamic properties of liquid alkalis [49,50]. Of course, account must be taken of screening of the ion-ion interactions by the conduction electrons in calculating liquid structure and properties. Starting from the structure factor of the OCP and treating screening by an optimized perturbational approach, one finds that the direct correlation function of the liquid metal is given by [51]

$$c(k) = c_0(k) - \frac{nu^*(k)}{k_B T} \tag{18}$$

where $c_0(k)$ is the direct correlation function of the OCP, which was shown in Fig. 2, and $u^*(k)$ is a perturbationally optimized form of the indirect ion-ion attraction via electronic screening. The results of such a calculation of the liquid structure for potassium near freezing are shown in Fig. 4 in comparison with diffraction data [52].

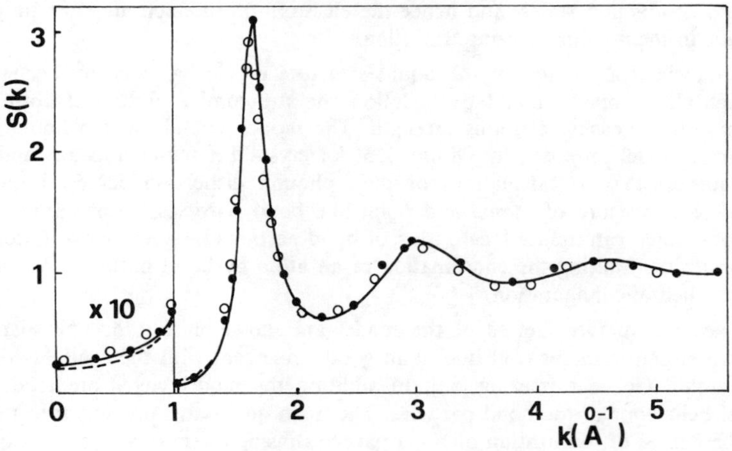

Figure 4. Structure factor $S(k)$ for liquid potassium near freezing, from theory (curves) and *X*-ray diffraction experiments (dots and circles). The magnitude of $S(k)$ in the small angle scattering region has been amplified by a factor of ten. From G. Pastore and M.P. Tosi [51].

Screening is indeed crucial at long wavelengths in order to recover a finite compressibility for the liquid metal and hence a finite value of $S(k)$ in the limit $k \to 0$. However, the structural consequences of screening are limited to the small-angle scattering region. All the structural features that are seen in Fig. 4 above $k \approx 1.5$ Å^{-1} are essentially unaffected by electronic screening and therefore closely follow those of the OCP at the appropriate value of the bare ion-ion coupling strength. Since $c(k) \approx c_0(k)$ in this whole region of wavenumber, a detailed microscopic theory of freezing in the alkalis should closely parallel the theory discussed in section 2.2 for the OCP, with the important difference that attention should be shifted from the interfacial drop to the volume change across the phase transition in treating the $k \to 0$ contributions to the coexistence and equilibrium conditions.

3.2. Elemental and Polar Semiconductors

Group IV elemental and group III - group V polar semiconductors crystallize at standard pressure in tetrahedrally coordinated open structures. Melting occurs with a *negative* volume change and a change of electronic transport character to metallic-type conductivity. The first neighbour coordination number increases on melting to a value of order 7. Yet the liquid structure of Ge and Si is quite distinct from that of other liquid metals [53]. In particular, the first neighbour coordination is still relatively low, but a second shell of neighbours is seen to lie at a short distance beyond the first shell, in a region where the pair distribution function in other liquid metals shows its main minimum. Similarly, the liquid structure factor shows a distinctive shoulder on the large - k side of its main peak.

Although the above structural characteristics can be approximately accounted for by means of effective interatomic pair potentials constructed by pseudopotential theory [54], one is clearly dealing with liquid metals which in a sense are at an opposite extreme relative to the alkalis. The drastic changes in atomic and electronic structure of these systems on melting are evidently associated with a release of valence electrons from interatomic bonds into conduction states, and hence the electron-ion interactions must be playing a crucial role in the melting/freezing transition.

A pseudoclassical model for the liquid structure of Ge has recently been proposed [55], which allows one to qualitatively follow the structural evolution of both ions and electrons with increasing coupling strength. The model extends to the liquid state the bond charge model proposed by Phillips [56] for crystalline semiconductors and used by various authors [57] in calculations of their phonon dispersion curves. Liquid Ge is described as a mixture of atoms and point-like bond particles, with mutual attractive interactions which can induce localization of bond particles between pairs of atoms under steric constraints limiting the coordination of an atom by bond particles to a maximum of 4 in tetrahedral configuration.

The partial structure factors of the model are shown in Fig. 5. The shape of the atom-atom structure factor (full line) is in good agreement with the available diffraction data on liquid Ge near freezing [53]. In addition the model yields predictions on the structural behaviour of the bond particles. The main qualitative prediction in this regard is that the degree of localization of bond particles inside interatomic bonds is quite high in the liquid near freezing, though not complete. This result is consistent with molten semiconductors being rather poor metallic conductors and should be further tested by full quantal calculations.

The strength of the atom-bond particle attractions, which are responsible for bond particle localization and for the consequent distinctive features of liquid structure, can be measured either through the valence-conduction band gap in the crystal (in units of $k_B T$) or through the 'plasma parameter' $z^2 e^2/(r_s k_B T)$, where z is the charge on a bond particle and $r_s = (4\pi n)^{-1/3}$ with n the atomic number density. In either case a melting criterion emerges from the theory. In particular, the characteristic atomic structure shown in Fig. 5 arises in the model for $z \approx 0.5$, which is the value estimated by Phillips [56] for the bond charge in Ge as $z = 2/\sqrt{\epsilon_0}$, with ϵ_0 the static dielectric constant of the crystal. This value yields the melting criterion

$$4e^2/(\epsilon_0 r_s k_B T_m) \approx 20 \tag{19}$$

which is indeed found [55] to be empirically satisfied in Si, Ge and the III-V compounds. The corresponding melting criterion involving the band gap E_g is $k_B T_m \approx E_g/10$.

Referring again to Fig. 5, we have marked at its top the locations of the allowed Bragg reflections of the diamond crystal structure, which is based on the fcc lattice. We see that there is good correspondence between the (220) and (311) Bragg diffraction spots and the shoulder in $S_{GeGe}(k)$. Furthermore, these RLV stars match the strong peak in

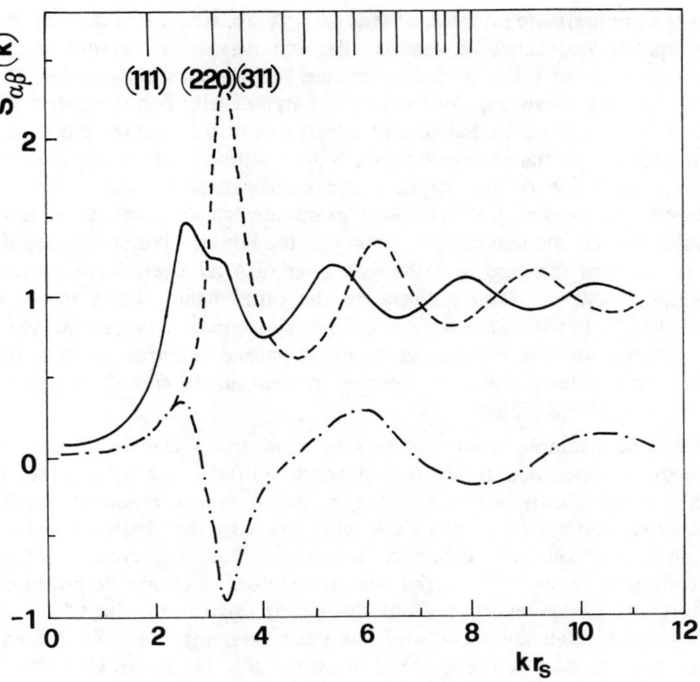

Figure 5. Partial structure factors $S_{\alpha\beta}(k)$ for a bond particle model of liquid Germanium near freezing. The various curves show the atom-atom (full line), bond-bond (dashed) and atom-bond (dotted-dashed) structure factors. The vertical bars mark the location of the Bragg diffraction spots from the diamond structure. From results of *A*. Ferrante and M.P. Tosi [55].

$S_{BB}(k)$ and the deep valley in $S_{AB}(k)$. The interpretation of the latter structural features is that in the liquid near freezing there is strong short-range order in the subsystem of bond particles and in their alternation with atoms in space.

Thus the Figure suggests that crystallization in Ge should be primarily looked upon as 'freezing of bonds' driven by tetrahedrally constrained attractions between ionic cores and valence electrons, rather than as a direct freezing of atoms into a crystal structure. It remains a challenge whether this qualitative suggestion can be developed into a microscopic theory of the phase transition, which should also account for the uncommon sign of the volume change on melting. It is also interesting to note that in the observed diffraction pattern of amorphous Ge [58] the shoulder of the liquid structure factor has evolved into a main peak at essentially the same wavenumber, whereas the main peak of the liquid structure factor has become a prepeak in correspondence with the (111) star of RLV's.

3.3. Ionic Systems

We shall pursue our main theme here in relation to freezing of ionic systems by a brief discussion of two classes of molten salts [59] and by an illustration of two further examples taken from areas outside conventional condensed-matter physics.

It was noticed by Reiss et al. [60] that standard-pressure melting of alkali halides satisfies the approximate criterion $e^2/(ak_BT_m) \approx 70$, where a is the sum of ionic radii of the two species and hence, in essence, the first neighbour distance in the crystal. The observed structure of a liquid such as molten NaCl is most simply described as a superposition of strong short-range order in the charge distribution (i.e. alternation of Na and Cl ions in space) onto excluded-volume effects due to the almost rigid ionic cores. Charge ordering yields a partial structure factor with a strong peak in correspondence with the (111) Bragg reflection of the crystal, whereas the partial structure factor associated with species-averaged density fluctuations is essentially gas-like except at low k, where its small value reflects the low compressibility of the liquid. Charge ordering drives crystallization at standard pressure with the assistance of a relatively large volume change [61]. The melting curve of alkali halides, on the other hand, shows in its shape a rapid decrease of $\Delta V/V$ with increasing pressure. The liquid structure evolves with pressure from Coulomb ordering to monatomic dense-packed ordering, so that freezing at very high pressure and temperature is expected to yield an fcc crystal on which the two ionic species are disordered [62].

Our second example from molten salts concerns a class of fluorite-type materials which acquire superionic conduction properties in the hot solid phase below melting through a rapid continuous disordering of the anionic component. $SrCl_2$ is the most extensively studied system in this class, when one considers both the solid and the liquid phase. The most prominent feature in the observed liquid structure of $SrCl_2$ near freezing is the main peak in the Sr-Sr partial structure factor, which can be put in correspondence with the (111) Bragg reflection from the fluorite structure. The Cl-Cl structure factor shows instead a relatively broad and low peak overlying the (200) reflection. Approximate account can be given [63] of the measured $S_{SrSr}(k)$ by an OCP model at coupling strength $\Gamma \approx 70$, as estimated from $\Gamma = 4e^2/(\epsilon_\infty r_s k_B T_m)$ where ϵ_∞ is the electronic dielectric constant accounting for screening of the Sr-Sr Coulomb repulsions by the 'background' of chlorines. Freezing of molten $SrCl_2$ can accordingly be described at an elementary level as an instance of classical Wigner crystallization of the Sr ions into an fcc sublattice, with the Cl component being strongly disordered though modulated by the Sr sublattice [61]. The superionic-to-normal transition which occurs in the crystal with further cooling can in turn be described as a process of continuous freezing of the modulated Cl component, primarily associated with a continuous growth of the (200) Cl order parameter from its finite value near melting [64]. The calculated behaviour of this order parameter with temperature across the continuous transition is in approximate agreement with diffraction data and directly yields the characteristic peak in the heat capacity which marks the superionic transition in fluorite materials.

After the foregoing examples from molten salts, it is of interest to briefly discuss the ordering exhibited by assemblies of charged macromolecules or macroparticles in aqueous suspensions. Solid-like structures can be formed for sufficiently strong interactions and short-range ordering can persist at relatively low ionic strength. These ordering effects are known for both biological and inorganic materials and have been studied in considerable quantitative detail in aqueous dispersions of charged spherical polystyrene particles. Screening by ions in solution leads to an effective Coulomb interaction between charged macroparticles which can be described by a Debye-Hückel screened repulsion under conditions that the Debye screening length is much larger than the particle diameters. To give an illustrative example, in the experiments of Brown et al. [65] the particle diameter is of order 250Å, the Debye screening length of order 3000Å and the mean interparticle distance of order 5000Å, yielding an effective repulsion between first neighbours of order $20k_BT$ which is responsible for liquid-like order in the macroparticle system as observed by light scattering. The phase diagram under screened Coulomb repulsions has been evaluated by Shih et al. [66], who find the fcc structure to be dominant at high densities while the bcc structure can become stable in dilute suspensions when the particle charge is high and the screening length is large. There is broad agreement between these theoretical results and observations, although the latter also show [67] a reduction of interparticle spacing in colloidal crystals of highly charged particles. It is also remarkable that recent studies of 2D suspensions of highly charged particles by optical microscopy and digital imaging [68] have given evidence for a two-step melting process occurring through an intermediate hexatic-type phase as predicted by the KTHNY theory.

Finally, to mention a recent example with relevance to astrophysics, freezing in a mixture of carbon and oxygen plasmas has been evaluated by Barrat et al. [69]. The phase diagram of the binary plasma is obtained by linearized density-wave theory with exclusion of the order parameter associated with the (200) star of the RLV's of the bcc lattice (see the discussion in section 2.2). In contrast with earlier theoretical predictions, the two nuclear species are found to be miscible in the solid phase, which is slightly richer in oxygen than the fluid phase. The authors also discuss the astrophysical consequences of this result with special attention to the white dwarf luminosity function.

4. Summary

We have emphasized in this brief review the special features of Coulomb interactions in driving phase transitions from disordered to ordered states. The discussion of the formal theory has been focussed for simplicity on the case of jellium, stressing (i) the existence of freezing/melting criteria in the form of specific values for the coupling strength at the phase transition and (ii) the relevance of the liquid-state density response to the microscopic description of the phase transition. The usefulness of these concepts has been illustrated by a number of examples regarding systems of increasing complexity. Two main conclusions can be drawn. Firstly, melting/freezing criteria apply for classes of similarly bonded materials, which are 'universal' as far as this expression can be used in relation to a first-order phase transition. Secondly, the relation between liquid structure and freezing in a classical system can be usefully exploited even at a level of qualitative understanding. As a matter of fact, much more abundant and accurate information is available on the structure of liquids near freezing than on solids near melting.

Acknowledgements

This review was prepared during the 1988 Research Workshop in Condensed Matter, Atomic and Molecular Physics at the International Centre for Theoretical Physics, Trieste. One of us (G.S.) wishes to thank Professor Abdus Salam, the International Atomic Energy Agency and UNESCO for hospitality at the Centre. We are grateful to Dr. A. Ferrante and Dr. G. Pastore for their collaboration and useful discussions. Finally, we wish to acknowledge the constant support given to our work by the Ministero della Pubblica Istruzione and the Consiglio Nazionale delle Ricerche of Italy.

Appendix. Electrochemical Potential in the Jellium Model

Here we consider the thermodynamic limit for the jellium model in the canonical ensemble, in relation to its electrochemical potential. The temperature is taken as fixed and therefore not explicitly considered. The potential energy function is

$$U(N,N_b) = \frac{1}{2}\sum_{i\neq j}^{N}\frac{e^2}{|\mathbf{r}_i-\mathbf{r}_j|} - \sum_{i}^{N}\frac{N_b}{V}\int_V d\mathbf{r}\frac{e^2}{|\mathbf{r}-\mathbf{r}_i|} + \frac{N_b^2}{2V^2}\int_V d\mathbf{r}\int_V d\mathbf{r}'\frac{e^2}{|\mathbf{r}-\mathbf{r}'|} \quad \text{(A.1)}$$

with the background density given by $n_b = N_b/V$, in terms of the number N_b of smeared background particles.

We recall that for any finite volume V it is possible to consider systems that are not neutral [70], provided that in the thermodynamic limit ($N, V \to \infty$ with $N/V \to n$) the condition $(N-N_b)/V^{2/3} \to 0$ is satisfied. Thus, for any finite volume we can define electrochemical potentials for both electrons and background particles according to

$$\mu = \frac{\partial F(N,N_b,V)}{\partial N}\Big|_{N_b,V} \quad \mu_b = \frac{\partial F(N,N_b,V)}{\partial N_b}\Big|_{N,V}. \quad \text{(A.2)}$$

From the expression of F in terms of the canonical partition function and exploiting the explicit form of the potential function of eqn. (A.1), it is immediately found that the background electrochemical potential is given by

$$\mu_b = <\frac{\partial U}{\partial N_b}> = <\frac{1}{V}\int_V d\mathbf{r}\sum_i^N\frac{e^2}{|\mathbf{r}-\mathbf{r}_i|} + \frac{N_b}{V^2}\int_V d\mathbf{r}\int_V d\mathbf{r}'\frac{e^2}{|\mathbf{r}-\mathbf{r}'|}> \quad \text{(A.3)}$$

$$= -\frac{1}{V}\int_V d\mathbf{r}\int_V d\mathbf{r}'\frac{e^2}{|\mathbf{r}-\mathbf{r}'|}[n(\mathbf{r})-n_b] \equiv -V_M$$

with $<...>$ denoting the canonical average and V_M being the spatial average of the Madelung potential.

Once the thermodynamic limit is taken, the distinction between density of electrons and density of background is lost because of charge neutrality. Thus, the Helmholtz free energy density $f(n)$ only depends on a single density n. This implies that knowledge of the bulk free energy density $f(n)$ only allows the calculation of

$$\mu^* \equiv \frac{\partial f(n)}{\partial n} = \mu + \mu_b = \mu - V_M. \quad \text{(A.4)}$$

References

1. E.P. Wigner, *Phys. Rev.* **46**, 1002 (1934) and *Trans. Faraday Soc.* **34**, 678 (1938).
2. C.M. Care and N.H. March, *Adv. Phys.* **24**,101 (1975).
3. D.M. Ceperley and B.J. Alder, *Phys. Rev. Lett.* **45**, 567 (1980).
4. S.G. Brush, H.L. Sahlin and E. Teller, *J. Chem. Phys.* **45**, 2102 (1966).
5. E.L. Pollock and J.P. Hansen, *Phys. Rev.* **A8**, 3110 (1973); W.L. Slattery, G.D. Doolen and H.E. DeWitt, *Phys. Rev.* **A21**, 2087 (1980) and **A26**, 2255 (1982); S. Ogata and S. Ichimaru, *Phys. Rev.* **A36**, 5451 (1987).
6. For a general introduction see N.H. March and M.P. Tosi, *Coulomb Liquids*, Academic, London (1984).
7. J.W. Gibbs, *Elementary Principles in Statistical Mechanics*, Dover, New York (1960).
8. P. Hohenberg and W. Kohn, *Phys. Rev.* **136**, B864 (1964).
9. N.D. Mermin, *Phys. Rev.* **137**, A1441 (1964).
10. M. Rovere and M.P. Tosi, *J. Phys.* **C18**, 3445 (1985).
11. G. Senatore and M.P. Tosi, *N. Cimento* **B56**, 169 (1980).
12. R. Lovett, C.Y. Mou and F.P. Buff, *J. Chem. Phys.* **65**, 570 (1976).

13. W. Kohn and L.J. Sham, *Phys. Rev.* **140**, 1133 (1965). See also W. Kohn and P. Vashishta, in *Theory of the Inhomogeneous Electron Gas*, ed. S. Lundqvist and N.H. March, Plenum, New York (1983).
14. G. Pastore and G. Senatore, to be published.
15. T.V. Ramakrishnan and M. Yussouff, *Phys. Rev.* **B19**, 2775 (1979).
16. A.D.J. Haymet and D.W. Oxtoby, *J. Chem. Phys.* **74**, 2559 (1981).
17. See, for instance, K.S. Singwi and M.P. Tosi, *Solid State Phys.* **36**, 177, ed. H Ehrenreich, F. Seitz and D. Turnbull, Academic, New York (1981).
18. K.S. Singwi, M.P. Tosi, R.H. Land and A. Sjölander, *Phys. Rev.* **176**, 589 (1968).
19. Tabulation of Fixed Node pair correlation functions for both unpolarized and polarized electron liquids were kindly provided to one of us (GS) by D.M. Ceperley.
20. D.M. Ceperley, *Phys. Rev.* **B18**, 3126 (1978).
21. A. Baldereschi, *Phys. Rev.* **B7**, 5212 (1973); D.J. Chadi and M.L. Cohen, *Phys. Rev.* **B8**, 5747 (1973).
22. H.M. van Horn, *Phys. Lett* **28A**, 706 (169).
23. D.K. Chaturvedi, G. Senatore and M.P. Tosi, *N. Cimento* **B62**, 375 (1981).
24. F.J. Rogers, D.A. Young, H.E. DeWitt and M. Ross, *Phys. Rev.* **A28**, 2990 (1983).
25. J.L. Barrat, *Europhys. Lett.* **3**, 523 (1987); J.L. Barrat, J.P. Hansen and G. Pastore, *Mol. Phys.* **63**, 747 (1988).
26. H. Iyetomi and S. Ichimaru, *Phys. Rev.* **B38**, 6761 (1988).
27. C.C. Grimes and G. Adams, *Phys. Rev. Lett.* **42**, 795 (1979).
28. V.B. Shikin, *JETP Lett.* **19**, 335 (1974).
29. D.S. Fisher, B.I. Halperin and P.M. Platzman, *Phys. Rev. Lett.* **42**, 798 (1979).
30. K.J. Strandburg, *Rev. Mod. Phys.* **60**, 161 (1988).
31. R.E. Peierls, *Ann. Inst. Henri Poincaré* **5**, 177 (1935).
32. L.D. Landau, *Phys. Z. Sowjetunion* **11**, 26 (1937).
33. N.D. Mermin, *Phys. Rev.* **176**, 250 (1968).
34. R.W. Hockney and T.R. Brown, *J. Phys.* **C8**, 1813 (1975).
35. R.C. Gann, S. Chakravarty and G.V. Chester, *Phys. Rev.* **B20**, 326 (1979).
36. R.K. Kalia, P. Vashishta and S.W. de Leeuw, *Phys. Rev.* **B23**, 4794 (1981).
37. P. Ballone, G. Pastore, M. Rovere and M.P. Tosi, *J. Phys.* **C18**, 4011 (1985).
38. R.H. Morf, *Phys. Rev. Lett.* **43**, 931 (1979) and *Helv. Phys. Acta* **56**, 743 (1983).
39. B.I. Halperin and D.R. Nelson, *Phys. Rev. Lett.* **41**, 121 (1978); D.R. Nelson and B.I. Halperin, *Phys. Rev.* **B19**, 2457 (1979).
40. A.P. Young, *Phys. Rev.* **B19**, 1855 (1979).
41. J.M. Kosterlitz and D.J. Thouless, *J. Phys.* **C6**, 1181 (1973).
42. D.J. Thouless, *J. Phys.* **C11**, L189 (1978).
43. F. Gallet, G. Deville, A. Valdes and F.I.B. Williams, *Phys. Rev. Lett.* **49**, 212 (1982).
44. R. Mehrotra, B.M. Guenin and A.J. Dahm, *Phys. Rev. Lett.* **48**, 641 (1982).
45. C.J. Guo, D.B. Mast, Y.-Z. Ruan, M.A. Stan and A.J. Dham, *Phys. Rev. Lett.* **51**, 1461 (1983).
46. P.M. Platzman and H. Fukuyama, *Phys. Rev.* **B10**, 3150 (1974).
47. F.M. Peeters and P.M. Platzman, *Phys. Rev. Lett.* **50**, 2021 (1983).
48. V.A. Ivanov, I.N. Makarenko, A.M. Nikolaenko and S.M. Stishov, *Phys. Lett.* **45A**, 18 (1973).
49. H. Minoo, C. Deutsch and J.P. Hansen, *J. Phys. Lettres* **38**, L191 (1977).
50. M. Ross, H.E. DeWitt and W.B. Hubbard, *Phys. Rev.* **A24**, 1016 (1981); K.K. Mon, R. Gann

and D. Stroud, *Phys. Rev.* **A24**, 2145 (1981).

51. D.K. Chaturvedi, M. Rovere, G. Senatore and M.P. Tosi, *Physica* **111B**, 11 (1981); G. Pastore and M.P. Tosi, *Physica* **124B**, 383 (1984).
52. A.J. Greenfield, J. Wellendorf and N. Wiser, *Phys. Rev.* **A4**, 1607 (1971); M.J. Huijben and W. van der Lugt, *Acta Cryst.* **A35**, 431 (1979).
53. Y. Waseda, *Structure of Non-Crystalline Materials*, McGraw-Hill, New York (1980).
54. J. Hafner and G. Kahl, *J. Phys.* **F14**, 2259 (1984).
55. A. Ferrante, Wang Li and M.P. Tosi, *Phil. Mag.* **A58**, 13 (1988); A. Ferrante and M.P. Tosi, *J. Phys.: Cond. Matter* **1**, 1679 (1989).
56. J.C. Phillips, *Covalent Bonding in Crystals and Molecules*, University of Chicago Press (1970).
57. R.M. Martin, *Phys. Rev.* **186**, 871 (1969); W. Weber, *Phys. Rev.* **B15**, 4789 (1977).
58. G. Etherington, A.C. Wrigth, J.T. Wenzel, J.C. Dore, J.H. Clark and R.N. Sinclair, *J. Non-Cryst. Solids* **48**, 265 (1982).
59. For a recent review see M. Rovere and M.P. Tosi, *Rep. Progr. Phys.* **49**, 1001 (1986).
60. H. Reiss, S.W. Mayer and J.L. Katz, *J. Chem. Phys.* **35**, 820 (1961).
61. B. D'Aguanno, M. Rovere, M.P. Tosi and N.H. March, *Phys. Chem. Liquids* **13**, 113 (1983).
62. J.L. Barrat, *J. Phys.* **C20**, 1031 (1987).
63. G. Pastore, P. Ballone and M.P. Tosi, *J. Phys.* **C19**, 487 (1986).
64. M. Rovere and M.P. Tosi, *Solid State Commun.* **55**, 1109 (1985).
65. J.C. Brown, P.N. Pusey, J.W. Goodwin and R.H. Ottewill, *J. Phys.* **A8**, 664 (1975).
66. W.Y. Shih, I.A. Aksay and R. Kikuchi, *J. Chem. Phys.* **86**, 5127 (1987).
67. N. Ise, T. Okubo, M. Sugimura, K. Ito and H.J. Nolte, *J. Chem. Phys.* **78**, 536 (1983); T. Yoshiyama, I. Sogami and N. Ise, *Phys. Rev. Lett.* **53**, 2153 (1984).
68. C.A. Murray and D.H. Van Winkle, *Phys. Rev. Lett.* **58**, 1200 (1987).
69. J.L. Barrat, J.P. Hansen and R. Mochkovitch, *Astron. Astrophys.*, in press.
70. E.H. Lieb and H. Narnhofer, *J. Stat. Phys.* **12**, 291 (1975).

PINNED CHARGE DENSITY WAVES AS MODELS OF GLASSY BEHAVIOR

L. PIETRONERO

Dipartimento di Fisica
Università di Roma 'La Sapienza'
P.le A. Moro 2, 00185
Roma, Italy

We discuss a simple model for the dynamics of a pinned Charge Density Wave inspired by the properties of $NbSe_3$. We show this model to be useful to investigate general questions related to the dynamics of glassy systems like the stretched exponential relaxation. The possibility that this behavior may be due to a hierarchical coupling between degrees of freedom is analyzed from a microscopic point of view.

1. Introduction

A remarkable series of experiments on quasi one dimensional metals like $NbSe_3$ and similar systems have provided evidence for a contribution to non linear transport due to the coherent sliding of a charge density wave (CDW) [1]. Above a threshold electric field the CDW seems to depinn and to contribute to the electric current. Below this threshold the system shows a number of properties that are characteristic of glassy systems like hystheresis effects [1] and streched exponential relaxation [2].

A model Hamiltonian for these systems consists of the overdamped dynamics of the CDW pinned by random impurities [1,3,4]. This model is able to explain various properties of the depinning transition and the noise spectrum above threshold. In addition it seems able to give some clues on the glassy properties. In view of its simplicity and the fact that it contains some basic ingredients common to glassy systems [5,6] like frustration effects, the existence of many metastable states with very similar energies etc., one can use it to ask questions of more general nature. In particular we consider here the microscopic nature of the dynamics that gives rise to streched exponential relaxation. It has been proposed, at a purely phenomenological level, that this may be due to a hierarchical coupling between degrees of freedom [7,8]. This model indeed shows this type of relaxation [9] and its theoretical analysis gives a clue on the microscopic origin of the hierarchical couplings. For this class of questions concerning dynamical properties the present model appears to be more suitable than spin glass models [6].

Part of the work discussed in this paper was done in collaboration with A. Erzan, A.G. de Vries, M. Versteeg and R. Heijungs.

In Section 3 we discuss some general questions related to the dynamics of metastable states in glasses and we introduce the concept of hierarchically coupled degrees of freedom.

In Section 4 we present a simple theoretical treatment of the model that shows the possible microscopic origin of hierarchical couplings that lead to a streched exponential relaxation.

R. Girlanda et al. (eds.), Progress on Electron Properties of Solids, 239–246.

In Section 5 we summarize the main results and outline the possible developments.

2. Model Hamiltonian for Pinned Charge Density Waves

In this section we show the origin of a model Hamiltonian for a pinned charge density wave as it has been proposed to describe the properties of the one dimensional conductor $NbSe_3$ [1]. We are then going to show how this Hamiltonian represents also an interesting model to study some questions of general interest for the properties of metastable systems and glasses.

A one dimensional metal is unstable with respect to the Peierls distortion. This gives rise to a charge density wave of the form [3,4]

$$\rho(x) = e\rho_0[1+C\cos(q_0x+\phi(x))] \tag{2.1}$$

where $\rho(x)$ is the charge density, $\phi(x)$ is the phase, $e\rho_0$ is the charge, q_0 is the Fermi wavevector and C the amplitude. For a Peierls distortion in a system with density of states N_0 the amplitude C can be related to the microscopic parameters by [4]

$$C = \frac{N_0\Delta}{a\lambda\rho_0} \tag{2.2}$$

where 2Δ is the energy gap, a the lattice constant and λ the electron phonon coupling.

We are now going to consider the function of the phase $\phi(x)$ due to the effect of pinning impurities. The amplitude fluctuations are neglected because they would involve larger energies. A charged impurity produces an electrostatic potential that tends to attract (or repel, depending on its sign) the regions with charge excess. The sliding of the CDW is then subject to the periodic potentials of the N impurities located at random positions. The total pinning potential is given by

$$V_p = V_0\sum_{j=1}^{N}\rho(x)\delta(x-x_j). \tag{2.3}$$

The phase dynamics can be described by the Hamiltonian of an elastic string [4]

$$H = \frac{1}{q_0^2}\int dx\left[\frac{1}{2}\rho_m\left[\frac{\partial\phi}{\partial t}\right]^2+\frac{1}{2}K\left[\frac{\partial\phi}{\partial x}\right]^2\right] \tag{2.4}$$

where the first term represents the kinetic energy (that we finally neglect) and the second one the elastic energy. In addition we assume a damping mechanism that, for mathematical convenience (this is not a serious restriction however), acts only at the impurity positions. This gives rise to a force

$$F_d = -\frac{1}{q^2}\frac{\rho m}{\tau}\sum_{j=1}^{N}\left[\frac{\partial\phi}{\partial t}\right]\delta(x-x_j) \tag{2.5}$$

where τ determines the strength of the damping. Finally we allow an external field E to applied, which shifts the whole CDW. This produces a potential

$$U_E = -\frac{e\rho_0}{q_0}E\phi. \tag{2.6}$$

The dynamics of this system is in general overdamped because of the small mass density with respect to damping effects. In this limit the equation of motion, including all terms, is [4]

$$-\frac{K}{q_0^2}\left[\frac{\partial^2\phi}{\partial x^2}\right]+\sum_{j=1}^{N}\delta(x-x_j)\left\{-\frac{\rho m}{q_0^2\tau}\left[\frac{\partial\phi}{\partial t}\right]+\right.$$
$$\left.+e\rho_0 CV_0\sin[q_0x_i+\phi(x)]\right\}+\frac{e\rho_0}{q_0}E=0.$$

This equation can be integrated analytically between two neighbouring impurity sites $(x_j;x_{j+1})$. This leads to a very simple equation for the phases at impurity sites only $(\psi_j\equiv\phi(x_j)/2\pi)$ [4]

$$\frac{\partial\psi_j}{\partial t}=B[\psi_{j+1}-2\psi_j+\psi_{j-1}]-\sin[2\pi(\psi_j+\phi_j^0)]+E. \tag{2.8}$$

The first term on the right represents the elastic coupling between the phases ψ_j at neighbouring impurity sites. The parameter $B\propto K/V_0$ represents the coupling constant of the model and can be related to the microscopic parameters. The phases $\phi_j^0=q_0x_j/2\pi$ are purely random (uniform distribution) and represent the quenched disorder due to the random position of the impurities.

In summary the model of Eq. (2.8) corresponds to an elastic string pinned by impurity potentials with quenched random phases and subject to an external field E as schematically shown in Figure 1.

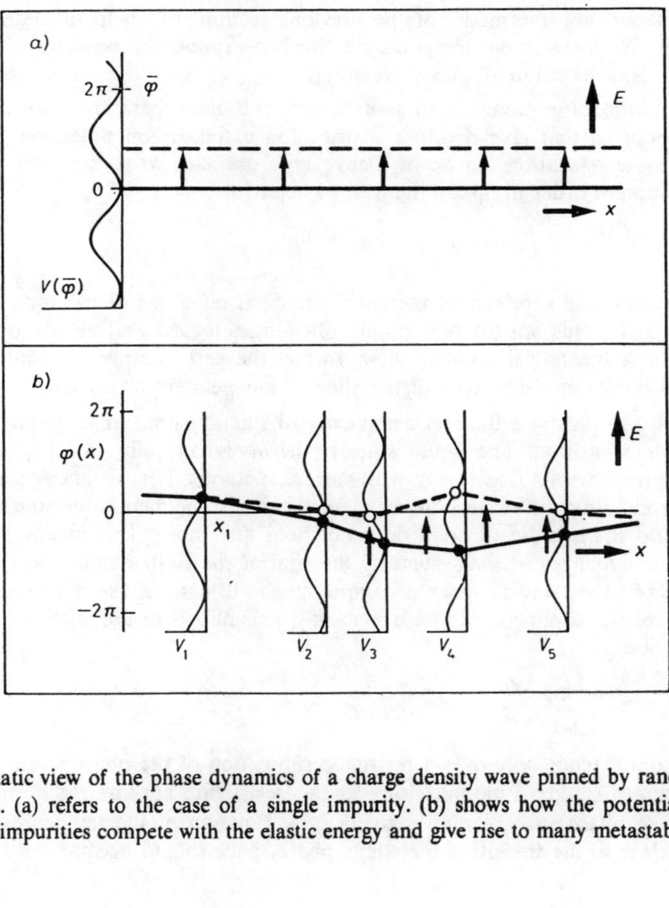

Figure 1. Schematic view of the phase dynamics of a charge density wave pinned by randomly located impurities. (a) refers to the case of a single impurity. (b) shows how the potentials of the various pinning impurities compete with the elastic energy and give rise to many metastable states.

This model has been used to describe a number of properties of $NbSe_3$ like for example the depinning transition [4,10]. For small values of the applied field E the CDW relaxes in a local minimum and it does not move asymptotically. By increasing E there will be a critical field E_c above which the potentials of the impurities are not able any more to localize the CDW that begins to slide contributing to the d.c. conductivity of the system. This dynamics gives rise to a peculiar noise in the power spectrum of the current that was also observed [1,3]. Other experiments in these systems show effects that are characteristic of all glassy systems like hystheresis and streched exponential relaxation [2]

$$P(t) = P_0 e^{-(t/\tau)^{\nu}}; \quad (0<\nu<1). \tag{2.9}$$

We are going to show that the model represented by Eq. (2.8), in addition to describing pinned CDW, it also represents an interesting Hamiltonian model to study various properties of glassy systems. In fact it has many metastable states of almost comparable energy, it shows hystheresis effects and the relaxation dynamics consist in a collective motion from one metastable state to another.

3. Hierarchical Dynamics in Glasses

Here we briefly discuss some concepts related to the dynamics of glassy systems and we will point out how the model of the previous section may help in understanding these properties. We focus in particular on the streched exponential behavior of relaxation (Eq. (2.9)) that is observed in all glassy systems [5,11].

The traditional approach is to assume that in a glass there are many types of relaxation phenomena that give rise to a distribution of relaxation times $W(\tau)$ [5]. Assuming each of these relaxations to be of Debye type one can write the total relaxation as a superposition in order to obtain the desired behavior.

$$\frac{P(t)}{P_0} = \int d\tau W(r) e^{-t/\tau} = e^{-(t/\tau)^{\nu}}. \tag{3.1}$$

Given the streched exponential therefore one determines the distribution $W(\tau)$ that may give rise to it. This approach is purely phenomenological and clearly does not explain much from a theoretical point of view. In fact the very complex dynamics of the glassy system in simply modelled by a distribution of independent relaxations.

In order to illustrate the real complexity of this dynamics it is useful to consider the glass-liquid transition. The liquid state is thermodynamically well behaved and one is able to write down its free energy with various methods. For the glassy state the problem is completely different because the barriers between some metastable states are practically infinite and the ergodic theorem does not hold any more. This means that one cannot replace the dynamics of the system by the sum of the Boltzmann factors over the whole phase space. One way to make a compromise is to assume the dynamics to be ergodic within a restricted portion of phase space [5]. This allows to use again to the usual partition function

$$Z = \sum_{\{\tilde{x}\}} e^{-\beta H\{x\}}; \quad (B = \frac{1}{k_B T}) \tag{3.2}$$

but the sum extends only over a restricted subsection of the phase space $\{\tilde{x}\}$. The problem is then of standard nature. However this restriction appears rather arbitrary. In fact the allowed phase space should actually be a function of the temperature $\{\tilde{x}\} = f(T)$ because close to the transition the whole phase space should became fully available. Also from a microscopic point of view one may think that, as the temperature in increased, some excitations may occur that open the way to more accessible phase space. This suggests that the main effect of the temperature is not just to appear in the Boltzmann factor but to give rise to a sort of invasion of phase space of dynamical nature [8].

These ideas are closely related to the proposal that relaxations in glasses are coupled in a hierarchical way [7]. This is also reminiscent of the ultrametric structure of phase space in spin glasses [6]. Consider for example a hierarchy of degrees of freedom with constraints only between successive levels. Relaxation of $a+1$-the level degree of freedom can only take place if μ_n n-th level degrees attained one particular state of all those possible. This implies the relation

$$\tau_{n+1} = 2^{\mu_n}\tau_n \tag{3.3}$$

between the corresponding relaxation times. Given that N_n is the number of degrees of freedom of the n-th level the total relaxation function is then

$$P(t) = \frac{1}{N}\sum_{n=0} N_n e^{-(t/\tau_n)}. \tag{3.4}$$

If one assumes furthermore that

$$\mu_n = \mu_0 \cdot n^{-p}; \ (p>1) \tag{3.5}$$

and

$$N_{n-1} = N_n/\lambda \tag{3.6}$$

one can see that Eq. (5.2) leads to a streched exponential relaxation [7]

$$P(t) = P_0 e^{-(t/\tau)^{\nu}} \tag{3.7}$$

with

$$\nu = (1+\mu_0 \ln 2)^{-1}. \tag{3.8}$$

Clearly this scheme is also phenomenological but it suggests a picture for the relaxation dynamics that is non trivial and that may eventually be supported by microscopic arguments. We are going to see in the next section that, using an approximate theoretical treatment, one can actually derive a hierarchical relaxation from Eq. (2.8) even though not exactly of the type we have discussed here.

4. A Simple Theory

If we consider the model system described by Eq. (2.8) (see also Figure 1) one can see that the relaxation dynamics is coupled in a hierarchical way. For example suppose the phase at the j-th impurity crosses the maximum of the potential giving rise to a change of configuration. After this event it is much easier, in view of the elastic energy, to change configuration in the same direction also for two neighbouring impurities.

Let us consider now the system relaxed in a local metastable state and we study how the system relaxes if an external field E is applied. It is useful to introduce a local threshold field χ_j (acting only at the j-th impurity site) that is the minimum field necessary to produce a change of configuration for the phase at this point. The values of these local threshold field will have a broad distribution in view of the different elastic forces acting on each point. For example this field is rather small at point 4 of Figure 1 before relaxation (black line) while it becomes rather large after relaxation (dashed lines). The elastic force due to the neighbours act in fact in opposite directions in the two cases. However, after the relaxation at point 4 the local threshold fields at the neighbouring points become smaller by an amount that is about half (for each) of the total increase of the threshold field at point 4.

We can try to cast this discussion in a mathematical form. First we introduce the distribution $W(\chi_i)$ for the values of the total threshold fields and we assume it to be flat over a range of size W. This means that

$$W(\chi_i) = h \quad (0 \leqslant \chi_i \leqslant W) \tag{4.1}$$

and it is zero otherwise. Since each point will have some local threshold field we have

$$h \cdot W = N. \tag{4.2}$$

If a field E is now applied to the whole system there will be a number of impurity sites in which this field is larger than the local threshold. In view of Eqs. (4.1) and (4.2) this number is

$$N_1 = h \cdot E. \tag{4.3}$$

The phases at these points will now change their configuration with a characteristic time τ_0. After these events take place however the distribution of local threshold fields is modified. In particular the N_1 sites that have just relaxed will have a large value of χ_j while the threshold fields of their neighbours (that are about $N' = 2N_1$ if $N_1 << N$) will be much lower and some of them will become also lower then the applied field E. One can argue that before the relaxation of the first N_1 sites the neighbouring N_2 sites had local thresholds distributed homogeneously between $W/2$ and W. After the relaxation of N_1 they get shifted to lower values by an average amount $W/2$. Their distribution is then constant $W(\chi) = h_1$ and shifted between zero and $W/2$. We have therefore

$$h_1 \frac{W}{2} = N' = 2N_1 = 2h \cdot E \tag{4.4}$$

that gives

$$h_1 = \frac{4h \cdot E}{W}. \tag{4.5}$$

Among these N' sites some of them will now have a local threshold field smaller than E. These are

$$N_2 = h_1 E \left[\frac{4E}{W} \right] \cdot (hE). \tag{4.6}$$

These will now relax with a characteristic time

$$\tau_2 = 2\tau_0 \tag{4.7}$$

because first they have to wait a time τ_0 for the relaxation of the sites N_1 and then it will take a similar time for their own dynamics.

It is now easy to see that the relaxations of these N_2 sites will also give rise to a rearrangement of all the local fields and this will induce a new generation of sites whose local field may become lower then E only after the relaxation of these N_2 sites. This effect gives rise to a hierarchy of relaxations whose generic term corresponds to a number of relaxing sites

$$N_n = (Eh)(\frac{4E}{W})^n \tag{4.8}$$

with a characteristic time

$$\tau_n = n\tau_0. \tag{4.9}$$

The total relaxation of the system is then given by the sum over all these hierarchies

$$P(t) = \sum_{n=1}^{\infty} N_n e^{-t/\tau_n} = \int dn e^{-(t/n \cdot \tau_0)} \tag{4.10}$$

where

$$K = \ln(\frac{W}{4E}). \tag{4.11}$$

The behavior of the integral of Eq. (4.10) can be easily analyzed by a saddle point method

$$P(t) = \int dn e^{-F(n,t)} = e^{-t^{1/2}} \tag{4.12}$$

because

$$\frac{\partial F}{\partial n} = K - \frac{t}{n^2} \tag{4.13}$$

leading to

$$n = t^{1/2} \tag{4.14}$$

and therefore to the stretched exponential behavior with $\nu = \frac{1}{2}$ (Eq. (4.12)).

We have seen therefore than the simple Hamiltonian system corresponding to Eq. (2.8) seems to be an appropriate model to study the hierarchical dynamics of glasses. Note however, that the type of hierarchy we have derived here is of a different type than those proposed on a purely phenomenological basis [7].

There is of course a well known Hamiltonian system that appears to contain the right features of glasses and this is the spin glass model [6]. With respect to the problem addressed here, namely the possibility of hierarchical couplings in the dynamics, the present model appears however more direct. In fact in the spin glass each degree of freedom (a single spin) has only two energy states without a barrier. Potential barriers develop only if one considers collective degrees of freedom that involve many spins. Since we are interested in the possible couplings between the dynamics across different barriers it is easy to realize that this is extremely difficult to achieve starting from a spin glass model while this question can be addressed in a more direct way within the present model.

We have also performed computer simulations on this model that show indeed streched exponential relaxation [9]. At the moment however it is difficult to extract from the numerical data some evidence for the specific hierarchical mechanism that we have discussed here. More work is needed in order to clarify this important point.

5. Summary

We have discussed a model Hamiltonian formulated to describe the dynamics of a charge density wave pinned by random impurities. This model is able to explain various features of the electronic properties of $NbSe_3$ that closely resemble the general dynamical behavior of glassy systems. In view of this similarity we point out that this model may provide an interesting example to investigate general questions about glassy dynamics. A particularly relevant feature in this respect is the possibility that the stretched exponential relaxation observed in glasses may be due to hierarchical couplings in the dynamics. In order to study this problem the present model appears to be more suitable then spin glass models. In fact it is possible to show with simple arguments that hierarchical coupling between degrees of freedom should naturally occur for such a system. This also leads to the possibility of deriving a stretched exponential relaxation from microscopic arguments.

References

1. H. Fukuyama and H. Takayama, in: Electronic Properties of Inorganic Quasi One Dimensional Materials, Ed. by P. Moncean, D. Reidel Publ. (1985), pp. 41; see also P. Monceau, *ibid.* pp. 139.
2. G. Kriza and G. Mihály, *Phys. Rev. Lett.* **56**, 2529 (1986).
3. L. Pietronero and S. Strassler, *Phys. Rev. B* **28**, 5863 (1983).
4. L. Pietronero in: Highlights of Condensed Matter Theory, Ed. by F. Bassani et al., North-Holland (1985), pp. 442.
5. G.S. Grest and M.H. Cohen, *Adv. Chem. Phys.* **48**, 55 (1981).
6. M. Mezard, G. Parisi and M.A. Virasoro Eds, Spin Glass Theory and Beyond, World Scientific, Singapore (1987).
7. R.G. Palmer, D.L. Stein, E. Abrahams and P.W. Anderson, *Phys. Rev. Lett.* **53**, 958 (1984).
8. L. Pietronero in: Fractals in Physics, Ed. by L. Pietronero and E. Tosatti, North-Holland (1986), pp. 147.
9. A. Erzan, R. Heijungs and L. Pietronero, unpublished.
10. L. Pietronero and M. Versteeg, preprint.
11. M.F. Shlesinger and E.W. Montroll, *Proc. Nat. Acad. Sci.*, USA, **81**, 1280 (1984).

RENORMALIZED BOLTZMANN EQUATION AND FERMI LIQUID THEORY FOR STRONGLY DISORDERED ELECTRON SYSTEMS

C. CASTELLANI AND C. DI CASTRO

Dipartimento di Fisica
Università di Roma 'La Sapienza'
I-00185 Roma
Italy

G. STRINATI

Scuola Normale Superiore
I-56100 Pisa
Italy

We review the localization phenomena for both noninteracting and interacting electrons in the presence of a random potential. Emphasis will be placed in the description of quantum transport via an appropriate kinetic equation with effective (renormalized) scattering term and self-consistent field, in the spirit of the Landau-Boltzmann approach.

1. Introduction

The classical approach to the electronic transport problem is governed by the Boltzmann-Drude theory [1], in which the presence of (weak) disorder leads to a diffusive motion of the particles. This motion is controlled by a diffusion coefficient $D_0 = v_F^2\tau / d$, where v_F is the Fermi velocity, τ is the impurity scattering time in the Born approximation, and d is the dimensionality of the system. At this level, many-body effects are introduced through the Landau theory of normal Fermi liquids [2]. The interaction among the electrons is thus taken into account by introducing effective parameters in the one-particle description via the quasiparticle concept.

In some classes of materials (e.g., doped semiconductors like $Si:P$ and amorphous alloys of a metal and a semiconductor like Nb_xSi_{1-x}) one observes peculiar features which cannot be interpreted within the Landau-Boltzmann scheme. Already in the weak-disorder regime ($k_Fl \gg 1$, where $l = v_F\tau$ is the mean free path), the temperature dependence of the conductivity shows a $T^{1/2}$ contribution in three dimensions and a $\ln T$ behavior in two dimensions, which usually leads to an increase of the resistivity by decreasing temperature. At larger disorder ($k_Fl \approx 1$), the system eventually undergoes a metal-insulator transition with the ($T=0$) conductivity vanishing at a critical value of disorder [3].

These experimental results indicate that, with increasing disorder, the electrons tend to localize within finite regions of space. This, in turn, can be interpreted by assigning to the system an effective diffusion coefficient which vanishes at the transition.

R. Girlanda et al. (eds.), Progress on Electron Properties of Solids, 247–253.

Additional anomalies appear in other quantities like the electronic specific heat and the spin susceptibility [4], whose interpretation further requires a modification of the Landau interaction parameters.

Theoretically, corrections to the Boltzmann-Drude theory appear within the one-particle picture by including quantum-interference effects in the scattering by the impurities [5]. This leads to the so-called Anderson localization, where the diffusion coefficient only becomes critical by acquiring scale-dependent corrections [6]. When the interaction is included, additional corrections to the diffusion coefficient [7] are accompanied by new corrections in the specific heat and in the spin susceptibility. The latter can be included as a renormalization of the Landau interaction parameters; the resulting scaling theory leads to various universality classes of the metal-insulator transition [8,9].

2. Noninteracting Electrons

Following a common practice of solid state physics, the localization problem was originally approached by the one-electron model with the underlying assumption that, once this problem would have been understood, the interaction could be taken care of by modifying the parameters of the theory in a trivial manner.

We start by considering the Drude result, where the zero-temperature conductivity is given by

$$\sigma_0 = 2e^2 N_0(\epsilon_F) D_0 = \frac{e^2}{\hbar}\frac{2}{d}\frac{S_d}{(2\pi)^d} k_F^{d-1} l. \tag{2.1}$$

Here $N_0(\epsilon_F)$ is the free-electron density of states at the Fermi level (per spin component), S_d is the solid angle in d dimensions, and e is the electronic charge. As the disorder is increased, we have to introduce corrections of successive orders in the expansion parameter t_0, which in two dimensions is proportional to $(k_F l)^{-1}$. t_0 can be expressed in terms of the measurable conductance G by

$$t_0 = \frac{1}{g}\frac{4}{d}\frac{S_d}{(2\pi)^d} \tag{2.2}$$

where

$$g = \frac{G}{e^2/\hbar} = \frac{\sigma_0 L^{d-2}}{e^2/\hbar}, \tag{2.3}$$

L being a characteristic linear dimension of the system. We notice that in the metallic regime the conductivity σ_0 is size independent, and therefore the expansion parameter t_0 has bare scale dimensionality $\epsilon = d-2$. This remark, in turn, suggests that $d = 2$ is in this case a special dimensionality since the coupling constant becomes scale independent, and further that logarithmic perturbative corrections can be expected at $d = 2$. Explicit calculations in the presence of nonmagnetic impurities only indeed give [5]

$$D = D_0[1-\frac{1}{\pi}t_0 \ln(L/l)+ \cdots]. \tag{2.4}$$

The dominant sequence of scattering events which leads to this singular correction is identified by the coherent superposition of pairs of patterns that remain equivalent under time reversal. The resulting quantum interference leads to a backscattering enhancement and yields a reduction of the diffusion coefficient at a macroscopic level $L > l$, as can be seen from Eq. (2.4).

In the presence of magnetic impurities, the lack of time-reversal invariance causes the first order correction in Eq. (2.4) to disappear. The relevant leading correction is now obtained at order t_0^2 [5]:

$$D = D_0[1-\frac{1}{8\pi^2}t_0^2 \ln(L/l)+ \cdots]. \tag{2.5}$$

However, a direct physical interpretation of the rather involved scattering processes contained in the diagrammatic structure cannot be easily visualized in this case.

The renormalization group equation can be invoked at this point to sum the logarithmic singularities of Eqs. (2.4) and (2.5) into a power-law behavior at $d = 2+\epsilon$, yielding the one-parameter scaling theory in both cases [6].

Instead of reproducing here the results of the scaling theory, we prefer to rephrase the localization theory in terms of a kinetic description, thus linking the localization problem to the more conventional treatment of transport phenomena and providing at the same time a physical interpretation of the mechanism underlying the localization [10,11].

The non-ohmic behavior of the quantum corrections to the diffusion coefficient in two dimensions (which in three dimensions amount instead to a square-root term) leads to a violation of the conventional Boltzmann approach. However, having in the back of one's mind the ideas of the renormalization group approach, one can evisage recovering the Boltzmann equation by suitably renormalizing the parameters of the theory. By using standard diagrammatic techniques, one can in fact show the following equation to hold [11]:

$$\begin{aligned}&[-i\Omega+v_F\hat{k}\cdot\nabla_{\mathbf{R}}-\nabla_R\Phi_{\text{ext}}(\mathbf{R},\Omega)\cdot\nabla_{\mathbf{k}}]n(\mathbf{k};\mathbf{R},\Omega)\\&= -\int\frac{d\hat{k}'}{S_d}\int d\mathbf{R}' W(\hat{k},\hat{k}';\mathbf{R}-\mathbf{R}',\Omega)[n(\mathbf{k};\mathbf{R}',\Omega)-n(\mathbf{k}';\mathbf{R}',\Omega)].\end{aligned} \tag{2.6}$$

Here $n(\mathbf{k};\mathbf{R},\Omega)$ is the time Fourier transform of the phase-space distribution function, Φ_{ext} is an external potential that drives the system, W is the scattering kernel that included quantum interference effects, and $|\mathbf{k}| = |\mathbf{k}'|$. In the hydrodynamic limit we can consider only the two leading moments

$$W_0(\hat{k},\hat{k}';\Omega) = \int d\mathbf{R}' W(\hat{k},\hat{k}';\mathbf{R}-\mathbf{R}',\Omega) \tag{2.7a}$$

$$\mathbf{W}_1(\hat{k},\hat{k}';\Omega) = \int d\mathbf{R}' W(\hat{k},\hat{k}';\mathbf{R}-\mathbf{R}',\Omega)(\mathbf{R}'-\mathbf{R}). \tag{2.7b}$$

For the two cases of interest they are given by:
(i) nonmagnetic impurities [10,11]

$$W_0(\hat{k},\hat{k}';\Omega) = \frac{1}{\tau}\{1+2t_0\mathbf{I}_d(\Omega)[S_d\delta(\hat{k}+\hat{k}')-1]\} \tag{2.8a}$$

$$\mathbf{W}_1(\hat{k},\hat{k}';\Omega) = 0; \tag{2.8b}$$

(ii) magnetic impurities [11]

$$W_0(\hat{k},\hat{k}';\Omega) = \frac{1}{\tau}\{1-\frac{t_0^2}{2}\mathbf{I}_d^2(\Omega)[1+2(\hat{k}\cdot\hat{k}')]\} \tag{2.9a}$$

$$\mathbf{W}_1(\hat{k},\hat{k}';\Omega) = -\frac{t_0^2}{2\tau}\mathbf{I}_d^2(\Omega)[1+\frac{\epsilon}{8} + 2(\hat{k}\cdot\hat{k}')(1+\frac{3}{8}\epsilon)]l(\hat{k}+\hat{k}'), \tag{2.9b}$$

at leading order in ϵ, where

$$\mathbf{I}_d(\Omega) = \int\frac{d\mathbf{q}}{(2\pi)^d}\frac{1}{\mathbf{q}^2-i\Omega} \tag{2.10}$$

is logarithmically singular in two dimensions. Eqs. (2.8) show that, in the presence of nonmagnetic impurities only, the time-reversal invariance implies an enhancement for W_0 in the backscattering channel and $\mathbf{W}_1 \equiv 0$, the latter result being due to the combined symmetries of the scattering kernel under the interchange of $\hat{k} \Leftrightarrow \hat{k}'$ and $\hat{k} \Leftrightarrow -\hat{k}'$. Eqs. (2.9) show instead that, when (quenched) magnetic impurities are present and thus time-reversal invariance does not hold, both moments survive with strong $\ln^2$ corrections in two dimensions. This feature would seemingly imply a non-renormalizability of W_0 and $\mathbf{W}_1$ when taken separately. Moreover, the nonlocality of the scattering kernel is in this approach the signal that, contrary to the nonmagnetic impurity case, one cannot identify a sequence of scattering events at a time leading to interference. Rather, a superposition of sequences results in a building up of the reflected waves over an extended region of space with $|\mathbf{W}_1|/(W_0 - \tau^{-1}) \propto l$.

The $\ln^2$ singularities of W_0 and $\mathbf{W}_1$ can be combined together and shown to give rise to a standard ln singularity by reducing the Boltzmann equation to the ordinary diffusion equation via an averaging procedure over all possible directions on the Fermi surface. In both cases one obtains in Fourier space [11]

$$[-i\Omega + D\mathbf{Q}^2]\delta\bar{n}(k;\mathbf{Q},\Omega) = -2\delta(\epsilon_k) D\mathbf{Q}^2 \Phi_{\mathrm{ext}}(\mathbf{Q},\Omega), \tag{2.11}$$

where

$$\delta\bar{n}(k;\mathbf{Q},\Omega) = \int \frac{d\hat{k}}{S_d} \delta n(\mathbf{k};\mathbf{Q},\Omega) \tag{2.12}$$

to first order in Φ_{ext}, with D given by Eqs. (2.4) and (2.5) in the two cases. Eq. (2.11) further leads to the diffusive form of the density-density correlation function

$$\chi_{nn}(\mathbf{Q},\Omega) = -2N_0(\epsilon_F)\frac{D\mathbf{Q}^2}{D\mathbf{Q}^2 - i\Omega}, \tag{2.13}$$

the Einstein relation then yielding the value $2e^2 N_0(\epsilon_F) D$ for the renormalized conductivity. The same diffusion coefficient also controls the spin and heat transport phenomena in the noninteracting case [12].

Experimentally, one can verify in many different ways that the framework so far described is far from being exhaustive. Theoretically, one realizes that the effects of the electron-electron interaction cannot be included in a trivial manner because there arise additional singular corrections to the conductivity, which are of the same order of those already described [7]. Moreover, other quantities besides σ (namely, the electronic specific heat and the spin susceptibility) acquire singular corrections and a critical behavior [13,14].

3. Interacting Electrons

The standard approach for treating the effects of the interaction on top of weak disorder relies on the Landau theory of Fermi liquids. All results of this theory can be embodied in a kinetic (Landau-Boltzmann) equation for quasiparticles. When the electronic system couples to an external scalar potential, this equation reads [15]

$$[-i\Omega + iv_F\hat{k}\cdot\mathbf{Q}]\delta n_\sigma(\mathbf{k};\mathbf{Q},\Omega) \tag{3.1}$$

$$+ iv_F\hat{k}\cdot\mathbf{Q}\delta(\epsilon_k)[\Phi_{\mathrm{ext}}(\mathbf{Q},\Omega) + F_0^s \frac{1}{2}\sum_{\sigma'}\int d\epsilon_{k'}\int \frac{d\hat{\mathbf{k}}'}{S_d}\delta n_{\sigma'}(\mathbf{k}';\mathbf{Q},\Omega)$$

$$= -\frac{1}{\tau}\left[\delta n_\sigma(\mathbf{k};\mathbf{Q},\Omega) - \int \frac{d\hat{\mathbf{k}}'}{S_d}\delta n_\sigma(\mathbf{k}';\mathbf{Q},\Omega)\right],$$

where $\delta n_\sigma(\mathbf{k};\mathbf{Q},\Omega)$ is the Fourier transform of the quasiparticle distribution function with spin σ (to linear order in the external field) and F_0^s is the symmetric part of the Landau interaction function. This equation holds provided the quasiparticles are well-defined excitations, i.e., whenever their inverse lifetime vanishes faster than the relevant distance from the Fermi surface.

The density-density correlation function can be readily retrieved from Eq. (3.1). One obtains

$$\chi_{nn}(\mathbf{Q},\Omega) = \sum_\sigma \int \frac{d\mathbf{k}}{(2\pi)^d}\delta n_\sigma(\mathbf{k};\mathbf{Q},\Omega)/\Phi_{\text{ext}}(\mathbf{Q},\Omega) \tag{3.2}$$

$$= -\frac{2\nu_{qp}}{1+F_0^s}\frac{D_{qp}\mathbf{Q}^2}{D_{qp}\mathbf{Q}^2 - i\Omega/(1+F_0^s)},$$

where $\nu_{qp}(=N_0(\epsilon_F)m^*/m)$ stands for the quasiparticle density of states (with effective mass m^*) and D_{qp} for the quasiparticle diffusion coefficient. From Eq. (3.2) one recovers, in particular, the Fermi-liquid corrections to the adiabatic compressibility and to the density diffusion coefficient:

$$\frac{\partial n}{\partial \mu} = \frac{2\nu_{qp}}{1+F_0^s}, \quad D_n = D_{qp}(1+F_0^s). \tag{3.3}$$

The Einstein relation in this case reads

$$\sigma = e^2\frac{\partial n}{\partial \mu}D_n = 2e^2\nu_{qp}D_{qp}. \tag{3.4}$$

For obtaining the spin-spin correlation function it is sufficient to replace Φ_{ext} in Eq. (3.1) with the spin-dependent coupling to a magnetic field and to recall that $\delta n_\sigma(\mathbf{k};\mathbf{Q},\Omega) = \sigma\delta n(\mathbf{k};\mathbf{Q},\Omega)$ in this case:

$$\chi_{SS}(\mathbf{Q},\Omega) = -\frac{2\nu_{qp}}{1+F_0^a}\frac{D_{qp}\mathbf{Q}^2}{D_{qp}\mathbf{Q}^2 - i\Omega/(1+F_0^a)}. \tag{3.5}$$

From this expression one gets that the Pauli susceptibility and the spin diffusion coefficient are modified into

$$\chi_s = \frac{2\nu_{qp}}{1+F_0^a}, \quad D_s = D_{qp}(1+F_0^a). \tag{3.6}$$

The diffusive form of the heat-heat correlation function is finally recovered from Eq. (3.1) by suppressing the driving term at its left-hand side (the self-consistent field being of order T^2 at low temperature), and by letting δn describe the deviation from local thermodynamic equilibrium. One obtains

$$\chi_{EE}(\mathbf{Q},\Omega) = -c_V T\frac{D_{qp}\mathbf{Q}^2}{D_{qp}\mathbf{Q}^2 - i\Omega} \tag{3.7}$$

where $c_V = (\pi^2/3)2\nu_{qp}T$ is the specific heat. The thermal conductivity κ is thus related to the electrical conductivity σ via the Einstein relation

$$\kappa = c_V D_{qp} = \frac{\pi^2}{3}\frac{\sigma}{e^2}T \tag{3.8}$$

that expresses the Wiedemann-Franz law. Notice from Eq. (3.7) that the heat diffusion coefficient coincides with the quasiparticle diffusion coefficient; three distinct diffusion coefficients are, however, assigned to the particle, spin, and heat transport.

Thus, in the presence of weak disorder, three quantities (namely, the quasiparticle density of states and the symmetric and antisymmetric Landau interaction parameters) besides the quasiparticle diffusion coefficient specify the modifications introduced by the interaction in the transport theory and in the equilibrium properties (as characterized by the static values of the correlation functions). When the disorder is increased, one expects that the effective interaction is enhanced owing to the slow diffusive motion of the quasiparticles. Explicit calculations [8,9,13] show that singular (logarithmic in two dimensions) corrections modify ν_{qp}, F_0^s, and F_0^a which become scale-dependent functions and generalize the single function $D(L)$ required in Section 2 for the noninteracting case. We do not provide here the explicit form of this scale dependence, but proceed instead on the assumption that the quantities $(\nu_{qp})_R$, $(F_0^s)_R$, and $(F_0^a)_R$ (where the suffix R stands for renormalized) are the only relevant parameters, besides the coupling t specifying the disorder, that allow us to evaluate all physical quantities of interest. On the basis of this assumption, which is in fact verified by renormalized perturbation theory, we can write a kinetic equation for the energy distribution of quasiparticles that holds in the presence of strong disorder, in the form [16]

$$\begin{aligned}&[-i\Omega+(D_{qp})_R\mathbf{Q}^2]\delta n_\sigma(\epsilon;\mathbf{Q},\Omega) \\ &= -\delta(\epsilon)(D_{qp})_R\mathbf{Q}^2[\Phi_{\rm ext}(\mathbf{Q},\Omega)+(F_0^s)_R\frac{1}{2}\sum_{\sigma'}\int d\epsilon'\delta n_{\sigma'}(\epsilon';\mathbf{Q},\Omega)],\end{aligned} \tag{3.9}$$

where $(D_{qp})_R$ is the renormalized diffusion coefficient.

The density-density correlation function can be directly obtained from Eq. (3.9):

$$\chi_{nn}(\mathbf{Q},\Omega) = -2(\nu_{qp})_R\frac{(D_{qp})_R\mathbf{Q}^2}{(D_{qp})_R\mathbf{Q}^2[1+(F_0^s)_R]-i\Omega}, \tag{3.10}$$

from which it follows that

$$\frac{\partial n}{\partial\mu} = \frac{2(\nu_{qp})_R}{1+(F_0^s)_R}, \quad D_n = (D_{qp})_R[1+(F_0^s)_R], \tag{3.11}$$

$$\sigma = 2e^2(\nu_{qp})_R(D_{qp})_R.$$

Eq. (3.11) generalizes Eqs. (3.3) and (3.4) in the presence of strong disorder. Similarly, for the spin-spin and heat-heat correlation functions one obtains

$$\chi_s = \frac{2(\nu_{qp})_R}{1+(F_0^s)_R}, \quad D_s = (D_{qp})_R[1+(F_0^a)_R], \tag{3.12}$$

$$c_V = \frac{\pi^2}{3}2(\nu_{qp})_RT, \quad D_E = (D_{qp})_R, \tag{3.13}$$

$$\kappa = c_VD_E = \frac{\pi^2}{3}\frac{\sigma}{e^2}T,$$

respectively.

The formal general structure given previously must be implemented by the renormalization group equations for the effective couplings in order to acquire physical content. To this end, the explicit form of the scale-dependent singular corrections to the disorder parameter t and to ν_{qp}, F_0^s, and F_0^a must be evaluated in a loop expansion in $2+\epsilon$ dimensions.

At leading order in ϵ, one finds a qualitatively different behavior depending on the relative relevance of $(F_0^a)_R$ with respect to $(F_0^s)_R$. This leads to various universality classes for the metal-insulator transitions, which have previously been discussed in the framework of the correlation functions [8,9,13].

References

1. See, e.g., N.W. Ashcroft and N.D. Mermin, *Solid State Physics*, Saunders College, Philadelphia, 1976, Chapter 16.
2. See, e.g., P. Nozières, *Theory of Interacting Fermi Systems*, W.A. Benjamin, New York, 1964.
3. For a recent review, see *Anderson Localization*, T. Ando and H. Fukuyama, Eds., Springer, New York, 1988.
4. S. Ikehata and S. Kobayashi, *Solid St. Comm.* **56**, 607 (1985); M.A. Paalanen, S. Sachdev, R.N. Bhatt, and A.E. Ruckenstein, *Phys. Rev. Lett.* **57**, 2061 (1986).
5. See, e.g., *Anderson Localization*, Y. Nagaoka and H. Fukuyama, Eds., Springer, New York, 1982.
6. E. Abrahams, P.W. Anderson, D.C. Licciardello, and T.V. Ramakrishnan, *Phys. Rev. Lett.* **42**, 673 (1979).
7. B.L. Altshuler, A.G. Aronov, and P.A. Lee, *Phys. Rev. Lett.* **44**, 1288 (1980); B.L. Altshuler, D. Khmelnitskii, A.I. Larkin, and P.A. Lee, *Phys. Rev. B* **22**, 5142 (1980).
8. A.M. Finkelstein, *Sov. Phys. JEPT* **57**, 97 (1983).
9. C. Castellani, C. Di Castro, P.A. Lee, and M. Ma, *Phys. Rev. B* **30**, 527 (1984).
10. S. Hershfield and V. Ambegaokar, *Phys. Rev. B* **34**, 2147 (1986).
11. G. Strinati, C. Castellani, and C. Di Castro, *Physica C* **153-155**, 697 (1988), and unpublished.
12. C. Castellani, C. Di Castro, and G. Strinati, *Europh. Lett.* **4**, 91 (1987); G. Strinati and C. Castellani, *Phys. Rev. B* **36**, 2270 (1987).
13. C. Castellani, C. Di Castro, P.A. Lee, M. Ma, S. Sorella, and E. Tabet, *Phys. Rev. B* **33**, 6169 (1986).
14. C. Castellani and C. Di Castro, *Phys. Rev. B* **34**, 5935 (1986).
15. O. Betbeder-Matibet and P. Nozières, *Ann. Phys.* **37**, 17 (1966), provide actually the most general form of the Landau-Boltzmann equation in the presence of weak disorder, by retaining the full angular dependence of the Landau interaction function and of the impurity collision probability. The simplified form (3.1) proves, however, sufficient for our purposes.
16. C. Castellani, C. Di Castro, G. Kotliar, P.A. Lee, and G. Strinati, *Phys. Rev. Lett.* **59**, 477 (1987); and G. Strinati, C. Castellani, C. Di Castro, and G. Kotliar (unpublished).

NONLINEAR MAGNETOOPTICS IN SOLIDS

D. FRÖHLICH

Insitut für Physik
Universität Dortmund
4600 Dortmund 50
FRG

Nonlinear magnetooptic experiments are discussed on different levels of complication. In the first chapter two-photon transitions to excitons and Landau levels are investigated, which are not simultaneously one-photon allowed. Within this chapter semiconductors with simple band structure as *ZnO* and Cu_2O (two-fold valence and conduction band) are discussed first. A more complicated band structure (four-fold valence band) leads to a complicated *P* exciton spectrum in a magnetic field as will be shown for typical zinc blende semiconductors as *ZnSe* and *CdTe*. Excitonic effects clearly show up in transitions to Landau levels as will be discussed for two-photon transitions in *CdTe* and multiple quantum wells. In a second chapter two-photon and three-photon transitions are investigated which are simultaneously one-photon allowed. In order to explain the observed *K*-dependent splitting the polariton character of the excited states has to be taken into account. This will be demonstrated for transitions in *CuCl* and alkali halides.

1. Introduction

Nonlinear spectroscopy has proven to be a powerfull tool to study electronic properties of solids. There are several motivations to do optical experiments, where more than one photon is necessary to induce an electronic transition. The general point of cause is, that there are additional 'degrees of freedom' in nonlinear spectroscopy as compared to linear spectroscopy, which can be used to get specific information about electronic properties which are not (easily) accessible to classical spectroscopic experiments (one-photon absorption: OPA). This point can be demonstrated by considering two-photon absorption (TPA) as an example. The selection rules are different for one and two-photon transitions. In systems with inversion symmetry two-photon and one-photon absorption to the same level exclude each other. This is analogous to Raman and infrared absorption in molecules with inversion center. The pronounced exciton structure of Cu_2O as measured by OPA and TPA [1] clearly shows the different selection rules. *P* excitons are one-photon allowed whereas only *S* and *D* excitons are seen in two-photon absorption. Besides symmetry selection rules which determine the polarization dependence, there is conservation of momentum. Since the photon energies add as scalars and the appropriate *K*-values as vectors, one can change in TPA the total *K*-vector by varying the angle between the individual *K*-vectors without changing the total energy. One can thus do 'spectroscopy in *K*-space'. Although one reaches through a rather small part of the Brillouin zone (1 / 1000), one can deduce interesting results on polaritons and even the longitudinal exciton. In order to determine the polariton dispersion by TPA the corresponding transition has to be simultaneously one- and two-photon allowed. It has to be one-photon allowed in order show polariton dispersion, it has to be two-photon allowed in order to allow the detection of the resonances by TPA. Again there is a close analogy

R. Girlanda et al. (eds.), Progress on Electron Properties of Solids, 255–274.

between Raman effect and TPA. The first phonon polariton dispersion was reported by Henry and Hopfield [2] in *GaP*, the first exciton polariton dispersion by Fröhlich, Wiesner and Mohler in *CuCl* [3]. The restriction to systems without inversion center was recently overcome by stepping on to three-photon spectroscopy (TPS). Resonances on the upper polariton branch and the longitudinal exciton of alkali halides were seen by Beerwerth and Fröhlich [4].

It is natural to ask the question if it is of any use to introduce external perturbations like magnetic or electric fields in nonlinear spectroscopy. The application of an electric field leads to a mixing of even and odd parity states, since parity is no longer a good quantum number. Transitions to these states are then simultaneously allowed in OPA and TPA. Mixing of e.g. *P* and *S* excitons in *AgI* was observed by Dinges and Fröhlich [5] in TPA. This leads to the unique possibility of 'polariton tuning', since the former pure *P* excitons become one-photon allowed by the electric field induced admixture of *S* excitons. The polariton character of the *S* excitons can thus be 'tuned' into the *P* excitons. By far more information is expected from TPA experiments in high magnetic fields.

The following two chapters will discuss magnetooptical experiments on different levels of complication. We will distinguish between transitions which are not simultaneously on-photon allowed (chapter II) and those which are one-photon allowed (chapter III). In the last case the polariton character has to be taken into account, which leads to unexpected effects as will be discussed later.

2. Two-Photon Transitions without Polariton Effect

This chapter is divided into three paragraphs. We will first consider excitons in crystals with simple band structure as e.g. ZnO and Cu_2O, where the relevant valence and conduction band are only two-fold degenerate. In most cases of zinc-blende structure (T_d), however, one has to deal with a four-fold valence band (Γ_8), which leads to additional terms in the Hamiltonian. The resulting rather complicated exciton structure can be analyzed to yield detailed information on band parameters, as will be shown in the second paragraph of this chapter. In the last paragraph of this chapter we will present recent two-photon measurements of Landau transitions in II-VI compounds and multiple quantum wells (MQW).

2.1. Excitons with Simple Valence Band Structure

The first two-photon magnetoopic experiments on excitons were reported by Dinges et al. [6] in *ZnO*. The authors determined the splitting of the 2*P* exciton lines (A_{2P} and B_{2P}) which correspond to transitions from the upper valence bands $\Gamma_7(A)$ and $\Gamma_9(B)$ to the lowest conduction band of Γ_7 symmetry. In Figure 1 the one- and two-photon spectrum of the two lowest exciton series are shown [from Ref. 6]. The splitting of the A_{2P} and B_{2P} excitons in a field of $4.2T$ is presented in Figure 2. The fact that *P* excitons are excited in TPA as compared to *S* excitons in OPA allows to determine the envelope *g* value $g_\mu = m_0(m_h^{-1} - m_e^{-1})$ which is directly connected to the effective masses of the corresponding bands. Although the splitting between the two valence bands is only $7\,meV$, nondiagonal terms between *P* excitons of the *A* and *B* series were not considered in the first analysis. The rather low field data of Figure 2 seem to justify this procedure. For measurements in higher fields ($B > 8T$) one would have to consider a more complicated valence band Hamiltonian by including the H_d-term. The formalism which is very

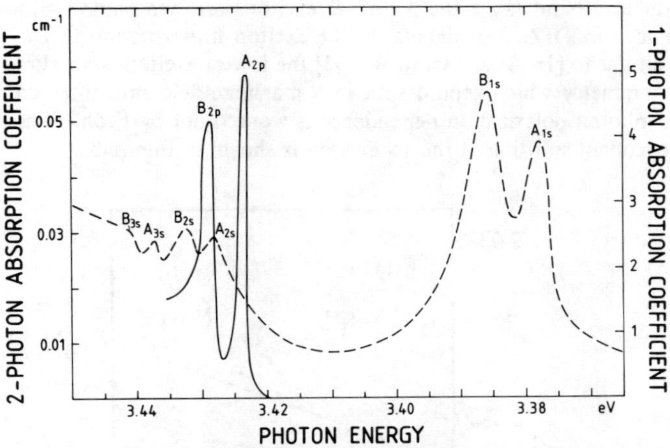

Figure 1. Dashed line, one-photon spectrum of ZnO at $4.2K$, polarization $\perp c$-axis. Solid line, two-photon spectrum of ZnO at $1.6K$, polarization $\perp c$-axis.

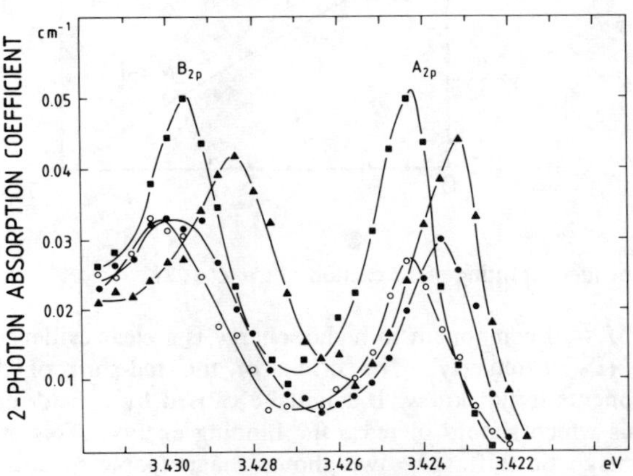

Figure 2. Two-photon spectrum of $2P$ excitons of ZnO at $4.2K$ in a magnetic field of $4.2T$, measured with circularly polarized light: squares (+ +), triangles (--), open circles (+-), closed circles (-+). The first and second sign refer to the polarization of the YAG laser and the tunable light source, respectively.

well known for zinc-blende structure, as will be discussed in the next paragraph, should be extended to wurtzite structure. The experimental results as presented in Figure 2 could be improved by the use of a tunable dye laser instead of the flashlamp and monochromator. It would certainly be of interest to measure the splitting of the C_{2P} exciton, which was first observed in TPA by Kaule [7]. The C exciton series stems from the $\Gamma_7(C)$ valence band which is about $40\,meV$ below the $\Gamma_7(A)$ valence band. CdS, which also crystallizes in wurtzite structure, has a similar valence band structure as ZnO. Two-photon magneto photoconductivity of A and B excitons up to $10T$ is reported by Seiler, Heiman and Wherrett [8].

As another rather simple example we have chosen TPA experiments on the lowest $1S$ exciton in Cu_2O. Because of the even parity valence band (Γ_7^+) and an even parity lowest conduction band (Γ_6^+) the S and D excitons are two-photon allowed and the P exciton are seen in OPA. For details on the exciton fine-structure in Cu_2O and further literature we refer to [1]. As shown in Ref. [1] the lowest exciton is an almost pure S exciton of Γ_5^+ symmetry which should split in a magnetic field into three components. The detailed two-photon polarization dependence is worked out by Fröhlich and Kenklies [9]. The field dependent splitting of the $1S$ exciton is shown in Figure 3.

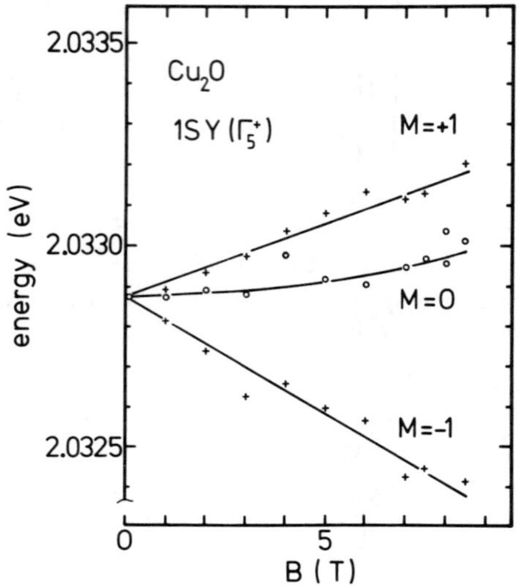

Figure 3. Field dependent splitting of $1S$ exciton of Cu_2O at $2K$.

The shift of the $M = 0$ component to higher energy is a clear evidence of its mixing with the paraexciton (Γ_2^+ symmetry). The origin of the red-shift of the average of the $M = \pm 1$ components is not know. It might be caused by a field-induced shrinkage of the exciton radius which should increase the binding energy. This effect even overcompensates the diamagnetic shift. The two-photon magnetoopic spectra of the higher excitons are expected to be much more difficult to analyze, since the mixing of S and D excitons has to be taken into account as discussed in Ref. [1]. A detailed study of the exciton fine structure in Faraday and Voigt configuration by TPA would certainly yield further interesting information on the model exciton system Cu_2O. It is expected, that one has to go beyond the spherical approximation and include a weak cubic H_d term to explain all details of the magnetooptical spectra quantitatively.

2.2. Excitons with Complicated Valence Band-Structure

In this paragraph we will discuss two-photon magnetoabsorption in zinc-blende semiconductors where we have to deal with a four-fold valence band (Γ_8-symmetry). Due to the high degeneracy of the valence band the Hamiltonian of the hole kinetic energy gets more complicated, as was first discussed by Luttinger and Kohn [10]. Altarelli and Lipari [11] and Baldareschi and Lipari [12] have written the Hamiltonian in the formalism of irreducible tensor operators which allows to calculate the angular part of the corresponding matrix elements in an elegant way. The remaining radial differential equations which include the Coulomb term have then to be solved numerically. As discussed in detail by

Hölscher, Nöthe and Uihlein [13] and Neumann, Nöthe and Lipari [14] two-photon magnetoabsorption is specially suited to study the P exciton finestructure, which yields a complete parameter set (band masses, g values, binding energy). The coupling of the threefold P envelope ($L=1$) to the fourfold hole states ($J=3/2$) leads to 12 wavefunctions ($F=5/2,3/2,1/2$) which have to be coupled to the spin functions of the conduction band ($S=1/2$). The resulting 24 states can be classified by their total momentum quantum number ($F_{tot}=3,2,1,0$). For P excitons the coupling between the electron spin **S** and the angular momentum **F** is negligible. This has to be contrasted to S exciton in Cu_2O where electron-hole exchange interaction has to be taken into account [1]. For the P excitons the electron spin doubles the degeneracy of the excitonic states ($B=0T$). In a magnetic field the Zeeman splitting due to the electron g value has to be included. In an axial model the two-photon polarization selection rules are easily derived since M_{tot} is still a good quantum number. The selection rules for Faraday (**K**||**B**) and Voigt configuration (**K**⊥**B**) are listed in Table I [13,14].

TABLE I

Selection rules for two-photon absorption. The polarizations (helicity) are referred to the magnetic field direction.

	Polarization state	M_{tot}
K||**B**	$\sigma_+ \; \sigma_+$	2
	$\sigma_+ \; \sigma_-$	0
	$\sigma_- \; \sigma_+$	0
	$\sigma_- \; \sigma_-$	-2
K⊥**B**	$\sigma \; \sigma$	$0,\pm 2$
	$\sigma \; \pi$	± 1
	$\pi \; \sigma$	± 1
	$\pi \; \pi$	0

Depending on the polarization of the two photon beams one can excite components up to $M_{tot} = \pm 2$. Experimental results on $ZnSe$ from Ref. 13 are shown in Figure 4 and Figure 5. At $B = 0T$ the envelope-hole coupling causes a splitting of the $2P$ excitons into 3 components $P_{5/2}$, $P_{3/2}$, $P_{1/2}$ in the spherical approximation. In cubic symmetry the $P_{5/2}$ splits into Γ_7 and Γ_8. This splitting is rather small as was first discussed by Sondergeld and Stafford [15]. The magnetic field dependence for the $M = 0$ components as measured in $(\sigma_+\sigma_-)$ polarization is shown in Figure 4. At $B = 8T$ the field-induced splitting (about $5\,meV$) is already much larger than the initial splitting due to the envelope-hole coupling ($1.7\,meV$). As discussed in detail in Ref. [13] this can be considered as an excitonic Paschen-Back effect. At high fields the splitting is determined by the electron-hole motion ($M_L=0,\pm 1$). The polarization dependence for a fixed field is summarized in Figure 5. On a first glance it looks rather puzzling that the spectra for $(\sigma_+\sigma_-)$ and $(\sigma_-\sigma_+)$ in Faraday configuration and the spectra for $(\pi\sigma)$ and $(\sigma\pi)$ in Voigt configuration look so different. The first polarization refers to the tunable dye laser

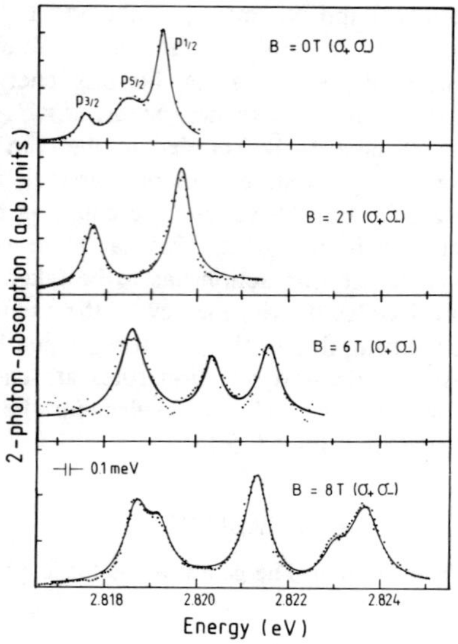

Figure 4. Two-photon absorption of $2P$ exciton structure of $ZnSe$ at $2K$, measured in Faraday configuration up to $8T$ with (σ_+, σ_-) polarization. The two polarizations refer to the dye laser and YAG laser respectively.

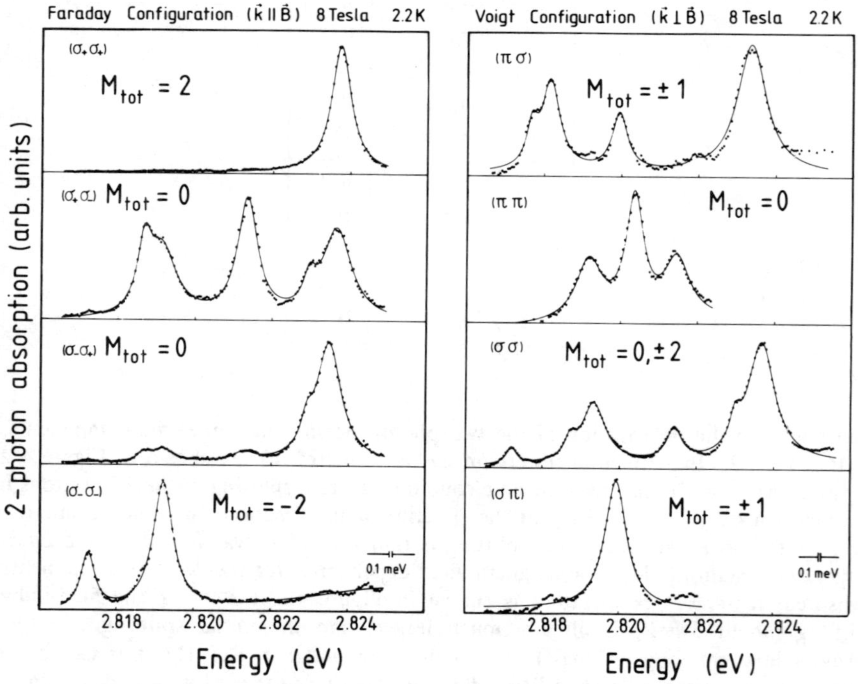

Figure 5. $2P$ exciton finestructure at $8T$ for different polarizations in Faraday (left) and Voigt configuration (right).

(about 2.2 *eV*) whereas the second refers to the high power Raman-shifted YAG laser of fixed photon energy (0.64966 *eV*). The difference in oscillator strength for equivalent polarizations (e.g. $(\sigma_+\sigma_-)$ and $(\sigma_-\sigma_+)$) is due to the rather large difference in photon energy of the two laser beams. This leads to different resonance denominators in second order perturbation theory, which has to be applied to determine the two-photon oscillator strength. In addition the two-photon oscillator strength depends strongly on the strength of the applied magnetic field, as was recently shown for *P* excitons in *GaAs* and *CdTe* by Neumann and Nöthe [16]. This increase is caused by a compression of the exciton wave-function, which leads to an increase of the derivative of the envelope wave-function at the origin. In a two-band model [17] one sees immediately, that the two-photon oscillator strength is proportional to the square of the derivative of the *P* envelope function at the origin. The oscillator strength of *S* excitons depends much less on the magnetic field, since it is proportional to the square of the wave function at the origin, rather than its derivative.

Besides *ZnSe* there are detailed two-photon mangetoabsorption measurements of *ZnTe*, *CdTe*, and *GaAs* published recently [14]. Because of the rather small exciton binding energy in the case of *CdTe* and *GaAs* (9.5 *meV* and 3.9 *meV*) the theoretical analysis has to include the coupling to higher envelope states ($L=5,7\cdots$).

2.3. Two-Photon Absorption to Landau Levels

Besides a magnetic field induced splitting of exciton states one also expects Landau levels in two-photon spectroscopy. First measurements of two-photon photoconductivity on rather low gap materials (*PbTe* and *InSb*) are reported by Button et al. [18]. A theoretical analysis of multiphoton magnetoabsorption with the Keldysh technique was given by Weiler, Reine and Lax [19]. Bassani and Girlanda [20] discuss in detail different cases of two-photon selection rules. They derive explicit expressions for the two-photon absorption constant for $\Delta n = 0$ and $\Delta n = \pm 1$ transitions (Δn: difference of Landau quantum numbers of conduction and valence band). Seiler, Littler and Heiman [21] have recently reported the first two-photon magneto-photoconductivity data of *GaAs*. They were able to resolve Landau transitions. They interprete their results in terms of a modified Pidgeon-Brown energy band model [22]. In this model the influence of Coulomb correlation is neglected. As was shown recently by Becker et al. [23] and Nöthe and Spata [24] the influence of the electron-hole correlation can be clearly extracted from the experimental data. A quantitative theoretical treatment of excitonic effects on Landau transitions has still to be done. It would certainly be of interest to compare one- and two-photon Landau transitions in the same material (e.g. *CdTe*), since they extrapolate below the bandgap to different excitons (*S* and *P* excitons, respectively).

In the following we want to present two-photon magnetoabsorption data in *CdTe* of Nöthe and Spata [24] and in multiple quantum wells (MQW) of Fröhlich and Wille [25]. In Figure 6 two-photon spectra of *CdTe* for different magnetic fields are shown [24]. In addition to the pronounced *P* exciton structure below 1.61 *eV*, which is discussed in detail in Ref. [14], there are very clearly resolved Landau transition. The authors [24] compare the experimental values with calculations using the quasi germanium model of Pidgeon and Brown [22]. As an example we show their results for the (σ_+, σ_-) polarization configuration in Figure 7. The solid lines are the results of the theoretical analysis. The energy difference between the experimental points and the lines is attributed to the effect of Coulomb correlation. Although one cannot expect, that this procedure leads to quantitative results, it can be considered as a first step to get information on the Coulomb correlation in the presence of a magnetic field. The authors find a strong dependence of

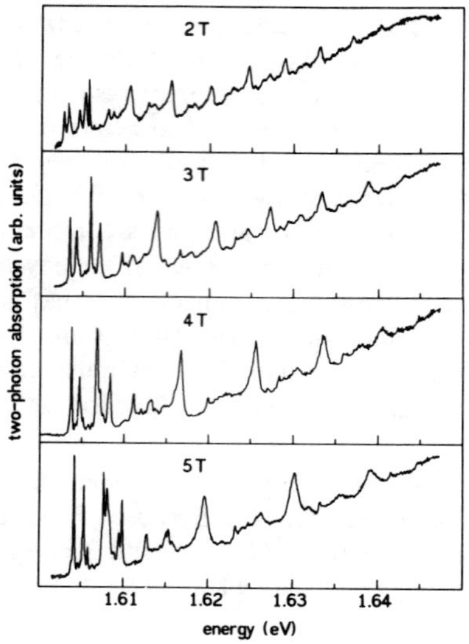

Figure 6.Two-photon absorption spectra of *CdTe* for (σ_+,σ_-) in magnetic field from $2T$ to $5T$. The polarization (σ_+,σ_-) refer to dye and CO_2 laser, respectively.

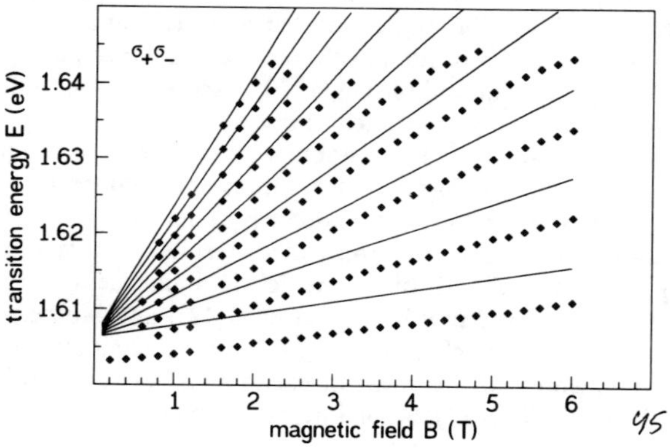

Figure 7. Magnetic field dependence of the calculated (solid lines) and measured (full squares) transition energies. Polarization of dye and CO_2 laser are σ_+ and σ_-, respectively.

the Coulomb correlation on the Landau quantum number, whereas the energy difference for a fixed Landau transition is nearly constant for a varying magnetic field at least in the low field regime ($B<3T$). Becker et al. [23] find in a first analysis of the *CdTe* data and one-photon results in *GaAs* as an empirical rule for the deviation at a fixed magnetic field:

$$\delta E_n / E_B \sim -C_1/(n+C_2) \qquad (1)$$

where C_1 and C_2 are in the order of one. E_B is the exciton binding energy and n the Landau quantum number. With the use of a two-band model, where the Coulomb potential is included as a perturbation, the authors are able to derive a dependence on n as given by eq. (1). A detailed theoretical analysis of one and two-photon magnetoopic spectra in the dominant exciton (below gap) and dominant Landau type region (above gap) would have to distinguish the different selection rules for the linear (one-photon) and nonlinear (two-photon) processes. It is an open question if in the Landau regime different envelope quantum numbers have to be assigned for one- and two-photon excited states, as has to be done in the exciton regime (*S* or *P* envelope). Besides the Coulomb correlation there is clear indication of a magneto-polaron effect in the one-photon and two-photon magnetooptic data as shown in Ref. [23]. The increasing deviation of the experimental points from the solid line at higher field strength, as marked in Figure 7, is attributed to the electron-phonon coupling, which leads to a pronounced pinning effect, if the conduction band Landau energy is close to the optical phonon energy.

In the last part of this paragraph we want to discuss two-photon magneto-absorption in multiple quantum wells (MQW). First measurements are reported by Fröhlich et al. [25]. It is not possible to detect two-photon absorption in MQW by the conventional technique, where one measures a small change in transmission for a tunable dye laser, which is induced by a high power laser (e.g. CO_2 laser). For the MQW with a typical effective thickness of less than a micron the sensitivity has to be increased by at least a factor of 10^3 as compared to the previous results on *CdTe* where samples of about 1*mm* thickness were used. The increase of sensitivity was achieved by measuring the two-photon excitation spectrum instead of the direct two-photon absorption. The experimental setup is shown schematically in Figure 8.

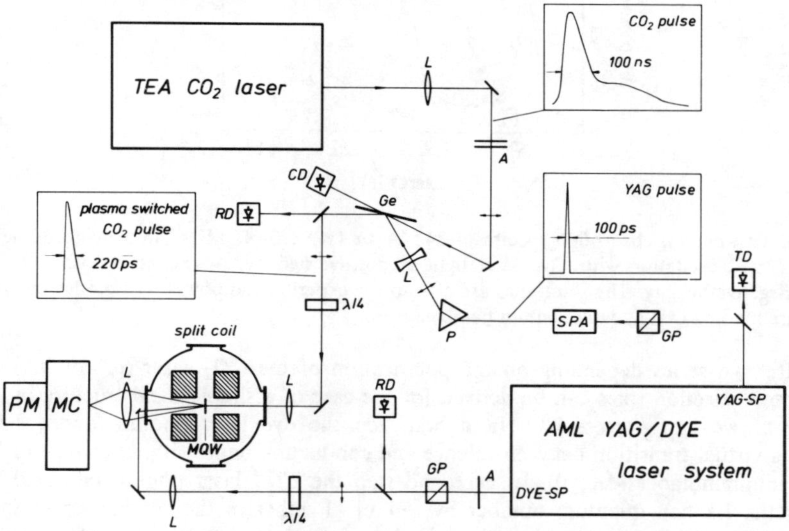

Figure 8. Block diagram of experimental setup. AML YAG/DYE, active mode-locked YAG-pumped dye laser system; TEA CO_2, transverse excited atmospheric CO_2 laser; *SP*, single pulse; SPA, single pulse amplifier; *GP*, glan polarizer; λ/4, quarter-wave plate; *P*, prism; *L*, lense; *Ge*, germanium plate; *TD*, trigger diode; *CD*, control diode; *RD*, reference diode; *A*, attenuator; *MC*, monochromator; *PM*, photomultiplier.

The setup is similar to the one described by Fröhlich et al. [26]. The central part is an active-mode-locked yttrium-aluminium-garnet (YAG)-dye laser system, which is described in detail by Fröhlich, Reimann, and Wille [27]. The laser provides 60*ps* pulses tunable between 780 and 865*mm* with the use of Styril 9 as a dye solution, and YAG pulses (1.06μm) of about 150*ps* which are perfectly synchronized to the pulses of the dye laser.

The $1.06\mu m$ pulses are further amplified to about $10mJ$ and then used for plasma-switching of a TEA CO_2 laser. The plasma-switching by the YAG pulse guarantees a perfect synchronization of the dye pulse and the CO_2 laser pulse which is necessary for two-photon processes. Both lasers are focussed to perfect overlap in the MQW which is mounted in Faraday configuration in a split-coil superconducting magnet at about $2K$; fields up to $6.6T$ are available. The helicity of both laser beams can be set independently by the use of quarter-wave plates. In order to separate the weak luminescence, which is induced by TPA, from scattered light of the high power dye laser a prism double monochromator and suitable interference filters are used.

Measurements on a MQW structure with *GaAs* layers of 84.5 Å (60 wells *As* with barrieres of 230 Å Al_xGa_{1-x} and an *Al* content of $x = 0.39$) show clearly resolved Landau transitions (Figure 9).

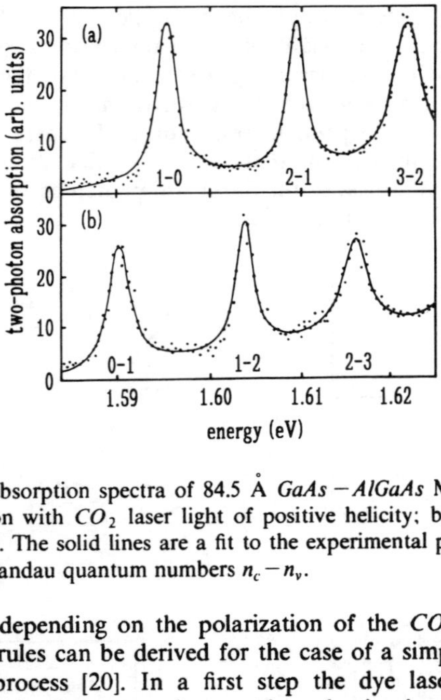

Figure 9. Two-photon absorption spectra of 84.5 Å $GaAs-AlGaAs$ MQW in a magnetic field at $5.5T$ at $2K$: a) excitation with CO_2 laser light of positive helicity; b) excitation with CO_2 laser light of negative helicity. The solid lines are a fit to the experimental points by Lorentzians. Resonances are marked by Landau quantum numbers $n_c - n_v$.

There are two series depending on the polarization of the CO_2 laser (σ^+ or σ^-). The two-photon selection rules can be derived for the case of a simple bandstructure by considering a two-step process [20]. In a first step the dye laser photon (about $1.5\,eV$) induces a virtual transition between valence and conduction band without change of Landau quantum number ($\Delta n = 0$). In a second step the CO_2 laser photon (about $0.1\,eV$) changes the Landau quantum number by $+1$ or -1 either in the conduction or in the valence band. The reverse process, which is also possible as seen by 2nd order perturbation theory is much weaker, because the CO_2 laser is too far off resonance to induce in the first step a virtual transition from the valence band to the conduction band of comparable strength. The field dependence is shown in Figure 10. Within the rather small range of magnetic field strength the resonances show a linear dependence. Neglecting again Coulomb correlation, non parabolicity in the conduction band and complications in the valence band, which are due to interaction between heavy and light hole levels [28], one can easily calculate cyclotron frequencies of the hole and electron bands separately. This is certainly an advantage as compared to one-photon spectroscopy, where in the simple case of a $\Delta n = 0$ selection rule one can derive only a reduced cyclotron frequency for the hole and electron band. The fact that the lines do not intersect a zero field is a

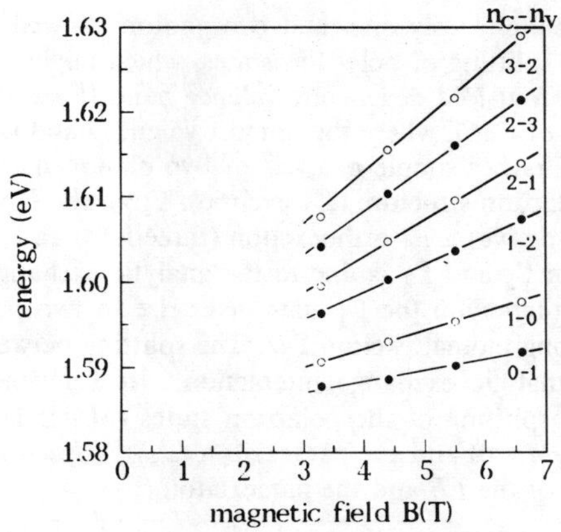

Figure 10. Field dependence of Landau transitions of 84.5 Å MQW as measured by TPA. Solid and open circles represent experimental results for negative and positive helicity of CO_2 laser, respectively. Slopes of solid lines are calculated from $m_e = 0.081$ and $m_{HH} = 0.27$. The numbers $n_c - n_v$ refer to Landau resonances of Figure 9.

clear indication, that excitonic effects have to be taken into account. Unfortunately it is not possible to resolve the *P* exciton structure as e.g. in *CdTe* [14], because of inhomogeneous broadening of the exciton lines in the MQW. TPA experiments on a single very homogeneous quantum well would certainly be of great interest to study excitons (*P* excitons) in a confined structure.

3. Nonlinear magnetoabsorption of polaritons

As outlined in the introduction nonlinear spectroscopy of transitions which are simultaneously allowed for one-photon transitions (linear spectroscopy) show additional complications. The polariton character of the exciton states has to be taken into account which leads in some cases to drastic *K*-dependent effects. For a recent review on polaritons we refer to an article by Bassani and Altarelli [29]. In nonlinear spectroscopy one has the unique possibility to tune energy and total *K*-vector within certain limitations independently of each other, thus doing *K*-space spectroscopy. As mentioned in the introduction two-photon transitions to polaritons can only be seen in systems without inversion center (e.g. copper halides). The polariton dispersion in crystals with inversion center (e.g. alkali halides) can be investigated by three-photon spectroscopy (TPS). This chapter will be divided in two paragraphs. In the first paragraph we discuss two-photon magnetoabsorption in *CuCl*. In the second paragraph three-photon magnetoabsorption of polaritons and excitons in alkali halides are presented.

3.1. Two-Photon Magnetoabsorption in *CuCl*

Two-photon transitions to excitons and biexcitons in copper halides (*CuCl*, *CuBr*) are extensively studied by two-photon absorption. For literature we refer to a review by Hönerlage et al. [30]. The copper halides are specially suited for two-photon spectroscopy of polaritons, because the strong admixture of *p* and *d* states in the valence band leads to

S excitons which are simultaneously one- and two-photon allowed. The application of a magnetic field leads to a splitting of polariton states, which might be rather complicated, if one has to deal with a four-fold degenerate valence band (Γ_8 at $K = 0$) as in the case of *CuBr* [30]. In the case of *CuCl*, where the upmost valence band is only two-fold degenerate (Γ_7), one expects a rather simple analysis of two-photon magnetoopic experiments on the lowest exciton-polariton structure (Z_3 exciton). From the Γ_7 valence band and the Γ_6 conduction band one derives a Γ_5 orthoexciton (three-fold) and a Γ_2 paraexciton. The energy difference between Γ_5 and Γ_2 is due to the analytic exchange interaction. Including the exciton photon interaction the Γ_5 state gives rise to two polariton branches Γ_{5T} (two-fold) and a single longitudinal exciton Γ_{5L}. The splitting between the lower Γ_{5T} and Γ_{5L} is due to the nonanalytic exchange interaction. In a magnetic field in Faraday configuration one gets a splitting of the polariton states ($M = \pm 1$) and a mixing of the longitudinal exciton ($LE, M = 0$) and the paraexciton (Γ_2). As seen in Figure 11 this mixing results in a repulsion of the LE and the paraexciton (Γ_2).

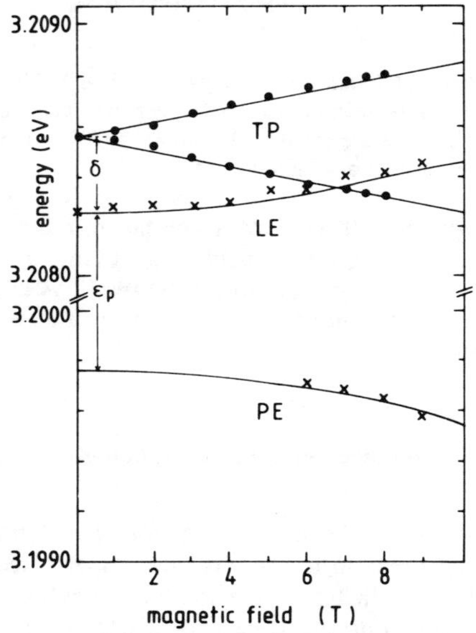

Figure 11. Energy values of Z_3 exciton components in *CuCl* as function of magnetic field. The solid lines are calculated using as parameters: $\delta = 0.31\,meV$, $\epsilon_p = 8.5\,meV$, $g_c + g_v = 1.02 \pm 0.05$, $g_c - g_v = 4.5 \pm 0.2$ (Ref. [31]).

The corresponding off-diagonal matrix element is proportional to the difference of conduction band and valence band g-values ($g_c - g_v$) as is discussed by Fröhlich, Hölscher, and Nöthe [31]. As expected the polariton shows a splitting into two components. The splitting depends linearly on the magnetic field. In a naive analysis this splitting is expected to be proportional to $g_c + g_v$ [31]. Further measurements by Fröhlich, Hölscher, and Mohler [32] showed that the polariton splitting depends strongly on the total K-vector (Figure 12). As discussed in detail in Ref. [32] the K-dependent splitting can be explained by taking into account the field induced mixing of the Z_3 (Γ_7 valence band) and the $Z_{1,2}$ (Γ_8 valence band) excitons and is given by:

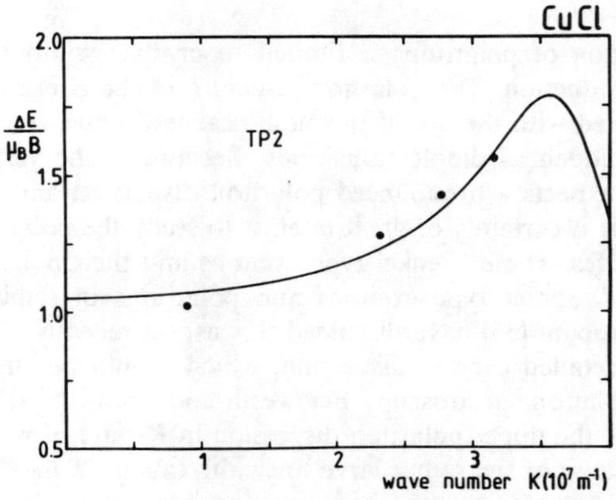

Figure 12. *K*-dependence of splitting of the upper polariton Z_3 in *CuCl*; full dots: experimental results, solid line: theoretical results (Ref. [32]).

	Z_3	$Z_{1,2}$
Z_3	$E_{z3}(0)\pm\frac{2}{3}(\tilde{K}+1)\mu_B B$	$\mp\frac{\sqrt{2}}{6}(2-\tilde{K})\mu_B B$
$Z_{1,2}$	$\mp\frac{\sqrt{2}}{6}(2-\tilde{K})\mu_B B$	$E_{1,2}(0)\pm\frac{1}{6}(5\tilde{K}+2)\mu_B B$

The upper and lower signs hold for the $M = +1$ and $M = -1$ components, respectively. As shown by Suga, and Mitani [33] $\tilde{K}$ is related to the coefficients of hybridization of the halogen p levels (a_p) and copper d levels (b_d) by $\tilde{K} = 2a_P^2 - 1$ with $a_P^2 + b_d^2 = 1$. The diagonal terms lead to a splitting linear in the magnetic field strength B. The off-diagonal terms cause a mixing of the Z_3 and $Z_{1,2}$ excitons, which changes the oscillator strength of the $M = \pm 1$ components. For the oscillator strengths of the excitons one obtains to first order in B:

$$f_{z_3} = f_{z_3}(0) \pm 2\sqrt{f_{z_3}(0)\cdot f_{z_{1,2}}(0)}\,\frac{\frac{\sqrt{2}}{6}(2-\tilde{K})\mu_B B}{E_{z_{1,2}}(0) - E_{z_3}(0)} \tag{3}$$

$$f_{z_{1,2}} = f_{z_{1,2}}(0) \mp 2\sqrt{f_{z_3}(0)\cdot f_{z_{1,2}}(0)}\,\frac{\frac{\sqrt{2}}{6}(1-\tilde{K})\mu_B B}{E_{z_{1,2}}(0) - E_{z_3}(0)}.$$

The upper and lower sign refer to the $M = +1$ and $M = -1$ components, respectively. With the use of a two-oscillator equation the *K*-dependent splitting can be calculated and compared with experiments. The increase of oscillator strength for the $M = +1$ and the decrease of the $M = -1$ component of the Z_3 exciton ($\tilde{K} = -0.54$ [32]) are responsible for the drastic K dependence of the polariton splitting. These rather unexpected findings show, that even in the case of simple band structure (two-fold valence and conduction band) two-photon magnetoabsorption experiments on polaritons cannot be analyzed without taking the polariton mixing by the external magnetic field into account.

3.2. Three-Photon Magnetoabsorption in Alkali Halides

Two-photon absorption of polaritons is limited to crystals wihtout inversion center as discussed in the introduction. The polariton structure of the S excitons in alkali halides can thus not be studied with the use of this nonlinear technique, since two-photon transitions are parity forbidden as dipole transitions. Because of the very strong one-photon absorption [34] one expects a pronounced polariton dispersion and a large longitudinal-transverse splitting. It is certainly of great interest to study the polariton and longitudinal exciton in alkali halides. These Frenkel type excitons and their polariton dispersion have to be compared to Wannier type excitons and polaritons in semiconductors. Bassani, Ruggiero and Quattropani [35] have discussed this aspect recently. They predict interesting features of the detailed exciton dispersion, which should be studied experimentally. By the use of three-photon spectroscopy Beerwerth and Fröhlich [4] were able to present first measurements of the upper polariton dispersion in *KI* and of resonances of the longitudinal exciton. Because of the rather large linewidth (about 20 *meV*) of these resonances it is difficult to study the Zeeman splitting as was done in *CuCl*. Beerwerth et al. [36] were able to resolve the splitting of the upper polariton.

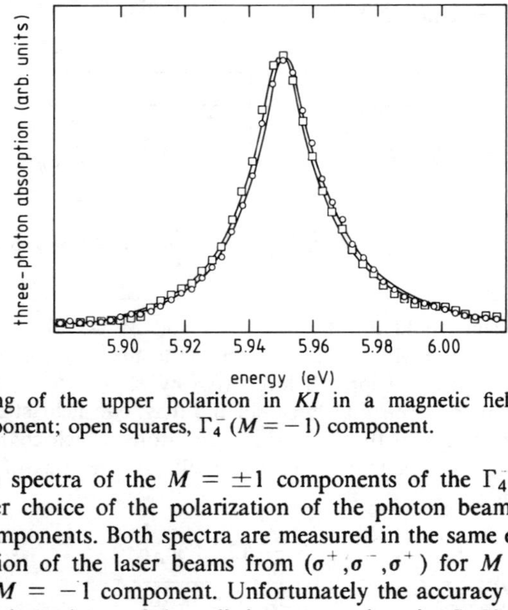

Figure 13. Splitting of the upper polariton in *KI* in a magnetic field of 7*T*; open circles, Γ_4^- $(M = +1)$ component; open squares, Γ_4^- $(M = -1)$ component.

In Figure 13 the spectra of the $M = \pm 1$ components of the Γ_4^- polariton in *KI* are shown. By proper choice of the polarization of the photon beams, one can selectively excite the two components. Both spectra are measured in the same experiment, by switching the polarization of the laser beams from $(\sigma^+, \sigma^-, \sigma^+)$ for $M = +1$ component to $(\sigma^-, \sigma^+, \sigma^-)$ for $M = -1$ component. Unfortunately the accuracy is not high enough to investigate the K-dependence of the splitting as was done in *CuCl*. Besides the orthoexcitons of Γ_4^- symmetry, which give rise to the transverse polaritons (*TP*) and the longitudinal exciton (*LE*), there are paraexcitons of Γ_5^- and Γ_3^- symmetry, which stem from the same valence to conduction band transition $(\Gamma_8^- \rightarrow \Gamma_6^+)$. Optical transition to these excitons are spin forbidden. In high magnetic fields ortho- and paraexcitons are mixed, which allows spectroscopy of paraexcitons, as was shown for *KI* by Beerwerth and Fröhlich [37] and for *RbI* in Ref. [36]. As an example we present in Figure 14 a spectrum of paraexcitons $(\Gamma_5^-, \Gamma_3^-$ symmetry) and the upper polariton $(\Gamma_4^-$ symmetry) in *KI*. Although the oscillator strength of the paraexcitons is very low as compared to the polariton, they are clearly resolved. The paraexcitons show a narrow linewidth of about 0.2 *meV*, which allows the determination of g values with high accuracy. In Figure 15 we show the field dependence of the splitting of the paraexciton in *KI*. As discussed in more detail in Ref. [36] the analysis of the polariton dispersion yields the longitudinal-

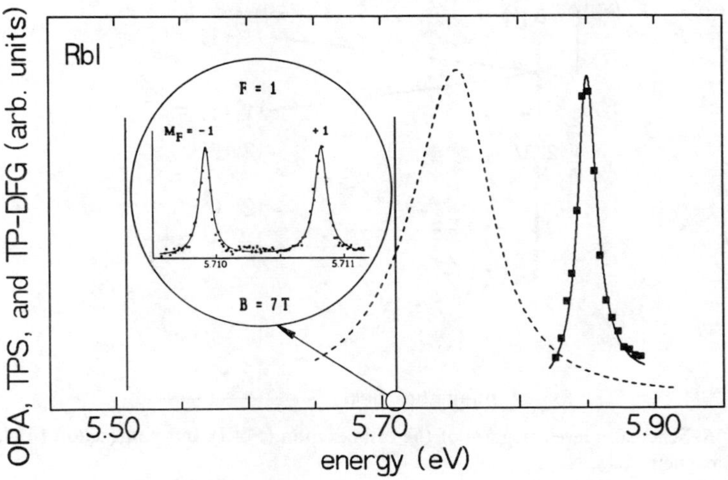

Figure 14. Three-photon spectra of lowest exciton structure of *KI* at 4.5*K*. Closed circles represent the strong resonances on the longitudinal exciton ($F=1, M_F=0$). The inset shows components of the paraexciton in a magnetic field of 7*T*.

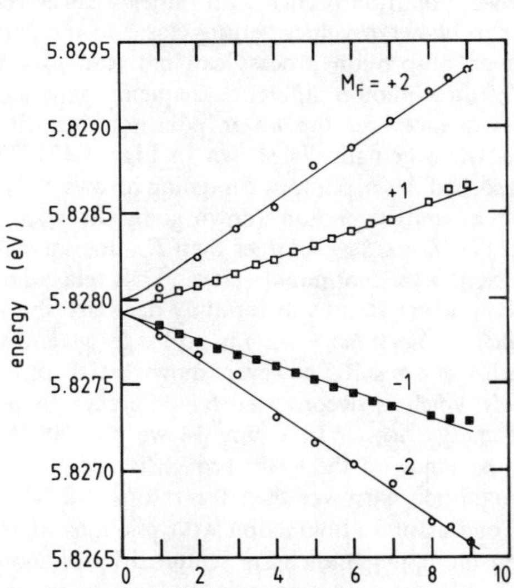

Figure 15. Field dependence of splitting of the paraexciton in *KI*. The solid lines correspond to a fit using $g_p = 1.35$ as *g* value (Ref. [37]).

transverse splitting Δ_{LT} and the energy of the longitudinal (*LE*) and transverse exciton (*TE*). From the magnetic field dependence of the paraexciton (Γ_5^-, Γ_3^- symmetry) one can than derive the exchange energy Δ_{ex} and even a value for the crystal field splitting ϵ of the $F = 2$ state into Γ_5^- and Γ_3^-. The schematic level diagram is shown in Figure 16. The accuracy of Δ_{LT} and Δ_{ex} is mainly dependent on the accuracy of the energy of the transverse exciton E_T, which is a fit parameter in a two-oscillator dispersion formula [36].

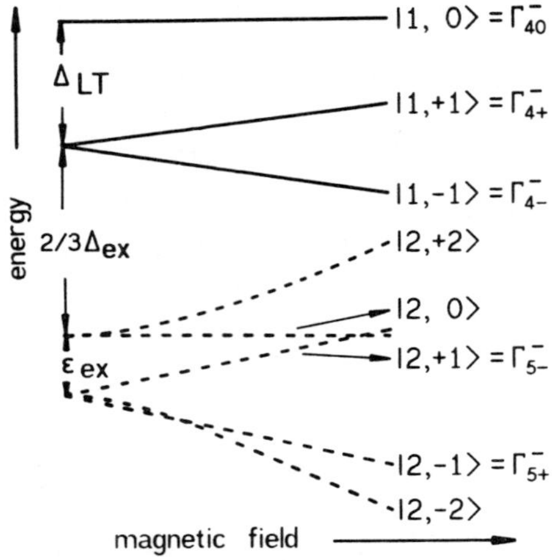

Figure 16. Schematic level diagram of the orthoexciton ($F = 1$) and paraexciton ($F = 2$) as a function of magnetic field.

The accuracy of this fit could certainly be improved considerably if resonances on the lower polariton branch close to the transverse exciton could be excited. This is not possible in three-photon absorption experiments, because the exciting laser photons give rise to polaritons on the lower polariton branch with rather small K-vectors. Since the polariton branch bends over to higher K-values getting closer to the first resonance, the fusion of three polaritons in an absorption process can not lead to resonances on the lower branch. With the use of three-photon difference frequency generation (TP-DFG) we have succeeded to excite resonances on the lower polariton branch of *KI* and *RbI* [38]. $TP - DFG$ to polaritons is schematically shown in Figure 17. The maximum K-vector $\mathbf{K}_1 + \mathbf{K}_2$ of the two 'absorbed' laser photons (up going arrows K_1 parallel K_2) is achieved if the K-vector $\mathbf{K}_3$ of the 'emitted photon' (down going arrow) is antiparallel to $\mathbf{K}_1$ and $\mathbf{K}_2$. It should be noted that $\hbar\omega_1 + \hbar\omega_2$ is larger than E_g, one therefore expects two-photon absorption to higher exciton or continuum states. Fast relaxation processes from these real intermediate levels to lower states will certainly deminish the $TP - DFG$ signal, since energy conservation ($\hbar\omega_{DF} = \hbar\omega_1 + \hbar\omega_2 - \hbar\omega_3$) is no longer given for these relaxed excited states. The *DFG* experimental results, however, show that there is a large coherent (non relaxed) component left which is deconverted by difference frequency generation to an outgoing polariton of energy $\hbar\omega_{DF}$. In Figure 18 we present first results [38] for *RbI*. The linewidth of the resonance on the lower branch is about $0.1\,meV$ which is by more than two orders of magnitude narrower then the resonances on the upper branch. This can be understood, if one assumes interaction with phonons as the main reason for the damping. Polaritons on the upper branch are scattered by phonon *emission* to the lower branch. This is apparently very efficient even at low temperatures. Polaritons on the lower branch can only be scattered efficiently by phonon *absorption* to polaritons at larger K-vectors on the same branch. If this mechanism is dominant, one should derive the phonon energy (probably LO phonon) from the temperature dependence of the linewidth. In Figure 19 resonances on the upper polariton branch, the longitudinal exciton and the lower polariton branch are shown.

The very small linewidth of the resonances on the lower polariton branch allows to determine the Zeeman splitting with high accuracy. In Figure 19 we show the splitting at $B = 7T$ for *RbI*. The two components can be selectively excited with the use of circularly polarized light. Due to the high accuracy we were able to derive a significant K-

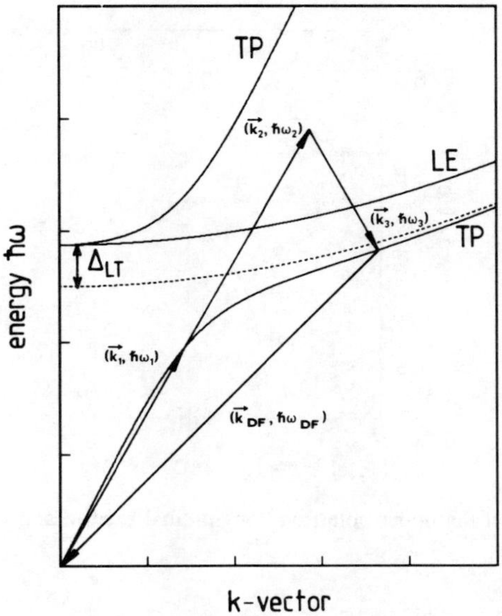

Figure 17. Schematic polariton diagram to demonstrate $TP-DFG$. TP, transverse polariton; LE, longitudinal exciton; Δ_{LT}, longitudinal-transverse splitting; $(\mathbf{K}_i, \hbar\omega_i)$ $i=1,2,3$ refer to incoming polaritons; $(\mathbf{K}_{DF}, \hbar\omega_{DF})$ refers to the polariton at the difference frequency.

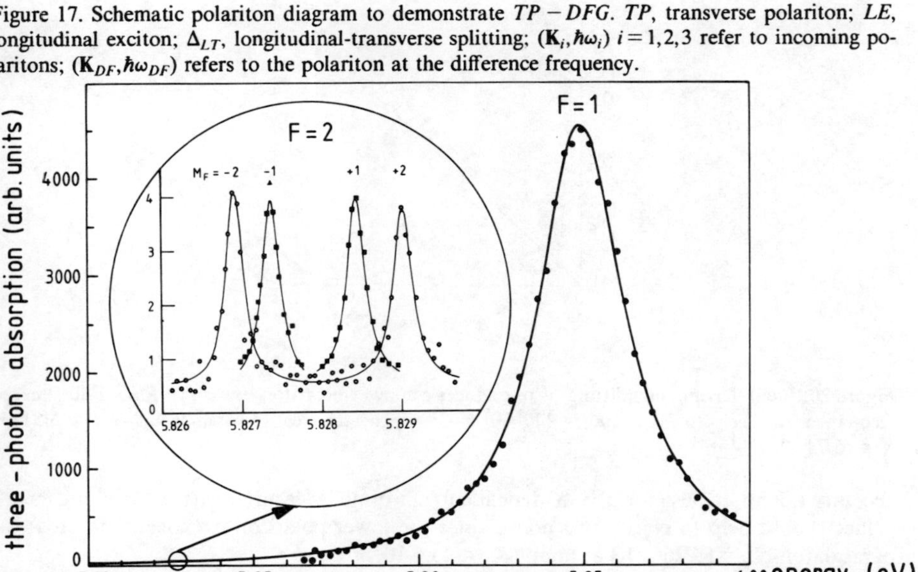

Figure 18. $TP-DFG$ spectra of two resonances on the lower polariton branch (5.51 eV and 5.71 eV) in RbI. For comparison one resonance on the upper polariton (5.85 eV) is also shown. The one-photon spectrum (dashed line) is taken from Ref. [34]. The inset shows the splitting of the 5.71 eV line in a magnetic field of $7T$.

dependence of the splitting as shown in Figure 20 for two K-values ($6.74 \cdot 10^7$ and $11.9 \cdot 10^7 m^{-1}$). From the splitting one should be able to derive accurate g-values for the orthoexciton. At the present it is not clear if a field-induced mixing of the ortho- and paraexciton or a mixing between the orthoexciton of the Γ_8^- valence band ($j=3/2$) and the orthoexciton of the Γ_6^- valence band ($j=1/2$) has to be considered in order to

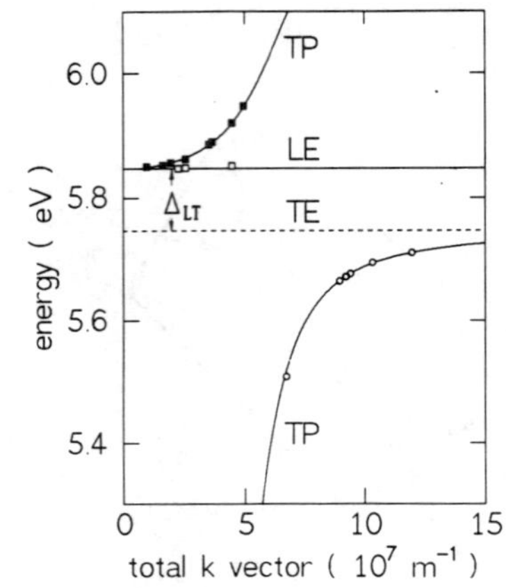

Figure 19. Resonances on the upper polariton, longitudinal exciton and lower polariton in *RbI* at 4.2*K*.

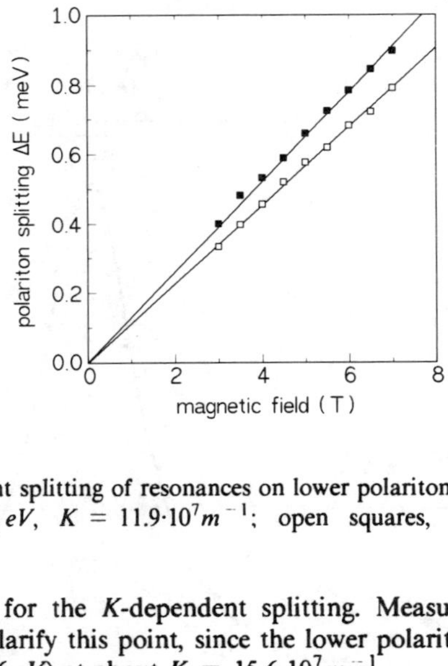

Figure 20. Field-dependent splitting of resonances on lower polariton branch in *RbI*. Full squares, resonance at $\hbar\omega = 5.71\,eV$, $K = 11.9{\cdot}10^7 m^{-1}$; open squares, resonance at $\hbar\omega = 5.51\,eV$, $K = 6.74{\cdot}10^7 m^{-1}$.

account quantitatively for the *K*-dependent splitting. Measurements at still higher *K*-values should help to clarify this point, since the lower polariton is expected to cross the paraexciton ($E_p = 5.7266\,eV$) at about $K = 15.6{\cdot}10^7 m^{-1}$.

The experimental technique of nonlinear magnetooptics can certainly be applied to many other interesting substances. Further increase of sensitivity should allow to study exciton splitting in MQW structures. Excitons and Landau levels in superlattices should be investigated. It would certainly be of interest to vary the angle between the direction of the magnetic field and the superlattice structure, which might lead to coupling effects between Landau levels and subbands.

References

1. Ch. Uihlein, D. Fröhlich and R. Kenklies, *Phys. Rev. B* **23**, 2731 (1981).
2. C.H. Henry and J.J. Hopfield, *Phys. Rev. Lett.* **15**, 964 (1965).
3. D. Fröhlich, E. Mohler and P. Wiesner, *Phys. Rev. Lett.* **26**, 554 (1971).
4. F. Beerwerth and D. Fröhlich, *Phys. Rev. Lett.* **55**, 2603 (1985).
5. R. Dinges and D. Fröhlich, *Solid State Commun.* **19**, 61 (1976).
6. R. Dinges, D. Fröhlich, B. Staginnus and W. Staude, *Phys. Rev. Lett.* **25**, 922 (1970).
7. W. Kaule, *Solid State Commun.* **9**, 17 (1971); *Z. Physik* **256**, 97 (1972).
8. D.G. Seiler, D. Heiman and B.S. Wherrett, *Phys. Rev. B* **27**, 2355 (1983).
9. D. Fröhlich and R. Kenklies, *Phys. Stat. Sol. (b)* **111**, 247 (1982).
10. M. Luttinger and W. Kohn, *Phys. Rev.* **97**, 869 (1955); J.M. Luttinger, *ibid* **102**, 1030 (1956).
11. M. Altarelli and N.O. Lipari, *Phys. Rev. B* **7**, 3798 (1973).
12. A. Baldereschi and N.O. Lipari, *Phys. Rev. B* **8**, 2697 (1973).
13 H.W. Hölscher, A. Nöthe and Ch. Uihlein, *Phys. Rev. B* **31**, 2379 (1985).
14. Ch. Neumann, A. Nöthe and N.O. Lipari, *Phys. Rev. B* **37**, 922 (1988).
15. M. Sondergeld and R.G. Stafford, *Phys. Rev. Lett.* **35**, 1529 (1975); M. Sondergeld, *Phys. Stat. Sol. (b)* **81**, 253 (1977); *ibid* **81**, 451 (1977).
16. Ch. Neumann and A. Nöthe, *Europhys. Lett.* **4**, 351 (1987).
17. G.D. Mahan, *Phys. Rev.* **170**, 825 (1968).
18. K.J. Button, B. Lax, M.H. Weiler and M. Reine, *Phys. Rev. Lett.* **17**, 1005 (1966).
19. M.H. Weiler, M. Reine and B. Lax, *Phys. Rev.* **171**, 949 (1968).
20. F. Bassani and R. Girlanda, *Opt. Commun.* **1**, 359 (1970); R. Girlanda, *Il Nuovo Cimento* **6B**, 53 (1971).
21. D.G. Seiler, C.L. Littler and D. Heiman, *J. Appl. Phys.* **57**, 2191 (1985).
22. C.R. Pidgeon and R.N. Brown, *Phys Rev.* **146**, 575 (1966); M.H. Weiler, R.L. Aggarwal and B. Lax, *Phys. Rev. B* **17**, 3269 (1978).
23. W. Becker, B. Gerlach, Th. Hornung, A. Nöthe, G. Spata and R.G. Ulbrich, *Proceedings of the 19th International Conference on the Physics of Semiconductors*, edited by W. Zawadzki, Polish Academy of Sciences, p. 1505 (1988).
24. A. Nöthe and G. Spata, *Phys. Rev. B* **39**, 8301 (1989).
25. D. Fröhlich, R. Wille, W. Schlapp and G. Weimann, *Phys. Rev. Lett.* **61**, 1878 (1988).
26. D. Fröhlich, R. Wille, W. Schlapp and G. Weimann, *Phys. Rev. Lett.* **59**, 1748 (1987).
27. D. Fröhlich, K. Reimann and R. Wille, *Europhys. Lett.* **3**, 853 (1987).
28. M. Altarelli, U. Ekenberg and A. Fasolino, *Phys. Rev. B* **32**, 5138 (1985).
29. F. Bassani and M. Altarelli, Handbook on Synchrotron Radiation, Vol. 1, edited by E.E. Koch, North-Holland Publ. Cy., p. 463-605 (1983).
30. B. Hönerlage, R. Lévy, J.B. Grun, C. Klingshirn and K. Bohnert, *Phys. Rep.* **124**, 162 (1985).
31. D. Fröhlich, H. Hölscher and A. Nöthe, *Solid State Commun.* **48**, 217 (1983).
32. D. Fröhlich, H. Hölscher and E. Mohler, Proceedings of the 17th International Conference on the Physics of Semiconductors, editors J.D. Chadi and W.A. Harrison, Springer-Verlag, New York, Berlin, Heidelberg, Tokyo, 1255 (1985).
33. S. Suga, T. Koda and T. Mitani, *Phys. Stat. Sol. (b)* **48**, 753 (1971).
34. K. Teegarden and G. Baldini, *Phys. Rev.* **155**, 896 (1967).

35. F. Bassani, F. Ruggiero and A. Quattropani, *Nuovo Cimento* **7D**, 700 (1986).

36. F. Beerwerth, D. Fröhlich, P. Köhler, V. Leinweber and A. Voss, *Phys. Rev. B* **38**, 4250 (1988).

37. F. Beerwerth and D. Fröhlich, *Phys. Rev. B* **36**, 6239 (1987).

38. D. Fröhlich, St. Kirchhoff, P. Köhler and W. Nieswand, *Phys. Rev. B* **50**, 1976 (1989).

DYNAMICS OF THE COHERENT AND NONCOHERENT NONLINEAR SUSCEPTIBILITY IN *CuCl*

B. HÖNERLAGE, R. LÉVY AND J.B. GRUN

Institut de Physique et Chimie
des Matériaux de Strasbourg
Groupe d'Optique Nonlinéare et d'Optoélectronique
Unité Mixte 380046-CNRS-ULP-EHICS
5, rue de l'Université
67084 Strasbourg Cédex
France

We discuss experimental methods to determine the dynamical properties of the nonlinear susceptibility in *CuCl*. The time-dependent nonlinear absorption gives information on the exciton and biexciton dynamics and therefore on the imaginary part of the noncoherent dielectric susceptibility. Information on its real part may be obtained from dynamical measurements of the refractive index. The coherent part of the nonlinear susceptibility gives rise to a parametric generation and its competition with the noncoherent part governs the generated signal intensity in wave-mixing processes.

1. Introduction

The study of the nonlinear dielectric susceptibility of semiconductors has been of increasing interest these recent years because of the possibility to develop optical devices from these materials [1,2]. Apart from lasers and light emitting diodes, optical bistable devices, logic gates, differential amplifiers and memories are important for optical computing and image processing. For the latter applications, the dynamical properties of the nonlinearities have to be known, since they limit the speed and the bandwidth of the devices.

Nonlinearities are governed by the real and imaginary parts of the susceptibilities. It is also important to distinguish between the coherent and noncoherent parts of these functions [3]: the first ones (coherent parts) are important for pulse propagation and signal generation processes; the second ones, for induced absorptions and changes of the refractive index.

If we consider a pulse propagating through a two-level system with quasi-stationary states $|1>$ and $|2>$, inducing transitions between the states, the transverse dephasing time T_2 is related to the coherence times τ_i of the states through the relation:

$$\frac{1}{T_2} = \frac{1}{2}\left[\frac{1}{\tau_1}+\frac{1}{\tau_2}\right]. \tag{1}$$

For a general system with several states, transverse damping constants Γ_{ij} are similarly defined through:

$$\Gamma_{ij} = \frac{1}{2}(1/\tau_i+1/\tau_j) \text{ for } i \neq j. \tag{2}$$

The coherence times τ_i of the quasi-stationary states are limited by elastic and inelastic scattering processes. In semiconductors, under low excitation conditions, they are in the

R. Girlanda et al. (eds.), Progress on Electron Properties of Solids, 275–290.

picosecond time range. Furthermore, they usually decrease with increasing occupation of excited states. Therefore, they depend on the shape and on the maximum intensity of the impinging light pulse and are different at its beginning and at its end.

This behaviour is due to the fact that the quasi-particles, coherently created by the light field, relax through elastic collisions, exciting other states which are no longer coherently driven by the light field. In this case, the energies of the quasi-particles remain the same but their phase, polarization and/or wave vector become different from their initial ones.

The population of quasi-particles with a definite energy decays with the longitudinal relaxation time T_{ii}. These time constants are usually on a nanosecond time scale in semiconductors and characterize the decay of the incoherent susceptibility.

In the following, we will decompose the noncoherent susceptibility into coherent and incoherent contributions which play different roles in the actual experiments performed.

We will first study the time dependence of the absorption and of the refractive index at the test frequency ω_t for a *CuCl* sample which is excited by a picosecond pump pulse of frequency ω_p. These experiments determine the real and imaginary parts of the noncoherent susceptibility. Then, we will discuss degenerate ($\omega_p = \omega_t = \omega$) wave-mixing processes using picosecond pulses which will allow us to determine the coherence times of the states.

Let us first discuss the propagation of the light fields in order to get a better insight into the role the coherence of the states is playing.

2. The Transient Propagating Light Pulse

Let us consider a classical electric field given by the complex expression $E_i(z,t)$ where the index i denotes its polarization. It varies in one space coordinate z and in time t, its transverse spatial dependence (x,y) being neglected:

$$E_i(z,t) = \frac{1}{2}[E_i^0(z,t)e^{i(\omega_i t - k_i z)} + \text{c.c.}] \tag{3}$$

$E_i^0(z,t)$ is the field envelope which varies slowly with respect to the frequency ω_i and the wave vector k_i of the light field, c.c. denotes the complex conjugated expression.

As explained in more details in Ref. [4], the wave equation for the propagating electric field $E_i(z,t)$ reads:

$$\frac{\delta^2}{\delta z^2}E_i(z,t) - \mu_0\sigma\frac{\delta}{\delta t}E_i(z,t) - \mu_0\epsilon_L\frac{\delta^2}{\delta t^2}E_i(z,t) = \mu_0\frac{\delta^2}{\delta t^2}P_{NL}^i(z,t) \tag{4}$$

where σ is the conductivity and μ_0 the permeability of the (nonmagnetic) system. $\epsilon_L = \epsilon_0(1+\chi_L)$ denotes the linear dielectric function, χ_L being the linear susceptibility. $P_{NL}^i(z,t)$ is the nonlinear polarization of the ith beam. It is defined as usual as a convolution product between the nonlinear susceptibility tensor χ_{NL}^{ij} and the jth electric field $E_j(z,t)$. Assuming a local spatial response for simplicity, we obtain:

$$P_{NL}^i(z,t) = \Sigma_j \int_{-\infty}^{t} \chi_{NL}^{ij}(z,t,t')E_j(z,t')dt'. \tag{5}$$

Here, again, $\chi_{NL}^{ij}(z,t,t')$ can be separated into fast and slowly varying contributions:

$$\chi_{NL}^{ij}(z,t,t') = \Sigma_{nn'}\chi_{NL}^{ij(n,n')}(z,t,t')e^{in(\omega t - kz)}e^{in'(\omega_j t - k_j z)} \tag{6}$$

where $(\omega_i,\mathbf{k}_i)$ denotes the frequency and wave vector of the field considered and $(\omega_j,\mathbf{k}_j)$ of the fields inducing the process. $(\omega,\mathbf{k})$ are the fundamental frequency and wave vector of the susceptibility. It will be the frequency and the wave vector of the pump beam in the coherent case, and that of the excited quasi-particles in the incoherent processes.

We now introduce the Fourier-transformed terms of $\chi_{NL}^{ij(n,n')}(z,t,t')$ [5]:

$$\chi_{NL}^{ij(n,n')}(z,t,\omega) = \int_0^\infty \chi_{NL}^{ij(n,n')}(z,t,t')e^{-i\omega t'}dt'. \tag{7}$$

The development presented here supposes that the pump beam is monochromatic and the other fields are weak.

With these definitions and using the relation

$$\chi(z,t,\omega) = \chi(z,t,-\omega)^* \tag{8}$$

we obtain in the slowly varying envelope approximation:

$$P_{NL}^i(z,t) = \frac{1}{2}[\Sigma_{jnn'}\chi_{NL}^{ij(n,n')}(z,t,\omega_j)E_j^0(z,t)e^{i[n\omega+(n'+1)\omega_j]t-i[nk+(n'+1)k_j]z} \tag{9}$$
$$+\chi_{NL}^{ij(-n,-n')^*}(z,t,\omega_j)E_j^0(z,t)^*e^{i[n\omega+(n'-1)\omega_j]t-i[nk+(n'-1)k_j]z}.$$

We introduce Eq. (9) into Eq. (4) using the dispersion relation [6] $\mu_0\epsilon_0\omega_j^2 = k_j^2$. We also assume that all time derivatives of E_j^0 and $\chi_{NL}^{ij(n,n')}$ are small compared to $E_j^0\omega_j$ and $\omega\chi_{NL}^{ij(n,n')}$, respectively, and that $\delta^2/\delta z^2\, E_j^0 << k_j\delta/\delta z\, E_j^0$. We then obtain a set of coupled nonlinear differential equations for the field amplitudes:

$$[-2ik_i\frac{\delta}{\delta z}E^0(z,t)-i\omega_i\mu_0\sigma E_i^0(z,t)]e^{i(\omega_i t-k_i z)}+\text{c.c.} \tag{10}$$
$$=-\mu_0\{\Sigma_{n,n'j}[n\omega+(n'+1)\omega_j]^2\chi_{NL}^{ij(n,n')}(z,t,\omega_j)\}E_j(z,t)e^{i[n\omega+(n'+1)\omega_j]t-i[nk+(n'+1)k_j]z}$$
$$\left.+\Sigma_{n,n'j}[n\omega+(n'-1)\omega_j]^2\chi_{NL}^{ij(-n,-n')^*}(z,t,\omega_j)E_j^{0*}(z,t)e^{i[n\omega+(n'-1)\omega_j]t-i[nk+(n'-1)k_j]z}\right\}.$$

The coupling arises via the different field amplitudes E_j^0 and the nonlinearity through the fact that the χ_{NL} terms are field dependent. If $\chi_{NL}^{ij(n,n')}$ and E_j^0 were independent of time and space, we could apply a Fourier analysis to Eq. (10). In the more general case, we obtain the following system of equations:

$$-2ik_i\frac{\delta}{\delta z}E_i^0-i\omega_i\mu_0\sigma E_i^0 = \tag{11}$$
$$-\mu_0[\Sigma_{nn'j}(n\omega+(n'+1)\omega_j)^2\chi_{NL}^{ij(n,n')}(z,t,\omega_j)E_j^0(z,t)\times$$
$$\times\delta(n\omega+(n'+1)\omega_j-\omega_i)\delta(nk+(n'+1)k_j-k_i)$$
$$+\Sigma_{nn'j}(n\omega+(n'-1)\omega_j)^2\chi_{NL}^{ij(-n,-n')^*}(z,t,\omega_j)E_j^0(z,t)^*\times$$
$$\left.\times\delta(n\omega+(n'-1)\omega_j-\omega_i)\delta(nk+(n'-1)k_j-k_i)\right]$$

where the conditions

$$\omega_i = n\omega+(n'+1)\omega_j \text{ and } k_i = (nk+(n'+1)k_j)$$

or

$$\omega_i = n\omega+(n'-1)\omega_j) \text{ and } k_i = (nk+(n'-1)k_j)$$

arise from energy and momentum conservations, respectively. They have to be strictly fulfilled in the stationary case. In the case of the slowly varying envelope approximation applied here, Eq. (11) remains approximately valid, but a small phase mismatch or frequency chirp may occur.

3. The Coherent and Noncoherent Transient Susceptibility

In order to discuss Eq. (11) further, we will specify the system we are studying. We will consider the case of *CuCl*, which is very well described by the three-level system given in Figure 1.

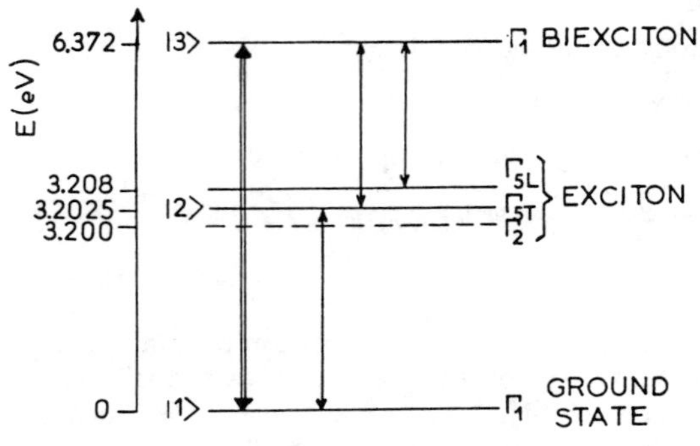

Figure 1. Three-level system describing *CuCl* together with the allowed optical transitions.

It consists of the crystal ground state $|1>$, the exciton state $|2>$, and the biexciton one $|3>$. One-photon transitions are allowed between the crystal ground state and the transverse exciton state, and between exciton and biexciton states [7]. The one-photon transition between the crystal ground state and the biexciton state is forbidden but the two-photon transition is allowed if the two photons have parallel polarizations. This system gives rise to strong nonlinearities which show up in absorption, reflection and emission spectroscopy.

Let us first specify Eq. (11) for the transmission measurements of a test beam (index t) with the generation of a signal beam (index s). We obtain the following equation for the test field:

$$-2ik_t\frac{\delta}{\delta z}E_t^0 - i\omega_t\mu_0\sigma E_t^0 = -\mu_0[\omega_t^2\chi_{NL}^{tt(0,0)}(z,t,\omega_t)E_t^0(z,t) \qquad (12)$$

$$+(2\omega-\omega_s)^2\chi_{NL}^{ts(-2,0)^*}(z,t,\omega_s)E_s^{0*}(z,t)\delta(2\omega-\omega_t-\omega_s)\delta(2\mathbf{k}-\mathbf{k}_t-\mathbf{k}_s)]$$

We obtain in Eq. (12) two nonlinear contributions which are of different origins: the first term $\chi_{NL}^{tt(0,0)}$ describes all the nonlinear absorption and dispersion changes for the test beam induced by the pump beam; the second term $\chi_{NL}^{ts(2,0)*}$ represents the generation of a signal beam at a frequency such that $2\omega = \omega_s+\omega_t$ and a wave vector given by $2\mathbf{k} = \mathbf{k}_s+\mathbf{k}_t$. Both terms have their own dynamics since they depend explicitly on the time t.

We may further decompose both terms into a coherent χ_{NLCO} and an incoherent χ_{NLINC} contribution with $\chi_{NL} = \chi_{NLCO}+\chi_{NLINC}$. This means that we are considering susceptibilities where scattering processes have been present or not before the measurements.

The induced absorption, obtain from $\mathrm{Im}(\chi_{NL}^{tt}(\omega_t))$, is proportional to the distribution of polaritons created by the pump beam [8,9]. The states can be coherently or incoherently occupied and their dynamics shows up as gain (negative absorption) or absorption in the measurement of the transmission. As discussed above, the coherent (χ_{NLCO}) and incoherent (χ_{NLINC}) parts have different relaxation times.

Let us now discuss our experimental results with Eq. (12).

4. Experimental Results for *CuCl*

4.1. Experimental Set-Up

We have performed time-resolved reflection [10] and transmission [11] measurements, using a 'pump and probe' technique, on *CuCl* monocrystalline platelets and on films, epitaxially grown on CaF_2 substrates.

The experimental set-up used is shown in Figure 2 [12].

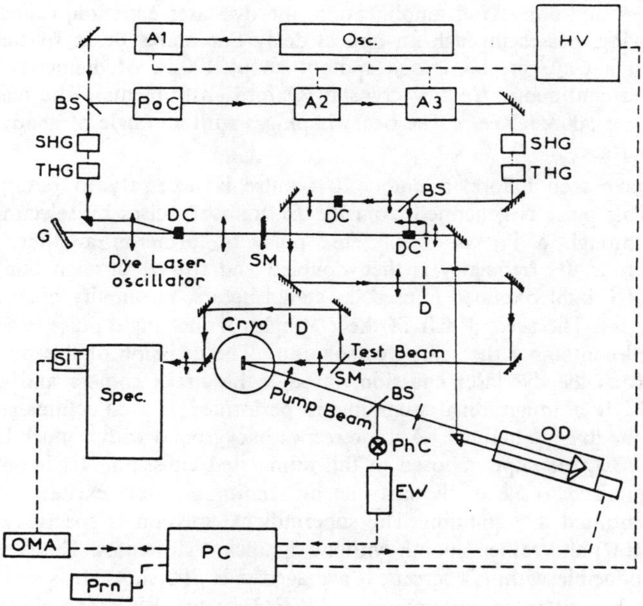

Figure 2. Experimental set-up:
Osc: Nd^{3+}: Yag Laser Oscillator. A1, A2, A3: three Nd^{3+}: Yag Amplifiers. *HV*: High Voltage generator for flash lamps. *PoC*: Pockels Cell. *SHG*, *THG*: Second and Third Harmonic Generator. *G*: Grating. *DC*: Dye Cells. *SM*: Semi-Transparent mirror. *D*: Diaphragm. *BS*: Beam Splitter. *OD*: Optical Delay line. *Cryo*: Continuous flow Cryostat. *Spec*: Spectrograph. *SIT*: Vidicon Camera. *PhC*: Photo-electric Cell. *EW*: Energy discriminator. *OMA*: Optical Multichannel Analyser. *PC*: Personal Computer. *Prn*: Printer. Optical paths: — Electrical wiring: —.

An active/passive mode-locked $Nd^{3+}:YAG$ laser oscillator (Quantel) emits a train of eleven pulses of 1064*nm* wavelength with a repetition rate of about 4*Hz*. The pulses have a duration of 35*ps* (*FWHM*) and are separated from one another by about 7*ns*. The laser emission is amplified in a $Nd^{3+}:YAG$ amplifier. After passing through two KDP^* crystals which respectively doubles (532*nm*) and triples (354*nm*) the light frequency, the *UV* light is separated from the green and *IR* by a chromatic dielectric mirror. Then, the *UV* train of pulses pumps synchronously a dye cell containing *BiBuQ* in toluene [10]. The dye laser cavity consists of an output mirror ($R = 60\%$) and a grating, mounted in a Littrow configuration. By inserting a diaphragm inside the cavity, we control the beam quality (its mode structure) and obtain uniform spots with homogeneous (Gaussian) profile. Under these conditions, we have at our disposal a laser emission consisting in a train of four pulses. Its characteristics have been studied in great details using an *ARP* streak camera system.

The emission of the dye laser passes through a second dye cell, containing the same dye solution as the oscillator. It is transversally pumped by a part of the *UV* light of a single pulse coming from the $Nd^{3+}:YAG$ laser (as will be described later). One of the four dye-laser pulses is amplified by a factor of 40 to 80. The amplifier works in its region of saturation and thus stabilizes the laser emission. Since our laser is not transform-limited, the total duration of the dye-laser pulses can be diminished to about 17*ps* when only the central part of the laser spot is selected through several diaphragms. In these experiments, we usually take the full dye-laser emission and thus work with pulses of 30*ps* duration. After amplification, the dye-laser emission, called 'pump beam' in the following, passes through an optical delay line and a beam splitter before being focused onto a *CuCl* crystal into a spot of about 100μm of diameter. The crystal is mounted in a continuous *He*-flow cryostat (Oxford Instruments). The maximum excitation intensity is $500 MW/cm^2$. The beam impinges with an angle of about 12 degrees on the crystal surface.

As we have seen before, a unique *UV* pulse is necessary to pump the dye-laser amplifier. This pulse is generated from the *IR* train of pulses by selecting one of these *IR* pulses through a Pockels cell. This pulse is furtheron amplified through two $Nd^{3+}:YAG$ rods. Its frequency is then doubled and tripled through nonlinear crystals. Half of the *UV* light obtained is used, as stated before, to amplify one of the dye-laser oscillator pulses. The second half of the *UV* light of the single pulse is focalized into a third dye cell containing the same dye solution. The emission of this dye cell has been characterized as the dye-laser emission, through the streak camera and a 65*cm HUET* spectrograph. If a longitudinal pumping is performed, a well-collimated superradient emission of the dye is obtained. A fluorescence background with a much longer duration (200*ps* for *BiBuQ*) is superimposed to this stimulated emission. Its intensity can, however, be minimized to 5% of the last one by sending a strong excitation intensity on a highly concentrated dye solution. The superradient emission is spectrally broad (about 40 *meV FWHM*) and rather smooth without significant structures. This 'continuum emission' is reproducible within 3 percent if averaged over 100 shots. It is very interesting to notice that it has the same duration (30*ps FWHM*) as the dye-laser emission. This stimulated emission can therefore be used as a test beam in our time resolved experiments. The jitter in time between the 'pump' and the 'test' pulses is about 5*ps*. The test beam is then focused onto a pinhole, which is imaged on the crystal inside the spot of the pump beam. Its diameter is about 40μm. Its intensity is kept low enough ($100 kW/cm^2$) so that it cannot induce renormalization effects alone. It impinges perpendicularly onto the surface of the sample, cooled down to 4*K*. The test and pump beams are linearly polarized parallel to each other. The test beam is either reflected by or transmitted through the sample, diaphragmed and focused onto the entrance slit of the spectrograph. The spectrum is analyzed by a *SIT* vidicon and an optical multichannel analyzer, (*PAR* 1205*D*). The spectra are transferred from the *OMA* to a microcomputer (Olivetti *M*24) [12].

Since we are studying nonlinear optical effects, the pump pulses have to be of nearly constant intensity. This is achieved by selecting them through an energy window as follows: if the pump energy measured by a fast photocell *UVHC* 20(*RTC*) is inside a fixed window (±10% around a mean value), the microcomputer accumulates in its memory the spectra measured by the vidicon, thus improving the signal to noise ratio. Then, when the transfer process from the *OMA* to the microcomputer is achieved, the latter triggers the $Nd^{3+}:YAG$ laser in order to obtain the next shot. Its repetition rate lies around $4Hz$.

4.2. Transmission and Reflection Measurements on *CuCl*.

Figure 3a shows the spectrum of the test beam transmitted through a *CuCl* sample of $30\mu m$ thickness, cooled down to $3K$, without any other light beam excitation.

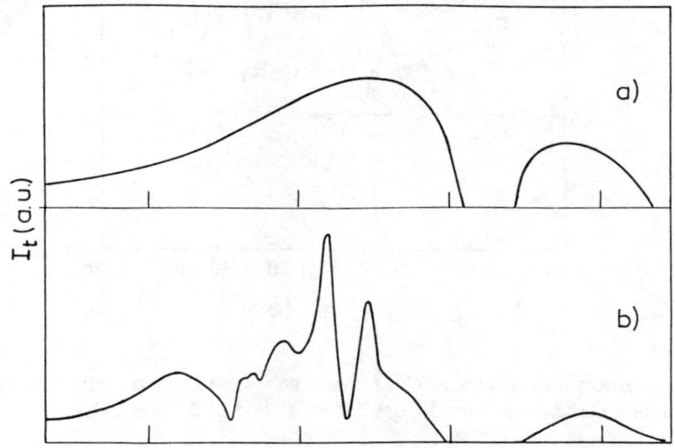

Figure 3. Transmission spectra of a *CuCl* sample without a) and with b) simultaneous excitation by a pump beam at $\hbar\omega_p = 3.186\,eV$ and $I_p = 300MW/cm^2$.

The strong linear absorption observed around $3.207\,eV$ is due to dipole active excitons with a Γ_5 symmetry. The energies of their longitudinal and transverse states are $E_L = 3.208\,eV$ and $E_T = 3.2025\,eV$, respectively.

When the sample is also excited by a strong pump beam at half the biexciton energy, $\hbar\omega_p = 3.186\,eV$, the transmission spectrum changes considerably as shown in Figure 3b. The maximum of both pulses coincide in time here.

From the spectra given in Figure 3a and b, we calculate the induced optical density K defined as follows [11]:

$$K(\omega_t,\Delta) = -\log_{10}[I_t(\omega_t,\Delta)/I_t(\omega_t,\Delta=\infty)] \tag{13}$$

where $\Delta = t_p - t_t$ is the time delay between the intensity maximum of the pump and the test pulse, respectively; $\Delta<0$ when the test pulse arrives after the pump pulse on the surface of the sample. The spectrum of the induced optical density K, when both pulses arrive simultaneously on the sample surface ($\Delta=0$), is drawn in Figure 4a. Several new absorption structures (A_p, A_N, A_B) appear as well as spectral regions of gain (G_B, R_T^+, R_T^-). As shown in Figure 4b, corresponding to $\Delta = -100ps$, these structures change in magnitude and spectral shape with time. Figure 5 gives a general view of the time evolution of K.

As shown in Figure 3b, strong induced absorption tails called A_X arise on both sides

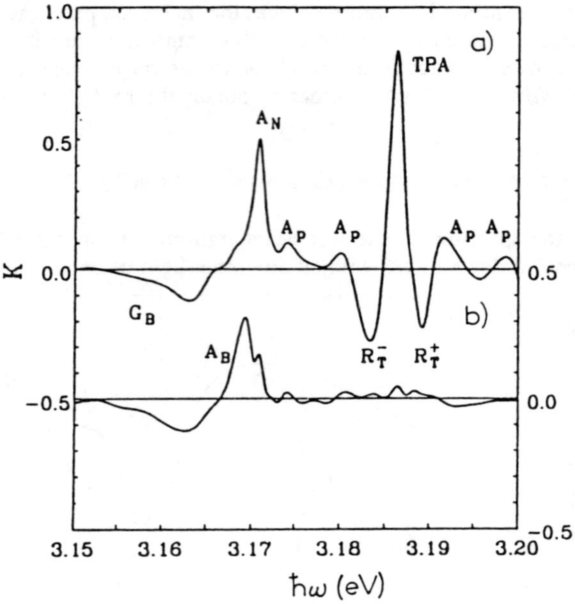

Figure 4. Induced absorption spectra of *CuCl* in a time-resolved pump and probe experiment, obtained at a) temporal coincidence from Figure 3, $\Delta = 0$, and b) $\Delta = -100ps$.

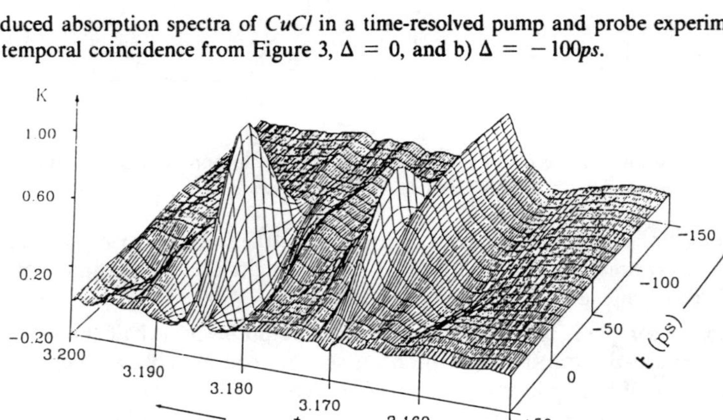

Figure 5. Induced optical density as a function of photon energy and time delay Δ.

of the exciton absorption line. In these regions K obeys approximately to [13]:

$$K = K_0 + K_1 \exp(-(\hbar\omega - E_0)/\gamma) \tag{14}$$

where E_0 is an arbitrary constant. The parameters K_0, K_1 and γ have been determined separately below and above the exciton absorption. γ and K_1 are time dependent showing rather long time constants (of several *ns*).

It should be noticed that this induced absorption has been substracted before plotting Figures 4 and 5.

This enhancement of the exciton absorption has been explained by collision processes between the photon-like polaritons from the test beam and exciton-like polaritons generated through the decay of biexcitons initially created by the pump beam [9]. These excitons have thermalized rapidly through elastic and inelastic collision processes. a theoretical fit of the exciton absorption line has led to the conclusion that the excitons obey to a Boltzmann distribution with a temperature around $50K$. This mean value depends on time but not appreciably on the excitation intensity [9].

If we go back to our propagation equations (10) and (12) for the photons from the test pulse, this collision process shows up as a resonance in the imaginary part of $\chi_{NLINC}^{tt(0,0)}(\omega_t = \omega)$. It leads to an induced absorption which is responsible for the transmission decrease of the test beam. The same type of discussion can be held for the absorption lines labelled A_N, A_p, A_B. The polaritons involved there have been created by the pump pulse through different nonlinear processes.

In the case of the A_p lines, they populate the photon-like branch of the polariton dispersion. They are observed by the photons of the test pulse through a transition towards biexciton states. In the case of the A_N line, this transition to biexciton states involves a population of excitonic polaritons with small kinetic energy, which are created immediately after the pump beam excitation. The shape of this line indicates that these excitonic polaritons are not yet fully thermalized in the bottleneck region. Furthermore, as shown in Figure 4 and 5, these absorption lines (A_N and A_p) have completely disappeared for $\Delta = -100ps$. On the contrary, the broad absorption line A_B reaches its maximum value (as shown in Figure 5) only after the end of the pump pulse. As the A_N line, it is due to a transition to biexciton states from excitons on the lower branch but having a higher kinetic energy. From the inverse Boltzmann shape of the A_B line, we deduced [14] that they are thermalized but have a temperature of about $30K$ which decreases with a time constant of $2.5ns$.

This population of secondary excitonic polaritons is obtained by the coupling of excitons from the bottleneck region to create (after emission of LO phonons) biexcitons with high kinetic energy [10]. They recombine, giving rise to photon-like polaritons (which in turn may contribute to the A_p line) and to exciton-like polaritons with large wave vectors. These excitons thermalize rapidly by mutual collisions. Their subsequent cooling down is then tested by the probe beam.

All these absorption lines involve elastic and inelastic scattering processes and contribute therefore to the incoherent part of the susceptibility χ_{NLINC}^{tt}. The two-photon absorption line (TPA), on the contrary, involves essentially the coherent part of the nonlinear susceptibility (χ_{NLCO}^{tt} in this notation), the singularities of which are induced by the pump and the test pulses. Furthermore, it strongly depends on the relative polarization of the two beams [15] and vanished in accordance with the selection rules for crossed polarizations [8].

Concerning the Hyper-Raman lines, labelled R_T^+, R_T^-, the same arguments (time dependence and polarization properties [15]) lead us to conclude that they appear through the coherent process governed by $\chi_{NLCO}^{ts(2,0)^*}\ \delta(2\omega_p - \omega_t - \omega_s)\delta(2\mathbf{k}_p - \mathbf{k}_t - \mathbf{k}_s)$. They are due to a two-photon absorption of the pump beam, creating coherent biexcitons, whose recombination is induced under energy and momentum conservation by the test beam, giving rise to a signal emission. The field of the signal (index s) shows up as a source term in the propagation equation (12) for the test pulse.

In addition to these coherent processes which show up as gain in the transmission spectrum, incoherent processes can also lead to gain: we attribute the G_B-band to a resonance in $\chi_{NLINC}^{ts(2,0)^*}$ since it is due to the induced recombination of biexcitons with high kinetic energy and wave vector which have been created through collision processes. They recombine into a longitudinal exciton and a photon in the G_B - band region. The longitudinal excitons scatter rapidly to the transverse branch, thus maintaining the population inversion between biexcitons and longitudinal excitons, which is necessary for the observation of gain. In the case of the coherent processes R_T^+ and R_T^-, the population inversion is obtained by the direct creation of coherent biexcitons through the pump pulse and thermalisation of the final state polaritons.

When we study the time-resolved reflection of a test beam by a $CuCl$ sample highly excited by a pump beam, we obtain, as in transmission measurements, information on the time evolution of the nonlinear susceptibility, but of its real part now. The samples investigated here are $CuCl$ films of a few microns thickness, epitaxially grown on CaF_2 substrates. As shown in Figure 6, we have recorded the spectra of the interference fringes of a test beam reflected by a sample excited or not by a pump beam at $3K$.

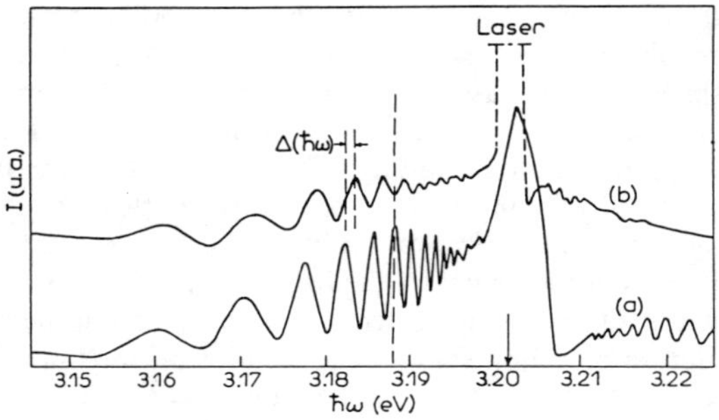

Figure 6. Reflection spectra of an epitaxial layer of $CuCl$ at $4K$, a) without and b) with pump beam excitation at $\hbar\omega_p = 3.2025\,eV$ and $I_p = 250MW/cm^2$. The test pulse is impinging onto the sample $50ps$ after the pump pulse ($\Delta = -50ps$).

The shift $\Delta(\hbar\omega)$ of the maximum $\hbar\omega$ of an interference fringe is related to the index change $\Delta n(\hbar\omega)$ at this photon energy of the test beam through the relation:

$$\Delta n(\hbar\omega) = \frac{n(\hbar\omega)}{\hbar\omega}\Delta(\hbar\omega). \tag{15}$$

Δn is a direct measurement of the real part of the nonlinear susceptibility χ_{NL} since they are related by the following expression:

$$\Delta n = \frac{1}{2n(\hbar\omega)}\mathrm{Re}\,\chi_{NL} \tag{16}$$

which is valid for small values of χ_{NL}/n.

We have analyzed the time dependence of Δn for two different photon energies of the pump beam and for different excitation intensities. The results are summarized in Tables I and II. We conclude that χ''_{NLINC} shows up in these processes and that its measured time constant is mainly due to the same polariton population which shows up in the A_p line in Figure 4. The coherent contributions around the pump laser line are not accessible in this type of measurements because of the presence of a strong laser stray light. Incoherent gain processes are observed, however, when decay times smaller than $40ps$ are obtained (Tables I and II).

The long-living exciton population in the bottleneck region shows up if we excite directly the sample in the exciton resonance. This is again the manifestation of the

TABLE I

Decay times of the change of the index of refraction at different photon energies for different pump beam intensities and a photon energy $\hbar\omega_p = 3.2083\,eV$.

	The spectral positions of interference fringes (eV)				
$I_p(MW/cm^2)$	3.163	3.176	3.182	3.185	3.188
100	56	53	68	60	-
200	35	59	61	60	90
250	35	52	36	59	70

TABLE II

As Table I, but for $\hbar\omega_p = 3.186\,eV$. (The measurements have been performed on different sample positions.)

	The spectral positions of interference fringes (eV)						
$I_p(MW/cm^2)$	3.159	3.170	3.173	3.177	3.179	3.189	3.191
100	-	-	53	-	32	27	36
200	-	54	-	52	-	54	-
300	42	56	-	83	-	90	-

resonance in χ^{tt}_{NLINC} which influences the complete spectral range below the exciton resonance. The complicated temporal behaviour of the χ^{tt}_{NLINC} is shown in Figure 7. It indicates the competition between different incoherent processes. Only at time delays longer than about - $400ps$ the excitons have thermalized. Their population then decays with time constants of several ns. The structures observed in Figure 7 during the first $100ps$ correspond quite well to the evolution of the absorption line A_N and A_B in Figures 4 and 5.

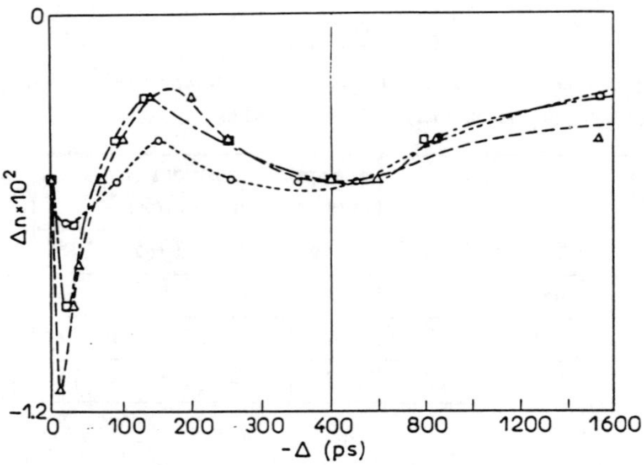

Figure 7. Time dependence of the index change Δn as a function of the time delay Δ under a pump beam excitation at $\hbar\omega_p = 3.2025\, eV$ and with $I_p = 250 MW/cm^2$ for different photon energies: $3.177\, eV$ (-.-); $3.182\, eV$ (...); $3.188\, eV$ (---).

4.3. Degenerate Wave-Mixing Processes

Up to now, we have discussed the nonlinear susceptibility acting on a test beam in the presence of a pump beam. In the degenerate case ($\omega = \omega_p = \omega_t = \omega_s$) the situation is simpler. Eq. (12) can be directly integrated [3,4], assuming that the nonlinearities are only induced by the pump pulse and not by the test and/or the signal pulses. We are then interested in the time-resolved generation of the signal pulse, emitted in the direction given by the coherently driven fields:

$$2\mathbf{k}_p - \mathbf{k}_t = \mathbf{k}_s. \tag{17}$$

These studies are very useful to determine the coherent lifetimes τ of the different states. Since we measure only the signal pulse emission, we perform a background-free spectroscopy.

Using a mean-field approximation, the intensity $I_s(l,t)$ of the signal pulse is given after a propagation through a crystal of thickness l by:

$$I_s(l,t) = E_t^2(t) e^{-2a'l} \frac{(b'^2 + b''^2)}{|\lambda^2|} snh^2(|\lambda| l) \tag{18}$$

with

$$a' = \frac{\omega^2 \mu_0}{2k} \mathrm{Im}(\chi_l - \chi_{NL}^{ss(0)}(t)) \tag{19}$$

$$a'' = -\frac{\omega^2 \mu_0}{2k} \mathrm{Re}(\chi_{NL}^{ss(0)}(t))$$

$$b' = -\frac{\omega^2 \mu_0}{2k} \mathrm{Im}(\chi_{NL}^{st(2)}(t))$$

$$b'' = -\frac{\omega^2 \mu_0}{2k} \mathrm{Re}(\chi_{NL}^{st(2)}(t))$$

and

$$\lambda^2(t) = -a''^2(t)+b'^2(t)+b''^2(t). \tag{20}$$

Eq. (18) is valid for $\lambda^2>0$; in the case of $\lambda^2<0$, one has to replace $snh(|\lambda|l)$ by $\sin(|\lambda|l)$. In the first case, the signal intensity increases exponentially with l (gain); in the second case, it oscillates and is simultaneously attenuated. As can be seen from Eqs. (18) and (19), $I_s(l,t)$ is a function of time. It depends on the instantaneous pulse shape $E_t(t)$ as well as on the time dependence of $\chi(t)$, introduced through the different relaxation processes of the material. As shown in Eq. (12), $\chi_{NL}^{tt(0)}$ and $\chi_{NL}^{st(2)}$ are different functions. They can be assumed to be equal only when we are working outside the resonances and in an adiabatic approximation. This point has often been neglected in the literature.

In order to discuss the time dependence of the signal generation in more details, we have calculated $I_s(l,t)$ from Eq. (18) for *CuCl* films of $10\mu m$ thickness, using again the three-level model together with the selection rules of Figure 1.

We assume radiative lifetimes (energy relaxation times) $T_{22} = T_{33} = 660ps$. The coherence times of the exciton and biexciton states are $\tau_2 = 6.6ps$ and $\tau_3 = 17ps$, respectively. The first time constant gives rise to a linear absorption $\alpha_L = 200cm^{-1}$, as measured in this spectral region. The second is the result of our recent work [17]. The damping of the ground state is neglected $(\tau_1 >> (\tau_2,\tau_3))$. We assume a Gaussian shape of the pump pulse of $20ps$ duration (*FWHM*) to excite the crystal. Then a test pulse, having the same characteristics as the pump pulse, impinges on the sample after the variable delay Δ. If the frequency of the light field is outside the biexciton resonance ($\hbar\omega_{Bi}/2 = 3.186\,eV$), a signal is generated only when test and pump pulses overlap appreciably. Then, the correlation trace is almost symmetrical, centered around $\Delta = 0$. The decay of the signal is not exponential. The temporal position of the maximum of the signal emission with respect to the pump pulse τ_M obeys to:

$$\tau_M = -1/3\Delta. \tag{21}$$

This behaviour is expected [17] for a process, in which the signal generation follows classical nonlinear optics, i.e. in which no time delay is introduced by the material and only the third order nonlinear susceptibility tensor $\chi^{(3)}$ is responsible for the signal generation. In this case, the signal intensity is proportional to:

$$I_s(t) \propto |\chi^{(3)}|^2 I_p(t)^2 I_t(t-\Delta) \tag{22}$$

which, for Gaussian pulses, leads to the result of eq. (21).

If the exciting light field is tuned into the biexciton resonance, the correlation trace of the generated signal shows a different time evolution: it lasts longer and is asymmetric, showing an exponential decay. Its time constant is given by the lifetime of the coherent biexciton states τ_3. As shown in Figure 8, the maximum of the signal emission τ_M does no longer follow Eq. (21): it turns from a slope of -1/3 to -1 at a delay $\Delta = -10ps$. This turnover happens later for less intense pulses. We conclude from these results that only the rising part of the correlation trace is due to the test-pump overlap, the decreasing part being due to the population of the coherent states alone.

In our experiment, the pulses used have a duration of $20ps$ and are tuned to the biexciton resonance at $3.186\,eV$. The spectral width is about $0.3\,meV$ and the excitation intensity $120MW/cm^2$. The generated signal is spatially filtered and registered by a streak camera. In order to obtain a time precision of $\pm 3ps$, part of the pump pulse is also detected by the streak camera and thus the jitter due to the trigger pulse is eliminated. Single shots are recorded and selected to be inside the energy window of $\pm 10\%$ around the mean value. Measurements are then averaged over five shots for each delay.

Figure 8 shows the temporal positions of the maximum of the signal emission τ_M together with its position calculated from our model. The change in the slope is

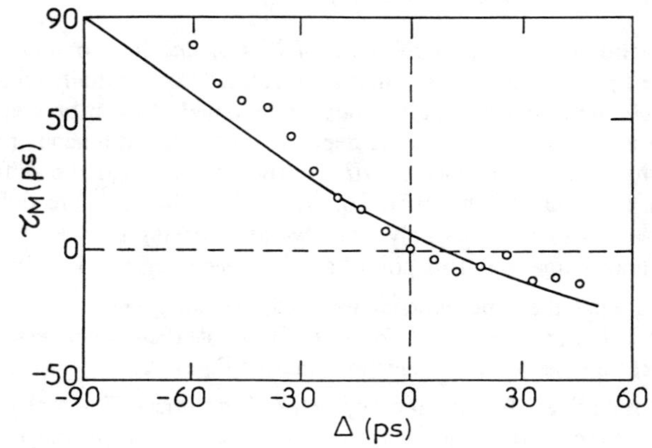

Figure 8. Temporal position of the maximum of the signal emission τ_M as a function of Δ.

spectacular. From the foregoing discussion, we conclude that, for $\Delta < -20ps$, only the biexciton decay is measured. In Figure 9, we give the corresponding correlation trace.

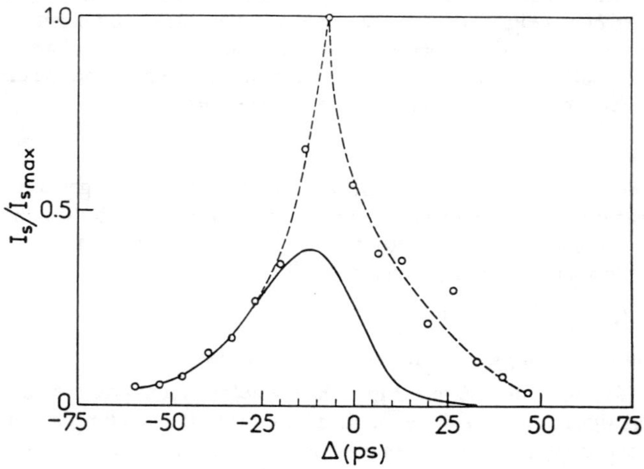

Figure 9. Experimental correlation trace as a function of Δ together with a theoretical fit (full line) for $I_p = 120MW/cm^2$ and $\hbar\omega_p = 3.186\,eV$.

By fitting it with our theoretical model (full line) we find a coherence time of the biexcitons τ_3 of $17ps$. It is important to notice that τ_3 increases with decreasing pump intensity. At $I_p = 5MW/cm^2$, we obtain e.g. $\tau_3 = 27ps$.

As it can be seen in Figure 9, our theoretical model reproduces quite well the decrease of the signal generation, but not its increasing part. This may have different reasons: the model used to obtain χ in Eq. (18) uses the approximation that the electric field varies slowly on a timescale given by τ_i. This approximatioon is bad around $\Delta = 0$, where the full propagation of the polarization should be considered. Another explanation may be that, in our experiments, the exciting light field is not monochromatic, but has a spectral width of $0.3\,meV$. This implies that even at resonance, biexcitons are created virtually by the absorption of two photons from the wings of the laser line. This process contributes only to the test-pump convolution part of the correlation trace. In addition, test and pump pulse create a periodic variation of the nonlinear susceptibility due to interference. On this grating, the pump pulse is self-diffracted, giving also rise to a signal generation [18].

5. Conclusion

We have shown that time-resolved pump and probe experiments can be used to determine energy relaxation times and transverse relaxation times even if they are shorter than, or comparable to, the coherence time T_c of the exciting light field. This is possible since the test-pump convolution strongly competes with the induced recombination of the coherent states. Since the time dependences of both processes are different, they can easily be distinguished in time-resolved experiments. These different time constants show up separately in the nonlinear susceptibility terms which describe absorption and gain. In addition, we have shown that it is quite useful to distinguish coherent and incoherent terms in both cases.

Acknowledgements

The authors are grateful to R. Leonelli, A. Manar, F. Tomasini, J.C. Mathae, M.J.M. Gomes and B. Kippelen for their assistance during the experimental part of this work. They have appreciated useful discussions with Professors H. Haug, J.M. Hvam, J. Miletic and S. Montasser. The work has been performed in the frame of a twinning program between the Universities of Frankfurt/Main and Strasbourg, supported by the European Communities.

References

1. E. Abraham and S.D. Smith, *Rep. Prog. Phys.* **45**, 815 (1982).
2. H.M. Gibbs, Optical Bistability, Academic Press, New York (1984).
3. B. Hönerlage and J. Miletic, *Europhysics Letters* **7**, 19 (1988).
4. B. Hönerlage, M. Frindi and J. Miletic, *Phys. Stat. Solidi (b)* **149**, 775 (1988).
5. J.Y. Bigot and B. Hönerlage, *Phys. Stat. Sol. (b)* **121**, 649 (1984).
   B. Hönerlage and J.Y. Bigot, *Phys. Stat. Sol. (b)* **123**, 201 (1984).
   J.Y. Bigot, J. Miletic and B. Hönerlage, *Phys. Rev.* **B32**, 6478 (1985).
6. A. Yarif, Quantum Electronics, Wiley and Sons, New York, 2nd edition (1975).
7. B. Hönerlage, R. Lévy, J.B. Grun, C. Klingshirn and K. Bohnert, *Phys. Rep.* **124**, 161 (1985).
8. A. Bivas, Vu Duy Phach, B. Hönerlage and J.B. Grun, *Phys. Stat. Sol. (b)* **84**, 235 (1977).
   Vu Duy Phach, A. Bivas, B. Hönerlage and J.B. Grun, *Phys. Stat. Sol. (b)* **84**, 731 (1977).
9. H. Haug, J. Liebler, R. Leonelli, A. Manar and J.B. Grun, *Phys. Rev. B* **38**, 10903 (1988).
10. A. Manar, M.J.M. Gomes, B. Hönerlage and R. Lévy, *Phys. Stat. Sol. (b)* **151**, 747 (1989).
11. R. Leonelli, J.C. Mathae, J.M. Hvam and R. Lévy, Proc. 18th Int. Conf. Phys.

Semiconductors, Stockholm (Sweden) 1986, pp. 1437.

12. J.C. Mathae, F. Tomasini and J.B. Grun, *J. Optics,* **19**, 69 (1988).
13. R. Leonelli, J.C. Mathae, J.M. Hvam, F. Tomasini and J.B. Grun, *Phys. Rev. Lett.* **58**, 1363 (1987).
14. R. Lévy, B. Hönerlage and J.B. Grun, *Phys. Rev. B* **19**, 2326 (1979).
15. R. Lévy, F. Tomasini, J.Y. Bigot and J.B. Grun, *J. Luminescence* **35**, 79 (1986).
16. R. Lévy, B. Hönerlage and J.B. Grun, *Phys. Stat. Sol. (b),* **150**, 825 (1988).
17. M.J.M. Gomes, B. Kippelen, B. Hönerlage and R. Lévy, *J. de Physique C2,* **49**, 263 (1988).
18. R. Lévy, H.J.M. Gomes, B. Kippelen and B. Hönerlage, to be published.

MULTIPHOTON SPECTROSCOPY IN LARGE GAP MATERIALS

I.M. CATALANO AND A. CINGOLANI

Dipartimento di Fisica dell'Università degli Studi
Unità GNEQP-CNR, Bari, Italy

The spectral behaviour of the direct and indirect interband multiphoton absorption (MPA) coefficient $\alpha_{d,i}^{(n)}$ for a number of crystals has been investigated in a large excitation energy range. Experimental results can be described well by a parametric formula with mixed spectral dependence, indicating the simultaneous contribution of different transition mechanisms to MPA amplitude. A qualitative and quantitative comparison between experimental and various theoretical MPA line-shapes is discussed.

1. Introduction

Multiphoton absorption (MPA) processes in crystalline solids have been the subject of extensive theoretical and experimental investigations over the last two decades (1,2). In particular, the different selection rules which govern MPA processes (when one-photon absorption is forbidden by selection rules, a higher order MPA may be allowed) have made non-linear spectroscopy very useful in determining the crystalline solids optical and electronic properties related to both excitonic (3) and interband (4) transitions. In this respect, it is worth noting that for multiphoton interband transitions, different absorption mechanisms, which depend on selection rules (hence on the appropriate choice of matrix elements), have been predicted theoretically. In fact, the electron-photon interaction matrix elements, which couple the valence (v) and the conduction (c) band to the intermediate states and also couple the latter to each other can be expanded as a function of **k** near critical points, assuming the general form:

$$\mathbf{P}_{ij}(\mathbf{k}) = \mathbf{P}_{ij}(0)+\mathbf{k}\cdot\nabla\mathbf{P}_{ij}+... \tag{1}$$

For coupled bands of opposite parity [allowed transitions (a)] the constant term of the matrix element power series (right hand-side of eq. 1) is dominant and the **k**-dependent terms are usually neglected. Conversely, for transitions involving bands of the same parity [forbidden transitions (f)], only the linearly **k**-dependent term of eq. 1 is considered since $\mathbf{P}_{ij}(0) = 0$. Therefore for direct interband MPA transitions, the following cases may occur (5):

(a)when the multiphoton transition amplitude ($M^{(n)}$) is **k**-independent (in this case all matrix elements $|P_{ij}|^2$ involved in $\alpha_d^{(n)}$ evaluation by means of the perturbation theory are assumed to be **k**-independent) the $\alpha_d^{(n)}$ theoretical frequency dependence is nearly equal to $(n\hbar\omega - E_{gd})^{1/2}$.

(b)when $M^{(n)}$ is a linear function of **k** (i.e. only one of $|P_{ij}|^2$ matrix elements is assumed to be linearly **k**-dependent) the $\alpha_d^{(n)}$ frequency dependence is roughly equal to $(n\hbar\omega - E_{gd})^{3/2}$.

(c)when $M^{(n)}$ is proportional to $|\mathbf{k}|^2$ (i.e. two of the $|P_{ij}|^2$ are considered as **k**-dependent) then the $\alpha_d^{(n)}$ frequency dependence is nearly equal to $(n\hbar\omega - E_{gd})^{5/2}$.

(d)when $M^{(n)}$ is proportional to $|\mathbf{k}|^3$ (i.e. three of the $|P_{ij}|^2$ are assumed to be **k**-

R. Girlanda et al. (eds.), Progress on Electron Properties of Solids, 291–301.

dependent), the typical $\alpha_d^{(n)}$ spectral dependence is nearly equal to $(n\hbar\omega - E_{gd})^{7/2}$.

Consequently, for direct two-photon absorption (DTPA) processes, three different absorption mechanisms [allowed-allowed $(a-a)$, allowed-forbidden $(a-f)$ and forbidden-forbidden $(f-f)$], depending on selection rules, have been theoretically predicted (6). This is because DTPA transitions occur through only one intermediate state hence two $|P_{ij}|^2$ matrix elements are involved in $\alpha_d^{(2)}$ evaluation as specified by the time dependent perturbation theory.

Similarly, for phonon assisted two-photon absorption processes (ITPA), two different transition mechanisms have been theoretically considered: $a-a$ and $a-f$. In the former mechanism, only **k**-independent matrix elements have been used for electron-photon interaction, giving an $\alpha_i^{(2)}$ frequency dependence nearly equal to $(2\hbar\omega - E_{gi} \pm \hbar\Omega)^2$ (7). Instead in the latter mechanism, **k**-dependent dipole matrix elements have also been taken into account and $\alpha_i^{(2)}$ proportional to $(2\hbar\omega - E_{gi} \pm \hbar\Omega)^3$ has been obtained (8) (where E_{gi} is the indirect gap).

Lastly for direct three-photon absorption (DThPA) processes, four possible absorption mechanisms ($a-a-a$, $a-a-f$, $a-f-f$ and $f-f-f$) have been predicted in accordance with the fact that DThPA transitions develop through two intermediate states so that three $|P_{ij}|^2$ matrix elements are involved in $\alpha_d^{(3)}$ evaluation (5).

Since MPA processes normally involve all the energy bands of the solid as intermediate states, terms with different **k** dependences are likely to contribute simultaneously to MPA transition amplitude (5) hence a mixed $\alpha_{d,i}^{(n)}$ frequency dependence can be expected.

This work aims to present our last multiphoton spectroscopy results in crystalline solids which enabled us to demonstrate experimentally the simultaneous contribution, in large direct and indirect gap materials, of all the mechanisms predicted on the grounds of the MPA selection rules to the two- and three-photon interband transition amplitude.

2. Experimental

Two- and three-photon absorption coefficients were measured in different direct and indirect gap crystals. Specifically DTPA processes were studied in ZnO, ZnS, $CuCl$ and $Bi_4Ge_3O_{12}$, ITPA processes in $AgCl$ and CdI_2 and DThPA processes in RbI, KI, CsI and NaI. These materials are especially well suited for our studies because: a) their high luminescence efficiency enables the MPA processes to be studied by utilizing nsec dye laser sources, b) due to their large gaps, dyes can be used which emits radiation in the visible range and which have - as is well-known - the highest available peak power. High quality crystals were selected to prevent damage from high power density excitation and to avoid dependence on the level of sample purity. The $\alpha_{d,i}^{(n)}$ spectral dependence in all the materials and for both non-linear processes considered was measured by means of the nonlinear luminescence technique (9). This consists in the detection of a luminescence signal $L^{(n)}$ excited by simultaneous absorption of n identical photons. The detected luminescence signal at a frequency $\hbar\omega_L$ is given by (9,10):

$$L^{(n)} = q^{(n)} f^{(n)} \alpha_{d,i}^{(n)} I_0^n z (1 - \alpha_L z/2) \quad (2)$$

where $q^{(n)}$ is the quantum efficiency for excitation at $n\hbar\omega$, $f^{(n)}$ is the fraction of collected luminescence, which is a function of the geometry, I_0 is laser intensity, z is sample thickness and α_L is the linear absorption coefficient at $\hbar\omega_L$. It is worthwhile noting that this technique is the most suitable one for measuring the very low $\alpha_i^{(2)}$ and $\alpha_d^{(3)}$ values when using nsec dye laser sources, hence to obtain different order MPA coefficients by means of the same experimental technique. The experimental set-up was similar to that reported in Ref. (9).

The $\alpha_{d,i}^{(n)}$ spectra were obtained by means of a Quantel Datachrome dye laser (pulse duration: 9 nsec, maximum peak power for Rhodamine 6G approx. equal to 30 MW/cm^2) pumped by the second and third harmonics of a Nd:YAG laser. Four efficient dyes were used: Coumarine 480 and 500 for the high energy side of the spectra and Rhodamine 6G and B for low energies. Emitted radiation was detected by a 60 ER response photomultiplier tube, while the incident beam was monitored by a fast response photodiode. Both signals were stored and processed by a computer data acquisition system. To reduce the effects of fluctuations in the input beam intensity, a ratio of the photomultiplier signal to the $n\,(n=2,3)$ power of the monitor signal was used. Moreover, the quadratic and cubic behaviour of the detected luminescence signal versus excitation intensity was checked at each experimental point and, for each point, different measurement runs were carried out. Experimental accuracy was within about 25% for each run.

Results and discussion

a - Direct two-photon absorption

Fig. 1a shows the DTPA line shape (dots) of $Zn\,0$ as a function of $(2\hbar\omega - E_{gd})$. The $\alpha_d^{(2)}$ values were obtained by normalizing experimental data with the absolute value $\alpha_d^{(2)} = 0.021$ cm/MW at $(2\hbar\omega - E_{gd}) = 0.13$eV measured in Ref. (11). The experimental results clearly show that $\alpha_d^{(2)}$ augments in this large gap semiconductor as the difference $(2\hbar\omega - E_{gd})$ increases. Specifically $\alpha_d^{(2)}$ energy dependence is sublinear near the energy gap, and becomes superlinear away from it. An attempt to give a general interpretation of the $\alpha_d^{(2)}$ line shape was made by fitting the experimental data with a parametric formula which takes into account all the three possible transition mechanisms mentioned in the introduction, i.e.

$$\alpha_d^{(2)}(\hbar\omega) = C_1(2\hbar\omega - E_{gd})^{1/2} + C_2(2\hbar\omega - E_{gd})^{3/2} + C_3(2\hbar\omega - E_{gd})^{5/2}. \tag{3}$$

The result of this fit is shown by the continuous curve in Fig. 1a. Agreement is good as shown by the x^2 value and by the fit probability, P, reported in the figure caption. The $C_m(m=1,2,3)$ coefficients in eq. 3 have the following values: $C_1 = 0.89\,10^{-3} cm/MW/(meV)^{1/2}$, $C_2 = 0.29\,10^{-5} cm/MW/(meV)^{3/2}$, $C_3 = 0.28\,10^{-7} cm/MW/(meV)^{5/2}$. These results show that all the terms in eq. 3 contribute to the DTPA spectral dependence, which actually proves to be of the mixed type. This fact suggests that all the three DTPA mechanisms ($a-a$, $a-f$ and $f-f$), which depend on the selection rules, contribute simultaneously to the DTPA interband transition amplitude in the considered energy range, though with different weights at different $2\hbar\omega$ values. The analysis of the experimental $\alpha_d^{(2)}$ line shape, which was carried out qualitatively by means of the functional form of eq. 3, was further improved by a quantitative comparison between the $\alpha_d^{(2)}$ experimental and theoretical data.

Fig. 1b shows the theoretical $\alpha_d^{(2)}$ line-shapes and the experimental points for this crystal. In the same figure the continuous curves a,b,c (calculated by Eqs. (1), (2) and (3) in Table I) are related respectively to: i) a three band model (12) with parabolic and non-degenerate bands and with both $|P_{ij}|^2$ matrix elements being **k**-independent ($a-a$ transitions), ii) a two-band model (13) with one of the bands degenerate and only one linearly **k**-dependent $|P_{ij}|^2$ matrix element ($a-f$ transitions) and iii) a three-band model (6) with both $|P_{ij}|^2$ matrix elements linearly **k**-dependent ($f-f$ transitions). It should be noted that the eq. (3) in Table I, does not consider band non-parabolicity and degeneration effects, which were shown to be important with excitation energy $2\hbar\omega > E_g$ (14).

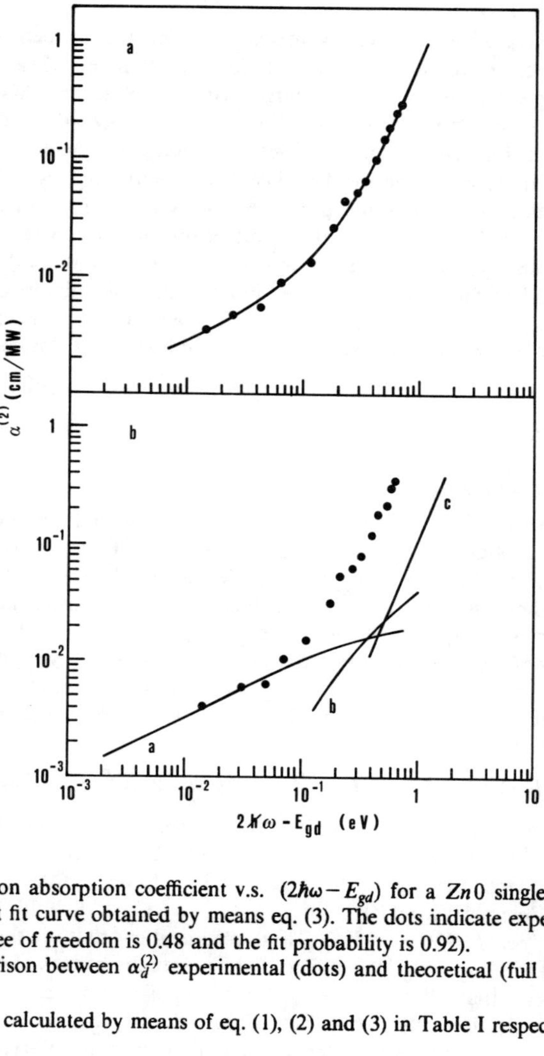

Figure 1. a) two-photon absorption coefficient v.s. $(2\hbar\omega - E_{gd})$ for a $Zn0$ single crystals at 80k. The full line is the best fit curve obtained by means eq. (3). The dots indicate experimental results. (The x^2 value for degree of freedom is 0.48 and the fit probability is 0.92).
b) quantitative comparison between $\alpha_d^{(2)}$ experimental (dots) and theoretical (full line) line shapes for $Zn0$.
Curve a), b) and c) are calculated by means of eq. (1), (2) and (3) in Table I respectively.

Thus, to take these effects into account in the numerical comparison given in Fig. 1b, a rough estimate was made as indicated in ref. 13. From Fig. 1b it appears that in the energy region just above the direct gap ($10\text{meV} < 2\hbar\omega - E_g < 120\text{meV}$), where choosing constant matrix elements is a reasonable assumption, the experimental data are in good qualitative and quantitative agreement with the theoretical model which considers both $|P_{ij}|^2$ matrix elements to be **k**-independent (spectral dependence: $(2\hbar\omega - E_{gd})^{1/2}$). In the intermediate energy region, for $(2\hbar\omega - E_{gd})$ ranging between 120meV and 400meV, an accurate proportionality is found between experimental data and the model assuming one of the matrix elements to be constant and the other one to be linearly **k**-dependent (spectral dependence $(2\hbar\omega - E_{gd})^{3/2}$). The $\alpha_d^{(2)}$ theoretical values underestimate the experimental data by a factor of 5. In the range beyond 400meV, good qualitative agreement is found only with the theoretical model assuming both matrix elements to be linearly **k**-

TABLE I.

Theoretical model calculations for two-photon direct transitions. Symbols have been defined in Refs. 12, 13 and 6.

Author	Model	Band structure	Transitions type	Result of calculation
Hassan (Ref. [12])	Three bands	Parabolic anisotropic bands	Allowed-allowed	(1) $\alpha_d^{(2)} = \frac{2^{5/2}e^4\pi}{n^2c^2m^{5/2}} \frac{\lvert P_{cn}\rvert^2 \lvert P_{nv}\rvert^2}{(\hbar\omega)^3}(\alpha_c+\alpha_v)^{-1} \times \frac{1}{M(BM)^{1/2}}\left[\frac{X}{1+X^2}+\arctan X\right]$; $B=\left[(\beta_v+\beta_n)-\frac{\alpha_n+\alpha_v}{\alpha_c+\alpha_v}(\beta_c+\beta_v)\right]$; $X=\left[\frac{B(2\hbar\omega-E_g)}{M(\beta_c+\beta_v)}\right]^{1/2}$
Vaidyanathan et al. (Ref. [13])	Two bands	Nonparabolic bands	Allowed-forbidden	(2) $\alpha_u^{(2)} = \frac{2^3\pi e^4}{3\epsilon_\infty C^2} \frac{E_g^{3/2}}{(m_{cv}^*)^{1/2}} \frac{1}{(\hbar\omega)^4}\left[\left[\frac{2\hbar\omega}{E_g}\right]^2-1\right]^{3/2}$
Braunstein et al. (Ref. [6])	Three bands	parabolic bands	Forbid.-Forbid.	(3) $\alpha_d^{(2)} = \frac{2^{13/2}\pi n^2 e^4 m^{7/2}}{9c^2(m_{nc}^*)^2(m_{vn}^*)^2(\alpha_c+\alpha_v)^{7/2}} \frac{1}{(\hbar\omega)^3} \frac{(2\hbar\omega-E_g)^{5/2}}{\lvert \Delta E-\hbar\omega+\frac{\alpha_n+\alpha_v}{\alpha_v+\alpha_c}(2\hbar\omega-E_g)\rvert^2}$

dependent, with a spectral dependence approximately equal to $(2\hbar\omega - E_{gd})^{5/2}$. In this case again, the theoretical values underestimate the experimental ones by a factor of 10.

It is worth noting that a similar $\alpha_d^{(2)}$ spectral behaviour was also observed in other large direct gap semiconductors (ZnS and CdI_2) and alkalide halide crystals ($CuC1$ and $Bi_4Ge_3O_{12}$) (15).

These results, obtained for different crystals in the energy range from a few meV up to a few hundred meV above the direct energy gap, clearly show that the $\alpha_d^{(2)}$ dispersion curve cannot be fully described by any one of the available theoretical models. In fact, each model predicts a good frequency dependence and satisfactory $\alpha_d^{(2)}$ numerical values only within a restricted energy range. Moreover, the complete $\alpha_d^{(2)}$ line shape can be described only by assuming that the three different DTPA mechanisms all contribute simultaneously to the transition amplitude with different weights at different $2\hbar\omega$ values.

b - Indirect two photon absorption

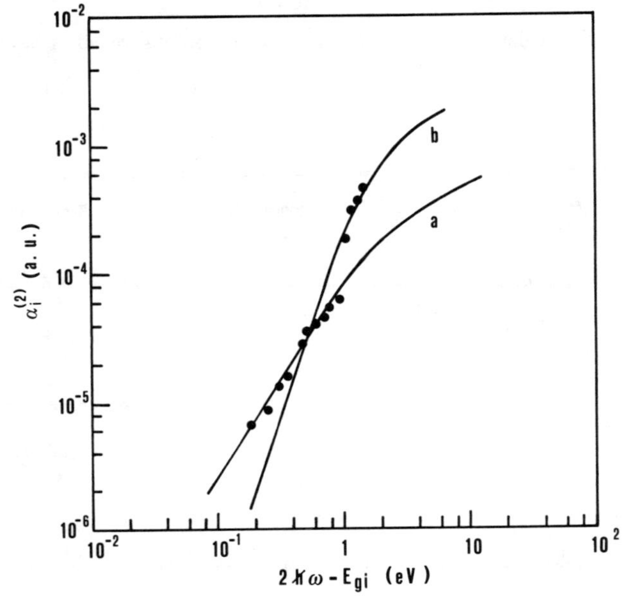

Figure 2. Indirect two-photon absorption coefficient *vs*$(2\hbar\omega - E_{gi})$ for an $AgC1$ sample. Full lines a and b are calculated by means of eq. (1) and (2) in Table II. The symbols indicate experimental results.

Fig. 2 gives the ITPA spectrum of $AgC1$. In this figure, curves a and b (calculated by eq. 1 and 2 in Table II) are due to a four-band (7) and a three-band model (8), respectively. In particular, the four band model assumes that the main contributions for electron-photon interaction are given by allowed virtual transitions, while the three-band model uses the forbidden virtual ones. It is evident that, in the region nearer to the indirect gap (174 meV $< 2\hbar\omega <$ 800meV), the experimental points show a good fit with the curve related to the four-band model, which is characterized by a spectral dependence roughly equal to $(2\hbar\omega - E_{gi} \pm \hbar\Omega)^2$. Beyond 800meV above the indirect gap, the $\alpha_i^{(2)}$ values fall along the three-band model curve. The latter is characterized by a spectral dependence approximately equal to $(2\hbar\omega - E_{gi} \pm \hbar\Omega)^3$.

A similar $\alpha_i^{(2)}$ spectral behaviour was obtained for another indirect gap semiconductor: CdI_2 (15).

This finding confirms that, for indirect gap materials also, the transitions of the allowed-forbidden type dominate the two-photon processes for energies above the indirect band gap whereas it is the allowed-allowed transitions which dominate just above this gap giving way to a mixed $\alpha_i^{(2)}$ spectral dependence in this case too.

c - Direct three-photon absorption

The three-photon absorption line shape (dots) of RbI versus $(3\hbar\omega - E_{gd})$ is given in Fig. 3a. The $\alpha_d^{(3)}$ values were obtained by normalizing experimental data with the $\alpha_d^{(3)}$ value measured at $(3\hbar\omega - E_{gd})$ = 0.91eV in ref. (16). The DThPA line shape of the RbI was

TABLE II.

Same as Table I for two-photon indirect transitions. Symbols have been defined in Refs. 7 and 8.

Authors	Model	Band structure	Transitions type	Result of the calculations
Bassani (Ref. [7])	Four bands	Parabolic isotropic bands	Allowed-allowed	(1) $\alpha_i^{(2)} = \frac{2e^4}{n^2c^2m\hbar^3}\frac{\lvert P_{mn}\rvert^2\lvert P_{nv}\rvert^2\lvert Q_{cm}\rvert^2}{(\hbar\omega)^3}(\alpha_c\cdot\alpha_v)^{-3/2}$ $\times\sum_{+,-}\frac{(2\hbar\omega-E_i\mp\hbar\Omega)^2}{\mp(e^{\mp\hbar\Omega/kT}-1)}$ $\times\left[\left(\frac{1}{(\Delta m-E_i\mp\hbar\Omega)^2}\right)\left(\frac{1}{(\Delta n-\hbar\omega)^2}+\frac{1}{(\Delta n-E_i+\hbar\omega\mp\hbar\Omega)^2}\right)\right.$ $\left.+\frac{1}{(\Delta n-E_i\mp\hbar\Omega)^2}\left(\frac{1}{(\Delta m-\hbar\omega)^2}+\frac{1}{(\Delta m-E_i+\hbar\omega\mp\hbar\Omega)^2}\right)\right].$
Yee (Ref. [8])	Three bands	Parabolic isotropic bands	Allowed-forbidden	(2) $\alpha_i^{(2)} = \frac{2^5e^6\pi m^2}{3n^2\hbar}\frac{\hbar\Omega}{(\hbar\omega_F)^2}\left(\frac{1}{\epsilon_\infty}-\frac{1}{\epsilon_0}\right)\frac{\lvert P_{cv}\rvert^2}{(\hbar\omega)^3}(\alpha_c\alpha_v)^{-3/2}$ $\times\left[\left(\frac{1}{\alpha_c}\lvert g_1^+\rvert^2+\frac{1}{\alpha_v}\lvert g_2^+\rvert^2\right)n_{kf}(2\hbar\omega+\hbar\Omega-E_i)^3\right.$ $\left.+\left(\frac{1}{\alpha_c}\lvert g_1^-\rvert^2+\frac{1}{\alpha_v}\lvert g_2^-\rvert^2\right)(n_{kf}+1)(2\hbar\omega-\hbar\Omega-E_i)^3\right]$ $g_1^{\pm}\approx\left[\frac{\alpha_c}{m\hbar\omega}\left(\frac{1}{E_{k_i'}-E_{k_v'}\mp\hbar\Omega}+\frac{1}{E_{k_v'}-E_{k_i'}+\hbar\omega}\right)\right.$ $\left.-\frac{\alpha_v}{m}\frac{1}{(E_{k_i'}-E_{k_v'}-\hbar\omega)(E_{k_i'}-E_{k_v'}\mp\hbar\Omega)}\right],$ $g_2^{\pm}\approx\left[\frac{\alpha_v}{m\hbar\omega}\left(\frac{1}{E_{k_v'}-E_{k_i'}\pm\hbar\Omega}+\frac{1}{E_{k_i'}-E_{k_v'}-\hbar\omega}\right)\right.$ $\left.+\frac{\alpha_c}{m}\frac{1}{(E_{k_v'}-E_{k_i'}\pm\hbar\Omega)(E_{k_v'}-E_{k_i'}+\hbar\omega)}\right]$ $n_{kf}=\frac{1}{e^{\hbar\Omega/k_BT}-1}$

fitted with a parametric formula which takes all the four possible transition mechanisms mentioned above into account; i.e.

$$\alpha_d^{(3)} = C_1(3\hbar\omega-E_{gd})^{1/2}+C_2(3\hbar\omega-E_{gd})^{3/2}+ \\ C_3(3\hbar\omega-E_{gd})^{5/2}+C_4(3\hbar\omega-E_{gd})^{7/2}. \tag{4}$$

The results of this fit are shown by the continuous curve in Fig. 3a. Agreement is good as shown by the x^2 values and by the fit probability P reported in the figure caption. $C_m(m=1,2,3,4)$ values for *RbI* are reported in ref. (16). The results of the fit show that all the terms in eq. 4 play a part in DThPA spectral dependence. This suggests that - again with direct three-photon absorption - all four DThPA mechanisms depending on the selection rules ($a-a-a$, $a-a-f$, $a-f-f$ and $f-f-f$), contribute simultaneously to the DThPA interband transition amplitude in the considered energy range, though with different weights at different $3\hbar\omega$ values.

Fig. 3b shows a comparison between $\alpha_d^{(3)}$ theoretical and experimental line-shapes for

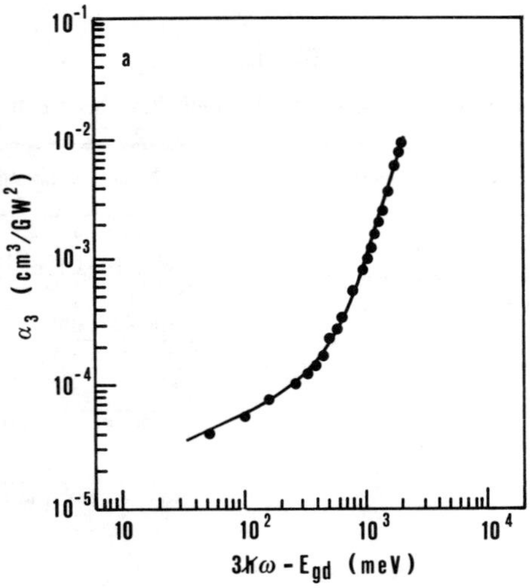

Figure 3a. Three-photon absorption coefficient versus ($3\hbar\omega - Egd$) for *RbI* crystals at 80°*k*. The full line is the best fit curve obtained by means of equation (4). The dots indicate experimental results. The x^2-value for degree of freedom is 0.13 and the fit probability is 0.999.

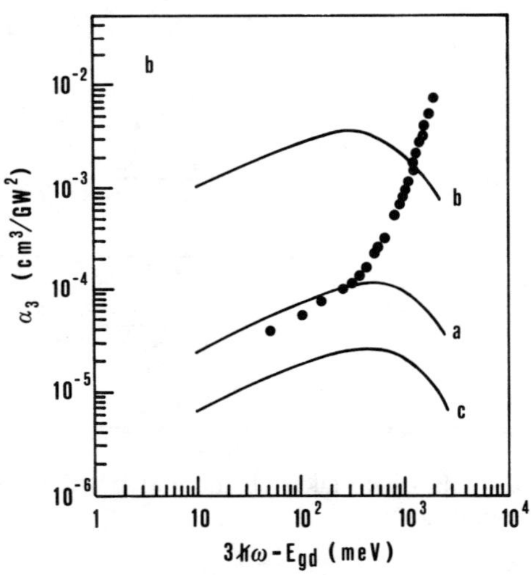

Figure 3b. Quantitative comparison between the $\alpha_d^{(3)}$ experimental (dots) and theoretical (full lines) line-shapes for *RbI*. Full line a and b and c have been calculated by means of equation (1), (2) and (3) in Table III.

RbI. The theoretical curves (*a*,*b* and *c* calculated by equations (1), (2) and (3) in Table III) are related to the three different theoretical approaches used to calculate $\alpha_d^{(3)}$: a four band model ($a-a-a$-transitions) (7), a two-band model ($a-f-f$ transitions) (5) and a non-perturbative approach (17).

In the first model, calculation was done by using the third order time-dependent perturbation theory and by assuming two higher conduction bands for the intermediate states, so that all three $|P_{ij}|^2$ matrix elements are assumed to be **k**-independent; in the second

TABLE III.

Theoretical model calculation for three-photon direct transitions. Symbols have been defined in Refs. 5 [Bobrisheva-Maskalenko (BM)], 7 [Bassani-Hassan (BH)], and 17 [Keldysh (k)], $a-a-a$ denotes allowed-allowed-allowed; $a-f-f$ denotes allowed-forbidden-forbidden (parabl. denotes parabolic; perturb. denotes perturbative; and quant. denotes quantization).

Author	Model	Calc. scheme	Transitions type	Result of calculation
BH	Four parabolic bands	3rd-order perturb. theory	$a-a-a$	$(1)\alpha_d^{(3)} = \frac{2^4\sqrt{2}\pi^2\hbar^2 e^6}{m^{9/2}c^3 n^3}\frac{1}{(\alpha_c+\alpha_v)^{3/2}}\frac{(3\hbar\omega-E_g)^{1/2}}{(\hbar\omega)^5} \times \frac{\lvert P_{cm}\rvert^2\lvert P_{mn}\rvert^2\lvert P_{nv}\rvert^2}{\left[E_n-\hbar\omega+\frac{\alpha_n+\alpha_v}{\alpha_c+\alpha_v}(3\hbar\omega-E_g)\right]^2\left[E_m-2\hbar\omega+\frac{\alpha_m+\alpha_v}{\alpha_c+\alpha_v}(3\hbar\omega-E_g)\right]^2}$ $\alpha_c = m_c/m_0,\ \alpha_v = m_v/m_0$
BM	Two parab. bands	Perturb. approach with 2nd-quant. method	$a-f-f$	$(2)\alpha_d^{(3)} = \frac{2^4 3^2\sqrt{2}\pi^2 e^5}{\epsilon^{5/2}\omega^7\hbar^5(m_{cv}^*)^{1/2}}\lvert P_{cv}\rvert^2 \times \left[\frac{4}{5}(3\hbar\omega-E_g)^{5/2} + \lvert P_{cv}\rvert^4\frac{(m_{cv}^*)^2}{4}(3\hbar\omega-E_g)^{1/2} - \frac{2}{3}\lvert P_{cv}\rvert^2(m_{cv}^*)(3\hbar\omega-E_g)^{3/2}\right]$
K	Two parab. bands	Nonperturb. approach		$(3)\alpha_d^{(3)} = \frac{2^2}{3\pi}\hbar\omega^2\left[\frac{8\pi}{n}\right]^3\left[\frac{m^*\omega}{\hbar}\right]^{3/2} \times\phi\left[\left[6-\frac{2\Delta}{\hbar\omega}\right]^{1/2}\right]\exp\left[6\left[1-\frac{e^2E^2}{4m^*\omega^2E_g}\right]\right]\left[\frac{e^2}{16m^*\omega^2cE_g}\right]^3$

model $\alpha_d^{(3)}$ calculation has been developed by using the S-matrix and the second quantization formalism; while, in the third instance, the transition rate was calculated between Stark-shifted levels, taking all the orders of the perturbation expansion into account. These models are all characterized by a dominant $(3\hbar\omega - E_{gd})^{1/2}$ spectral dependence. It is clearly apparent that for *RbI* a satisfactory qualitative agreement between the $\alpha_d^{(3)}$ experimental line-shape and those obtained by the three theoretical models considered here was only found for $(3\hbar\omega - E_{gd})$ ranging between approximately 50 *meV* and 200 *meV* above the direct band gap. In this range, the four-band model also shows a good quantitative agreement with the experimental data which, conversely, are greatly overestimated by the two-band model and underestimated by the nonperturbative one.

On the other hand, in the range beyond 200 *meV* above the direct energy gap, the experimental points do not agree - qualitatively or quantitatively - with the theoretical models considered here. This is not surprising because each of the perturbative models takes into account just one of the four possible DThPA mechanisms predicted on the grounds of the selection rules.

A similar $\alpha_d^{(3)}$ spectral behaviour was observed on three other direct gap materials (*KI*, *CsI* and *NaI*). The values of related $C_m (m = 1,2,3,4)$ coefficient are reproted in ref. [16], [18].

Overall, the results obtained for interband two- and three-photon transitions in large gap crystals allow us to conclude that for this class of materials:

(i) the spectral behaviour of the $\alpha_{d,i}^{(n)}$ direct multiphoton absorption coefficient can be qualitatively and quantitatively described in the energy range considered here by the parametric formula:

$$\alpha_d^{(n)} = \sum_{m=1}^{n+1} C_m (n\hbar\omega - E_{gd})^{m-\frac{1}{2}}$$

indicating the simultaneous contribution of all possible transition mechanisms, that depend on the selection rules, to the MPA transition amplitude, though with different weights at different values of $n\hbar\omega$.

(ii) for both direct and indirect MPA processes, none of the available theoretical models can fully describe the $\alpha_{d,i}^{(n)}$ dispersion curve in the whole energy range considered. The inadequacy of the theoretical models in describing the $\alpha_{d,i}^{(n)}$ dispersion curve - both qualitatively and quantitatively - must be ascribed not only to the approximations made in choosing the optical matrix elements but also to the fact that each model takes into account a limited number of intermediate states and thus does not verify the gauge $(-\mathbf{A}\cdot\bar{p}(e/mc), -\mathbf{E}\cdot\mathbf{x}e)$ invariance of the electron-photon interaction.

In order to improve the qualitative agreement between the experimental and the theoretical $\alpha_{d,i}^{(n)}$ line shapes a matrix element $\mathbf{k}$-dependence of the type: $P_{ij}(\mathbf{k}) = \mathbf{P}_{ij}(0) + \mathbf{k}\cdot\nabla\mathbf{P}_{ij}$ would have to be considered. This would result into a mixed spectral dependence of the $\alpha_{d,i}^{(n)}$ absorption coefficients. Instead, quantitative agreement could be improved by making sure that the gauge invariance will in any case be satisfied even though the complete set of states of the crystal bands cannot be considered. According to ref. [19], this could be done: a) by resorting to appropriate approximations which would employ the sum rule on the matrix elements, discovered by Robinson [20] precisely by imposing the gauge invariance; and b) by taking into account the non-parabolicity and degenerate band effects for excitation energies greater than the energy gap.

Acknowledgements

The authors acknowledge their gratitude to Prof. F. Bassani for the many years devoted to a fruitful scientific interaction on these topics.

The experimental support of Dr. M. Lepore and Dr. R. Cingolani was proved essential to carry out this work.

References

1. V. Nathan, A.H. Guenther and S.S. Mitra, *J. Opt. Soc. Am. B* **2**, 294 (1985).
2. P. Liu, W.L. Smith, H. Lotem, J.H. Bechtel, N. Bloembergen and R.S. Adhav, *Phys. Rev. B* **17**, 4260 (1978).
3. F. Beerwerth and D. Frolich, *Phys. Rev. Lett.* **55**, 2603 (1985).
   F. Beerwerth and D. Frolich, *Phys. Rev. Lett.* **57**, 1344 (1986).
4. J.P. van der Ziel, *Phys. Rev.* **16B**, 2775 (1977).
   D.G. Seiler, C.L. Littler and D. Heiman, *J. Appl. Phys.* **57**, 2191 (1985).
   R. Pizzoferrato, M. Casalboni, R. Francini, U. Grassano, F. Antonangeli, M. Piacentini, M. Zena and F. Bassani, *Europhys. Lett.* **2**, 571 (1986).
5. A.I. Bobrysheva and S.A. Moskalenko, *Sov. Phys. Semic.* **3**, 1347 (1970).
6. R. Braunstein and N. Ockman, *Phys. Rev.* **134A**, 449 (1964).
7. F. Bassani and A.R. Hassan, *Nuovo Cimento* **7B**, 313 (1972).
8. J.H. Yee, *J. Phys. Chem. Solids* **33**, 643 (1972).
9. I.M. Catalano and A. Cingolani, *J. Appl. Phys.* **50**, 5638 (1978).
10. I.M. Catalano, A. Cingolani, M. Ferrara and M. Lugarà, *Optica Acta* **27**, 625 (1980).
11. G. Kobbe and C. Klingshirn, *Z. Phys. B* **37**, 9 (1980).
12. A.R. Hassan, *Nuovo Cimento B* **70**, 21 (1970).
13. A. Vaidyanathan, A.H. Guenther and S.S. Mitra, *Phys. Rev. B* **22**, 6480 (1980).
14. I.M. Catalano and A. Cingolani, *Phys. Rev.* **15B**, 1130 (1983).
15. I.M. Catalano, R. Cingolani and M. Lepore, *Solid State Comm.* **60**, 385 (1986).
    I.M. Catalano, R. Cingolani and M. Lepore, *Nuovo Cimento* **9D**, 1313 (1987).
16. I.M. Catalano, A. Cingolani and M. Lepore, *Phys. Rev.* **38B** 13460 (1988).
17. L.V. Keldysh, *Zh. Eksp. Teor. Fiz.* **47**, 1945 (1964); *Sov. Phys. - JETP* **20**, 1307 (1965).
18. I.M. Catalano, A. Cingolani, R. Cingolani and M. Lepore, *Phys. Rev.* **38B**, 3438 (1988).
19. A. Quattropani and F. Bassani, *Interaction of radiation with matter*, Quaderni della Scuola Normale Superiore, Pisa 481 (1987).
20. G.W. Robinson, *Phys. Rev.* **26A** 1482 (1982).

SUPERCONDUCTIVITY ENERGY GAPS IN THE HIGH-T_c LAYERED CUPRATES

J.C. PHILLIPS

AT&T Bell Laboratories
Murray Hill, New Jersey 07974

A wide range of energy gaps has been measured for high-T_c superconductors, such as $YBa_2Cu_3O_{7-x}$, by optical, tunneling and other methods. I comment briefly on these results, allowing for inhomogeneous oxidation and cation disorder. I conclude that in better samples $E_g / kT_c \geqslant 6$, which places these materials in the strong-coupling limit. This limit excludes superconductive mechanisms of the Little type based on electron-exciton exchange and requires that the superconductivity be mediated either by conventional electron-phonon exchange or by an exotic mechanism involving electron-magnon exchange.

1. Introduction

The subject of high-temperature superconductivity in oxides, $T_c \geqslant 30K$, has recently attracted much attention. The highest transition temperatures are found in the layered cuprates, $T_c \geqslant 80K$, and it is even possible that (so far unobserved) layering of oxygen vacancies explains $T_c \sim 30K$ in $(Ba,K)BiO_{3-y}$ alloys. This is a very large subject, which I have treated extensively elsewhere [1].

For the present occasion, rather than discussing the crucial role played by layering, which would require a lengthy discussion drawing extensively on my book [1], I will discuss briefly an important subject which has generated widespread confusion, namely various attempts to measure the superconductive energy gap. Some of this confusion stems from failure to recognize that all oxides, but especially superconductive cuprates which must, after initial formation as insulators, be annealed in O_2 at high temperatures to render them metallic and superconducting, are not single-phase materials, until this has been proven explicitly. This point is discussed briefly in Chapter X of ref. [1], but much still remains to be done before we can describe the multi-phase nature of these materials accurately. The physical mechanisms behind these inhomogeneities are the internal stresses and strains of a ferroelastic nature which generate domains of varying composition. Even small variations in composition, of order a few percent in the oxygen concentration and as little as one percent in the cation site ordering, are sufficient to reduce T_c by one third. These small variations associated with domains are extremely difficult to measure, even if one has single crystals large enough for accurate neutron studies, which has so far not been the case.

Adding to these difficulties are the almost certain increase in defect concentrations near surfaces, where most optical and tunneling experiments probe. At present therefore, one should analyze the data assuming that the samples *are* inhomogeneous, that their measured properties are the result of some kind of percolation, and that different experiments probe different kinds of percolation. Qualitatively we may distinguish two kinds of percolation. The first or more familiar kind is scalar percolation, where a dendritic path

R. Girlanda et al. (eds.), Progress on Electron Properties of Solids, 303–304.

connects one sample boundary with another. Scalar percolation requires the smallest fraction of superconductive material, and it determines the threshold or onset temperature for electrical resistivity measurements of T_c.

The second kind of percolation is vector percolation, which permits planewave like propagation (acoustic waves, optical waves) from one sample boundary to another. This generally requires a larger active fraction of material, and hence will be associated with an apparently smaller value of T_c.

Finally, the third and fourth kinds of percolation are the blocking out of scalar or vector percolation by a normal metallic phase, which may be present in addition to the superconductive phase. In *ac* measurements normal-phase percolation may give rise to resistivity tails and leakage resistances in electrical and optical measurements.

From these comments it will be apparent that many things can happen which can cause the energy gap as measured near the surface to be reduced from its largest bulk value. I will now list the leading optical and tunneling experiments which give the largest values of E_g / kT_c which are of greatest theoretical interest.

The leading optical experiment [2] used a small, partially oriented polycrystalline mosaic sample and obtained $E_g = 2\Delta \sim 500 cm^{-1}$. This result is in good agreement with an estimate [3] by Raman scattering which gave $E_g = 2\Delta \sim 470 cm^{-1} \sim 700K \sim 7.5kT_c$.

The leading tunneling experiments [4,5] which have obtained the largest gaps have used an *STM* probe which has been driven vertically into the sample. It seems likely to me that in these experiments the sample has cracked internally and that the junction studied is actually a native $S-I-S$ junction, i.e., the measured gap is 2Δ, not Δ, which is what one would measure if the junction were between the sample and *STM* tip. This reduces E_g / kT_c from the quoted values [4,5] of 11 and 12, to $\geqslant 5.5$, but we should still note that kT_c at the crack may be less than its bulk value. Thus the value 5.5 is lower bound. This value is fully consistent with break junction measurements carried out with epitaxial films [6].

In conclusion, at present the best available experimental data show that E_g / kT_c is far above the weak-coupling value of 3.5, which means that the energy E_b of the boson exchanged with electrons to produce the attractive interaction responsible for superconductivity cannot be as large as an exciton energy ($\sim 1eV$), and therefore the Little mechanism [7] (quite popular at the Berkeley conference [8]) is excluded. Other experimental data (especially diffuse scattering of spin-polarized neutrons) firmly exclude magnetic mechanisms [1]. This leaves only the conventional electron-phonon mechanism, modified in a complex way [1] in layered cuprates to produce high T_c's.

References

1. J.C. Phillips, Phys. of High-T_c Superconductors, Academic Press, Boston, 1989.
2. Z. Schlesinger, R.T. Collins, D.L. Kaiser and F. Holtzberg, *Phys. Rev. Lett.* **59**, 1958 (1987).
3. S.L. Cooper, M.V. Klein, B.G. Pazol, J.P. Rice and D.M. Ginsberg, *Phys. Rev. B* **37**, 5920 (1988).
4. M.D. Kirk et al., *Appl. Phys. Lett.* **52**, 2071 (1988).
5. A.L. de Lozanne, K.W. Ng, S. Pan, R.N. Silver and A. Berezin, *J. Mic. Soc.* **151** (1988).
6. J.S. Tasi et al., *Phys. C* **153-155**, 1385 (1988), unpublished.
7. W.A. Little, *Phys. Rev.* **A134**, 1416 (1964).
8. Novel Superconductivity, Ed. S.A. Wolf and V.Z. Kresin, Plenum, N.Y. (1987).

PHOTOEMISSION STUDIES OF HIGH TEMPERATURE SUPERCONDUCTORS

F.C. BROWN*)

Department of Physics, FM-15
University of Washington
Seattle, WA 98195
USA

1. Current Understanding

Since the discovery of high temperature superconductivity in doped La_2CuO_4 [1] and in $YBa_2Cu_3O_7$ [2], a large amount of work has been carried out in order to determine the essential properties of these compounds. See, for example, a recently published volume, which summarizes the physical properties of the high temperature superconductors [3]. From the beginning, the powerful experimental technique of photoemission spectroscopy has been utilized to probe the electronic structure of these materials.

Some of the first photoemission experiments on $YBa_2Cu_3O_{6+z}$ were carried out on sintered material in March, 1987 at the University of Illinois soft x-ray beam line at the Synchrotron Radiation Center in Stoughton. Wisconsin [4]. The Illinois group has continued work on this beam line [5], and, worldwide, many investigators have used photoemission, as well as inverse photoemission, to study the properties of high temperature superconductors [6]-[32]. Among these investigations, references [25]-[32] deal with experiments on single crystals cleaved in ultrahigh vacuum, including the new bismuth compound $Bi_2CaSr_2Cu_2O_8$ [26], [29].

Photoemission is known to be a surface sensitive probe, therefore questions about reconstruction at the surface, or loss of oxygen after preparation of the surface, are most relevant. It was determined very early that the common method of surface preparation for monovalent metals, argon ion bombardment, damaged the surface and, therefore, must be avoided. The emphasis then shifted from sintered samples to single crystals cleaved *in situ*. Unfortunately, the freshly cleaved surface of materials such as $EuBa_2Cu_3O_{6.7}$ is unstable in vacuum above $55K$, as recently emphasized by R.S. List, et al. [31], [32]. Properly oxygenated crystals cleaved and measured at $20K$ show substantial density of states near the Fermi level consistent with a metallic Fermi surface. This issue will be further discussed in Section 6.

The gradual availability of large high quality single crystals has been very important to the photoemission work. It has been even more important to neutron scattering studies, which so clearly demonstrate the essential nature of spin interactions in these materials. See, for example, the review by Birgeneau and Shirane [33].

The current theory of high temperature superconductivity is still in a formative state. Detailed band calculations for the various materials [34]-[39] have provided results for comparison with photoemission and other experimental work. The comparison between experiment and one electron band theory is not entirely satisfactory, as discussed below in Section 6. In fact, these results, and the importance of spin, suggest that one look

*) Emeritus Professor of Physics at the University of Illinois, Urbana

R. Girlanda et al. (eds.), Progress on Electron Properties of Solids, 305–318.

beyond the one electron approximation. Electrons and holes in the high temperature superconductors appear to be highly correlated, therefore a number of theories now specifically incorporate, or begin with, the local interactions [40], [44]. This point of view is reinforced by the experimental evidence from photoemission discussed in Section 4 and 5.

What mediates the pairing in high temperature superconductivity? Through isotope experiments, it has been learned that phonons play only a minor role in the superconductivity of doped La_2CuO_4, and no role at all in the superconductivity of $YBa_2Cu_3O_7$. Apparently, the main process does not involve the formation of Cooper pairs through interaction with phonons. The spin state of the system and dynamic spin fluctuations appear to be crucial. Undoped La_2CuO_4, as well as $YBa_2Cu_3O_7$ with low oxygen content, have long range magnetic order at low temperature, due to the antiferromagnetic interaction between Cu^{2+} ions. Upon doping or addition of oxygen, holes tend to localize on oxygen ions in $O^- 2p$ states. These holes interact ferromagnetically with adjacent coppers to frustrate [45] the $Cu^{2+} - Cu^{2+}$ antiferromagnetic interaction. Further, the holes localized on the oxygen are the carriers of electricity. In the theory of Emery [41] these $2p\sigma$ holes move as quasiparticles which carry both spin and charge. At low temperature, *s*-state pairing occurs produced by enhanced superexchange interactions and large zero point fluctuations in kinetic energy. In the cluster calculations of Goddard and coworkers [44], the holes on the oxygen are in $2p\pi$ states. Again, the holes interact ferromagnetically with electrons on the copper and a (triplet state) pairing interaction arises. Because of these pairing interactions (plus a possible admixture of other excitonic or phonon interactions), a kind of BCS [46] ground state arises in which the carriers of electricity, are not readily scattered by imperfections or lattice vibrations. Superconductivity occurs at intermediate and at low temperatures.

Let us briefly discuss crystal structure in the next Section, realizing the importance of the ionic lattice, as well as conduction in the sheets and chains. Experimental details concerning photoemission and sample preparation are given in Section 3. The Fermi edge as it is beginning to be revealed by photoemission is discussed in Section 6, and finally a summary with ideas for future work given in Section 7.

2. Crystal Structure and Reduced Dimensionality

Many similarities exist between the two superconducting systems $La_{2-x}Sr_xCuO_4$ and $YBa_2Cu_3O_{6+x}$. For example, both have a crystal structure involving $Cu-O$ planes separated by heavy metal or rare earth ions. In both, a kind of $2D$ conduction occurs, due to the hopping of holes between oxygens situated in sheets and planes within the structure. Also, when both materials are lightly doped or undoped ($x=0$) long range magnetic order occurs at low temperature. The lanthanum compound has been most extensively studied, especially by neutron scattering techniques [33], so we introduce the structure in this case.

Figure 1a shows the crystal and magnetic structure of the high temperature tetragonal phase (space group I4/mmm) of La_2CuO_4. Below $530K$ a superlattice forms involving a doubling of the unit cell due to transformation from tetragonal to orthorhombic phase. In this structural transformation the oxygen neighbors of the central oxygen rotate slightly.

The antiferromagnetic long range order indicated by arrows in the upper and lower $Cu-O$ planes of Fig. 1a sets in below about $250K$. here the arrows represent the unpaired electron spins in $d_{x^2-y^2}$ orbitals on Cu^{2+} ions, $3d^0$ configuration, assuming an

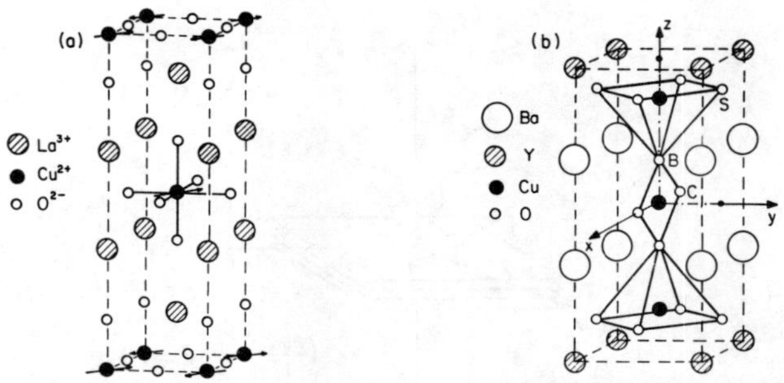

Figure 1. a) shows the crystal and magnetic structure of La_2CuO_4 below the Neel temperature. b) is the 15 site unit cell for $YBa_2Cu_3O_7$. The two dots indicate the position of order vacancies. Inequivalent oxygen ions are indicated in sheet (S), bridge (B) and chain (C) sites.

ionic model. To first order, the magnetic structure consists of stacks of planes each a square $2D$, $S = 1/2$, Heisenberg antiferromagnetic lattice [33], [47]. This long range magnetic order is rapidly lost by mild doping with Sr in $La_{2-x}Sr_xCuO_4$, which introduces holes localized on the oxygens. On the other hand, magnetic fluctuations persist. As mentioned above, this behavior appears to be due to competition between the antiferromagnetic order of $3d^9$ copper ions and a strong ferromagnetic interaction of holes localized on oxygen [45].

A large amount of work has been done on the high temperature superconductors to ascertain the role which oxidation plays in relation to the transition temperature T_c. For $YBa_2Cu_3O_{6+x}$ it turns out that the highest transition temperature $T_c \approx 90K$ occurs when $0.8 \leqslant x < 1.0$. Between $x \approx 0.5$ and 0.8, T_c has decreased to about $60K$ and, for smaller values of x, T_c deteriorates until below $x = 0.3$ the material becomes antiferromagnetic. Zaanen et al. [48] have discussed this phase diagram and argue for the importance of holes distributed between the sheets and chains in the crystal structure. Figure 1b shows the 15 site unit cell of orthorhombic $YBa_2Cu_3O_7$ in the so-called vacancy model (two oxygen vacancies are indicated by small dots in the figure). Three different oxygen sites are labelled by S (sheets), B (bridge) and C (chains). Notice that the structure is basically one containing layers of copper and oxygen atoms interleaved with layers containing only yitterium or barium. Conduction in the material is largely two-dimensional, and the $Cu-O$ groups in the sheets and in the chains play the major role in conduction.

Through core level studies, photoemission yields information on the charge of various ions. Resonance behavior allows one to separate different contributions to a band or level and to properly assign observed structure. We will illustrate this in Section 4, but first let us briefly discuss the experimental photoemission technique.

3. The Photoemission Technique

In general, photoemission measurements on solids are two types, angle integrated or angle resolved. Figure 2 illustrates angle resolved photoemission for photons of energy $h\nu$ incident onto the sample at a given polarization and angle α. Excited photoelectrons travel to the surface and penetrate into vacuum where they have a kinetic energy $E = \hbar k^2/2m$ and angular coordinates θ and ϕ, depending upon position of the angular resolved analyzer and detector. The figure shows schematically the clean top surface of a layered crystal produced, say, by cleaving in ultrahigh vacuum. This might be done

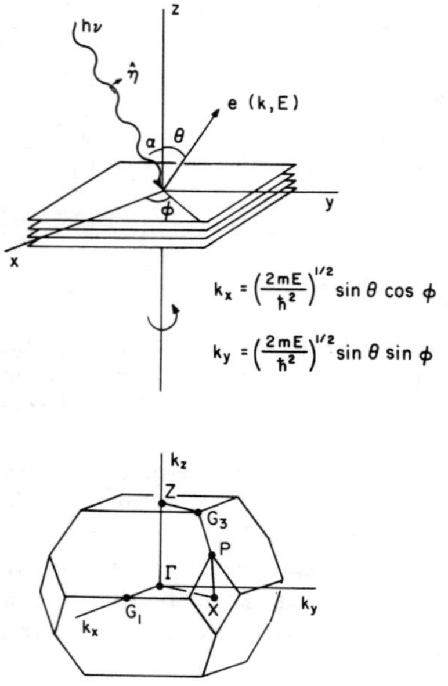

Figure 2. Illustrating the angle photoemission technique. The reduced zone for $Bi_2CaSr_2Cu_2O_8$ is shown in the lower part of the figure [39]. The equations show how k_x and k_y relate to the θ,ϕ and the free electron kinetic energy E, assuming that the component of wavevector parallel to the surface is conserved during photoemission.

either with a cleaving tool or by pulling apart a secured crystal to which a small tab has been cemented using low vapor pressure epoxy.

It turns out that the initial electron energy in the crystal can be determined as a function of wavevector using angle resolved photoemission. This is especially convenient to do for two dimensional layered crystals by varying the angle θ for given directions ϕ (along lines in the Brillouin zone shown in the lower part of the Fig. ..). Electron wavevector parallel to the surface is conserved; therefore, the relations shown in the Fig. apply when a photoelectron penetrates a well cleaved surface. Even in the case of three dimensional crystals, the E versus k relations can be explored at normal emission by using known final state bands or free electron final states. This type of work is just beginning to be done on the superconducting compounds [27]. A good review on the technique of angle resolved photoemission is that by Plummer and Eberhardt [49].

More often the emitted electrons are integrated over angle to obtain a photoelectron energy distribution curve $N(E,h\nu)$, a function of both kinetic energy E and photon energy $h\nu$. It is well known that $N(E,h\nu)$ depends upon both the initial density of states $\rho(E)$ and the photoionization cross section $\sigma(h\nu)$ for transitions between the initial bound state wave function and a final continuum wave function. For the valence band of a solid which is known to be a mixture of p and d derived states, as a first approximation one can write:

$$N(E,h\nu) = \rho_p(E)\sigma_p(h\nu)\rho_d(E)\sigma_d(h\nu).$$

The calculated atomic photoionization cross sections [50] are at least a guide to the behavior of photoemission from bands of mixed symmetry. By using an intense beam of photons from a synchrotron source, tunable over a wide energy range, it is possible to enhance one specific cross section or the other. Sometimes it is also possible to identify a photoemission feature by tuning the photon energy through a resonance.

We will now expand upon these remarks by describing the photoemission experiments of references [4] and [5], which were carried out in the ultrahigh vacuum photoemission chamber at the focus of an ellipsoidal mirror on the University of Illinois beam line [51] on the Aladdin storage ring in Stoughton, WI. Here an extended range grasshopper monochromator is used, which covers the range 55 to 1000eV. The photoelectrons were analyzed by a hemispherical energy analyzer. The overall energy resolution was 0.075 to 0.25eV depending upon the slit settings and photon energies used.

High quality pressed pellet superconducting samples were prepared with T_c's as high as 93K. The samples were pressed against indium on small stages inserted into the photoemission chamber through a vacuum interlock. They were then picked up on a sample manipulator, which could be cooled below 80K. The vacuum before and after cooling was in the mid 10^{-10} Torr range. The initial run was made after thorough outgassing in ultrahigh vacuum but without heating or further surface treatment. Later runs were all made after the surface had been abraded with a file in ultrahigh vacuum. Typical results are given in the next Section.

4. Valence Band and Outer Core Emission

In general, one likes to take photoemission data over a wide range of photon energies in order to exploit the dependence of photoionization cross section upon $h\nu$, as described above. The superconducting compounds have a number of different atoms and many core levels.

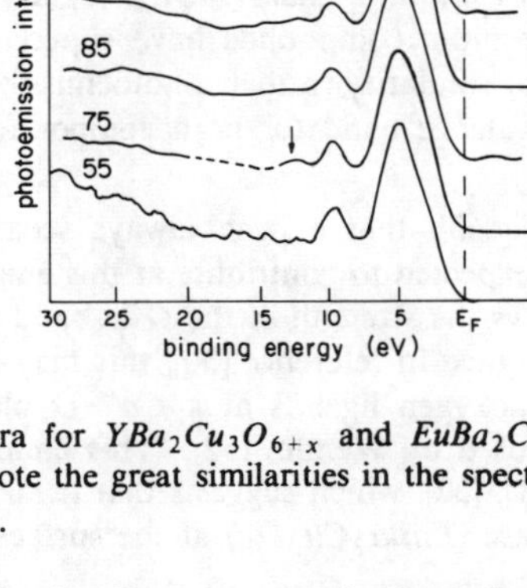

Figure 3a. Photoemission spectra for $YBa_2Cu_3O_{6+x}$ and $EuBa_2Cu_3O_{6+x}$ taken with different photon energies as indicated. Note the great similarities in the spectra for the two different compounds, from Samsavar et al. [5].

Figure 3a shows energy distribution curves of photoelectrons from a sample of $YBa_2Cu_3O_{6+x}$ taken at various photon energies from 55 to 180eV [5]. Except for details very close to the Fermi level, quite similar results have been reported on single crystals of this material [30].

The prominent peak at $5eV$ binding energy in Fig. 3a is due to emission from the valence band, which is largely composed of $Cu\,3d$ and $O\,2p$-derived levels. At the higher photon energies $Cu\,3d$-derived states have larger cross section than $O\,2p$, and the main band at $5eV$ is almost entirely due to $Cu3d$. The shoulder on this band at $2.5eV$ increases with decreasing photon energy since it involves oxygen $2p$-derived states whose cross section is large at the lower photon energies. A pair of core level photoemission lines appear in Fig. 3a at 14 and $16eV$. The barium $5p$ outer core levels lie at these energies, thus probably account for this structure. These assignments can be tested by carefully looking for resonance effects as a function of photon energy. Notice that the $Ba5p$ lines are enhanced around $100-130eV$, which is near the maximum of the $Ba4d$ to f absorption. A line at $30eV$, due to the $Ba5s$ core, is also enhanced by tuning in the range $100-130eV$.

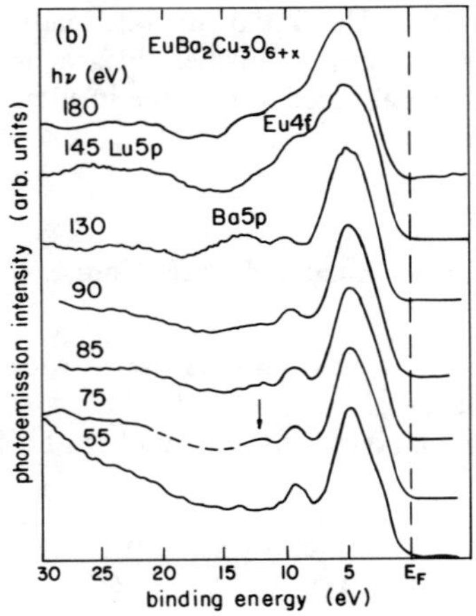

Figure 3b.

Entirely similar results were obtained on $EuBa_2Cu_3O_7$, except that emission from the $Eu4f$ levels appears near $7eV$ binding energy at the higher photon energies. See Fig. 3b. Both the yttrium and europium compounds have superconducting transition temperatures around $90K$. The great similarity in their photoemission spectra leads to the conclusion that emission in the valence band for both compounds is dominated by the $Cu3d$ and $O\,2p$ states.

Notice in Fig. 3a in Fig. 3b that a peak always seems to occur at $9.5eV$ binding energy. No core levels are expected to contribute at this energy for either compound. The intensity of this band follows the strength of the $O\,2p$ band at $2.5eV$, and appears to be a satellite of that band. As argued in reference [30], this may be due to a two hole satellite, with the holes on $3d^{10}L^2$ oxygen ligends at a $Cu-O$ plane on the surface. Such an explanation has been proposed by Wendin [52]. This band is absent in crystals cleaved at very low temperature [31], [32], which suggests that it may be characteristic of an oxygen depleted insulating phase ($EuBa_2Cu_3O_6$) at the surface of grains or crystals cleaved at high temperature.

Core level studies, both photoemission and photoabsorption, are useful in answering questions about the oxidation states of Cu and of O in these materials. In the early work of reference [4], emission from $Cu3p$ core levels was studied in detail. Results appeared in agreement with the idea that the copper was predominently in a Cu^{2+} state, although some Cu^{3+} could not be ruled out. Observation of the copper $2p$ and $3p$ absorption edges [8] show energy positions very close to those found in CuO, giving further confirmation of the Cu^{2+} state for most of the copper. Likewise, evidence has been given for the existence of $O\,2p$ holes, which in $YBa_2Cu_3O_7$ are localized primarily on the $Cu-O$ chains of the structure shown in Fig. 1. Additional evidence for these various oxidation states comes from the satellite structure observed in photoemission. This will be further discussed in Section 5.

Photoemission and inverse photoemission within about $14eV$ of the Fermi level is shown for the superconducting compounds in Fig. 4. These spectra were obtained by Meyer et al. [22] using sintered samples fractured in ultrahigh vacuum, except that the $Bi_2CaSr_2Cu_2O_8$ material was a single crystal cleaved in vacuum.

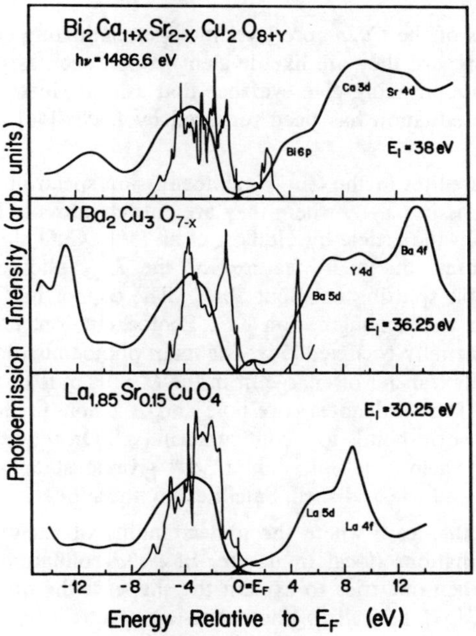

Figure 4. Showing photoemission and inverse photoemission for three superconducting materials compared with the one electron density of states. From Meyer, Hill, Weaver, and Nelson [22].

Also shown are the density of states calculated by one electron band theory [34]-[36]. Notice that the valence bands are all strikingly similar, and that a low density of states is observed near the Fermi level. In fact, the density of states observed by photoemission is disorted when compared with band theory, in that it always appears shifted downwards by about $1.7eV$. It has been pointed out by several authors [8]-[10] that this indicates localization of the $Cu\,3d$ and $O\,2p$ derived states along with an on-site Coulomb interaction U which is at least greater than about $2eV$. Again, the occurence of satellites supports this, as discussed in the next section.

5. Satellites and Local States

The arrows above the $75eV$ curves in Figs. 3a and 3b point to a satellite feature at $12eV$ which has been discussed at length in the literature [53]. It is derived from excitation of an electron on a Cu ion in a $3d^9$ ground state configuration, leaving two correlated holes on a copper site, namely the $Cu3d^8$ configuration. The fact that this feature is resonant at $75eV$ indicates transitions on the Cu, since this is the energy of the $Cu3p$ core. Two different photoemission processes are involved, direct photoemission:

$$3d^n \rightarrow 3d^{n-1}\epsilon_k$$

and the atomic process:

$$3p^6 3d^n \rightarrow 3p^5 3d^{n+1} \rightarrow 3p^6 3d^{n-1}\epsilon_k$$

where ϵ_k indicates the final photoelectron state. The two processes interfere and a kind of Fano resonance occurs.

Photoemission scans of the $Cu2p$ core level in superconducting compounds also show clear satellites. In this regard they are like divalent oxides, such as CuO and NiO, where strong core level satellite structure give evidence that a local cluster, rather than a band approach, applies. The situation has been reviewed by Davis [54] and also by Eglehoff [55].

The occurence of satellites in the Cu 2p photoemission spectrum for CuO is in sharp contrast to what happens in Cu_2O, where they are entirely absent. This is shown in Fig. 5, which was taken from the article by Hulbert et al. [56]. CuO shows satellites - Cu_2O does not. In both cases, the main features of the $L_{2,3}$ photoemission spectra are separated by a spin-orbit splitting of about $20eV$. The copper in CuO has a valence of $+2$, therefore an outer shell configuration $3d^9$. Photoexcitation, creating a hole, forms two final states, each partially occupied [54]. The main photoemission line can be thought of as core excitation plus transfer of charge from the $O2p$ ligands to the $Cu3d$ level, i.e., a configuration $c3d^{10}\underline{L}$. (Here c denotes core hole and $\underline{L}$ a hole in the $O2p$ ligands.) The observed $10eV$ satellite corresponds to a configuration cd^9. On the other hand, the copper in Cu_2O has a formal valence of $+1$, thus a $3d^{10}$ ground state, and there can be no charge transfer from ligand to Cu d-shell. Satellites do not appear.

Nickel oxide is another case where the understanding of excited state spectroscopy involves various combinations (local in nature) of $Ni3d$ configurations and oxygen $2p$ ligands. This is true when one tries to explain the shape of the observed Ni 3p absorption edge at $65eV$ [54], [57], as well as photoemission spectra from both core levels and valence bands [58].

The point of view taken by Sawatsky et al. [59] is that, whereas modern band theory may fairly well represent the ground state of the transition metal oxides, a different approach is needed to explain core level and valence band photoemission. For example, to explain the valence band photoemission spectrum of NiO, so-called three hole states (d^7,$d^8\underline{L}$,d^9, and $d^{10}\underline{L}^3$) are required. The one hole states needed to interpret inverse photoemission spectra are (d^9 and $d^{10}\underline{L}$). Important configuration interactions are the $d-d$ Coulomb interaction (NiO is a so-called high U material) and the $O2p-Ni3d$ hybridization. In other words, although the uppermost $Ni3d$-derived bands of the ground state of NiO are only partially filled, NiO is an insulator and a conductivity gap exists of the order of $4.3eV$. Like NiO, the high temperature superconductors appear to be high U materials, although they are not insulators in the same sense.

From photoemission binding energy data for the $Cu3p$ core levels, we assert that the

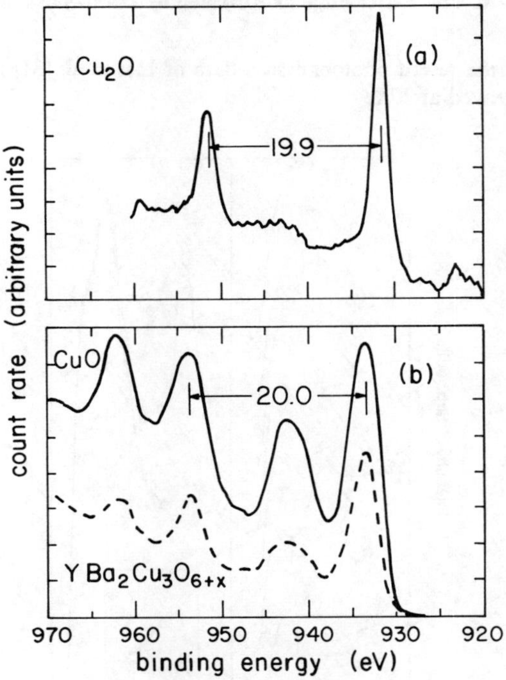

Figure 5. The $CuL_{2,3}$ photoelectron distribution curves for (a) Cu_2O and (b) CuO taken from Hulbert et al. [56]. The energy distribution curve for the $L_{2,3}$ core levels of Cu in $YBa_2Cu_3O_{6+x}$ is shown by a dotted line below the curve for CuO, according to Ihara and coworkers [12].

valence of Cu in $YBa_2Cu_3O_{6+x}$ is predominantly Cu^{2+}. This is confirmed by scans of the $Cu2p$ core level in the superconducting compound which show clear satellites, as for CuO. This was first reported by Ihara and coworkers [12], whose data are sketched as a dotted line below CuO in Fig. 5. Thus, both core and valence band photoemission spectra for YBa_2Cu3O_{6+x} show satellites.

To summarize, it appears that the assembled photoemission data on the new high temperature superconducting compounds are better compared with the local cluster model, as used for high U materials such as the transition metal oxides, than with the results of band theory. In this picture, the final states for valence band photoemission from the high temperature superconductor would be combinations of d^8, d^9L and $d^{10}L^2$, where the notation is similar to that used in the cases of CuO or NiO. It is not that $YBa_2Cu_3O_{7-\chi}$ is a Hubbard insulator in the usual sense, but similar features occur such as strong Coulomb $d-d$ interaction leading to high U, as well as charge transfer fluctuations (moderate or low $\Delta < U$). This localized ionic model leads to incomplete screening of the hole in photoemission, consequently the overall downward shift ($\sim 1.7eV$) of the main valence band density of states, and also to various satellite phenomena.

6. The Fermi Edge as Revealed by Photoemission

In Fig. 6, we show the recent photoemission data of List et al. [31] for a single crystal of $EuBa_2Cu_3O_{6+x}$ cleaved at $20K$.

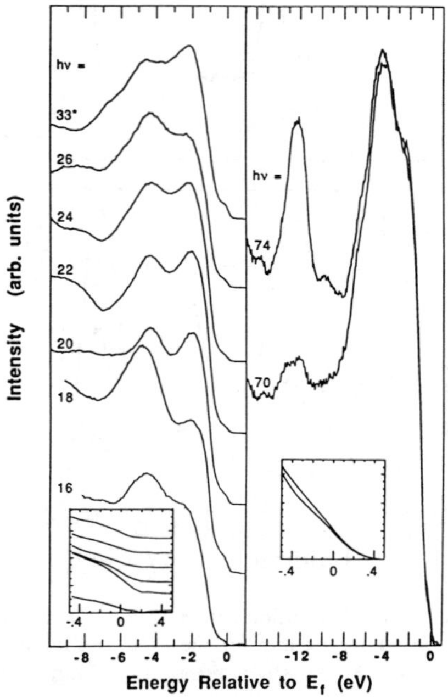

Figure 6. Photoemission from a $EuBa_2Cu_3O_{6+x}$ single crystal near the Fermi edge, across resonant thresholds for $Eu\,5p(18eV)$, $O\,2s(22eV)$ and $Cu\,3p(74eV)$. The 16 through $26eV$ spectra are scaled to equal peak intensity near $-4.5eV$. The crystal was cleaved *in situ* at $20K$. From R.S. List et al. [31].

Notice that a definite edge appears although it is broadened somewhat by instrument resolution (0.1-0.2 eV). The intensity in this edge, relative to the main band at 5 eV, is not unlike copper metal. Resonance effects are discernable in the $Cu3p$ to $3d$ region at $74eV$. Likewise, an antiresonance occurs at $22eV$ corresponding to the $O2$ to $2p$ threshold. Such results illustrate the mixed O2p, Cu 3d band-like character of states very close to E_F. These results are consistent with the existence of a Fermi surface in well-oxygenated material at low temperature.

Even though a higher density of states is now found near E_f, the main $5eV$ feature still has to be shifted down to greater binding energy by $1-2eV$ in order to bring it in agreement with the results of band theory. As suggested by Yarmoff et al. [8], it still appears that a local model for photoemission is appropriate.

The resonating valence bond RVB model of Anderson [40] has some specific consequences as far as photoemission is concerned. For example, Huber [60] shows that, as a consequence of phase space limitations, the photoemission intensity for the normal phase should decrease in the neighborhood of the Fermi level over an interval set by the spinon bandwidth. Recent observations on $Bi_2CaSr_2Cu_2O_9$ [29] and $EuBa_2Cu_3O_7$ [31] show a definite Fermi edge with decrease in photoemission as E_f is approached. The spinon energy would have to be $60meV$ or less, if the RVB model is correct. If the spinon energy is more like 100 to $200meV$, the recent photoemission results above Tc favor a Fermi liquid model.

Finally, the recent photoemission experiments on $Bi_2Ca_8Sr_2Cu_2O_9$ [29], [31] are consistent with the opening of a superconducting gap $\Delta = 20$ to $40meV$ upon cooling single crystals below T_c. Such work points the way for exciting developments on the experimental side within the coming months.

7. New Directions

To summarize, photoemission spectroscopy has contributed its share to a growing understanding of the new superconducting materials. Results range from a general view of the valence band density-of-states to detailed evidence for the dominance of the $Cu^{2+}\,3d^9$ configuration and localization of holes in O^{-2p} states. In order to observe states near the Fermi edge, the importance of properly oxygenated crystals at very low temperature has been recently demonstrated [31], [32]. Angle resolved photoemission experiments on good quality single crystals of $YBa_2Cu_3O_7$ have now been reported [27], and these show angular dispersion plus strong photon energy dependence near E_f. The angle resolved technique will certainly be extended to other crystals in the future and will benefit from improved surface preparation.

The oxide superconductors appear to be highly ionic systems. They should have interesting surface properties. Until now, the surfaces of strongly ionic compounds have been little studied by photoemission or other surface sensitive techniques, such as photoelectron or low energy electron diffraction. This situation is rapidly changing as new photoemissions results on oxides and other ionic crystals are reported [61].

It should be remarked that photoemission is a useful tool to survey the distribution of electronic levels as new superconducting materials are discovered. Investigations usually begin on sintered samples, which are fractured or abraided in high vacuum, but high quality single crystals should be used as soon as they become available. So far, the largest amount of photoemission work has been carried out on the $40\,K$ $La_{2-x}Sr_xCuO_4$ and the $80\,K$ rare earth $Ba_2Cu_3O_{6+z}$ systems.

The $85\,K$ non-rare earth system $Bi_2CaSr_2Cu_2O_8$ has been studied by photoemission, using both sintered [62] and single crystal samples [26]. A definite shoulder in density-of-states near E_f has been observed, especially in the single crystal work. Furthermore, there is now evidence for changes in this shoulder or edge below the critical temperature [29]. Recently, angle resolved photoemission has been utilized [63] to map energy bands in the Γ to X direction in momentum space (shown schematically in Fig. 2). Results close to E_f are in general agreement with the calculated electronic structure [39]. The resonance behavior of states near the Fermi level indicate a dominant $O\,2p$ character. This kind of work should be extended to other non-rare earth systems, such as the $125\,K$ material, $Tl_2Ca_2Ba_2Cu_3O_4$ [64], [65].

Finally, for comparison's sake, it would be interesting to carry out photoemission experiments on non-copper systems, such as the $30\,K$ material $Ba_{1-x}K_xBiO_3$ [66], [67]. This is especially true in light of the suggestion [68] that conventional phonon mechanisms may be operative here rather than magnetic pairing interactions.

Acknowledgements

The author would like to thank Profs. J. Bardeen, G. Margaritondo, C.G. Olson, M. Onellion and J. Weaver for helpful discussions. Support by the National Science Foundation under Grant DMR 88-96157, and also by the Eastman Kodak Company, is much appreciated.

References

1. J.G. Bednorz and K.A. Mueller, *Z. Physik* **64**, 189 (1986).
2. C.W. Chu, P.H. Hor, R.L. Meng, L. Gao, Z.J. Huang and Y.Q. Wang, *Phys. Rev. Lett.* **58**, 405 (1987).
3. D.M. Ginsberg ed., *Physical Properties of High Temperature Superconductors*, World Scientific Publishers (1989).
4. F.C. Brown, T.C. Chiang, T.A. Friedmann, D.M. Ginsberg, G.H. Kwawer, T. Miller and M.G. Mason, *J. Low Temp. Phys.* **69**, 151 (1987).
5. A. Samsavar, T. Miller, T.C. Chiang, B.G. Pazol, T.A. Friedmann and D.M. Ginsberg, *Phys. Dev. B* **37**, 5164 (1988).
6. M. Onellion, Y. Chang, D.W. Niles, R. Joynt, G. Margaritondo, N.G. Stoffel and J.M. Taruscon, *Phys. Rev. B* **36**, 819 (1987).
7. Y. Chang, M. Onellion, D.W. Miles, R. Joynt, G. Margaritondo, Y.G. Stoffel and M. Tarascon, *Solid State Commun.* **63**, 717 (1987).
8. J.A. Yarmoff, D.R. Clarke, W. Drube, V.O. Karlsson, V. Teleb-Ibrahami and F.J. Himpsel, *Phys. Rev. B* **36**, 3967 (1987).
9. P.D. Johnson, S.L. Piu, L. Jiang, M.W. Ruckman, M. Strongin, S.L. Hulbert, R.F. Garrett, B. Sinkovic, N.V. Smith, R.J. Cava, C.S. Jee, N. Nichols, E. Kaczanowicz, R.E. Salomon and J.E. Crow, *Phys. Rev. B* **35**, 8811 (1987).
10. Z.X. Shen, J.W. Allen, J.J. Yeh, J.S. Kang, W. Ellis, W.E. Spicer, I. Lindau, M.B. Maple, Y.D. Dalichaouch, M.S. Rorikachvih, J.Z. Sun and T.H. Geballe, *Phys. Rev. B* **36**, 8414 (1987).
11. P. Steiner, J. Albers, V. Kinsinger, I. Sander, B. Siegwart, S. Hüfner and C. Politis, *Z. Phys. B* **66**, 275 (1987); **67**, 494 (1987); **67**, 494 (1987).
12. H. Ihara, M. Hirabayashi, N. Teada, Y. Kimura, K. Senzalsi, M. Arimoto, K. Bushida, F. Kawashima and R. Uzuka, *Jpn. J. Appl. Phys.* **28**, L460 (1987).
13. E.R. Moog, S.D. Bader, A.J. Arko and B.K. Flandermeyers, *Phys. Rev. B* **36**, 5583 (1987).
14. R.L. Kurtz, R.L. Stockbauer, D. Mueller, A. Shi, L.E. Toth, M. Osofsky and S.A. Wolf, *Phys. Rev. B* **36**, 8818 (1987).
15. N.G. Stoffel, J.M. Tarascon, Y. Gang, M. Onellion, D.W. Niles and G. Margaritondo, *Phys. Rev. B* **36**, 3986 (1987).
16. F. Garcia-Alvarado, E. Moran, M. Vallet, J.M. Genzalez-Calbet, M.A. Alavior, M.T. Perez-Frias, J.L. Vicent, S. Ferner, E. Garcia-Michel and M.C. Asensio, *Solid State Commun.* **63**, 507 (1987).
17. T. Takahasi, F. Mueda, H. Arui, H. Kutagama-Yoshida, Y. Okube, T. Suzuki, S. Hosoga, A. Fujimori, T. Shidata, T. Koide, T. Miyahara, M. Onoder, S. Shamato and M. Soto, *Phys. Rev. B* **36**, 5686 (1987).
18. N. Nücker, J. Fink, B. Renker, D. Ewert, G. Politis, J.W.P. Weigs, J.C. Fuggle, *Z. Phys. B* **67**, 9 (1987).

19. B. Reihl, T. Rieslerer, J.G. Bednorz and K.A. Müller, *Phys. Rev. B* **35**, 8804 (1987).

20. A. Fujimori, E. Takayama-Muromachi, Y. Uchida and B. Okai, *Phys. Rev. B* **35**, 8814 (1987).

21. Y. Gao, T.J. Wagener, J.H. Weaver, A.J. Arko, B. Flandermeyer and D.W. Capone II, *Phys. Rev. B.* **36**, 3971 (1987); **36**, 3899 (1987).

22. H.M. Meyer III, D.M. Hill, J.H. Weaver and D.L. Nelson, Chap. 21 in *Physical Chem. of High Temperature Superconductors*, ed. by T.F. George and D.L. Nelson, Amer. Chem. Soc. (1987).

23. J.H. Weaver, H.M. Meyer III, T.J. Wagener, D.M. Hill and Y. Hu, in *Special Symposium Amer. Vac. Soc.*, AIP Conf. Proc. (1988).

24. H.M. Meyer III, D.M. Hill, T.J. Wagener, Y. Gao, J.H. Weaver, D.W. Capone and K.C. Goretta, *Phys. Rev. B* **38**, 6500 (1988).

25. H.M. Meyer III, D.M. Hill, J.H. Weaver, D.L. Nelson and C.F. Gallo, *Phys. Rev. B* **38**, 7144 (1988).

26. M. Onellion, M. Tang, Y. Chang, G. Margaritondo, J.M. Tarascon, D.A. Morris, W.A. Bonner and N.G. Stoffel, *Phys. Rev.* **38**, 881 (1988).

27. N.G. Stoffel, Y. Chang, M.K. Kelly, L. Dottl, M. Onellion, P.A. Morris, W.A. Bonner and G. Margaritondo, *Phys. Rev. B* **37**, 7952 (1988).

28. N.G. Stoffel, P.A. Morris, W.A. Bonner, D. La Graffe, Ming Tang, Y. Chang, G. Margaritondo and M. Onellion, *Phys. Rev. B* **38** (1988).

29. C. Chang, Ming Tang, R. Zanoni, M. Onellion, R. Joynt, D.L. Huber, G. Margaritondo, P.A. Morris, W.A. Bonner, J.M. Tarascon and N.G. Stoffel, to be published.

30. Ming Tang, N.G. Stoffel, Qi Biao Chen, D. La Graffe, P.A. Morris, W.A. Bonner, G. Margaritondo and M. Onellion, *Phys. Rev. B* **38**, 897 (1988).

31. R.S. List, A.J. Arko, Z. Fisk, S.W. Cheong, J.D. Thompson, J.A. O'Rourke, C.G. Olson, A.-B. Yang, Tun-Wen Pi, J.E. Shirber and H.D. Shinn, *J. Appl. Phys.*, to be published.

32. A.J. Arko, R.S. List, Z. Fisk, S.-W. Cheong, J.D. Thompson, J.A. O'Rourke, C.G. Olson, A.-B. Yang, Tun-Wen Pi, J. Magnetism and Magnetic Materials, **75**, L1 (1988).

33. R.J. Birgeneau and G. Shirane, in *Physical Properties of High Temperature Superconductors*, ed. by D.M. Ginsberg, World Scientific Publ. (1989).

34. L.F. Mattheiss, *Phys. Rev. Lett.* **58**, 1028 (1987).

35. L.F. Mattheiss and D.R. Hamann, *Solid State Commun.* **63**, 395 (1987).

36. S. Massidda, J.J. Yu, A.J. Freeman and D.D. Koelling, *Phys. Rev. Lett. A* **122**, 198 (1987); J. Redinger, A.J. Freeman, Yaejun Yu and S. Massida, *Phys. Lett. A* **124**, 469 (1987); S. Massiolda, J.J. Yu and A.J. Freeman, *Physica C*, in press.

37. J. Yu, A.J. Freeman and J.H. Xu, *Phys. Rev. Lett.* **58**, 1028 (1987).

38. T. Fujiwara and Y. Hatsugai, *Japan J. Appl. Phys.* **26**, L716 (1987).

39. F. Herman, R. Kasowski and W. Hsu, *Phys. Rev. B* **36**, 6904 (1987); *Phys. Rev.* **B38**, 204 (1988).

40. P.W. Anderson, *Science* **235**, 1196 (1987); P. Anderson, et al., *Phys. Rev. Lett.* **58**, 2790 (1987).

41. V.J. Emery, *Phys. Rev. Lett.* **58**, 2794 (1987); J. Emery and G. Reiter, *Phys. Rev. B* **38**, 4547 (1988).

42. J.E. Hirsch, *Phys. Rev. Lett.* **59**, 228 (1987).

43. J.R. Schrieffer, X.G. Wen and S.C. Zhang, *Phys. Rev. Lett.* **60**, 944 (1988).

44. Y. Guo, J.M. Langlois and W.A. Goddard III, *Science* **239**, 896 (1988), G. Chen and W.A. Goddard III, *Science* **239**, 899 (1988).

45. A. Aharony, et al., *Phys. Rev. Lett.* **60**, 1330 (1988); R.J. Birgeneau, M.A. Kastner and A. Ahrony, *Z. Phys. B* **71**, 57 (1988).

46. J. Bardeen, L.H. Cooper, J.R. Schrieffer, *Phys. Rev.* **108**, 1175 (1957).

47. R. Birgeneau, et al., *Phys. Rev. Lett.* **59**, 1329 (1987).
48. J. Zaanen, A.T. Paxton, O. Jepsen and O.K. Anderson, *Phys. Rev. Lett.* **60**, 2685 (1988).
49. E.W. Plummer and W. Eberhardt, *Adv. Chem. Phys.* **49**, 533 (1981).
50. T.J. Yeh and I. Lindau, *Atomic Data Nuclear Data Tables* **32**, 49 (1985).
51. S.L. Hulbert, J.P. Stott, F.C. Brown and H. Lien, *Nucl. Instrum. Meth. Phys. Rev.* **208**, 43 (1983).
52. G. Wendin, *Proc. 14 International Conf. on X-ray and Inner-Shell Processes*, J. Physique (France), in press.
53. M.R. Thuler, R.L. Benbow and Z. Hurych, *Phys. Rev. B* **26**, 669 (1982).
54. L.C. Davis, *J. Appl. Phys.* **59**, R25 (1986).
55. W.F. Eglehoff, *Surface Sci. Reviews* **6**, 6-8 (1987).
56. S.L. Hulbert, B.A. Bunker, F.C. Brown and P. Pianetta, *Phys. Rev. B* **30**, 2120 (1984).
57. F.C. Brown, C. Gahwiller and A.B. Kunz, *Solid State Commun.* **9**, 487 (1971).
58. A. Fujimori and F. Minami, *Phys. Rev. B* **29**, 5225 (1984); *Phys. Rev. B* **30**, 957 (1984).
59. G.A. Sawatsky and J.W. Allen, *Phys. Rev. Lett.* **53**, 2339 (1984); see also J. Zaanen, G.A. Sawatzky and J.W. Allen, *Phys. Rev. Lett.* **55**, 418 (1985); see also A. Fujimori and F. Minami, *Phys. Rev. B* **29**, 5225 (1984); **30**, 957 (1984).
60. D.L. Huber, *Solid State Commun.*, **68**, 459 (1988).
61. K.E. Smith and V.E. Henrich, *Phys. Rev.* **B38**, 5965 (1988); R.J. Lad and V.E. Henrich, *Phys. Rev.* **B38**, 10860 (1988).
62. Z.-X. Shen, P.A. Lindberg, I. Lindau, W.E. Spicer, C.B. Eom and T.H. Geballe, *Phys. Rev. B* **38**, 7152 (1988).
63. T. Takahashi, H. Matsuyama, H. Katayama-Yoshida, Y. Okabe, S. Hosoya, K. Seki, H. Fujimoto, M. Sato and H. Inokuchi, *Nature* **334**, 691 (1988).
64. Z.Z. Sheng and A.M. Hermann, *Nature* **332**, 138 (1988).
65. S.S. Parkin, V.Y. Lee, E.M. Engler, A.I. Nazzal, T.C. Huang, G. Gorman, R. Savoy and R. Beyers, *Phys. Rev. Lett.* **60**, 2539 (1988).
66. L.F. Mattheiss, E.M. Gyorgy and D.W Johnson, *Phys. Rev. B* **37**, 3745 (1988).
67. R.J. Cava, et al., *Nature* **332**, 814 (1988).
68. L.F. Mattheiss and D.R. Hermann, *Phys. Rev. Lett.* **60**, 2681 (1988).

UNDERSTANDING THE OXIDE SUPERCONDUCTORS

TRUMAN O. WOODRUFF

Michigan State University
East Lansing, MI 48824
and
Scuola Normale Superiore
I-56100 Pisa

Here you will find descriptions of the various classes of compounds in which bonding between metal atoms and oxygen is a prominent feature and which become superconducting at relatively high temperatures (often greater than $77K$). Compounds of this kind, much investigated since the epochal work of Bednorz and Müller in 1986, are high-temperature or high-transition-temperature superconductors - usually referred to as high-T_c superconductors. It is argued that the bonds referred to, which are affected by doping, give rise to resonances (somewhat narrow absorptions) in the spectral density function associated with the dielectric permittivity *DP* (as defined in one of the usual ways), and that these resonances (excitations) can lead to Cooper-pairing of the electrons and hence superconductivity. The resonances are excitations of many (coupled) small sub-systems, in each of which the distances between the metal and oxygen atoms are strongly sensitive to the state of excitation of the bonding electrons. It is argued further that analysis of such a system is greatly facilitated by including at an early stage consideration of its *DP*. *DP*'s for simple model systems are exhibited, special attention being paid to the way in which the resonances are incorporated in them and how these resonances enter into the determination of T_c as a functional of the *DP*. Finally, it is conjectured that Bose condensation of bipolarons formed form the self-trapped holes which give rise to the resonances - as well as Cooper-pairing of the electrons - can contribute to the superconductivity of these oxides.

This essay addresses the present state and a possible future evolution of our understanding of the several classes of oxide superconductors which have been the subject of the worldwide paroxysm of investigation triggered by the remarkable discovery of Bednorz and Müller published [1] in 1986. It begins with a brief survey of discoveries subsequent to [1] but prior to October 29, 1988 - the day I started writing this article for publication in a volume to honor Professor Franco Bassani on his sixtieth birthday, October 29, 1989. Next I discuss connections between the tensor function describing the response of a particular condensed matter system to an external space- and time-varying vector potential and the properties relating to the superconductivity of the system. I argue that this function of the wave vector $\mathbf{q}$ and the frequency ω of the perturbing charge distribution, usually denoted $\epsilon(\omega,\mathbf{q})$, can be made to serve as the central unifying entity in a complete theory of the superconductivity of all the copper- and bismuth-oxide- containing materials thus far found to be superconductors (often with $T_c>77K$, the approximate temperature at which nitrogen boils in the atmosphere.)

Soon after the publication by Bednorz and Müller [1] of their investigations of a mixed-phase sample containing *La*, *Ba*, *Cu* and *O*, it was established that compounds of the family $La_{2-x}M_xCuO_{4-\delta}$ where *M* may be *Ba*, *Sr* or *Ca*, are superconducting at temperatures as high as $40K$. (For references and a good description of the perovskite-related structure, see [2].) Early in 1987, superconductivity at $93K$ in a mixed phase containing *Y*, *Ba*, *Cu* and *O* was reported [3]. The superconductor in this mixture was quickly found to be $YB_2Cu_3O_{7-\delta}$ [4]. Again, the structure was shown to be related to that of

R. Girlanda et al. (eds.), Progress on Electron Properties of Solids, 319–329.

perovskite. There are in the literature a large number of figures depicting this structure; I found those given by Beno et al. [5] most helpful. The characteristic of both the *La*- and *Y*-containing structures which I believe to be central to understanding their superconductivity is the presence of planes (sometimes puckered) of *Cu* and *O* atoms linked like this:

```
      |          |          |
      0          0          0
      |          |          |
 -   Cu  -  0  - Cu  -  0  - Cu  -
      |          |          |
      0          0          0
      |          |          |
```

Figure 1.

(Each *Cu* linked to 4 0 atoms, each 0 atom between and linking 2 *Cu* atoms.) The state and behaviour of the charge carriers in each of these planes must of course be significantly influenced by the electrostatic fields of the ions lying above and below that plane, but we shall try to obtain a preliminary idea of the essence of the superconductivity by ignoring these effects.

A second wave of discoveries relating to cuprate superconductors originated in the observation by Michel et al. [6] of superconductivity at about 20K in the $Bi/Sr/Cu/0$ system for a compound now known to be $Bi_2Sr_2Cu0_6$. Key steps in this evolution were taken by Maeda et al. [7] who added *Ca* to this just mentioned system and by Sheng and Herman [8] who studied related systems with *Tl* in place of *Bi*. By July of 1988 it was understood that two homologous series of superconducting cuprate compounds with each member containing either *Bi* or *Tl* can be made; the highest T_c observed was 125K (see Parkin et al. [9] and references given there). To describe the two series, Haldar et al. [10] proposed the general formula

$$(B^{III})_y(A^{II})_2Ca_{n-1}Cu_n0_z \qquad \text{(F)}$$

in which $B^{III} = Ti$ or Bi, $A^{II} = Br$ or Sr, and (F)

$$z = (3/2)y + 2n + 1 + \delta,$$

y being 1 for one series and 2 for the other series. All of these compounds form layered crystals related to perovskite. Within each series the superconducting transition temperature (T_c) increases monotonically with n, the number of consecutive copper oxide layers or double layers (for the $y = 2$ case).

To summarize what has been learned from all of the discoveries mentioned: All copper-oxide related superconducting crystals are perovskite-related layered compounds; all of the high-T_c cuprate superconductors contain layers (planes or puckered planes) of *Cu* and 0 atoms arranged as in Figure 1. Between the layers or bilayers of $Cu-0$ planes there are layers containing charged ions; little has been written as yet about how the properties of the $Cu-0$ layers are modified by these intervening planes.

At the same time the experimenters were accomplishing what we have just described, hosts of theorists were trying to fit these discoveries into same coherent theory adequate to explain the properties of the layered cuprates, particularly their superconductivity. To describe this profusion of efforts would take us too far afield - and is unnecessary because of the availability of excellent reviews of the theoretical situation by T.M. Rice [11] and M. Schluter [12]. Note, however, that the $Cu-0$ cluster calculations of M.J. Rice and Wang [13], Guo, Langlois, Chen and Goddard III [14,15] and Stechel and

Jennison [16], which I regard as very significant and to which I shall return later in this essay, are mentioned either not at all or only cursorily, in most cases because the reviews were completed before publication of the theoretical studies in question.

Let us turn now to another stream of investigation which has as its source the discovery made in 1975 by Sleight, Gellson and Bierstedt [17] of another kind of 'high-T_c' superconducting oxide. This work will be shown later to complement the studies of layered copper-oxides. Sleight et al., aware of earlier discoveries that oxides could be superconductors [17a], searched for and found it in the system $BaPb_xBi_{1-x}0_3$, with a maximum T_c (for $x = 0.75$) of $13K$ - which up until 1986 was considered a very high temperature for superconductivity. It appears that interest in the $Pb(Bi)-0$ related systems, rekindled by the wave of advances relative to superconductivity in the cuprates which we have surveyed, led to the discoveries by Mattheiss et al. [18] and by Cava et al. [19] of superconductivity in the $20-30K$ range for $Ba_{1-x}M_xBi0_3$ with $M = K, Rb$ (three-dimensional perovskites). The significance for theory of this development was emphasized by T.M. Rice [20]. Everyone who has read this far will agree on the desirability of a single theory that would clarify the phenomena relating to superconductivity in both the $Cu-0$ and $Pb(Bi)-0$ systems. In what follows I outline some ideas on how to construct such a theory. (What follows can be regarded as a more complete explanation and an extension of my recent Brief Note [21].) These ideas are based on or depart from those developed over the last twenty years by Ginzburg, Kirzhnits, Dolgov, Maksimov and others at the P.N. Lebedev Physics Institute, USSR Academy of Sciences. (Hereafter I refer to this group of scientists as the Lebedev Group.) What encouraged me to believe in the possibility of a valid theory based on special features of the spectrum of electronic excitations of the dielectric medium (excitonic coupling), and hence to consult their extensive body of investigations was the paper by Varma, Schmitt-Rink, and Abrahams [21a].

As recapitulated in the recent book edited by Ginzburg [22], the principles on which the Lebedev Group base their work are the following:

1) An essential feature of the superconducting state (as discovered by Bardeen, Cooper and Schrieffer (BCS) [23]) is Cooper-pairing of the carriers.

2) Cooper-pairing is only possible if and when the interaction between the carriers has certain characteristics (discussed below). Considering for the time being the simplest case, we can say that there must be a net attractive interaction between carriers with oppositely-directed momenta and spins. For the simplest case, the Fourier transform of the (mediated) two-particle interaction $\mathcal{V}(\omega,\mathbf{q})$ can be written in the form $4\pi e^2/q^2\epsilon(\omega,\mathbf{q})$ where (still in the simplest case) ϵ can be identified with or related to the macroscopic permittivity function as defined by

$$\delta V_{\text{ext}}(\omega,\mathbf{q}) = \epsilon_M(\omega,\mathbf{q})\delta V_{\text{tot}}(\omega,\mathbf{q})$$

in which δV_{tot} is the change in the total electrostatic potential resulting from a change δV_{ext} in the external scalar potential (pp. 36-7 of [22]).

3) A BCS-type analysis of the system having $\mathcal{V}(\omega,\mathbf{q})$ for pair interaction establishes that the superconducting transition temperature T_c is a functional of $\epsilon(\omega,\mathbf{q})$:

$$T_c = \mathcal{F}\{\epsilon(\omega,\mathbf{q})\}.$$

Thus the Lebedev Group are led to the conclusion that the existence and characteristics of a superconducting state in any particular material - including those ceramic or crystalline materials just discussed, in which oxygen linked to copper or bismuth (lead) atoms is prominent - can be related to the behavior of the function $\epsilon(\omega,\mathbf{q})$ (hereafter referred to as the Dielectric Permittivity or DP), which is determined by the atomic/electronic structure of the material. Hence an efficient way to learn about the oxide superconductors is to calculate or otherwise devise expressions for the DP of each, and then to perform experiments the results of which are affected by the form of this function.

The rest of this essay will be devoted to:

1) discussing this point of view and its limitations - all of which I feel can eventually be removed or corrected for;
2) describing calculations which lead to forms for the DP function or, failing that, suggest constraints on, or conditions that must be satisfied by, such functions;
3) reporting and analyzing results of experiments that have been performed on the oxides of interest and suggesting other experiments, the results of which, if they were available, would provide information about the form of the DP for one or more of these oxides.

I shall conclude with some evidence and some speculations concerning the essential but unusual features of the electronic structure of these materials which make them good superconductors. These insights have emerged from efforts by others and myself to elucidate the structures of the various oxides.

My goal is to persuade the reader that a good (maybe the best) way to understand superconductivity in a particular system is to understand the (zero-temperature) dielectric permittivity of that system. (At the moment I am assuming the existence of a single well defined *DP*, ignoring subtle difficulties which are mentioned elsewhere in this essay.) One way to proceed is to collect examples of *DP*'s and study their analytic properties. Ginzburg and Kirzhnits (Chapters 1 and 2 of [24]) have made it easy for us to start a collection by assembling expressions for the *DP*'s describing some of the simple model systems which are discussed in first courses on condensed matter physics. They list the following examples:

1) for a medium consisting of noninteracting harmonic oscillators of charge e, mass m eigenfrequency ω_r

$$\epsilon(\omega,q) \equiv \epsilon_r(\omega) = 1 + \frac{\omega_r^2(\epsilon_0 - 1)}{\omega_r^2 - \omega^2},$$

in which ϵ_0 is the *DP* of the medium without the oscillators and

$$\omega_r^2(\epsilon_0 - 1) = 4\pi e^2 n_r / m$$

where n_r is the concentration of oscillators.

2) 'jellium' - a structureless electron-ion plasma. For this model, if $\omega << qvF$, damping neglected,

$$\epsilon(\omega,q) = 1 + \frac{\kappa^2}{q^2} - \frac{\omega_i^2}{\omega^2}$$

with

$$\omega_i^2 = \frac{4\pi e^2 Z^2 n_i}{M} = \frac{4\pi e^2 Zn}{M}$$

(eZ is the ion charge, M its mass, n_i the ion concentration and $\frac{1}{\kappa}$ the screening radius).

$$\kappa^2 = \frac{6\pi e^2 n}{E_F} = \frac{4e^2 m q_F}{4\hbar^2}$$

3) generalized jellium (jellium supplemented by a set of oscillators with eigenfrequency

$\omega_r = \Omega$ (it is assumed that $\Omega >> \omega_i$.)

$$\epsilon(\omega,q) = 1+\frac{\kappa^2}{q^2}-\frac{\omega_i^2}{\omega^2}+\frac{\Omega^2(\epsilon_0-1)}{\Omega^2-\omega^2}.$$

(To include damping replace $\Omega^2-\omega^2$ by $\Omega^2-\omega^2-i\delta$, $\delta>0$.) See also Eq. (9) below.

Obviously, good *ab initio* calculations (including contributions of both valence (conduction) electrons and ion displacements) of total *DP*'s for particular systems would be useful. One can hope that performing such calculations will become more popular among theorists.

In Chapter 2 of [24] Kirzhnits has shown how $\epsilon(\omega,\mathbf{q})$ can be incorporated in the Hamiltonian to be associated with a system of condensed atoms which can exhibit superconductivity and how to derive from the Hamiltonian the equations determining the critical temperature T_c and the energy gap Δ. He bases the dynamical description of the electron subsystem on the electron-electron interaction Hamiltonian

$$H' = \frac{1}{2}\int d^4xd^4x'\psi_\alpha^\dagger(x)\psi_\beta^\dagger(x')V(x-x')\psi_\beta(x')\psi_\alpha(x),$$

in which $V(x)$ is the Fourier transform of $\mathcal{V}(\omega,\mathbf{q}) = 4\pi e^2/q^2\epsilon(\omega,\mathbf{q})$ having as a result of the frequency dispersion of the permittivity a retarded character ($x=\mathbf{x},t,d^4x=d\mathbf{x}dt$).

Included in $\mathcal{V}(x)$ could be interactions between electrons resulting from the mutual coupling of lattice disortions which might 'clothe' the electrons. A limitation of this work by Kirzhnits is that the interaction between electrons is taken to be spin-independent. (It should be noted, however, that the superconducting transition may be well described if $V(x-x')$ gives correctly the interaction between electrons with oppositely directed spins.)

The goal of the Kirzhnits analysis is to determine T_c of a homogeneous and isotropic superconductor with an interaction between electrons arbitrary in form but weak. Weak here means $\alpha<1$, where

$$\alpha \equiv me^2/\pi p_F$$

which is equivalent to $r_s<1$, where r_s is defined by

$$\frac{1}{n} = \frac{4\pi}{3}(a_B r_s^3)$$

n being the electron density and a_B the first Bohr radius, m the mass of the electron and p_F the momentum of an electron at the Fermi surface. The results can be summarized as follows

$$T_c = \frac{\xi}{1.45}\exp\left(-\frac{1}{\lambda-\mu^*}\right) \tag{1}$$

with

$$\mu^* = \frac{\mu}{1+\mu\ln(E_F/\xi)} \tag{2}$$

$$\mu = \frac{\alpha}{2}\ln\frac{1}{\alpha} \tag{3}$$

$$\lambda = \mu[1-\frac{1}{\epsilon(0,\bar{q})}] \tag{4}$$

E_F is the Fermi surface energy. For more precise definition and discussion of $\bar{q}$ and ξ, I refer you to pp. 76-77, especially Eq. 2.66, of [24]; for our purposes it is enough to know that

$$\kappa \ll \bar{q} \ll q_F, \quad (q_F \equiv p_F/\hbar) \tag{5}$$

where κ is the inverse Debye-screening radius which at zero temperature in the weak coupling case is given by

$$\kappa = 2(p_F m e^2/\pi)^{1/2}$$

or

$$\kappa^{-1} = \frac{1}{2}\alpha^{1/2} q_F = \alpha^{1/2} p_F/2\hbar \tag{6}$$

and ξ is called the mean logarithmic frequency because it is determined by averaging the logarithm of the frequency over the real spectrum of the system:

$$\ln\xi = \frac{\int_0^\infty \frac{dE}{E} \ln E\sigma(E,\bar{q})}{\int_0^\infty \frac{dE}{E}\sigma(E,\bar{q})}. \tag{7}$$

Dolgov and Maksimov [25] discuss in detail the point that (as first noted by Cohen and Anderson [26]), some stringent limitations on the maximum transition temperature T_c are implied by the expressions just given for T_c, λ and μ together with the relation [25]

$$\mu - \lambda = N(0) < \frac{4\pi e^2}{q^2}\epsilon^{-1}(0,q)> \tag{8}$$

in which $N(0)$ is the density of electronic states at the Fermi surface and the angled brackets indicate an average over the momentum q. They conclude that a necessary condition for the existence of high T_c is

$$\lambda > \mu$$

which, according to (8) can only be true if $\epsilon^{-1}(0,q)$ is negative. These considerations make clear why the question of admissible signs of the *DF* is of such great importance and has stimulated so much [22-28] theoretical work. (Note in fact, that the title of [28], a fairly lengthy work, is 'On an admissible sign of the static dielectric function'.) The conclusions reached are summarized in Section 6 of [25] and Section 2.5 of [22]. Among these conclusions note the following:

1) A necessary condition for the effectiveness of a superconductivity mechanism, i.e., for the existence of values of T_c greater than about $1K$, is that the total (electron plus ion) static dielectric function $\epsilon_{tot}(0,q)$ be negative for some values of q which are large ($\sim q_F$).

2) That this condition is satisfied does not contradict any general requirements. In fact, it is not unusual for this condition to be satisfied; it is satisfied in many real systems.

3) The condition that *DF* be negative can be satisfied both by the total *DF* including any ionic contribution as well as by the electronic component of the *DF*.

4) For nonphonon superconductivity mechanisms to be effective, the electronic part of the *DF* must take on negative values.

I hope that my presentation of the Kirzhnits formula for T_c (valid asymptotically for $\alpha \to 0$ (weak coupling) and of the examples of *DP*'s for simple models of condensed systems will make it a little easier for the reader to profit from the final part of Chapter 2 - the Kirzhnits chapter - in [24], which is aimed at discovering what factors raise T_c. His most interesting conclusions arise from applying his formula (remember their limitations: weak coupling! etc.) to the generalized jellium model for which

$$\epsilon(\omega,\mathbf{q}) = 1-\frac{\omega_c^2}{\omega^2-\omega_c^2 q^2/\kappa^2+i\delta}-\frac{\omega_i^2}{\omega^2-\omega_0^2(\mathbf{q})+i\delta}. \tag{9}$$

It is assumed that the ions are bound and have (bare) frequency spectrum $\omega_0(\mathbf{q})$, $\omega_i^2 = 4\pi Ze\rho_i/M$ with ρ_i being the density of ions of charge Ze and mass M; $\omega_c^2 = 4\pi e\rho_c/m$ with ρ_c being the density of conduction electrons of mass m. κ is defined in Eq. (6) above. He finds

1) that the decisive influence on the critical temperature is exerted by the value of the static permittivity $\epsilon(0,\overline{q})$ of the superconducting material at large wavevector value $\overline{q}$ lying between the Debye wavevector κ and the Fermi wave vector values q_F. The smaller the value of $1/\epsilon(0,\overline{q})$ is, the higher the critical temperature is. So, for high T_c, a positive permittivity $\epsilon(0,q)$ should have the largest possible value; it is even better when $\epsilon(0,\overline{q})$ is negative. These requirements correspond to a material which has the largest possible cross section for scattering fast electrons through large angles; they follow immediately from Eqs. (1), (2) and (4).

2) The critical temperature depends also on the system's excitation spectrum which is determined by the equation $\epsilon(\omega,\mathbf{q}) = 0$. The spectrum affects the mean logarithmic frequency via Eq. (7).

3) Damping of the excitations reduces the critical temperature, as does the presence of many vibration modes. Highest T_c is associated with the case when the entire spectrum of the excitations reduces to a single narrowest possible line located at energy ξ and has the strongest possible interaction with the conduction electrons.

The goal of my Brief Note [21] was to begin the work of discovering the correct expression for the *DP* associated with a particular compound of the type $La_{2-x}Sr_xCuO_4$. My thought was to proceed by successive approximations, starting with one of the simple model forms described above and altering it as needed to make it accord with available experimental data. Repeating this mostly empirical fitting procedure for a) several different compounds in which $Cu-0$ layers are a main feature; b) several compounds of the $3D\ Pb(Bi)-0$ class, and comparing the prominent features of each and how they enter into the formula for T_c should enable us to see what special features of these compounds lead to their high T_c's. I shall not try to conceal from the reader my suspicion that what is special about the $Cu-0$ and $Pb(Bi)-0$ superconductors is the structure of the low lying dipole-allowed optical excitations. For both classes of oxides I believe one will find reasonably sharp pairs of absorptions, one member of the pair in the range of a few hundred cm^{-1}, the other in the range of a few thousand cm^{-1}. This pairing is the result of coupling of electronic and atomic-displacement degrees of freedom in each of many small regions of the crystal, each such region being the neighborhood of a self-trapped hole. The small regions are weakly coupled through the lattice degrees of freedom.

I hasten to admit that the details of the picture I have outlined came to me as a result of pondering the work of Rice and Wang [13] which provides a theoretical model for the interpretation of infrared absorption, a model leading to resonance-like expressions for the spectral density which I suggested in [21] should be used to determine the parameters of the localized oscillators to be included in the generalized jellium model to be fitted to the $La-Sr-Cu-0$ compound. (Note that a better expression for the *DP* would be the one given in Eq. (2.15) of [22] and quoted below.) The key result of [13] - referred to there as Eq. 4 - needed in our construction of $\epsilon(\omega,q)$ is

$$\sigma(\omega) = -in'\omega X_0(\omega)/[1-\lambda D_0(\omega)\tilde{X}_0(\omega)];$$

$n' = n_{P^+}$ is the number of trapped holes P^+ per unit volume. It would be interesting to compare this spectral density function with other forms that could no doubt be extracted from their calculations for small clusters by Goddard III et al. [14,15] and by Stechel and Jennison [16].

Readers who have come this far with me should now be convinced that a satisfactory frame of reference for questions and answers concerning the properties of the high-T_c superconductors is or will be provided by the theoretical structure which can be built up around the central concept of the dielectric permittivity (*DP*) of a system. (Perhaps the greatest virtue of an analysis relating the superconducting transition to a *DP* function is that it establishes a framework within which it is possible to take proper account of the fact that the interaction between the carriers of a pair is determined by the very complicated *medium* made up of all the other electrons and atom-cores of which the system is composed. Insight is gained better by devising and testing models of a *DP* describing such a medium than by trying to elucidate the interactions between the two particles of a pair starting from 1-particle calculations and adding effects of all possible intermediate as well as direct interactions. An effort of the latter kind which merits attention, though it does not do justice to the complexity stemming from the medium-like quality ('mediation?') of the interactions between Cooper-pair partners is [30].) Parts of this theoretical structure are in place but there are many gaps to be filled and blemishes to be removed. Even in its present unsatisfactory state, however *DP* theory offers some insights and suggests some conjectures concerning the unusual characteristics of the high T_c oxides to which I now turn.

1) The Kirzhnits (*K*) analysis (recapitulated above) of factors affecting T_c may have wider validity than has generally been appreciated; in particular the Coulombic coupling between electrons, as measured by the dimensionless number r_s and $\alpha = [(4/9\pi^4)^{1/3} r_s]$ may be sufficiently weak because the electron density n that should be entered in the formula may be larger than we have used in the past. Perhaps for the clusters around a trapped hole, n should include all of the valence (including *Cu d*-) electrons.

2) If, as I believe, the *K* weak coupling analysis is relevant it teaches us: a) that 'resonances' (narrow strong absorptions associated with electronic absorptions) in the spectral density $\sigma(\omega, \mathbf{q})$ can have a decisive effect in raising T_c; b) coupling of localized electronic excitations (of the kind modelled by Rice and Wang [13]) with 'lattice' (whole atom) displacements has an important influence on potentiating the 'resonances'. c) for both of the grand classes of oxide superconductors - those based on $Cu-0$ and those based on $Pb(Bi)-0$ - there is experimental evidence (some of it, admittedly, highly contested) of strong 'resonances' ($\sim 0.5\,eV$ in the $Cu-0$ layer, $\sim 1.0\,eV$ in the $Pb(Bi)-0$ skeleton). d) Electronic excitations of the kind that can be associated with these resonances will modify the coordinates of nearby atoms: i.e., there will be excitation-lattice coupling. For the $Pb(Bi)-0$ systems this coupling is reflected in the tendency for bipolarons to form [20,20a]. Recapitulating: Superconductivity of both the Cooper-pairing and Bose-condensation types [29,29a] can occur in the oxide systems surveyed here. What is characteristic of all these systems is the existence of bonds between the metal atoms $M(Cu, Bi$ or $Pb)$ and oxygen which can capture a hole when the system is doped by the addition of an electron-remover. The doped system with holes localized in a fraction of the $M-0$ bonds can be described as having mixed- or fluctuating- (as holes hop from one bond to a nearby one) valence [20a,b]. The self-trapped-hole centers (P^+ in the language of Rice and Wang [13]) possess excitations (in the $1\,eV$ range) having large oscillator-strength, which can lead to Cooper-pairing of the $0(2p)$ and $Cu(3d)$ electrons (polarization- or exciton-induced Cooper-pairing). On the other hand, the self-trapped holes in both $Cu-0$ and $Pb(Bi)-0$ compounds can be coupled by the small-polaron, space-localized distortion closely surrounding each of them to form bipolarons - which can lead to superconductivity by the Bose-

condensation route [29].

3) For a single $Cu-0$ layer and hence also in the multi-layer systems described by Eq. (F), T_c may be enhanced by $2D$ effects (see the reviews [11] and [12]) - which could help to explain why T_c's in the $Cu-0$ layer systems tend to be higher than in the $3D$ $Pb(Bi)-0$ systems.

4) The great weakness of the K analysis as it stands is its omission of spin dependencies. There is much evidence (some of it summarized in [31]) that spin interactions are important in determining the properties of the $Cu-0$ superconductors (which, however, is not true for the $Pb(Bi)-0$ superconductors). To improve the K analysis in this respect one can
a) include the spin operator in the basic Hamiltonian (equivalent to replacing the Schroedinger equation by an electron-spin enhanced modification often called the Schroedinger-Pauli equation); or
b) redefine the DP by letting the driving charge wave which disturbs the system be a wave of polarized electrons. Just as inelastic scattering of unpolarized electrons yields information about the DP as previously defined, scattering of polarized electrons [32] could be used to determine the form of a redefined spin-dependent DP. Accompanying these changes must be reconsideration and possibly revision of the relevant sum rules.

The goal of this essay has been to persuade you, the reader, that understanding superconductivity in general and the superconducting oxides in particular can be reduced to understanding the dielectric permittivity DP for the relevant system for some convenient temperature T. It is true that there remain significant obstacles to obtaining mastery over DP, including
1) Confirming and extending the validity of the Kirzhnits analysis given in [24] and finding how it is modified for almost 2-dimensional layers as in the $Cu-0$ materials;
2) Discovering how to calculate the difference between the DP (ϵ_M) determined from a linear-response or Kubo-type analysis of response of the system to an external vector potential, and the DP which is the Fourier transform of the $V(x)$ entering H' (written above);
3) Developing reliable means for deciding which electrons to include in the 'sea' which is analyzed à la Kirzhnits. There is hope that for the right subset, n is sufficiently large that $\alpha \ll 1$, so that the electrons can be treated as weakly coupled.

If eventually there is confirmation of what is conjectured above, viz., that the oxides which have been the main focus of investigation since 1986 can become superconducting via either or a combination of the two mechanisms: Cooper-pairing induced by (mainly) electronic excitations of self-trapped holes, or Bose-condensation of bipolarons formed from these holes, it will seem less strange that understanding the oxide superconductors has required so much effort.

Acknowledgements

This essay is dedicated to my dear friend, Franco Bassani, without whose stimulating example and encouragement it would never have been born. I hope he will forgive its imperfections.

References

1. J.G. Bednorz and K.A. Müller, *Z. Phys. B* **64**, 189 (1986).
2. R.J. Cava, A. Santoro, D.W. Johnson, Jr. and W.W. Rhodes, *Phys. Rev. B* **35**, 6716 (1987).
3. M.K. Wu, C.J. Ashburn, Jr., P.H. Torng, R. Hor, L. Meng, L. Gaol, Z.J. Huang, Y.Q. Wang and C.W. Chu, *Phys. Rev. Lett.* **58**, 908 (1987).
4. R.J. Cava, B. Batlogg, R.B. VanDover, D.W. Murphy, S. Sunshine, T. Siegrist, J.P. Remeika, E.A. Rietman, S.M. Zahurak and G.P. Espinosa, *Phys. Rev. Lett.* **58**, 1676 (1987).
5. M.A. Beno, D.W. Soderholm, D.W. Capone II, D.G. Hinks, J.D. Jorgensen, J.D. Grace, I.K. Schuller, C.U. Segre and Z. Zhang, *Appl. Phys. Lett.* **51**, 57 (1987).
6. C. Michel, M. Hervieu, M.M. Borel, A. Granding, F. Deslandes, J. Provost and B. Raveau, *Z. Phys. B* **68**, 421 (1987).
7. M. Maeda, Y. Tanaka, M. Fukutomi and T. Asano, *J. Appl. Phys. Pt 2* **27**, L209 (1988).
8. Z.Z. Sheng and A.M. Hermann, *Nature* **332**, 55 (1988).
9. S.S.P. Parkin, V.Y. Lee, A.I. Nazzal, R. Savoy, R. Beyers and LaPlaca, *Phys. Rev. Lett.* **61**, 750 (1988).
10. P. Haldar, K. Chen, B. Maheswaran, A. Roig-Janicki, N.K. Jaggi, R. S. Markiewicz and B.C. Giessen, *Science* **241**, 1198 (1988).
11. T.M. Rice, *Z. Phys. B* **67**, 141 (1987).
12. M. Schluter, *Int'l. J. Mod. Phys. B* **2**, 167 (1988).
13. M.J. Rice and Y.R. Wang, *Phys. Rev. B* **36**, 8794 (1987).
14. Y. Guo, J.M. Langlois and W.A. Goddard III, *Science* **239**, 896 (1988).
15. G. Chen and W.A. Goddard III, *Science* **239**, 899 (1988).
16. E.B. Stechel and D.R. Jennison, *Phys. Rev. B* **38**, 4632 (1988).
17. A.W. Sleight, J.J. Gillson and P.E. Bierstedt, *Solid State Commun.* **17**, 27 (1975).

17a. For an excellent recent survey of the chemistry of high-temperature superconductors, including a more complete history than given here of work on oxide superconductors, see A.W. Sleight, *Science* **242**, 1519 (1988).

18. L.F. Mattheiss, E.M. Gyorgy and D.W. Johnson, Jr., *Phys. Rev. B* **37**, 3745 (1988).
19. R.J. Cava, B. Batlogg, J.J. Krajewski, R. Farrow, L.W. Rupp, Jr., A.E. White, K. Short, W.F. Peck and T. Kometani, *Nature* **332**, 814 (1988).
20. T.M. Rice, *Nature* **322**, 780 (1988).

20a. L.J. DeJongh, *Physica C* **152**, 171 (1988).

20b. F. Parmigiani and G. Samoggia, *Europhys. Lett.* **7**, 543 (1988).

21. T.O. Woodruff, *Nuovo Cimento D* **10**, 1001 (1988).

21a. C.M. Varma, S. Schmitt-Rink and E. Abrahams, *Solid State Comm.* **62**, 681 (1987).

22. V.L. Ginzburg, (ed.), Superconductivity, Superdiamagnetism, and Superfluidity, MIR Publishers, Moscow, 1987.
23. J. Bardeen, L.N. Cooper and J.R. Schrieffer, *Phys. Rev.* **108**, 1175 (1957).
24. V.L. Ginzburg and D.A. Kirzhnits, (eds), High Temperature Superconductivity, Consultants Bureau, 1982; Edition in Russian published in 1977.
25. O.V. Dolgov and E.G. Maksimov, *Sov. Phys. Usp.* **25**, 688 (1982).
26. M.L. Cohen and P.W. Anderson in: AIP Conference on *d*- and *f*-Band Superconductivity, AIP Conf. Proc., New York, **17** (1972).
27. D.A. Kirzhnits, *Sov. Phys. Usp.* **19**, 530 (1976).

28. O.V. Dolgov, D.A. Kirzhnits and E.G. Maksimov, *Rev. Mod. Phys.* **53**, 81 (1981).
29. M.J. Rice and Y.R. Wang, *Phys. Rev. B* **37**, 5893 (1988).
29a. P. Nozières and S. Schmitt-Rink, *J. Low Temp. Phys.* **59**, 195 (1985).
30. M.D. Nuñez Regueiro and A.A. Aligia, *Phys. Rev. Lett.* **61** 1889 (1988).
31. V.J. Emery, *Nature* **328**, 756 (1987); *ibid* **333**, 14 (1988).
32. R.J. Celotta and D.T. Pierce, *Science* **234**, 333 (1986).

OPTICAL PROPERTIES OF SEMICONDUCTOR SURFACES

G. CHIAROTTI AND P. CHIARADIA

Department of Physics
University of Rome 'Tor Vergata'
Rome, Italy

1. Introduction

It is well known that surface states are special solutions of the Schrödinger equation for electrons in a solid, made possible by the breaking of translational symmetry at the surface [1]. This was first demonstrated by Tamm using a one-dimensional Kronig-Penney model [2]. Within this simple model surface states are characterized by an imaginary value of the momentum k (meaning exponential decay away from the surface) and energy in the forbidden gap. In real cases the energy position of surface states critically depends both upon the atomic rearrangement at the surface (reconstruction and/or relaxation) and the presence of adsorbates. Moreover in three dimensions surface states (which are regular Bloch states in the plane of the surface) can be either in a gap of the bulk band structure projected onto the surface [3] or degenerate with bulk states. In the latter case they are called surface resonances. From an intuitive point of view, surface states might be thought as associated to the dangling bonds occurring at the surface. Generally speaking, surface states are important in all interactions of the solid with the ambient (chemisorption and catalysis are significant examples). Furthermore, in semiconductors they usually control the surface position of the Fermi level and therefore play a major role in device physics.

A manifold of surface-sensitive techniques have been developed in the last decades to study surface states, especially in semiconductors. In this article the main results obtained with optical methods applied to semiconductor surface physics will be shortly reviewed.

2. Surface Differential Reflectance

Let us define the surface dielectric function (*SDF*) as the *DF* of the transition region near the surface where the optical constants are different both from the ambient and bulk values. This region is typically $5-10$ Å thick. It should be stressed that conceptually the *SDF* differs from the bulk *DF* for two reasons. The first is the peculiar electronic structure characteristic of the surface (surface states). The second is the fact that in general bulk-like optical transitions themselves are modified near the surface, for instance due to lack of k conservation in the direction perpendicular to the surface and to different periodicity along the surface.

Reflection of light from a crystal surface clearly depends on both the bulk and surface dielectric functions. The main task of surface optical spectroscopy is to separate the surface term from the bulk term, which is usually much larger. In Surface Differential Reflectance (*SDR*) this is accomplished by changing the surface conditions in a proper way, usually by saturating the dangling bonds of an atomically clean surface with oxygen. In this way the surface related term is changed while the bulk term is not affected. Under simplifying hypotheses, it can be demonstrated that the fractional change $\Delta R / R$ is

R. Girlanda et al. (eds.), Progress on Electron Properties of Solids, 331–336.

related to the surface contribution to reflectance. In fact the analysis developed by McIntyre and Aspnes [4] of the classical problem of light reflection in a three-media system [5] shows that $\Delta R/R$ at normal incidence is given by

$$\frac{\Delta R}{R} \cong -\mathrm{Im}\left[\frac{\hat{\epsilon}_1-\hat{\epsilon}_2}{\hat{\epsilon}_1-\hat{\epsilon}_3}\right] \tag{2.1}$$

where $\hat{\epsilon}_j = \epsilon_j' + i\epsilon_j''$ ($j=1,2,3$) is the dielectric function of the ambient, surface and substrate, respectively. Being d is the thickness of the surface layer, for the usual case of vacuum as first medium and non-absorbing substrate the previous equation becomes

$$\frac{\Delta R}{R} \cong \frac{4n_1}{n_1^2-n_3^2}\frac{\omega}{c}(\epsilon'' d) \tag{2.2}$$

showing that $\Delta R/R$ is simply proportional to the imaginary part of the *SDF*. The general case is fairly complicated since $\Delta R/R$ depends on the real and imaginary parts of the *DF* of surface, substrate and oxide layer. A detailed analysis must be carried out case by case [6]. However, it should be pointed out that in most cases the last two contributions are structureless in the energy range of interest, for example in *Si* below 3.5 *eV*.

The above macroscopic treatment of light reflection from a surface [4,5] is very crude [7]. For instance the exact definition of some parameters, like the *SDF* itself, is by no means obvious. However, more recent microscopic theories [8,9,10,11] yield essentially the same results as the Drude-McIntyre-Aspnes model. In particular, the phenomenological parameters appearing in the macroscopic theory are nothing but spatial averages (with respect to the surface normal) of the quantities related to the surface band structure which in principle can be calculated.

Experimentally, *SDR* spectra are currently measured with two distinct configurations: multiple internal (total) reflection [12,13] and external reflection [6] usually at normal incidence. The advantage of the former method is the amplification of the small surface signal, by the number of internal reflections. On the other hand internal reflection is intrinsically limited to the region of transparency of the substrate and does not allow to fully exploit light polarization [6]. Both these shortcomings are overcome in the external mode [6]. Historically clean cleaved surfaces of semiconductors have been studied first with multiple internal reflection [12,14] and later on with external reflection [6,15].

3. Results

In the following some recent results obtained with *SDR* on the cleavage faces of *Si* and *GaAs* [*Si*(111)2×1 and *GaAs*(110), respectively] will be discussed.

Figure 1 shows the *SDR* spectrum of the *Si*(111)2×1 surface in the near infrared, below the absorption threshold of the substrate. The peak at 0.45 *eV* has been interpreted as a surface state to surface state optical transition [12], proving that an optical gap exists between full and empty surface states. The gap originates from inequivalence of the two atoms in the 2×1 surface unit cell due to atomic reconstruction, the ideal unreconstructed surface being metallic. The *Ge*(111)2×1 surface behaves in a similar way [6,12] while in *GaAs*(110), instead, the energy separation of surface states is larger than the fundamental gap [16]. This is easily understood by keeping in mind that dangling bonds associated to *Ga* and *As* are already separated in the ideally terminated surface and atomic relaxation further increases the gap between them.

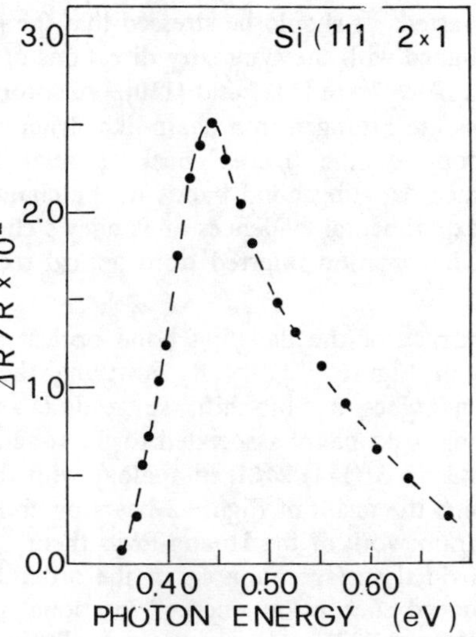

Figure 1. Surface Differential Reflectance spectrum of $Si(111)2\times1$ below the fundamental gap, at room temperature, with unpolarized light.

In the case of the 2×1 reconstruction of the (111) face of group IV semiconductors, three possible domains are allowed by symmetry. However, single domain surfaces can be obtained by suitable cleavage techniques. Figure 2 shows the polarization dependence of the *SDR* spectrum of a single domain $Si(111)$ 2×1 surface, in the visible and near infrared energy range. A sketch of the Low Energy Electron Diffraction pattern is reported in the inset.

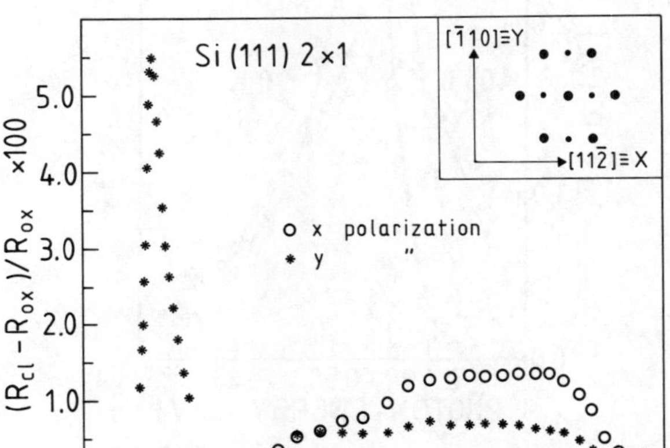

Figure 2. Surface Differential Reflectance for a single domain $Si(111)2\times1$ surface with polarized light, in the visible and near infrared energy range, at room temperature. The inset shows a sketch of the *LEED* pattern, with the two main directions $[11\bar{2}]$ and $[\bar{1}10]$ labelled x and y respectively.

The pronounced peak at 0.45 eV is completely anisotropic, while above the gap the anisotropy is incomplete and inverted. It should be stressed that the principal axes of the surface dielectric tensor are aligned with the symmetry directions of the 2×1 reconstruction, that in the case shown in Figure 2 are [112] and [110]. According to Pandey's model of Si(111) 2×1, surface atoms are arranged in a chain-like structure [17] along [$\bar{1}$10] directions. The strong anisotropy of the 0.45 eV peak is related to the (almost) one-dimensional nature of surface dangling bond bands in the chain model. Indeed this has been one of the soundest experimental evidences of Pandey's chain model and a remarkable example of structural information inferred from optical techniques using symmetry arguments [18].

The temperature dependence of the dangling bond optical absorption at 0.45 eV in Si(111) 2×1 is reported in Figure 3 [19]. By lowering the temperature, the peak becomes narrower and undergoes a blue-shift, suggesting a strong electron-phonon interaction [19]. This finding is probably associated to the (one-dimensional) localization of the surface dangling bonds in Si(111) 2×1, in analogy with the behaviour of localized centers in solids [20] to which the result of Figure 3 bears much resemblance. An analysis of the above result in the framework of the Huang-Rhys theory for localized centers [21] yields a satisfactory fit provided surface phonons of the order of 10 meV are excited in the optical process [19]. Indeed such a low energy vibrational mode has been observed recently with He-atom scattering [22] and also theoretically interpreted as an acoustic mode arising from the quasi one-dimensional nature of the chain model [23]. In the framework of the chain model, however, the distinction between optical and acoustic modes at the zone boundary is probably irrelevant.

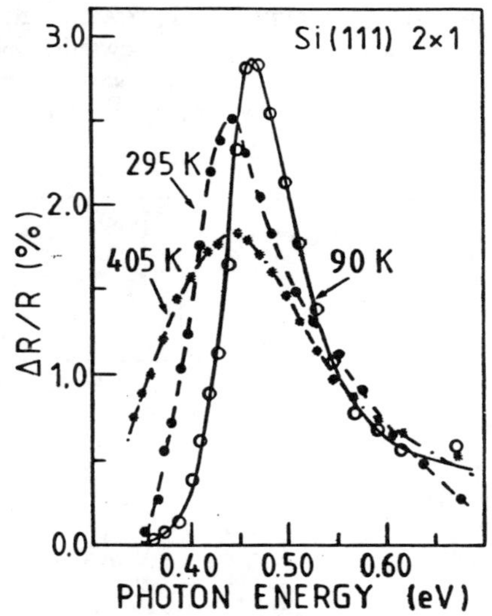

Figure 3. Temperature-dependence of the Surface Differential Reflectance spectrum of Si(111)2×1 in the near infrared energy range, with unpolarized light.

In $GaAs$(110), as well as in the cleavage faces of other III-V compounds like GaP and InP, surface contributions to reflectance have been detected only at energies larger than the gap [16,24] as already mentioned. Figure 4 shows the SDR spectrum of $GaAs$(110) in the visible range, for complete and partial oxidations. The small oscillatory behaviour which can be seen in both curves is due to Franz-Keldysh effect of the (bulk-like) optical transitions at about $3\,eV$. The *FK* effect is brought about by the variation of the surface electric field associated to the change of band bending induced by oxidation [25]. The curve corresponding to partial oxidation allows one to subtract the *FK* effect from the broad curve measured after full oxidation, which is essentially due to surface state. Unlike $Si(111)\ 2\times 1$, in the case of *GaAs* in the energy range of Figure 4 both the real and imaginary parts of the surface dielectric functions contribute to the *SDR* spectrum. As a consequence, the spectral shape of surface state optical absorption is somewhat different from the *SDR* spectrum [6]. A peak in the imaginary part of the surface dielectric function of *GaAs*(110) is found at $2.9\,eV$ [6], in agreement with theoretical calculations [26].

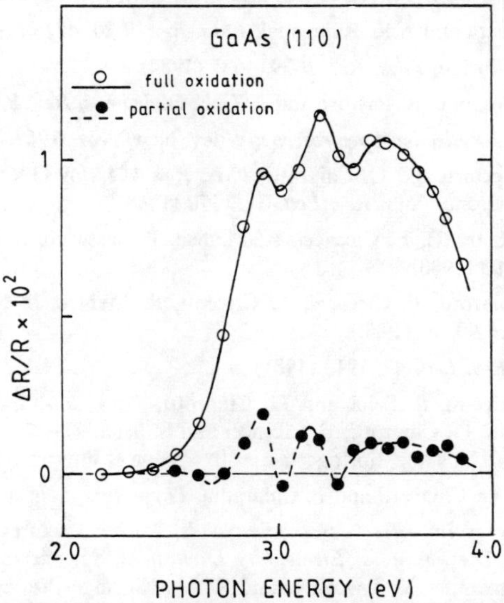

Figure 4. Surface Differential Reflectance spectrum of the *GaAs*(110) cleaved surface, from 2 to $4\,eV$, with unpolarized light, at room temperature. Upper and lower curves correspond to full and partial oxidations, of the order of 10^6 and 10^4 Langmuir respectively.

Acknowledgements

The authors wish to mention that the results described in the present article have been obtained over a period of two decades, with the contribution of many people. The complete list of coworkers would be too long. The collaborations of S. Nannarone, S. Selci and F. Ciccacci are especially acknowledged.

References

1. A. Many, Y. Goldstein and N.B. Grover, *Semiconductor Surfaces*, North-Holland, Amsterdam 1965.
2. I. Tamm, *Z. Physik* **76**, 849 (1932); *Phys. Zeit. Sov.* (in German) **1**, 732 (1932).
3. M. Schlüter, *Festkörperprobleme*, XVIII (1976) p. 155.

4. J.D.E. McIntyre and D.E. Aspnes, *Surf. Sci.* **24**, 417 (1971).
5. P. Drude, *The Theory of Optics*, Longmans, Green and Co., London, 1920, and references therein.
6. S. Selci, F. Ciccacci, G. Chiarotti, P. Chiaradia and A. Cricenti, *J. Vac. Sci. Techn. A* **5**, 327 (1987).
7. P. Chiaradia, *Semiconductor Interfaces: Formation and Properties*, ed. by G. Lelay, J. Derrien and N. Boccara, Springer Proc. in Physics 22, Springer-Verlag, 1987, p. 290.
8. M. Nakayama, *J. Phys. Soc. Japan* **39**, 265 (1975).
9. P.J. Feibelman, *Phys. Rev. B* **14**, 762 (1976) and *Prog. Surf. Sci.* **12**, 287 (1982).
10. A. Bagchi, R.G. Barrera and A.K. Rajagopal, *Phys. Rev. B* **20**, 4824 (1979).
11. R. Del Sole and E. Fiorino, *Phys. Rev. B* **29**, 4631 (1984).
12. G. Chiarotti, S. Nannarone, R. Pastore and P. Chiaradia, *Phys. Rev. B* **4**, 3398 (1971).
13. N.J. Harrick, *Internal Reflection Spectroscopy*, Wiley, New York, 1967.
14. G. Samoggia, A. Nucciotti and G. Chiarotti, *Phys. Rev.* **144**, 749 (1966); G. Chiarotti, G. Del Signore and S. Nannarone, *Phys. Rev. Lett.* **21**, 1170 (1968).
15. S. Nannarone, P. Chiaradia, F. Ciccacci, R. Memeo, P. Sassaroli, S. Selci and G. Chiarotti, *Sol. St. Comm.* **33**, 593 (1980).
16. P. Chiaradia, G. Chiarotti, F. Ciccacci, A. Cricenti, R. Memeo, S. Nannarone, P. Sassaroli and S. Selci, *Surf. Sci.* **99**, 70 (1980).
17. K.C. Pandey, *Phys. Rev. Lett.* **47**, 1913 (1981).
18. P. Chiaradia, A. Cricenti, S. Selci and G. Chiarotti, *Phys. Rev. Lett.* **52**, 1145 (1984); P. Chiaradia, A. Cricenti, G. Chiarotti, F. Ciccacci and S. Selci, *The Structure of Solids I*, ed. by M.A. Van Hove and S.Y. Tong, Springer-Series in Chemical Physics, 1985, p. 66.
19. F. Ciccacci, S. Selci, G. Chiarotti and P. Chiaradia, *Phys. Rev. Lett.* **56**, 2411 (1986).
20. G. Chiarotti, *Theory of Imperfect Crystalline Solids*, Trieste Lectures, 1970 (IAEA, Vienna, 1971); *Y. Toyozawa, The Physics of Elementary Excitations*, Springer Series in Solid State Science **12**, ed. by S. Nakajima, Y. Toyozawa and R. Abe, Springer-Verlag, Berlin, 1980, Chapter 7.
21. K. Huang and A. Rhys, *Proc. Royal Soc., London A* **204**, 403 (1950).
22. U. Harten, J.P. Toennies and C. Woell, *Phys. Rev. Lett.* **57**, 2947 (1986).
23. O.L. Alerhand and E.J. Mele, *Phys. Rev. Lett.* **59**, 657 (1987).
24. A. Cricenti, S. Selci, A.C. Felici, C. Goletti and G. Chiarotti, *Surf. Sci.* **211**, 552 (1989).
25. F. Ciccacci, S. Selci, G. Chiarotti, P. Chiaradia and A. Cricenti, *Surf. Sci.* **168**, 28 (1985).
26. D.J. Chadi, *Phys. Rev. B* **18**, 1800 (1978), F. Manghi, E. Molinari, R. Del Sole and A. Selloni, *Surf. Sci.* **189/190**, 1028 (1987).

PLASMONS OF A SINGLE GRAPHITE LAYER FLOATING ON A METAL SURFACE

G. CAMPAGNOLI

Dipartimento di Fisica 'A. Volta'
Università di Pavia
Pavia, Italy

E. TOSATTI

International School for Advanced Studies
Trieste, Italy
and
International Centre for Theoretical Physics
Trieste, Italy

We study the collective electronic excitations and the corresponding electron energy-loss spectrum of a single graphite layer, such as one finds floating on $Ni(111)$ and $Ru(001)$ *graphitic* surfaces. The spectrum is compared to that of three-dimensional ($3D$) graphite; similarities and differences are pointed out. Direct comparison of our single layer results with Electron Energy Loss Spectroscopy (*EELS*) data on $Ni(111)$ is very favorable, and demonstrates how in this geometry an extra peak of purely kinematic origin may arise when $\omega = \mathbf{q} \cdot \mathbf{v}$.

1. Introduction

The study of surfaces provides many examples of physical systems which are almost exactly two-dimensional ($2D$), particularly in the area of phase transitions [1]. It seems rarely, however, to provide systems which are $2D$ in their electronic excitations, especially the collective plasmon excitations.

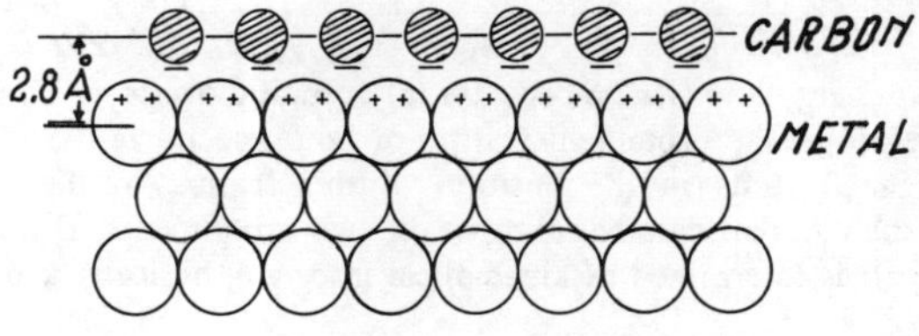

Figure 1. Sketch of a graphitic overlayer on a (111)*fcc* transition metal surface. The spacing shown applies for $C/Ni(111)$.

In this paper we suggest that examples of such systems are, to a good accuracy, the carbon monolayer that is found to float on the surface of transition metals such as $Ni(111)$ [2,3], as schematized in Figure 1.

Our starting point is the fact that, under some conditions, the carbon overlayer on

R. Girlanda et al. (eds.), Progress on Electron Properties of Solids, 337–352.

$Ni(111)$ is *graphitic*, i.e., it has the $2D$ structure of graphite with reasonable accuracy, and it floats some 2.8Å above the Ni surface. It is further argued from photoemission spectroscopy [3] that the main influence of the substrate consists in about 0.02 electrons acquired by each carbon atom at the expenses of the underlying Ni atoms. The charge transfer to this single graphite layer is strongly reminiscent of that of donor intercalation compounds, and suggests that the known physics of plasmons in graphite [4] and graphite intercalates [5] may be applied to the present case as well. An immediate important confirmation of this viewpoint comes from *EELS* data [3] which show a striking similarity between data of $C/Ni(111)$ (graphitic) and $3D$ graphite [6]. Both data are reproduced in Figure 2, for ready comparison. This means that the collective excitation spectra of the two systems are, in the $(\mathbf{q}, \omega)$ region relevant to these experiments, quite similar. There are also differences, the main one being a new peak around $3.3\,eV$ in the $C/Ni(111)$ case, together with a small energy shift of the $7\,eV$ π-electron plasmon.

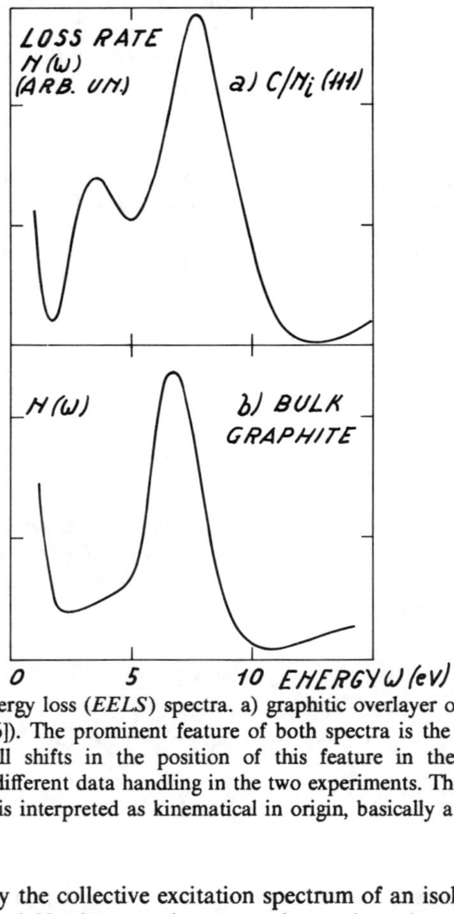

Figure 2. Experimental energy loss (*EELS*) spectra. a) graphitic overlayer on $Ni(111)$ (ref. [4]); b) crystalline graphite (ref. [6]). The prominent feature of both spectra is the π-electron plasmon of graphite near $7\,eV$. Small shifts in the position of this feature in the two spectra are not significant, in view of the different data handling in the two experiments. The smaller extra peak of $C/Ni(111)$ around $3.5\,eV$ is interpreted as kinematical in origin, basically a single layer effect (see text).

In this work we study the collective excitation spectrum of an isolated graphite monolayer, assumed to have 0.02 electrons in excess for each carbon atom, as found on $Ni(111)$. Besides this charge transfer, the two main additional effects of the Ni substrate on the graphite layer are a) *electronic* and b) *electromagnetic*. Of these two, the electronic couplings, though certainly not negligible, are not very large, as clearly indicated by photoemission data [3]. In this respect the situation is opposite to *carbidic* carbon, which is believed to be deeply implanted inside a trough of the $Ni(111)$ substrate, and whose states should mix heavily with *Ni* states, much in the same way as it happens on *C*/*Ru*(001) [7]. In summary, we may, in a first approximation, neglect electronic coupling of the graphite monolayer to the *Ni* substrate.

The electromagnetic effects on the other hand might be very roughly modeled by an image-charge-type screening of the electron-electron interaction within the graphite layer or, in a better approximation, by including the full microscopic screening of the semi-infinite nickel. This effect is in principle straightforward to include, possibly using a realistic dielectric function for the semiinfinite *Ni* substrate. Before moving on to this level of sophistication, however, it seems important and instructive to study the simpler case of a graphite monolayer suspended in vacuum, i.e. to neglect first *all* effects of the *Ni* substrate, other than charge transfer. This is what we propose to do here.

This paper is organized as follows. In Section 2 we illustrate the simple band model chosen to describe the graphite monolayer. It is a Hückel model, including π-electrons only, which has been previously tested in many instances in graphite [8] and in intercalated graphite [9] and which we have extended to yield a very good description of plasmon dispersion in 3*D* pristine graphite [10]. In Section 3 we calculate the $(\mathbf{q},\omega)$-dependent irreducible π-electron polarizability $\chi^{\pi}(\mathbf{q},\omega)$ of the graphite monolayer. Section 4 is devoted to the calculation of the energy-loss rate of a fast charged particle crossing a monolayer described by a general $\chi(\mathbf{q},\omega)$. Of course in the real *EELS* experiment the fast particles are electrons, and can, in principle, exchange with the monolayer electrons. Also, they can cross the monolayer twice, being reflected back by the *Ni* substrate. Our simple discussion neglects exchange - as is appropriate for fast electrons - and considers only one crossing of the monolayer, which is quite appropriate if one is interested only in a *single* energy loss scattering event. In comparison with the corresponding 3*D* energy-loss formulae, our monolayer results show that a new *kinematic EELS* peak should appear at $\omega = \mathbf{q}\cdot\mathbf{v}$, $\mathbf{v}$ being the fast electron velocity. This *kinematical resonance* peak is absent in 3*D* bulk *EELS*. The *EELS* spectrum of the graphite monolayer is calculated in Section 5, for different values of $\mathbf{q}$. Comparison with the experimental results of Figure 2a is, at this point possible, and is found to be very instructive. In particular it is concluded that the 3.3 *eV* peak is likely to be associated with the kinematical resonance typical of the monolayer geometry, which is then absent in 3*D* graphite. The overall agreement of the calculated spectrum with that observed for *C*/*Ni*(111) (graphitic) provides a direct confirmation that both electronic and electromagnetic perturbation of the *Ni* substrate on the graphite monolayer plasmon spectrum is in fact weak, as argued earlier.

Finally, we discuss in Section 6 a comparison between the full $(\mathbf{q},\omega)$-dependent calculated excitation spectrum of the graphite monolayer with that of 3*D* graphite. Although the two spectra are markedly different for $\mathbf{q}\to 0$, they rapidly become similar as $|\mathbf{q}|$ increases. It is this last fact that explains the close similarity between the two spectra of Figure 2, that are taken in a Cylindrical Mirror Analyser (*CMA*) geometry where $\mathbf{q}$ is integrated upon, and large momenta therefore dominate.

2. Band Model for the Graphite Monolayer

In the 2*D* graphite structure, shown in Figure 3a, there are two carbon atoms per unit cell at the positions

$$\mathbf{d}_1 = a\left[0, \frac{\sqrt{3}}{3}\right], \qquad \mathbf{d}_2 = a\left[0, \frac{2\sqrt{3}}{3}\right] \tag{1}$$

the primitive direct lattice vectors being

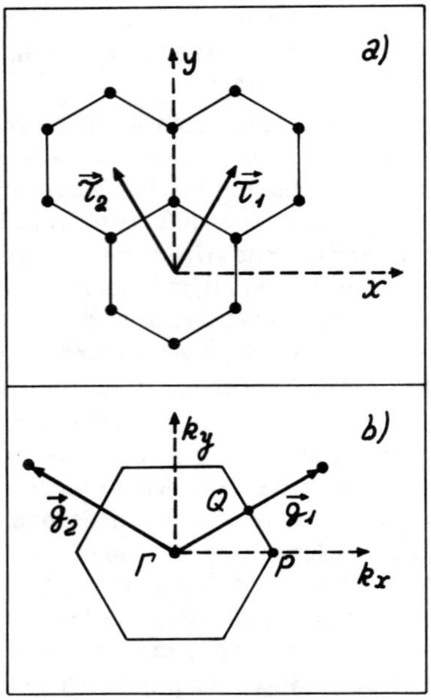

Figure 3. a) Honeycomb structure of a simple graphite sheet, b) its reciprocal lattice and 2*D* Brillouin Zone.

$$\tau_1 = a\left[\frac{1}{2},\frac{\sqrt{3}}{2}\right], \quad \tau_2 = a\left[-\frac{1}{2},\frac{\sqrt{3}}{2}\right] \tag{2}$$

with a $a = 2.46\text{Å}$; the primitive vectors of the reciprocal lattice are

$$g_1 = \frac{4\pi}{a}\left[\frac{1}{2},\frac{1}{2\sqrt{3}}\right], \quad g_2 = \frac{4\pi}{a}\left[-\frac{1}{2},\frac{1}{2\sqrt{3}}\right] \tag{3}$$

and the hexagonal first Brillouin Zone are shown in Figure 3b.

Following earlier work [8,9] we restrict to π bands alone, which we consider within tight binding. If $\varphi(\mathbf{r})$ is the ortho-normalized $2p_z$ wave function for a carbon atom in the lattice, the crystal wave function for π electrons can be written

$$\Psi(\mathbf{k},\mathbf{r}) = c_1(\mathbf{k})\Phi_1(\mathbf{k},\mathbf{r})+c_2(\mathbf{k})\Phi_2(\mathbf{k},\mathbf{r}) \tag{4}$$

where $\Phi_{1,2}$ are the Bloch sums ($\{R\}$ being direct lattice vectors)

$$\Phi_{1,2}(\mathbf{k},\mathbf{r}) = \frac{1}{\sqrt{N}}\sum_R e^{i\mathbf{k},\mathbf{r}}\varphi(\mathbf{r}-\mathbf{R}-\mathbf{d}_{1,2}). \tag{5}$$

Assuming a single particle hamiltonian operator

$$H = \frac{p^2}{2m}+\sum_{iR}V(\mathbf{r}-\mathbf{R}-\mathbf{d}_i), \tag{6}$$

neglecting three-centers integrals and considering the interaction between nearest neighbours only, the energies $E(\mathbf{k})$ and the coefficients c_1,c_2 are obtained as eigenvalues and eigenvectors of the matrix whose compatibility equation takes the simple form

$$\begin{vmatrix} E_0-E(\mathbf{k}) & -\gamma_0 M(\mathbf{k}) \\ \text{c.c.} & E_0-E(\mathbf{k}) \end{vmatrix} = 0 \tag{7}$$

where

$$M(\mathbf{k}) = 1+e^{-i\mathbf{k}\cdot\tau_1}+e^{-i\mathbf{k}\cdot\tau_2}$$

and $-\gamma_0$ is the hopping integral between nearest neighbours which we take as a parameter. Setting conventionally the energy zero as $E_0 = <\varphi(\mathbf{r})|H_0+V|\varphi(\mathbf{r})>$ we obtain

$$E(\mathbf{k}) = \mp\gamma_0|M(\mathbf{k})| \tag{8}$$

with the corresponding bonding (+) and antibonding (-) wave functions:

$$\psi^{\pm}(\mathbf{k},\mathbf{r}) = \frac{1}{\sqrt{2}}\left[\Phi_1(\mathbf{k},\mathbf{r})\pm\frac{M^*(\mathbf{k})}{|M(\mathbf{k})|}\Phi_2(\mathbf{k},\mathbf{r})\right]. \tag{9}$$

The energy bands (8) are schematically sketched in Figure 4, where we take $\gamma_0 = 3\,eV$.

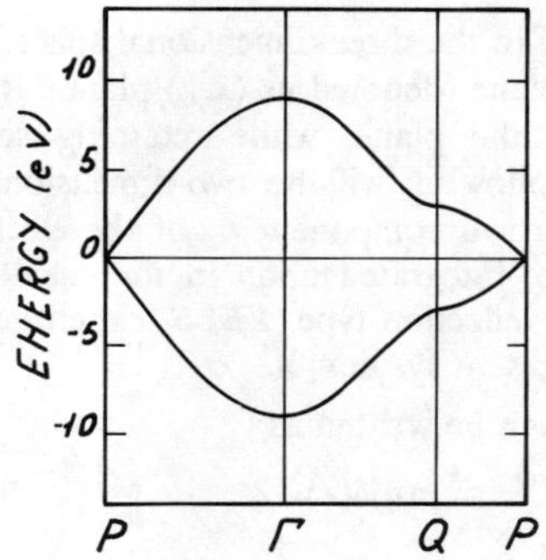

Figure 4. Simplified tight-binding π-electron band structure of a single graphite layer, obtained with the nearest neighbour hopping of magnitude $\gamma_0 = 3\,eV$. The Fermi level is at $E = 0$ without charge transfer, and at $E = 0.6\,eV$ for a transfer of 0.02 el/carbon, as discussed in the text.

For the purpose of calculating the dielectric response, we shall need an explicit form for the electronic orbitals. As an approximate localised orbital we may take a normalized function $\chi(\mathbf{r})$ on a carbon atom

$$\chi(\mathbf{r}) = Nre^{-\alpha^2r^2}\cos\theta \tag{10}$$

$$N = \left[\frac{2^7}{\pi^3}\right]^{\frac{1}{4}}\alpha^{\frac{5}{2}}$$

which mimics a p_z orbital, but has a Gaussian form which simplifies the algebra. The overlap S between the functions $\chi(\mathbf{r})$ on nearest neighbours atoms depends only from the relative distance, e.g.,

$$S = \int\chi^*(\mathbf{r})\chi(\mathbf{r}-\mathbf{d})d\mathbf{r} = e^{-\frac{1}{2}\alpha^2d^2}, \tag{11}$$

where $\mathbf{d} = \mathbf{d}_2-\mathbf{d}_1$. In order to orthogonalize, we consider the linear combination with coefficients λ,μ

$$\varphi(\mathbf{r}) = \lambda\chi(\mathbf{r})+\mu\{\chi(\mathbf{r}-\mathbf{d})+\chi(\mathbf{r}-\mathbf{d}+\tau_1)+\chi(\mathbf{r}-\mathbf{d}+\tau_2)\}; \tag{12}$$

the gaussian parameter α has been chosen in such a way that the maximum of the radial distribution probability density is the same as calculated from Clementi and Roetti's *single* ζ function for the $2p_z$ carbon orbital [11]. The coefficients λ,μ are determined from the normalization condition and orthogonality to n.n. carbons (neglecting the overlap beyond first neighbours), i.e.,

$$\int \varphi^*(\mathbf{r})\varphi(\mathbf{r})d\mathbf{r} = 1$$

$$\int \varphi^*(\mathbf{r})\varphi(\mathbf{r}-\mathbf{d})d\mathbf{r} = 0.$$

We find (with the carbon bond length $d = 1.42\text{Å}$), $\alpha = 1.482\text{Å}^{-1}$, $S \sim 0.109$, $\lambda = 1.0138$, $\mu = -0.056$.

3. Calculation of the Longitudinal Polarizability $\chi(\mathbf{q}, w)$ for a Graphite Monolayer

Consider a graphite monolayer in the three-dimensional space. Since translational invariance holds in the monolayer plane (denoted as (x,y) plane), it is convenient to introduce two-dimensional **k**-vectors in the plane, while retaining separately the z-coordinate. Hence, all **k** vectors in the following will be two-dimensional $\mathbf{k} \equiv k_x, k_y$. The present treatment assumes that the vertical component k_z of the electron momentum transfer is not measured, and is therefore integrated upon in the calculated energy-loss rate. This applies ideally to the usual reflection-type *EELS* experiments of the kind that are currently carried out for the present system [3].

The density operator can then be written as

$$\rho(\mathbf{q},z) = \delta(z)\sum_i e^{-i\mathbf{q}\mathbf{r}_i} = \rho(\mathbf{q})\delta(z) \tag{13}$$

where $\mathbf{r}_i = (x_i,y_i)$ are also two-dimensional. An external longitudinal perturbation of potential

$$V^{\text{ext}}(\mathbf{r},z;t) = \sum_{\mathbf{k},\omega} V(\mathbf{k},z;\omega)e^{i(\mathbf{k}\cdot\mathbf{r}-\omega t)} \tag{14}$$

will couple to the monolayer via a perturbing hamiltonian

$$\begin{aligned} H'(\mathbf{q},\omega) &= \int \rho(\mathbf{q},z')V(-\mathbf{q},z',\omega)dz' \\ &= \int dz\delta(z')\sum_i e^{-i\mathbf{q}\mathbf{r}_i} V(-\mathbf{q},z',\omega) \\ &= \sum_i e^{-i\mathbf{q}\cdot\mathbf{r}_i} V(-\mathbf{q},z=0,\omega) \\ &= V(-\mathbf{q},z=0,\omega)\rho(\mathbf{q}). \end{aligned} \tag{15}$$

The $2D$ π-polarizability of the monolayer in the single particle approximation is defined as

$$\chi^{\pi}(\mathbf{q},\omega) = \sum_{\mathbf{k}} |<v_{\mathbf{k}}|e^{-i\mathbf{q}\cdot\mathbf{r}}|c_{\mathbf{k}+\mathbf{q}}>|^2 \times \tag{16}$$

$$\times \left[\frac{1}{E_v(\mathbf{k})-E_c(\mathbf{k}+\mathbf{q})-\hbar\omega-i\delta}+\frac{1}{E_v(\mathbf{k})-E_c(\mathbf{k}+\mathbf{q})+\hbar\omega+i\delta}\right].$$

Here $|v_{\mathbf{k}}>$ denotes a filled state and $|c_{\mathbf{k}}>$ an empty state, at $T = 0$.

Inserting for $|v_k>$ and $|c_k>$ the pertinent π-electrons Bloch functions from the previous Section we have, for the matrix element $<v_k|e^{-i\mathbf{q}\cdot\mathbf{r}}|c_{\mathbf{k}+\mathbf{q}}>$,

$$<v_{\mathbf{k}}|e^{-i\mathbf{q}\cdot\mathbf{r}}|c_{\mathbf{k}+\mathbf{q}}> = \frac{1}{2}\left\{F_{11}(\mathbf{k},\mathbf{q})\pm F_{12}(\mathbf{k},\mathbf{q})\frac{M^*(\mathbf{k}+\mathbf{q})}{|M(\mathbf{k}+\mathbf{q})|}\right. \tag{17}$$

$$\left.\pm F_{21}(\mathbf{k},\mathbf{q})\frac{M(\mathbf{k})}{|M(\mathbf{k})|}\pm F_{22}(\mathbf{k},\mathbf{q})\frac{M(\mathbf{k})M^*(\mathbf{k}+\mathbf{q})}{|M(\mathbf{k})|\,|M(\mathbf{k}+\mathbf{q})|}\right\}$$

where

$$\begin{aligned}
F_{11}(\mathbf{k},\mathbf{q}) &= <\Phi_1(\mathbf{k},\mathbf{q})|e^{-i\mathbf{q}\cdot\mathbf{r}}|\Phi_1(\mathbf{k}+\mathbf{q},\mathbf{r})> \\
&= e^{-i\mathbf{q}_{\shortparallel}\mathbf{d}_1}<\varphi(\mathbf{r})|e^{-i\mathbf{q}_{\shortparallel}\mathbf{r}}|\varphi(\mathbf{r})> \\
F_{22}(\mathbf{k},\mathbf{q}) &= <\Phi_2(\mathbf{k},\mathbf{q})|e^{-i\mathbf{q}\cdot\mathbf{r}}|\Phi_2(\mathbf{k}+\mathbf{q},\mathbf{r})> \\
&= e^{-i\mathbf{q}_{\shortparallel}\mathbf{d}_2}<\varphi(\mathbf{r})|e^{-i\mathbf{q}_{\shortparallel}\mathbf{r}}|\varphi(\mathbf{r})> \\
F_{12}(\mathbf{k},\mathbf{q}) &= <\Phi_1(\mathbf{k},\mathbf{q})|e^{-i\mathbf{q}\cdot\mathbf{r}}|\Phi_2(\mathbf{k}+\mathbf{q},\mathbf{r}) \\
&= <\varphi(\mathbf{r}-\mathbf{d}_1)|e^{-i\mathbf{q}\cdot\mathbf{r}}|\varphi(\mathbf{r}-\mathbf{d}_2> \\
&+e^{-i(\mathbf{k}+\mathbf{q})\cdot\tau_1}<\varphi(\mathbf{r}-\mathbf{d}_1)|e^{-i\mathbf{q}\cdot\mathbf{r}}|\varphi(\mathbf{r}-\mathbf{d}_2-\tau_1) \\
&+e^{-i(\mathbf{k}+\mathbf{q})\cdot\tau_2}<\varphi(\mathbf{r}-\mathbf{d}_1)|e^{-i\mathbf{q}\cdot\mathbf{r}}|\varphi(\mathbf{r}-\mathbf{d}_2+\tau_2) \\
F_{21}(\mathbf{k},\mathbf{q}) &= <\Phi_2(\mathbf{k},\mathbf{q})|e^{-i\mathbf{q}\cdot\mathbf{r}}|\Phi_1(\mathbf{k}+\mathbf{q},\mathbf{r})> \\
&= <\varphi(\mathbf{r}-\mathbf{d}_1)|e^{-i\mathbf{q}\cdot\mathbf{r}}|\varphi(\mathbf{r}-\mathbf{d}_2)> \\
&+e^{i(\mathbf{k}+\mathbf{q})\cdot\tau_1}<\varphi(\mathbf{r}-\mathbf{d}_2)|e^{-i\mathbf{q}\cdot\mathbf{r}}|\varphi(\mathbf{r}-\mathbf{d}_1-\tau_1)> \\
&+e^{i(\mathbf{k}+\mathbf{q})\cdot\tau_2}<\varphi(\mathbf{r}-\mathbf{d}_2)|e^{-i\mathbf{q}\cdot\mathbf{r}}|\varphi(\mathbf{r}-\mathbf{d}_1-\tau_2).
\end{aligned}$$

The signs in eq. (17) depend on the initial and final states we are considering, i.e. if we are dealing with interband transitions (bonding $\rightarrow$ antibonding) or intraband transitions (bonding $\rightarrow$ bonding or antibonding $\rightarrow$ antibonding). The $\Phi_{1,2}$ are Bloch functions and φ are the (Wannier-like) localised functions (12) on the carbons. The real and the imaginary part of the π-polarizability are plotted in Figure 5 in the case in which we have 0.02 electrons per carbon in excess and for different values of the transferred momentum $\mathbf{q}$. A non zero phenomenological damping $\delta = 0.12\,eV$ has been added for the sole purpose of smoothing out the singularities.

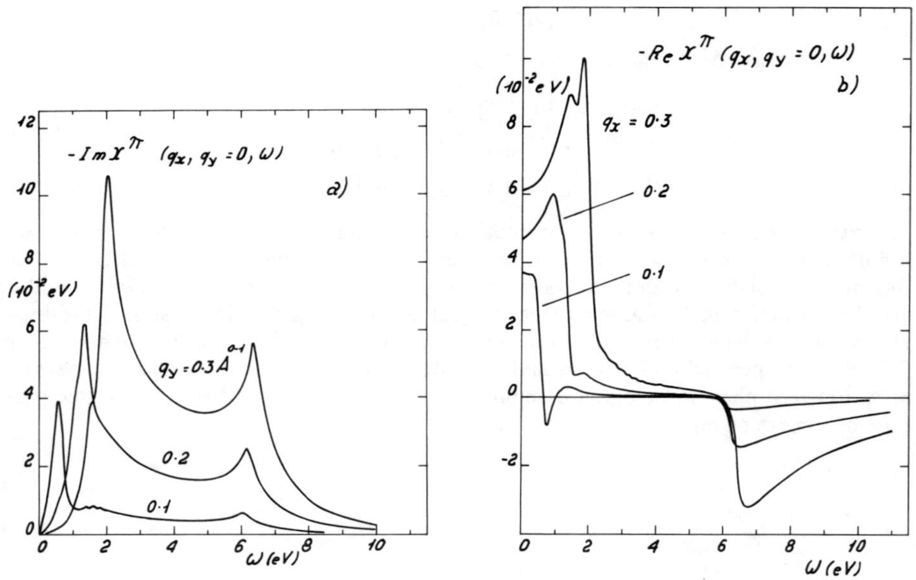

Figure 5. π-electron susceptibility of a single graphite sheet with 0.02 extra electrons per carbon atom. a. Imaginary part, $\mathbf{q}\|x$; b. real part, $\mathbf{q}\|x$; One can note a highly dispersive low-energy peak, which represents the edge for transitions near the P point, and a non-dispersive high-energy peak, representing the $\pi \rightarrow \pi^*$ transitions (which, in reality, occur near 4.6 eV at the Q-point. The susceptibility, as expected, is isotropic for small $|\mathbf{q}|$, and develops in-plane anisotropy as $|\mathbf{q}|$ increases (cf. Fig. 5c).

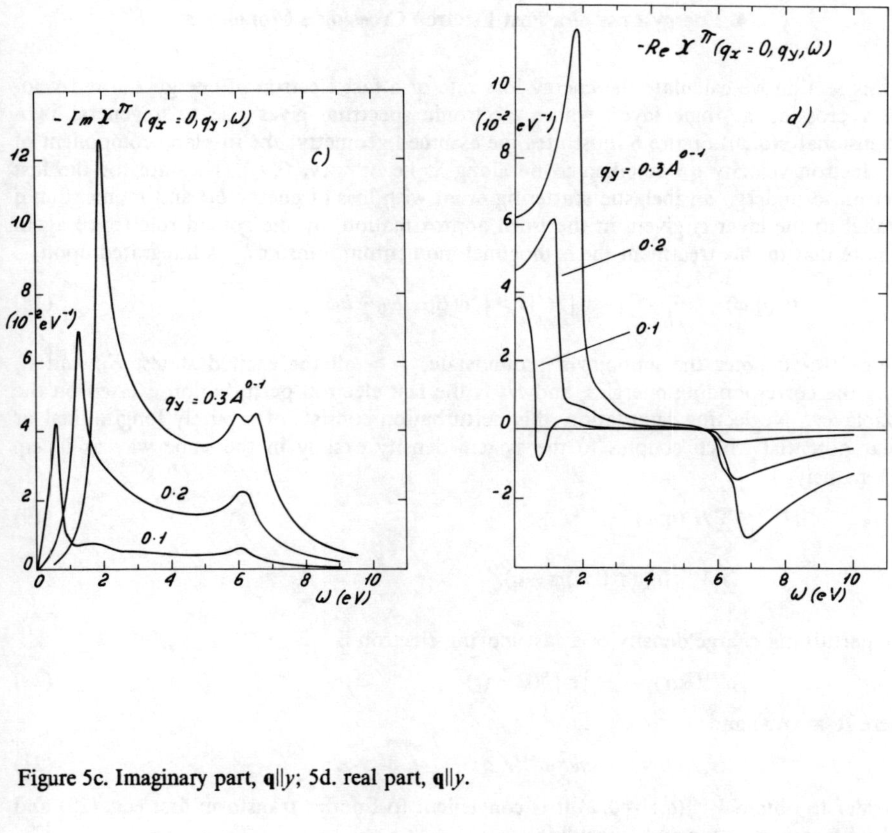

Figure 5c. Imaginary part, **q**||*y*; 5d. real part, **q**||*y*.

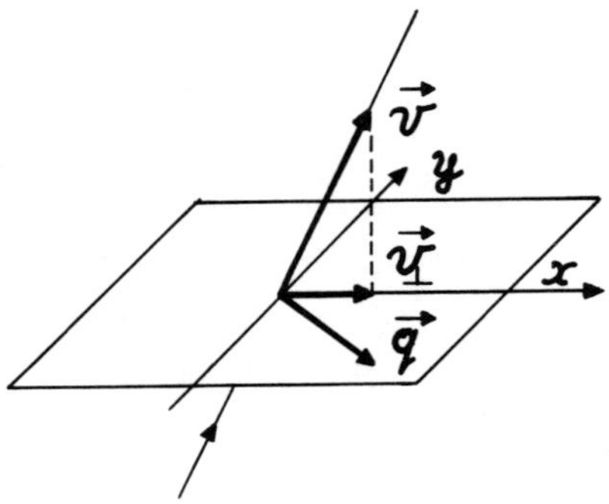

Figure 6. Geometry assumed in the calculations. Here **v** is the electron velocity, **v**⊥ its in-plane projection, and **q** the planar momentum transfer, of components (q_x, q_y).

4. Energy Loss of a Fast Electron Crossing a Monolayer

In this section we calculate the energy loss rate of a fast electron of energy E_0 and velocity **v** crossing a single layer, whose electronic spectrum gives rise to a general two-dimensional $\chi(\mathbf{q},\omega)$. Figure 6 illustrates the assumed geometry; the in-plane component of the electron velocity $\mathbf{q}\cdot\mathbf{v}$ is taken to be along x, i.e. $\mathbf{v} = (\mathrm{v}_x, 0, \mathrm{v}_z)$. The rate for the fast electron to undergo an inelastic scattering event with loss of energy $\hbar\omega$ and momentum **q** parallel to the layer is given, in the Born approximation, by the golden rule (once again we note that in this treatment the orthogonal momentum transfer k_z is integrated upon)

$$W(\mathbf{q},\omega) = \frac{2\pi}{\hbar}\sum_i |<0|H'|i>|^2 \delta(E_i - E_0 - \hbar\omega) \tag{18}$$

where $|0>$ denotes the monolayer groundstate, $|i>$ all the excited states, E_0 and E_i being the corresponding energies, and H' is the fast electron perturbation exerted on the monolayer. Neglecting retardation, this perturbation consists of a purely longitudinal or scalar potential which couples to the system density exactly in the same way as in eq. (15), namely

$$\begin{aligned} H' &= \sum_{\mathbf{q},\omega} H'(\mathbf{q},\omega) \qquad (19)\\ &= \sum_{\mathbf{q},\omega} V^{\mathrm{tot}}(\mathbf{q}, z=0, \omega)\rho(-\mathbf{q}). \end{aligned}$$

The perturbing charge density of a fast moving electron is

$$-|e|\rho^{\mathrm{ext}}(\mathbf{R},t) = -|e|\delta(\mathbf{R} - \mathbf{v}t) \tag{20}$$

where $\mathbf{R} = (\mathbf{r}, z)$ and

$$\nabla^2 V^{\mathrm{ext}}(\mathbf{r},z,t) = -4\pi e^2 \rho^{\mathrm{ext}}(\mathbf{r},z,t). \tag{21}$$

In order to obtain $V^{\mathrm{ext}}(\mathbf{q}, z=0, \omega)$ it is convenient to Fourier transform first eqs. (20) and (21) with respect to $\mathbf{r}, z$ and ω yielding

$$\rho^{\mathrm{ext}}(\mathbf{q},k_z,\omega) = \frac{2\pi}{T\Omega}\delta(\omega - \mathbf{q}\cdot\mathbf{v} - k_z \mathrm{v}_z), \tag{22}$$

$$V^{\mathrm{ext}}(\mathbf{q},k_z,\omega) = \frac{8\pi^2 e^2}{T\Omega}\frac{1}{q^2 + k_z^2}\delta(\omega - \mathbf{q}, \mathbf{v} - k_z v_z) \tag{23}$$

(where T is a conventional large collision time, later to be set equal to unity) and subsequently sum over k_z:

$$V^{\text{ext}}(\mathbf{q},z=0,\omega) = \sum_{k_z} V^{\text{ext}}(\mathbf{q},k_z,\omega) \tag{24}$$

$$= \frac{4\pi e^2}{TA\,|\mathrm{v}_z|} \frac{1}{q^2+\left[\dfrac{\omega-\mathbf{q}\cdot\mathbf{v}}{\mathrm{v}_z}\right]^2}.$$

Here, Ω denotes the total volume, and A the total area of the system so that $\Omega = AL_z$. We must now go from V^{ext} to $V^{\text{tot}}(\mathbf{q},z=0,\omega)$. This is done by introducing an induced potential in the usual way

$$V^{\text{tot}} = V^{\text{ext}} + V^{\text{ind}} \tag{25}$$

with

$$V^{\text{ind}}(\mathbf{q},\mathbf{k},\omega) = \frac{4\pi e^2}{L_z}\frac{1}{q^2+k_z^2}\rho^{\text{ind}}(\mathbf{q},\omega) \tag{26}$$

and

$$\rho^{\text{ind}}(\mathbf{q},\omega) = \chi(\mathbf{q},\omega)V^{\text{tot}}(\mathbf{q},z=0,\omega), \tag{27}$$

where $\chi(\mathbf{q},\omega)$ is the *total* layer polarizability.

Putting all this together we have:

$$V^{\text{tot}}(\mathbf{q},z=0,\omega) = V^{\text{ext}}(\mathbf{q},z=0,\omega)+\sum_{k_z}\frac{4\pi^2}{AL_z}\frac{1}{q^2+k_z^2}\rho^{\text{ind}}(\mathbf{q},\omega) \tag{28}$$

$$= V^{\text{ext}}(\mathbf{q},z=0,\omega)+\frac{2\pi e^2}{qA}\rho^{\text{ind}}(\mathbf{q},\omega)$$

$$= V^{\text{ext}}(\mathbf{q},z=0,\omega)+\frac{2\pi e^2}{qA}\chi(\mathbf{q},\omega)V^{\text{tot}}(\mathbf{q},z=0,\omega)$$

whence finally

$$V^{\text{tot}}(\mathbf{q},z=0,\omega) = \frac{1}{1-\dfrac{2\pi e^2}{qA}\chi(\mathbf{q},\omega)}V^{\text{ext}}(\mathbf{q},z=0,\omega), \tag{29}$$

i.e. the external potential is *screened* by the monolayer. This result bears of course a strong resemblance to the corresponding $3D$ formula

$$V^{\text{tot}}(\mathbf{Q},\omega) = \frac{1}{1-\dfrac{4\pi e^2}{Q^2\Omega}\chi(\mathbf{Q},\omega)}V^{\text{ext}}(\mathbf{Q},\omega) \tag{30}$$

except that the $2D$ Coulomb interaction $\dfrac{2\pi e^2}{qA}$ replaces the $3D$ interaction $\dfrac{4\pi e^2}{Q^2\Omega}$. The total polarizability of the monolayer can be written as

$$\chi(\mathbf{q},\omega) = \chi^{\pi}(\mathbf{q},\omega)+\chi^{\sigma}(\mathbf{q},\omega); \tag{31}$$

the screening due to the π-electrons is that given in Section 3 while for σ-electrons we consider a simple phenomenological expression, as explained in the Appendix.

We can now return to our energy loss rate (18). Since we are interested in one $(\mathbf{q},\omega)$ value only then the pertinent perturbation, in the linear approximation, is just $H'(\mathbf{q},\omega)$, so that in the one electron approximation we can write

$$W(\mathbf{q},\omega) = \frac{2\pi}{\hbar}|V^{\mathrm{tot}}(\mathbf{q},z=0,\omega)|^2\sum_{\mathbf{k}}|<v_{\mathbf{k}}|\rho_{\mathbf{q}}|c_{\mathbf{k}+\mathbf{q}}>|^2\delta(E_c(\mathbf{k}+\mathbf{q})-E_v(\mathbf{k})-\hbar\omega)$$

$$= -\frac{2\pi}{\hbar}\left|\frac{V^{\mathrm{ext}}(\mathbf{q},z=0,\omega)}{1-\frac{2\pi e^2}{qA}(\chi^{\pi}(\mathbf{q},\omega)+\chi^{\sigma}(\mathbf{q},\omega)}\right|^2 \mathrm{Im}\,\chi^{\pi}(\mathbf{q},\omega) \tag{32}$$

where $V^{\mathrm{ext}}(\mathbf{q},z=0,\omega)$ is given in eq. (24).

In comparison with the corresponding $3D$ result for an ideal crystal made of non overlapping graphite layers [9]

$$W^{3D}(\mathbf{q},\omega) = -\frac{2\pi}{\hbar T}\left|\frac{\frac{8\pi^2e^2}{q^2\Omega}}{1-\frac{4\pi e^2}{q^2\Omega}(\chi^{\pi}(\mathbf{q},\omega)+\chi^{\sigma}(\mathbf{q},\omega))}\right|^2 \mathrm{Im}\,\chi^{\pi}(\mathbf{q},\omega)\delta(\omega-\mathbf{q}\cdot\mathbf{v}), \tag{33}$$

we note a totally similar structure with a $2D$ screening replacing the $3D$ one and the *smearing* of the kinematical $\delta(\omega-\mathbf{q}\cdot\mathbf{v})$ into a broader function $[q^2+(\frac{\omega-\mathbf{q}\cdot\mathbf{v}}{\mathrm{v}_z})^2]^{-1}$. As we shall see this denominator is capable of introducing a kinematical peak (q-direction dependent) in the electron loss rate across a single layer, while no such peak exists in the corresponding $3D$ case.

5. Plasmons and *EELS* Spectra of a Graphite Monolayer

We have calculated the loss rate (32) with $\chi^{\pi}(\mathbf{q},\omega)$ given in (16) and $\chi^{\sigma}(\mathbf{q},\omega)$ given in A.2. The results obtained by further choosing the kinetic energy of the electron beam $E_0 = 100\,eV$ and a crossing angle of 45°, i.e.

$$\frac{\mathbf{q}\cdot\mathbf{v}}{|q||\mathbf{v}|} = \frac{1}{\sqrt{2}}, \quad v_z = \sqrt{\frac{2E_0}{m}}$$

are shown in Figure 7.

These calculated energy loss exhibit three main features: *1)* a low-energy narrow peak, *2)* a broad, very q-dependent intermediate peak present for $\mathbf{q}\cdot\mathbf{v} \neq 0$ ($q=q_x$) and absent for $\mathbf{q}\cdot\mathbf{v}$ ($q=q_y$) and *3)* a high-energy peak near $7\,eV$.
By direct inspection we find that peaks 1 and 3 arise due to minima in the real part of the $2D$ dielectric function $1-\frac{2\pi e^2}{1A}(\chi^{\pi}(\mathbf{q},\omega)+\chi^{\sigma}(\mathbf{q},\omega))$, and therefore can legitimately be referred to as *plasmons*, In fact the lower-energy plasmon 1 is a collective excitation of the extra 0.02 electrons/carbon which the graphite monolayer has acquired. Such *intra-band* plasmons are well known experimentally [12] and theoretically [9] in the $3D$ graphite intercalates. The high-energy plasmon 3 is easily interpreted as the $2D$ remnant of the ordinary $7\,eV$ π-plasmon of the $3D$ graphite.

The intermediate broad peak in the neighborhood of $3\,eV$, on the contrary, does not arise due to any features of the dielectric polarizability (16) but rather because of the minimum of $q^2-(\omega-\frac{\mathbf{q}\cdot\mathbf{v}}{\mathrm{v}_z})^2$ at $\omega = \mathbf{q}\cdot\mathbf{v}$. It is therefore kinematic in origin and does not carry any particular microscopic significance. It shows that, when dealing with *EELS*

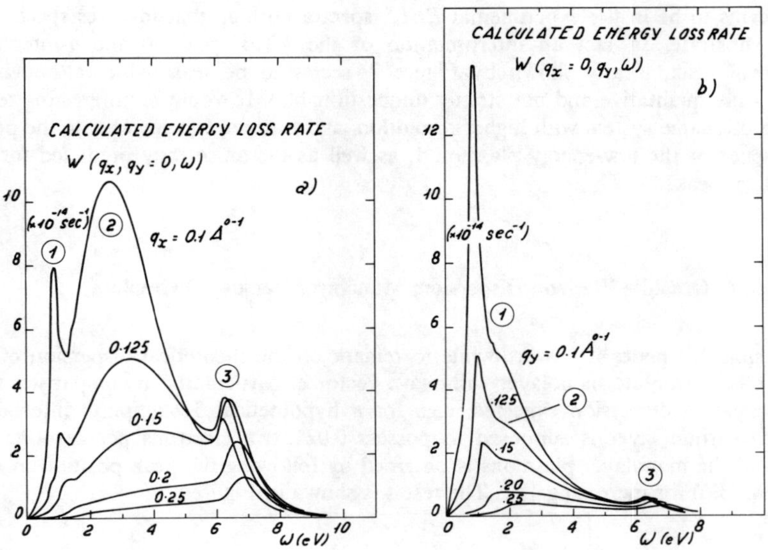

Figure 7. Electron energy loss rate for a single graphite sheet, with 0.02 extra electrons/carbon: a) when the **q**-vector lies in th electron incidence plane, $\mathbf{q} \| \mathbf{v}_\perp$.
b) when the **q**-vector is orthogonal to the incidence plane $\mathbf{q} \perp \mathbf{v}_\perp$. Each spectrum has three main peaks: 1,2 and 3. Peak 1 is the intraband plasmon, due to the charge transfer. Peak 3 is the overall π-electron plasmon. The broad peak 2 is kinematical, centered at $\omega = \mathbf{q} \cdot \mathbf{v}$, as also indicated by its anisotropy. The overall difference of the two spectra a) and b) decreases as $|\mathbf{q}|$ increases.

from monolayer problems, one must be cautious before identifying *all* features with some well-defined electronic excitations or other: there may also be kinematic features of this kind.

We can now proceed to compare our calculated loss rates of Figure 7 with the *EELS* findings for a graphitic monolayer on *Ni*(111) of Figure 2. This comparison must, by necessity, be only qualitative at this stage. This is because the experimental data represent *CMA* integrated intensity which implies some unspecified **q**-integration of our calculated rates and also because the experimental resolution was not sufficient to reveal any structure - such as our peak 1 - at or below 1 *eV*. Another reason is of course our total neglect of the *Ni* substrate influence in the calculations. It seems reasonable, however, to assume that *forward* scattering ($\mathbf{q} \cdot \mathbf{v} \sim \omega$) should dominate over *transverse* scattering ($\mathbf{q} \cdot \mathbf{v} \sim 0$) after integration. We also note that **q**-integration will eventually enhance the intensity of non-dispersive peaks, like the π-plasmon, over the dispersive ones, like the kinematical peak at lower energy. This observation provides a clue for reconciling the very different distribution of intensities calculated (Figure 7) and observed (Figure 2a).

With this in mind, the overall agreement of calculated and experimental spectra is good, and indicates two first interesting conclusions:

i) The π-plasmon spectrum of a graphite monolayer is very close to that of $3D$ graphite, particularly at medium-large momenta. This is the main reason why the two spectra of Figure 2 resemble one another so closely.

ii) There seems to be in the experimental *EELS* spectra nothing that involves specifically the *Ni* substrate. In fact an interpretation of the $3.3\,eV$ peak as the **q**-integrated kinematical peak, clearly shown by Figure 7, seems to be reasonable (although, at present, only qualitative and not strictly unquestionable). It would be interesting to re-examine this same system with higher resolution, as an attempt to investigate the possible presence of the low-energy plasmon 1, as well as the anisotropy predicted for the kinematical peak.

6. Graphite Plasmon Dispersion: Monolayer Versus $3D$ Graphite

Before closing this paper it is worthwhile to remark on the theoretical dispersion of the plasmons of the graphite monolayer with wave vector **q**, particularly in comparison with the corresponding dispersion expected, e.g., for a hypothetical $3D$ graphite intercalate, where each carbon layer is supposed to possess 0.02 extra electrons per carbon. The dispersion of the monolayer plasmons is obtained by following the peak position in ω of $W(\mathbf{q},\omega)$ (eq. (32)) for increasing $|\mathbf{q}|$. The result is shown in Figure 8.

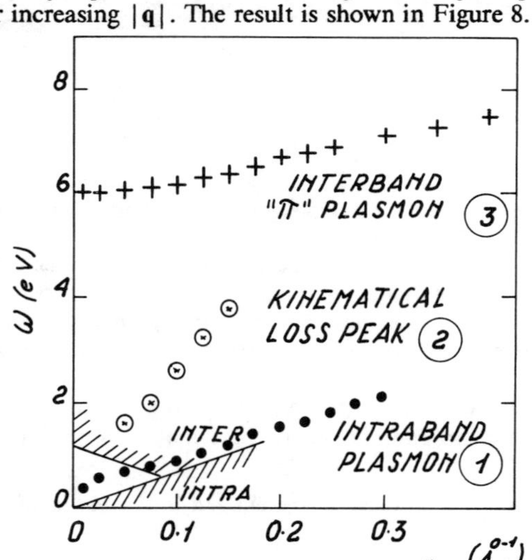

Figure 8. Summary of all calculated peak positions, in the particular case $\mathbf{q} = (q_x,0)$. For $|\mathbf{q}| \to 0$, the dispersion of the intraband plasmon is compatible with a $\sqrt{q}$ behaviour, that of the kinematical peak is linear with $|\mathbf{q}|$, and that of the interband plasmon tends to the interband $\pi \to \pi^*$ transition energy ω_0^π (here $6\,eV$, in reality $4.6\,eV$). The sharpness of the intraband plasmon peak 1 for small wavevector, seen in Figure 7, is due to its falling inside the *window* between the intraband and interband continua, as indicated by shaded regions.

This figure shows first of all the intra- and inter-band electron-hole pair continua (shaded). The intraband plasmon falls in a gap between intra- and inter-band continua, which explains its extraordinarily sharp calculated lineshape in Figure 7 (of course in the real $C/Ni(111)$ situation, the lifetime should be very substantially reduced by decay into Ni surface electron-hole pair). The dispersion of this intra-band plasmon is very strong for small $|\mathbf{q}|$-vectors and extrapolates to $\omega = 0$ in the long wavelength limit. A dispersion $\omega \sim \sqrt{q}$ is actually what one expects for a $2D$ electron gas with no retardation. The interband π-plasmon, at higher energy, is intrinsically broader since it falls inside an interband electron-hole continuum. It is, however, still rather well defined, because this continuum is of very low intensity. The dispersion of the π-plasmon is also interesting. It starts off at small $|\mathbf{q}|$ exactly at the frequency of the strong $\pi \rightarrow \pi^*$ saddle point interband transition ω_0^π, which occurs at point $\mathbf{Q}$ of the Brillouin zone [4,8]. The actual frequency of this transition is $4.6\,eV$ in $3D$ graphite, but $\omega_0^\pi = 6\,eV$ in the present crude model. For increasing $|\mathbf{q}|$ the plasmon energy should rise at first linearly, and only later quadratically as in $3D$ graphite. As Fig. 8 shows, the linear behaviour for small $\mathbf{q}$ is not visible in practice.

The above demonstrates that, while plasmon dispersion for a monolayer is very different from that of a $3D$ layered crystal for small $|\mathbf{q}|$, the modes of these two systems and their dispersion with $\mathbf{q}$ tend to become extremely similar for large $|\mathbf{q}|$. Clearly the crossover between these two different behaviours must occur in the neighbourhood of $q_c \sim \frac{2\pi}{c_0}$, where c_0 is the interlayer separation for the $3D$ case.

Acknowledgement

One of the authors (E.T.) would like to express his deep gratitude to Prof. F. Bassani who, twenty-two years ago, assigned him plasmons in graphite as a Ph.D. argument, and gave him guidance in his first steps in the field of solid state.

Appendix

In the present case we must also allow for the extra screening of V^{ext} due to the σ-electrons, which were not included in our band hamiltonian of Section 2, but whose screening contribution cannot be neglected. This is done by adding a σ-polarizability inside $\chi(\mathbf{q},\omega)$ of eq. (29), i.e.,

$$\chi(\mathbf{q},\omega) = \chi^\pi(\mathbf{q},\omega) + \chi^\sigma(\mathbf{q},\omega). \tag{A1}$$

Here $\chi^\pi(\mathbf{q},\omega)$ is that given in Section 3 while for $\chi^\sigma(\mathbf{q},\omega)$ we use a phenomenologial expression

$$\begin{aligned}\chi^\sigma(\mathbf{q},\omega) &= \frac{q^2\Omega}{4\pi e^2}\frac{\pi}{2\omega_\sigma}\omega_{p\sigma}^2\frac{1}{\pi}\left[\frac{1}{\omega_\sigma-\omega+i\delta}+\frac{1}{\omega_\sigma+\omega-i\delta}\right] \\ &= \frac{q^2\Omega}{4\pi e^2}\frac{\omega_{p\sigma}^2}{\omega_\sigma^2(\mathbf{q})-\omega^2+2i\omega\delta}\end{aligned} \tag{A2}$$

where

$$\omega_\sigma(\mathbf{q}) = \omega_\sigma(0) + \frac{q^2}{2m}. \tag{A3}$$

In the present context [10] the σ plasma frequency $\omega_{p\sigma}^2$ and the σ effective *Einstein frequency* $\omega_\sigma^2(0)$ are two adjustable parameters that can be chosen so that *a*) the total $\pi+\sigma$

f-sum rule is satisfied

$$\int_0^\infty \omega \, \mathrm{Im}\,\chi(\mathbf{q},\omega) d\omega = \frac{q^2 \Omega \pi}{4\pi e^2} \frac{\pi}{2} \left[\frac{4\pi n_{\pi+\sigma} e^2}{m} \right] \tag{A4}$$

where $n_{\pi+\sigma}$ is the volume density of $\pi+\sigma$ electrons; and b) the π-electron plasmon should be shifted from 12.5 eV to 7 eV, as suggested by the work of Taft and Philipp [13]. This yields, in the present case, $\omega_{p\sigma} = 22.94\,eV$ and $\omega_\sigma(0) = 14.42\,eV$.

References

1. See, e.g., Proceedings of ECOSS-9, *Surface Sci.* **189/190**, (1987).
2. D.W. Goodman, R.D. Kelley, T.E. Madey and J.W. White, *J. Catalysis* **64**, 479 (1980); D.W. Goodman, R.D. Kelley, T.E. Madey and J.T. Yates, *J. Catalysis* **63**, 226 (1980).
3. R. Rosei, S. Modesti, F. Sette, C. Quaresima, A. Savoia and P. Perfetti, *Phys. Rev. B* **29**, 3416 (1984).
4. E. Tosatti and F. Bassani, *Nuovo Cimento* **65B**, 161 (1970).
5. See *Physics of Intercalation Compounds*, L. Pietronero and E. Tosatti eds., Springer, Heidelberg (1981); *Intercalation Compounds of Graphite III*, A. Herold, D. Guerard and P. Lagrange eds., Elsevier-Sequoia, Lausanne (1983).
6. L. Papagno and L.S. Caputi, *Surface Sci.* **125**, 530 (1983).
7. P.J. Feibelman, *Surface Sci.* **103**, L149 (1981); *Phys. Rev. B* **26**, 5347 (1982).
8. F. Bassani and G. Pastori Parravicini, *Electronic states and electronic transitions in solids*, Pergamon Press, Oxford (1975).
9. J. Blinowski, C. Rigaux, Nguyen Hy Han, J.P. Vieren, R. Le Toullec, G. Furdin and J. Melin, *J. Phisique* **41**, 47 (1980); G. Campagnoli and E. Tosatti, *J. Phys. C* **15**, 1457 (1982).
10. G. Campagnoli, E. Tosatti and S. Leelaprute, unpublished; S. Leelaprute, Ph.D. Thesis, ISAS, Trieste (1981).
11. E. Cleenti and C. Roetti, *Atomic Data and Nuclear Data Tables* **14**, 177 (1974).
12. M. Zanini and J.E. Fischer, *Mat. Sci. Eng.* **31**, 169 (1977).
13. E.A. Taft and H.R. Philipp, *Phys. Rev.* **138**, A197 (1969).

ELECTRONIC STATES IN SEMICONDUCTOR SUPERLATTICES AND QUANTUM WELLS

M. ALTARELLI

European Synchrotron Radiation Facility
B.P. 220
38043 Grenoble Cedex
France

The theory of electronic states in semiconductor superlattices and quantum wells is briefly reviewed, with special emphasis on current problems and controversies. Applications of the envelope function method to subband levels, excitons and the effect of electric and magnetic fields are described.

1. Introduction

Theoretical calculations of electronic subbands in two-dimensional semiconductor systems have been performed in Si MOS structures [1] and later in III-V heterostructures. In the latter case, the tight-binding method [2-4], the pseudopotential method [5,6] or ab initio schemes based on the density functional formalism [7] were adopted, in addition to the envelope-function or effective-mass method.

Tight-binding methods have the remarkable advantage of treating the whole Brillouin zone on the same footing and of being computationally simple. On the other hand, they have difficulty in incorporating the effects of external fields (magnetic, electric and strain fields), the Coulomb field of impurity and exciton interactions, etc. The pseudopotential method is computationally more complex and may suffer of a conceptual difficulty in the non-uniqueness of pseudopotentials; again, external fields and excitons are not easily incorporated in such a method. Ab initio calculations are conceptually very appealing, but limited by computational complexity to short-period superlattices. In addition, the well known difficulties [8] of the local density approximation to reproduce the energy gaps correctly are also present. In spite of its limitations (most notably, the difficulty in handling states resulting from the admixture of different k-points in the bulk band structure) the envelope-function method has enjoyed a large popularity, because of its intuitive character and of its great flexibility.

2. Envelope-Function Description of Electronic States

The Effective-Mass Equation

The envelope-function approximation is an effective-mass theory, familiar from the treatment of shallow impurities [9]. We consider, in the one-electron approximation, the Schrödinger equation for the motion of an electron in a semiconductor in presence of some additional potential $U(\mathbf{r})$:

R. Girlanda et al. (eds.), Progress on Electron Properties of Solids, 353–372.

$$\left[\frac{p^2}{2m_0} + V_{per}(\mathbf{r}) + U(\mathbf{r}) - E\right]\psi(\mathbf{r}) = 0. \tag{1}$$

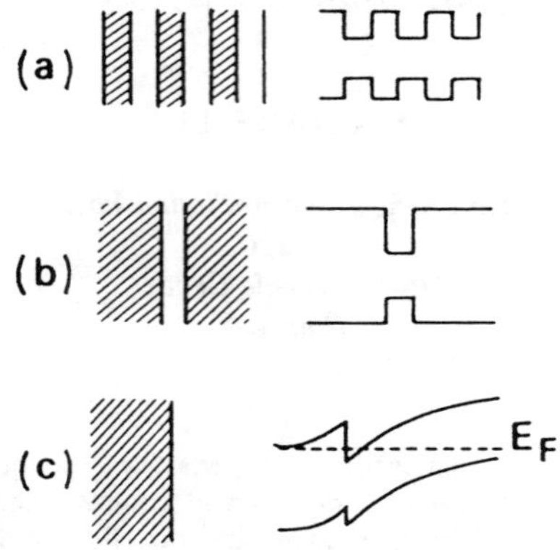

Figure 1. Schematic representation of three types of heterostructures and, on the right, the corresponding band edge profile. (a) Superlattice; (b) Single quantum well; (c) Single heterojunction between an n-type material (on the left side) and a p-type one (on the right side of the heterojunction).

Here m_0 is the free-electron mass, V_{per} is the periodic potential of the perfect bulk semiconductor and $U(\mathbf{r})$ is the additional potential, which we assume to be slowly varying and weak, in the sense precisely stated below. For example, $U(z)$ could be the constant potential in one of the quantum wells of Fig. 1(a), or the band bending potential on the right-hand side of Fig. 1(c), arising from the depleted acceptors. Eq. (1) would then describe the electron in the quantum well or on the right-hand part of the heterojunction. A similar equation would hold for one of the barrier layers in Fig. 1(a), or on the left-hand side of the heterojunction in Fig. 1(c), but with different V_{per} and U, and we shall worry later about the matching of the solutions across the interfaces. When $U(\mathbf{r}) = 0$, the solutions of Eq. (1) are the Bloch functions, $\psi_{n\mathbf{k}}(\mathbf{r})$, with:

$$\left[\frac{p^2}{2m_0} + V_{per}(\mathbf{r})\right]\psi_{n\mathbf{k}}(\mathbf{r}) = E_n(\mathbf{k})\psi_{n\mathbf{k}}(\mathbf{r}) \tag{2}$$

and

$$\psi_{n\mathbf{k}}(\mathbf{r}) = e^{i\mathbf{k}\cdot\mathbf{r}}u_{n\mathbf{k}}(\mathbf{r}),\ (u_n \text{ periodic}). \tag{2'}$$

In Eqs. (2), (2') $\mathbf{k}$ is restricted to the first Brillouin zone. For $U(\mathbf{r}) \neq 0$, we expand the solutions of Eq. (1) in terms of the complete set of Bloch functions:

$$\psi(\mathbf{r}) = \sum_n \sum_k \phi_n(\mathbf{k})\psi_{n\mathbf{k}}(\mathbf{r}). \tag{3}$$

It can be shown in detail [9,10] that, *if*:

(a) $U(\mathbf{r})$ is slowly varying on the scale of the lattice parameter a of the semiconductor, i.e. its Fourier transform $U(\mathbf{k})$ is appreciably non-zero only in a region around $k = 0$ of extension Δk, with $\Delta k \ll 2\pi/a$

(b) $U(\mathbf{r})$ is weak, i.e. if the matrix elements of U between Bloch functions are much smaller than the interband gaps at $k = 0$, $E_n(0) - E_{n'}(0)$, $(n \neq n')$

(c) we are considering eigenvalues in the neighborhood of a simple parabolic band edge n, well separated from all others:

$$E_n(\mathbf{k}) \approx E_n(0) + \frac{\hbar^2}{2m^*}k^2 \tag{4}$$

then the Fourier transform $F(\mathbf{r})$ of $\phi_n(\mathbf{k})$

$$F(\mathbf{r}) = (2\pi)^{-3/2} \int d^3k \Phi_n(\mathbf{k}) e^{i\mathbf{k}\cdot\mathbf{r}} \tag{5}$$

satisfies the effective-mass equation:

$$\left[\frac{\hbar^2}{2m^*}\nabla^2 + U(\mathbf{r})\right] F(\mathbf{r}) = (E - E_n(0))F(\mathbf{r}) \tag{6}$$

familiar from the theory of shallow donors.

Using the first order $k{\cdot}p$ expansion [11] of the periodic factor $u_{nk}(\mathbf{r})$ in Eq. (2'):

$$u_{n\mathbf{k}}(\mathbf{r}) = u_{n0}(\mathbf{r}) + \sum_{m \neq n} \frac{\mathbf{k}\cdot\langle u_{m0}|\mathbf{p}|u_{n0}\rangle}{m_0(E_n(0) - E_m(0))} u_{m0}(\mathbf{r}) \tag{7}$$

it is easy to show [10] that, in the approximation of Eq. (6) the total wavefunction $\psi(r)$ is related to the 'envelope' function $F(r)$ via the relationship:

$$\psi(\mathbf{r}) = F(\mathbf{r})u_{n0}(\mathbf{r}) + \sum_{m \neq n} \frac{-i(\nabla F(\mathbf{r})\cdot\mathbf{p}_{mn})}{m_0(E_n(0) - E_m(0))} u_{m0}(\mathbf{r}) \tag{8}$$

where p_{mn} is the momentum matrix element appearing in Eq. (7) as well.

Eq. (8) shows that, to lowest order, $F(\mathbf{r})$ is a slowly varying 'envelope' modulating the rapidly varying Bloch part $u_{n0}(\mathbf{r})$. The following correction term shows that there is a contribution from other $k = 0$ Bloch functions, proportional to the gradient of the envelope function $F(\mathbf{r})$. It is important not to forget this term if one is interested in the derivative of the wavefunction $\psi(\mathbf{r})$, which enters the boundary conditions for the effective-mass equation at the sharp boundary between two semiconductors.

Boundary Conditions for the Envelope Functions

The effective-mass equation (6) has the remarkable feature that all reference to the microscopic structure of the host semiconductor is condensed in the effective-mass m^* and the band edge energy $E_n(0)$. This is possible when the potential $U(\mathbf{r})$ is weak and slowly varying. The two parameters m^* and $E_n(0)$ assume different values in the two semiconductors, say A and B, making up an interface system. Given the high quality of state-of-the-art heterostructures the transition region includes only a few atomic layers. One could think of writing a more general form of Eq. (6), in which the effective-mass and the band edge vary as a function of z. The z-dependent $E_n(0)$ can be attached to $U(\mathbf{r})$ to form a new effective potential $U(\mathbf{r}) + E_n(0,z)$. The modified equation (6) would then read:

$$\left[-\frac{\hbar^2}{2}\nabla\cdot\left[\frac{1}{m^*(z)}\nabla\right] + U(\mathbf{r}) + E_n(0,z)\right] F(\mathbf{r}) = EF(\mathbf{r}). \tag{9}$$

The kinetic energy term has been rewritten, for a z-dependent mass, in a way which restores the hermitian character of the Hamiltonian, following Harrison [12] and Ben Daniel and Duke [13]. For z well to the left of the interface, $m^* = m_A^*$, $E_n(0) = E_n^A(0)$, and for z well on the right side $m^* = m_B^*$ and $E_n(0) = E_n^B(0)$.

We cannot take Eq. (9) seriously, however, because the variation in $E_n(0,z)$ between $E_n^A(0)$ and $E_n^B(0)$, which is typically as large as 0.1-1 eV, takes place over a few lattice distances. The potential term in Eq. (9), therefore, varies much too rapidly for the effective-mass formalism to be valid. Nevertheless, we can learn something about the boundary conditions from this differential equation. Assuming the continuity of the envelope-function (see Eq. (11) below), taking the limit in which the effective-mass and band edge energy variations occur over an infinitesimal thickness 2ϵ and integrating Eq. (9) between $z = -\epsilon$ and $z = +\epsilon$ we obtain:

$$F^A(-\epsilon) = F^B(+\epsilon) \tag{10}$$

$$\frac{1}{m_A^*}\frac{\partial}{\partial z}F^A|_{-\epsilon} = \frac{1}{m_B^*}\frac{\partial}{\partial z}F^B|_{+\epsilon}.$$

In order for the boundary conditions (10) on the envelope functions to make physical sense, we must see their implications on the total wavefunction as described by Eq. (8). In order for ψ to be continuous when F is, one must assume that:

$$u_{n0}^A \approx u_{n0}^B \tag{11}$$

and that the second term of the wavefunction (8) be small, i.e. that the k-dependence of the Bloch function u_{nk} about $k = 0$ be weak, as emphasized by Ben Daniel and Duke [13]. Eq. (11) is plausible, in the III-V semiconductor family, as long as we are considering the same band edge on both sides (e.g. the conduction band Γ minimum). Then by looking at pseudopotential wavefunctions one sees that Eq. (11) is reasonably verified.

Consider now the probability current operator. The existence of stationary states (probability density constant in time) implies that the z-component of the current be the same on all planes parallel to the interface, therefore, also that its average over a microscopic volume Ω, including one or few unit cells, be the same on both sides of the interface. Let us calculate this average on the A side. To do this, we use the wave-function as given by Eq. (8) and recall that

$$\int_{\text{cell}} u_{n0}^* \frac{\partial}{\partial z} u_{m0} = \frac{i}{\hbar} p_{nm} \tag{12}$$

and that this matrix element vanishes for $m = n$, in the conduction and valence band edges of cubic semiconductors, for symmetry reasons [14]. We find, for the average current J

$$J_A = \frac{\hbar}{m_0}\text{Im}\int_n d^3r \Psi_A^* \frac{\partial}{\partial z}\Psi_A = \frac{\hbar}{m_A^*}\text{Im}\left[F_A^*(0)\frac{\partial}{\partial z}F_A(0)\right] \tag{13}$$

with use of the well-known $k{\cdot}p$ expression for the effective-mass of band n:

$$\frac{1}{m^*} = \frac{1}{m_0} + \frac{2}{m_0^2}\sum_{m\neq n}\frac{p_{nm}^z p_{mn}^z}{E_n(0) - E_m(0)} \tag{14}$$

and of Eq. (8). Therefore, the continuity of J implies:

$$\frac{\hbar}{m_A^*}\text{Im}\left[F_A^*(0)\frac{\partial}{\partial z}F_A(0)\right] = \frac{\hbar}{m_B^*}\text{Im}\left[F_B^*(0)\frac{\partial}{\partial z}F_B(0)\right]. \tag{15}$$

It is now apparent that the boundary conditions (10) on the envelope-functions imply that the average of the probability current is constant (see Eq. (15)). They are, therefore, meaningful on physical grounds and compatible with the limit behaviour of the effective-mass equation for z-dependent mass and band edge, and we shall adopt them in our treatment of heterostructures. A further discussion of the boundary conditions is given at the end of this section.

Coupled Bands

The conditions for the validity of the simple-band approach are certainly violated in many situations of interest. This may happen because of various reasons:

a) Band degeneracy near an extremum, as in the case of the valence band maximum at Γ in all cubic semiconductors.

b) Coupling between bands producing deviations from parabolicity, as in the conduction band of direct gap semiconductors. For narrow-gap materials, like *InAs* or *InSb* the non-parabolicity of the conduction band, due to coupling with the valence bands, is quite large for energies very near the band minimum [11], but even in *GaAs* it has a sizeable effect on levels with energy $\geqslant 0.1\,eV$ above the band minimum.

c) There are situations specific to heterostructures in which the single band approach fails; if the two materials have a 'staggered' energy gap configuration, then, in a large and interesting energy range, the wavefunction has conduction band character on one side of the junction and valence band character on the other. $InAs-GaSb$ superlattices provide an example of this situation.

The simple-band case was treated in a way modelled on the theory of donor impurities; the case in which many bands contribute with comparable weight to the formation of the eigenfunctions [10] is modelled on the theory of acceptors [9]. We start by identifying the bulk bands which are to be treated on the same footing, and describe their behaviour near $k = 0$ by the generalization of Eq. (4), i.e. we write [11]

$$H_{\rm lm}(\mathbf{k}) = E_l(0)\delta_{\rm lm} + \sum_{\alpha=1}^{3} P^{\alpha}_{\rm lm} k_{\alpha} + \sum_{\alpha,\beta=1}^{3} D^{\alpha,\beta}_{\rm lm} k_{\alpha} k_{\beta} \tag{16}$$

where $\rm l,m = 1,2,\cdots,n$, and α,β run over the x,y and z directions. Given a k-vector, the n band energies $E_l(k)$ are given by the eigenvalues of the $n \times n$ matrix $H_{\rm lm}(k)$. The direct $k{\cdot}p$ coupling between the n bands is thus retained in the terms $P^{\alpha}_{\rm lm}k_{\alpha}$, where the matrix P^{α} is given by:

$$P^{\alpha}_{\rm lm} = \frac{\hbar}{m_0} \langle u_l | p^{\alpha} | u_m \rangle. \tag{17}$$

The k-quadratic terms proportional to the matrix $D^{\alpha,\beta}$, on the other hand, represent the indirect $k{\cdot}p$ coupling between two of the n bands via the other bands ($n+1$ to ∞) not included in the set. They have indeed an expression very similar to the r.h.s. of Eq. (14). A specific example is the 6×6 matrix which represents the conduction and the upper spin-orbit split component of the valence band. This is a very good description of these bands for materials with large spin-orbit splittings. We have, therefore, a conduction band with s-like character at Γ and two spin states, $s\uparrow$ and $s\downarrow$. The effective-mass m^* appearing in the corresponding diagonal terms originates from coupling to the bands not included in the set, and primarily from the split-off valence band [15]. The valence band has p-like character and the upper spin-orbit manifold corresponds to $J = 3/2$ states, classified by the four possible J_z values from -3/2 to 3/2. The P matrices are expressed entirely in terms of a single parameter P, defined as the interband momentum matrix element, $iP = \langle s | p^x | p_x \rangle$, where p_x indicates the p_x-like valence wavefunction and s the conduction wavefunction at Γ. Then it turns out that $m^{*-1} = 1 + 2P^2/3(E_c - E_v + \Delta)$, where Δ is the valence band spin-orbit splitting. If we wish to consider the energy region close to the valence band top, we can ignore the conduction band altogether. The resulting 4×4 matrix is the Luttinger Hamiltonian [16], which describes the valence band top and the splitting in light- and heavy-hole bands at $k \neq 0$. The D matrices are expressed in terms of the three Luttinger parameters, γ_1, γ_2 and γ_3, specific of each material.

We are now ready to generalize Eq. (6), (8) and (10) for the many-band case, in analogy with the many-band effective-mass equation written by Luttinger and Kohn [17] for acceptor impurities. We obtain a system of n differential equations (we assume $\hbar = 1$ in the following):

$$\sum_{m=1}^{n} \left[H_{lm}(-i\nabla) + U(\mathbf{r})\delta_{lm} \right] F_m(\mathbf{r}) = EF_l(\mathbf{r}) \tag{18}$$

for the n-component envelope-function $F_l(r)$, $l = 1,2, \cdots ,n$. The total wave-function, $\phi(r)$, is expressed, in analogy to Eq. (8), as:

$$\Phi(\mathbf{r}) = \sum_{l=1}^{n} \left[F_l(\mathbf{r})u_{l0}(\mathbf{r}) + \sum_{m>n} \frac{-i(\nabla F_l(\mathbf{r}))\cdot p_{lm}}{m_0(E_l(0) - E_m(0))} u_{m0}(\mathbf{r}) \right]. \tag{19}$$

Note that the 'kinetic energy' part is just obtained by replacing k with $-i\nabla$ in the $k\cdot p$ matrix (16) and the potential energy term is diagonal in the band index. This is a consequence of its slow spatial variation: as it can be taken as a constant in each unit cell, its off-diagonal matrix elements vanish by Bloch function orthogonality. It is also straightforward to generalize the boundary conditions. One finds

$$F_l^A(-\epsilon) = F_l^B(+\epsilon) \quad l = 1,2, \cdots ,n$$

$$\sum_{m=1}^{n} \left[P_{lm}^{z} - i \sum_{\alpha=1}^{3} \left(D_{lm}^{z\alpha} + D_{lm}^{\alpha z} \right) \nabla_\alpha \right] F_m \tag{20}$$

continuous at $z = 0$ for $l = 1,2, \cdots ,n$.

One must, of course, assume Eq. (11) for the n bands of interest. Let us write down these boundary conditions for an ideal planar interface between lattice matched semiconductors A and B and for the $n \times n$ coupled-band case of Eq. (16). In this situation the potential $U(\mathbf{r})$ depends only on z and k_x and k_y are good quantum numbers. That means that we can write

$$F^{A,B}(\mathbf{r}) = e^{ik_x x} e^{ik_y y} F^{A,B}(z) \tag{21}$$

so that the second Eqs. (20) reads

$$\sum_{m=1}^{6} \left[P_{lm}^{z} + \sum_{\alpha=x,y} \left(D_{lm}^{z\alpha} + D_{lm}^{\alpha z} \right) k_\alpha - 2iD_{lm}^{zz} \frac{\partial}{\partial z} \right] F_m \tag{22}$$

continuous.

A further simplification of the boundary conditions occurs by noticing that, if Eq. (11) holds for $l = 1, \cdots ,n$, then also the P_{lm} matrix elements between these Bloch functions must be equal, as they are momentum matrix elements connecting the same conduction and valence band functions. It is empirically known [18] that such matrix elements depend essentially only on the lattice constant and, therefore, are indeed about equal in two lattice-matched materials. Thus the continuity of the $P_{lm}^{z} F_m$ terms follows from the continuity of the F_m alone, and Eq. (22) further simplifies to:

$$\sum_{m=1}^{6} \left[\sum_{\alpha=x,y} \left(D_{lm}^{z\alpha} + D_{lm}^{\alpha z} \right) k_\alpha - 2iD_{lm}^{zz} \frac{\partial}{\partial z} \right] F_m \tag{22'}$$

continuous. This is the form of boundary conditions [10,19,20], used e.g. in the specific examples to be discussed soon.

DISCUSSION OF THE ENVELOPE-FUNCTION APPROXIMATION

In deriving the equations for the envelope-function approximation, the many and sometimes very restrictive assumptions which were needed were stated in detail. It is quite clear that there are situations in which they are untenable. Apart from the cases of strong or rapidly varying potentials $U(\mathbf{r})$, which cause the breakdown of effective-mass theories also for impurity states, the condition (11) cannot be satisfied for many interesting heterostructures. This occurs for all cases in which the eigenstates we want to compute are derived from the mixing of Bloch waves from different points of the Brillouin zone (e.g. the conduction-related states in $Si-Si_xGe_{1-x}$ heterostructures, or in $GaAs-Al_xGa_{1-x}As$ with $x>0.4$, where the alloy is indirect, etc.). In these situations, one must resort to other methods, such as the empirical tight-binding scheme [2-4], or the pseudopotential method [5,6,21]. In either approach, some of the attractive features of the envelope-function method are lost. Alternatively, one can devise an extension of the envelope-function method suitable for such heterostructures, but at the price of introducing some additional parameters in the theory. Ando [22] has proposed such an approach for *GaAs*-indirect *AlGaAs* heterostructures.

Also in the cases where the two semiconductors have relevant band edges which, because of their similarity in symmetry and chemical origin, can be paired to satisfy Eq. (11) approximately, a few words of caution on the accuracy of the envelope-function calculations are necessary. Our discussion of boundary conditions is based on the idea of perfect periodicity of both media up to a geometrical plane defining the interface. This is certainly an idealization and, in the very important case in which one of the components is an alloy (e.g. $GaAs-Al_xGa_{1-x}As$) it is impossible even to define such a plane. It would be more realistic to talk about an interface layer, comprising several atomic planes, separating the two semiconductors, and characterized by given reflection and transmission coefficients. These would be again parameters of unknown value to be used as input.

Finally, a word of caution concerning the use of $k{\cdot}p$ matrices to describe the bulk band edges was recently put forward by Schuurmans and 't Hooft [20]. They pointed out that a Hamiltonian like Eq. (16) can have unphysical solutions; in the case of the 8×8 description of *GaAs* including conduction, upper- and split-off valence bands, for example, they find bulk solutions with E in the gap and very large real k-values. Care must be exerted to prevent that such unphysical branches are included in the construction of the superlattice solution.

In spite of these caveats the envelope-function results are in overall good agreement with experiments, sometimes even more than one would expect. When comparisons with e.g. tight-binding calculations - which are free from the problem of boundary conditions - are performed in detail, the agreement is very good [20,2-3,22,23]. These facts can be regarded as valid a posteriori justifications for the approximations of the method.

There is, on the other hand, a conceptual need to justify the envelope-function method and the boundary conditions rigorously. An attempt in this direction is being made by Burt [24-25], but more work is clearly necessary.

Examples of Results for Some Systems

$GaAs-Al_xGa_{1-x}As$

The envelope-function method is applicable to this system for all states derived from Γ-point bulk Bloch functions. This excludes the conduction subbands high enough in energy to approach the secondary minima at X and L; this is the case at all energies for $x > 0.4$, where the alloy has an indirect gap. The valence subbands, on the other hand, are accessible to the method at all alloy compositions.

An important input parameter of the calculations is the band offset. Even for an extensively investigated system like $GaAs-AlGaAs$, this quantity was very controversial until recently. There seems to be an emerging consensus for a value of the conduction band discontinuity between 60% and 70% of the difference in band gap between $AlGaAs$ and $GaAs$ [26-30]. Theory can only be of limited help, given the uncertainty ($>0.1\,eV$) of state-of-the art band theory in bulk materials.

We start by some general considerations. The growth axis of MBE heterostructures is in a <001> direction [31]. In a superlattice, one has, therefore, a Brillouin zone which is very thin in the z direction ($2\pi/d$, where d is the period) and has the usual $\sim 2\pi/a$ size, where a is the lattice constant, in the x and y direction. One has, therefore, a very asymmetric band structure with a bulk-like bandwidth ($\sim 10\,eV$) in the k_x, k_y plane and $\sim (a/d)^2 \cdot 10\,eV$, typically $10\,meV$ bandwidth in the k_z-direction.

The dispersion in the k_z-direction is relatively simple to handle even in the 6×6 coupled-band model, because most of the off-diagonal terms vanish for $k_x, k_y = 0$. In detail, we see that $J_z = \pm 3/2$ states, corresponding to the heavy holes, completely decouple and behave as simple particles with effective mass $m_{hh} = 1/(\gamma_1 - 2\gamma_2)$. We have a Kronig-Penney type of eigenvalue problem, with the boundary condition (10), and, as shown e.g. by Bastard [32], one finds the implicit dispersion relation:

$$\cos(k_z d) = \cos(k_{zA} d_A)\cos(k_{zB} d_B) - \frac{1}{2}\left[\frac{m_{hh}^A k_{zB}}{m_{hh}^B k_{zA}} + \frac{m_{hh}^B k_{zA}}{m_{hh}^A k_{zB}}\right]\sin(k_{zA} d_A)\sin(k_{zB} d_B) \tag{23}$$

where

$$k_{zA,B}^2 = 2m_{hh}^{A,B}\left[E - E_v^{A,B}\right] \tag{23'}$$

and the superlattice period d is made up of thickness $d_A (d_B)$ of material $A(B)$.

As for light-holes and conduction electrons, one has the 2×2 matrix between $s\uparrow$ and $J_z = +1/2$, or an equivalent one for $s\downarrow$ and $J_z = -1/2$:

$$\begin{bmatrix} E_c + \frac{1}{2m^*}k_z^2 & -i\sqrt{\frac{2}{3}}Pk_z \\ i\sqrt{\frac{2}{3}}Pk_z & E_v - \frac{1}{2}(\gamma_1 + 2\gamma_2)k_z^2 \end{bmatrix}. \tag{24}$$

One could solve directly the superlattice problem for these coupled bands as discussed previously or, for energies very near the band edges, find the eigenvalues of (24), $\xi_c(k_z), \xi_v(k_z)$ and describe their non-parabolic dispersion via energy dependent effective-masses. For a complete discussion of k_z-dispersion in $GaAs-AlGaAs$ superlattices, we refer the reader to the treatment of Schuurmans and 't Hooft [20]. In general, the larger the energy distance from the band extremum, the more accurate the description of the non-parabolicity should be.

Consider now the dispersion for $k_x, k_y \neq 0$. This is much more complicated, but also very interesting. The main features can be investigated in a model in which the conduction band is uncoupled from light- and heavy-hole bands, but the latter are coupled to each other via the Luttinger Hamiltonian [16]. For the conduction subbands one has again a simple Kronig-Penney like problem with possible non-parabolic effects embodied in an energy-dependent effective-mass.

For the valence bands we have to deal with the 4×4 Luttinger Hamiltonian, which can be written in the form:

$$\begin{array}{c|cccc|} & 3/2 & 1/2 & -1/2 & -3/2 \\ 3/2 & a_+ & b & c & 0 \\ 1/2 & b^* & a_- & 0 & c \\ -1/2 & c^* & 0 & a_- & -b \\ -3/2 & 0 & c & -b^* & a_+ \end{array} \tag{25}$$

with

$$a_{\pm} = E_v - \frac{1}{2}(\gamma_1 \pm \gamma_2)(k_x^2+k_y^2) - \frac{1}{2}(\gamma_1+2\gamma_2)k_z^2 \tag{25'}$$

$$b = \sqrt{3}\gamma_3(k_x - ik_y)k_z$$

$$c = \frac{\sqrt{3}}{2}\gamma_2(k_x^2-k_y^2) - 2i\gamma_3 k_x k_y.$$

The bulk solutions of (25) give the spin-degenerate light- and heavy-hole bands. This degeneracy is actually lifted in non-inversion symmetric zincblende materials, like *GaAs*, by terms linear in the *k*-vector [33]. The terms are, however, extremely small and we neglect them here. The valence bands are anisotropic and, therefore, one has to expect that also for superlattices the band dispersion will be different for different directions in the k_x, k_y plane. However, this warping of the bands is small and a good approximation is obtained by the 'axial model' in which [34,35] γ_2 and γ_3 in the *c* matrix elements (see the last of Eq. (25')) are replaced by $\gamma = (\gamma_2+\gamma_3)/2$. With this replacement, Eq. (25) acquires cylindrical symmetry about the *z*-axis, and the bands are isotropic in the k_x, k_y plane.

The solution of the four-component equation with the boundary conditions was originally pursued numerically [35] but, as shown by Andreani, Pasquarello and Bassani [36], an analytic reduction to a search for the zeroes of a small determinant is possible. At $k_x = k_y = 0$ there is no mixing of the light and heavy components. It is therefore possible to identify each subband as heavy or light (this is a consequence sequence of the envelope function approximation. For a more complete description, see Ref. (21)). As we move out of the $k = 0$ axis, the mixing grows rapidly and produces the non-parabolic behaviour, first pointed out by Nedorezov [37], who solved the problem exactly in the limit of infinitely high barriers. A particularly striking feature is the positive in-plane effective mass of the first light-hole subband, which changes sign and becomes hole-like only for $k \sim \pi/L$, where L is the well thickness.

In the example of $GaAs-AlGaAs$, one has to deal with a system in which no significant charge rearrangement across or near the interfaces takes place, if we consider superlattices involving intrinsic semiconductors. In contrast, in modulation doped superlattices, filling of the quantum wells takes place, with carriers released from impurities located in different layers. whenever such transfers take place, they contribute to the definition of the one-electron potential $U(\mathbf{r})$ appearing in the envelope-function equations (6) or (18). The potential depends on the charge density of the states filled by the transferred carriers; and, because the corresponding wavefunctions depend in turn on the potential $V(\mathbf{r})$, the problem must be solved self-consistently, within some scheme of approximation, e.g. the Hartree approximation, the local density approximation, etc. These procedures are familiar from the treatment of *Si* MOS systems [38]. The relative ease of performing such self-consistent calculations is one of the advantages of the envelope-function methods.

From the experimental point of view, the complexity of the valence bands is best revealed by intra- and interband optical experiments. Analysis [39-41] of Raman scattering [42] results on *p*-type modulation doped wells and of polarized luminescence [43] in *n*-type structures confirms the strong light-heavy hole mixing and the upward curvature of the light-hole first subband. Further confirmation is offered by magneto-optical experiments, to be discussed later.

$InAs-GaSb$ Superlattices and Quantum Wells

The $InAs-GaSb$ system is particularly interesting [44] because of the peculiar band line-up of these materials. The valence band top of the $GaSb$ is estimated to lie $\sim 0.15\,eV$ higher than the bottom of the $InAs$ conduction band [44,45]. Electrons tend to transfer to the $InAs$ layers, unless the layers are so thin that the particle-in-a-box quantization energy reverses the level order, locating empty $InAs$ quantum well conduction states above the full $GaSb$ valence states. Therefore, charge transfer occurs only for $InAs$ layers thicker than $\sim$90Å, and in this case self-consistency plays a crucial role in determining the level structure. Such a calculation [46], was performed in the Hartree approximation and in the 6×6 band model with neglects the split-off valence band. The results show strong valence-conduction subband mixing, with small (few meV) hybridization gaps opening up at subbands crossing points. For a 120Å $InAs-80$ Å $GaSb$ superlattice, the density of transferred carriers is $\sim 0.9\times10^{12}\,cm^{-2}$, thus providing good justification to the Hartree approximation. Magneto-optical experiments, to be discussed later, indicate good agreement with the envelope function results. Single quantum wells of $InAs$ between $GaSb$ barriers, however, show a concentration of electrons in $InAs$ much larger than that of holes in the barriers [47]. The origin of these extrinsic charges is not clear, although a plausible explanation is that the well is in a strong band-bending region, deriving from the proximity of the $GaSb$ free surface of the outer barrier [48].

$CdTe-HgTe$ Superlattices

Superlattices of II-VI compounds have recently attracted much attention, and among them the $CdTe-HgTe$ system is especially interesting. This is a consequence of the zero-gap character of $HgTe$ and of the band line-up, which puts the $HgTe$ Γ_8 edge above the corresponding edge of $CdTe$. The value Δ of this band offset is controversial, the estimate $\Delta = 0.04\,eV$ from magneto-optics [49,50] being confirmed by Raman measurements [51], but contradicted by photoemission results [52,53] which suggest a much larger value $\Delta\approx0.35\,eV$.

These systems provide an excellent testing ground for the boundary conditions, Eqs. (10) and (22'). This is because the Γ_8 'light holes' are in fact Γ_8 'electrons' in $HgTe$, i.e. they form an unoccupied band with positive mass in this material, due to the interchange in the energy position of the Γ_6 and Γ_8 edges. There are therefore three types of Γ_8 states in these superlattices: Γ_8 'electrons', Γ_8 heavy holes, both confined in the $HgTe$ wells, and in addition unusual interface states [23,54] arising in the gap between Γ_8 light holes in $CdTe$ and Γ_8 electrons in $HgTe$ and decaying exponentially on both sides of an interface. If one ignores other bands for simplicity, then it is easy to see that the boundary conditions, Eq. (10), allow for such states if m_A, m_B have opposite signs.

The good agreement [23] between envelope-function calculations, in which the boundary conditions are so crucial in predicting the interface states, and tight-binding results, provides strong a posteriori confirmation of the envelope-function method. On the other hand, the claim that the band offset value carefully measured by photoemission has to be drastically reduced to account for infrared and Raman results leads to the conclusion that some aspects of the physics of this system are not understood. Jaros et al. [6] went as far as claiming that the envelope function method is not applicable to these systems, because of the strength of the *Hg* potential, and that pseudopotential calculations can reconcile the observed features of the electronic structure with a large band offset. This statement is however controversial [55,56] and very recently if was shown, by an envelope function calculation [57], that the properties of the electronic structure relevant to the interpretation of optical experiments (gap, effective mass, etc.) are strongly non-monotonic functions of the band offset, with similar values for $\Delta = 0.04\,eV$ or $0.35\,eV$. This would provide a resolution of the paradox, without giving up the applicability of few-band calculations.

External Fields

As indicated in the discussion of the envelope-function method, any external field varying slowly with respect to the lattice parameter of the constituents can be included in the envelope function equations. In particular:

- Electric fields [58], via the potential $e\mathbf{F}\cdot\mathbf{r}$ (diagonal in Eq. (18)).
- Magnetic fields, via the replacement $\nabla \rightarrow \nabla - \frac{e}{c}\mathbf{A}$ where **A** is the vector potential [59].
- Strain fields [60,61], via the potential $\bar{e}\cdot\bar{d}$, where *e* and *d* are the strain and deformation potential tensors, and via the strain dependence of the band gap discontinuities.
- Screened Coulomb fields of donor and acceptor impurities, via the potential $\pm e^2/\epsilon|\mathbf{r}-\mathbf{R}|$, where **R** is the position of the impurity [62,63].

The effect of an external electric field on a superlattice is interesting because it becomes possible to observe [64] some spectacular phenomena, such as the Wannier-Stark quantization and the Bloch oscillations which are not accessible in an ordinary solid. Let us consider the Hamiltonian for an electron subject to a periodic potential, $V_{per}(z)$ and to an electric field F:

$$H = \frac{p^2}{2m} + V_{per}(z) - eF_z. \tag{26}$$

It is easy to see that if $\psi(z)$ is an eigenstate of Eq. (26) with energy E, then $\psi(z+\nu d)$, where ν is an integer and d is the period of V_{per} is also an eigenstate, with energy $E - eF\nu d$. Eigenstates belong therefore to 'ladders' of equidistant levels with spacing eFd. This characteristic energy eFd (or frequency $f = eFd/h$) is also found in the semiclassical derivation of Bloch oscillations [65]. Let $E_n(\mathbf{k})$ be the dispersion of the n-th band (or subband) in the system, and let $F = (0,0,F_z)$ be the external field. If Zener tunneling (see Fig. 2) between different bands is negligible, the semiclassical equation of motion of an electron in subband n is (we restore h in Eqs. 27-29)

$$\hbar\frac{dk_z}{dt} = eF_z \tag{27}$$

with solution

$$k_z(t) = k_z(0) + \frac{e}{\hbar}F_z t. \tag{28}$$

The periodicity of the band structure for $k_z \rightarrow k_z + 2\pi/d$ implies that the particle executes a periodic motion, with the characteristic frequency f discussed above.

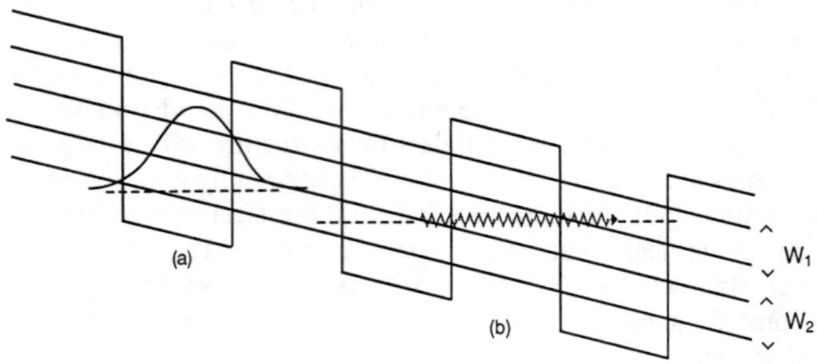

Figure 2. A superlattice in an external electric field (schematic): (a) sketch of the wavefunction of an electron executing oscillatory motion in the semiclassical picture; (b) Zener tunneling process between two states as in (a) belonging to different bands W_1 and W_2, separated in space, but with the same energy.

In a superlattice with $d = 200$ Å, $f = 4.8\times10^{12}s^{-1}$ for $F_z = 10^4 V/cm$. The amplitude of the periodic motion (see Fig. 2 and Eq. 27) in terms of the bandwidth W_n, is:

$$l = \int_0^{T/2} \frac{1}{\hbar}\frac{dE_n}{dk_z}\cdot dt = \int_0^{W} \frac{1}{eF_z} dE_n = \frac{W_n}{eF_n}. \tag{29}$$

In a usual solid, $W_n \sim 10\,eV$ and $d \sim 5$Å. Therefore, this 'Stark localization length' for $F_z = 10^4 V/cm$ is about $10\mu m$, a huge distance compared with the lattice constant or the mean free path; similarly the spacing of the Stark ladder is $\sim 0.5\,meV$. These figures account for the difficulty of observing the predicted effects in practice. In superlattice with $d \sim 100$Å, $W_n \sim 10\,meV$, on the other hand, we find for the same $F_z, l \sim 100$Å, corresponding to localization in one cell, and a ladder energy spacing of the same order as the bandwidth. Very recently, optical experiments [64] seem to have revealed energy features with separation linear in the applied electric field, in agreement with the Wannier-Stark ladder prediction.

Many revealing experiments on heterostructures are performed with an external magnetic field, which has a profound influence on the energy spectrum. It produces an enrichment and a sharpening of the optical properties and striking transport phenomena, like the quantum Hall effect.

To include a magnetic-field in the many-band envelope-function formalism [59], we follow the lines of the classical work of Luttinger [16] on the cyclotron resonance of holes in semiconductors. The field $\mathbf{B} = (0,0,B)$ is described by the vector potential $\mathbf{A}$. (It is convenient to choose a gauge with $A_z = 0$). In the $k.p$ bulk Hamiltonian, Eq. (16), k is to be replaced by $\mathbf{k}' = \mathbf{k} + (e/c)\mathbf{A}$. Then it is easy to see that the x and y components of this new operator do not commute, but instead

$$\left[k'_x, k'_y\right] = -i(e/c)B. \tag{30}$$

Also, new diagonal terms arise, representing the direct coupling of the electron and hole spins to the field. The conduction band diagonal elements in the 6×6 Hamiltonian now become

$$H_c = E_c + \frac{1}{2m^*}(\mathbf{k} + (e/c)\mathbf{A})^2 + \frac{e}{2c}g^* s_z B \tag{31}$$

where s is the electron spin, m^* is the effective-mass defined previously in the discussion of Eq. (16-17) and the effective g-factor [15] g^* is simply $2/m^*$. For the valence band diagonal elements, one must add the term

$$\frac{e}{c}\kappa J_z B + \frac{e}{c} q J_z^3 B \tag{32}$$

where J_z is the spin 3/2 operator and k and q are two material parameters [16]. Actually q is very small for the semiconductors of interest here and the second part of (32) can be neglected. It is easy to see that as a consequence of (30) we can define operators a, a^+

$$a = \sqrt{\frac{c}{2eB}}(k_x' - ik_y') \tag{33}$$

$$a^+ = \sqrt{\frac{c}{2eB}}(k_x' + ik_y')$$

with commutator

$$[a, a^+] = 1 \tag{33'}$$

so that all terms in k_x or k_y in the Hamiltonian can be expressed in terms of these harmonic oscillator raising and lowering operators.

The resulting energy level scheme (the Landau levels) is strongly dependent on the orientation of the field with respect to the layers. We will shortly discuss the cases in which the field is either perpendicular or parallel to them.

Landau Levels: Perpendicular Fields

It is possible to obtain a solution of the 6×6 Hamiltonian in closed form [59], if the axial approximation in the valence band portion is adopted (see the discussion following Eq. (25)). This procedure was applied to *GaAs − AlGaAs*, to *InAs − GaSb* and to *CdTe − HgTe* systems. The Landau levels obtained are directly comparable to intraband magneto-optical experiments in *GaAs − AlGaAs*, but not to interband ones, which are strongly affected by exciton effects, to be discussed in a later Section. For *InAs − GaSb* and *CdTe − HgTe* the distinction between the two types of experiments is ambiguous, because of the strong mixing induced by the band line-up.

The structure of the valence band Landau levels, in particular, can be directly probed with intraband experiments. Cyclotron resonance experiments [66] on p-type *GaAs − AlGaAs* single heterojunctions were interpreted by several groups [67-70] in terms of Landau level calculations in the $B = 0$ selfconsistent potential of the heterojunction. The agreement can be good, although this must involve some accident, because of the use of the Hartree approximation (inadequate for hole densities $\sim 5\times10^{11} cm^{-2}$), of the axial approximation, etc. Comparisons of cyclotron resonance in p-type quantum wells are less favourable [71], whereas satisfactory semiquantitative agreement is found for intraband Raman scattering experiments [72] on p-type quantum wells. In general, while theory seems to be correct in predicting the main features of the valence band Landau levels, more work is needed to establish quantitative agreement with interband magneto-optics.

In the *InAs − GaSb* superlattices, magneto-optical experiments [45] were crucial in establishing the basic properties of the electronic structure. The calculated [59] transition energies are in good agreement with experiment. More recently [73], the theory was shown to provide a reasonable account of magneto-optical experiments performed under hydrostatic pressure, which modifies the band offset; due to the peculiar nature of this system, the reduction of the overlap between the *GaSb* valence band and the *InAs* conduction band with hydrostatic pressure affects the electronic states very much, and produces a change in the character of prominent transitions from interband-like to intraband-like.

LANDAU LEVELS: PARALLEL FIELDS

Interesting information on the electronic properties of quantum wells and superlattices can be obtained by performing magneto-optical experiments in a field parallel to the layers of the heterostructure. In an intuitive picture, the field forces the particles to perform cyclotron orbits in a plane intersecting both the well and the barrier materials. If the radius of such orbits (i.e. the magnetic length $a_M = (c/eB)^{1/2}$) is large compared to the thickness of the layers, then the completion of the cyclotron orbit requires traversal of the barriers. This condition is easily met experimentally as, e.g., a $10T$ field corresponds to $a_M \sim 8nm$. Parallel field experiments probe therefore intriguing properties such as tunneling, perpendicular transport, etc., and are particularly fruitful in the investigation of superlattice band structure in the growth direction [74], which is more elusive than the in-plane dispersion discussed in the previous sections.

To keep the notation of the previous section, we now denote the growth direction by x, and choose the gauge $\mathbf{A} = (0, Bx, 0)$ for the field B parallel to z. Consider first an electron in a bulk semiconductor in a simple parabolic band with mass m^*. The Schrödinger equation can as usual be cast in the form

$$\left[\frac{p_z^2}{2m^*} + \frac{p_x^2}{2m^*} + \frac{1}{2}m^*\omega_c^2(x-x_0)^2\right]\psi = E\psi \tag{34}$$

where $x_0 = -a_M^2 k_y$, $\omega_c = eB/m^*c$.

If we now consider a heterostructure, e.g. a superlattice, we must add in Eq. (34) a periodic potential $V_s(x)$. It is then apparent that the degeneracy of the free-electron Landau levels with respect to x_0 is now lifted, because it makes a difference if x_0 is, e.g., in a barrier or in a well.

The general problem of a particle in a field plus a periodic potential cannot be solved exactly. However, if the magnetic length a_M is large compared to the superlattice period d, one can use the semiclassical quantization scheme [75]. The approach is discussed in great detail by Zilberman [76]. One must first consider the subband structure problem for the superlattice in the absence of the field, and construct the sections of the constant energy-surfaces in k-space with the k_z = constant planes. For the $k_z = 0$ case, these are schematically shown in Fig. 3. This figure reflects the fact that the bandwidth in the superlattice direction, x, is much smaller than in the in-plane directions. The closed curves correspond to the energies between the subband bottom and the bandwidth in the x direction. For energies exceeding this value, the constant energy surface touches the edge of the superlattice Brillouin zone, corresponding to the 'open orbit' case (one must think of a periodic repetition of the zone in the superlattice direction x).

When the field in the z direction is turned on, both open and closed curves correspond to k-space trajectories of the representative point in the usual semiclassical Bloch dynamics [75]. To get a deeper picture of the quantized spectrum, one can regard these k-space trajectories as phase-space trajectories of some kind. Indeed, Eq. (30) shows that k_x' and k_y' are like canonical variables (i.e. like 'y' and 'k_y'). We then construct the potential in the 'y' direction that would produce the classical phase space trajectories of Fig. 3. Clearly, it is a periodic potential, with potential wells in which closed phase space orbits occur for low energies and unbound orbits for energies higher than the bandwidth.

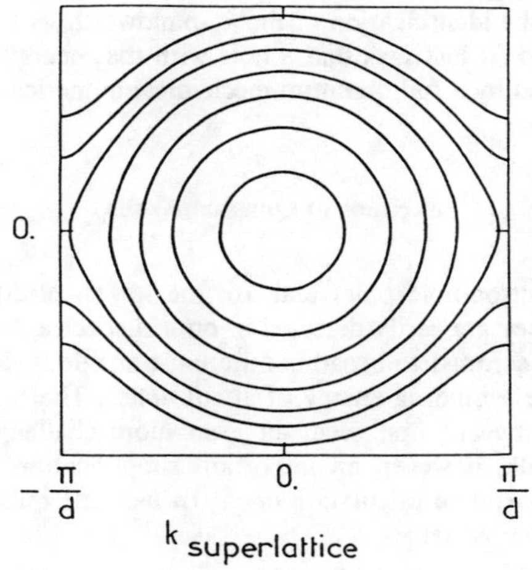

Figure 3. Constant energy surfaces for a subband in a superlattice of period d. The Brillouin zone in the $k_{/\!/}$ direction extends to much larger values (of order π/a, where a is the lattice constant of the constituent semiconductors).

The corresponding quantized spectrum is obtained by solving the motion in this potential, in a WKB-like approach.

Then, the energies within the x-direction bandwidth (closed orbit region) correspond to essentially discrete levels, or to very narrow bands, as the probability of k-space tunneling to the next periodically equivalent well is small [76]. The gap between these discrete levels is $\approx\omega_c$. Their energies are given by the Onsager prescription:

$$A_k(E) = \frac{2\pi eB}{c}(n+\frac{1}{2}) \tag{35}$$

where $A_k(E)$ is the k-space area of the closed orbit. As the energy gets close to the bandwidth, the orbits get larger and the tunneling probability increase. The discrete levels begin to broaden appreciably. At larger energies, the nearly flat open orbits correspond to nearly-free phase space motion, therefore to a weak periodic potential, in which wide bands separated by narrow gaps occur. Zilberman [76] shows that a 'band' extends approximately over the energy region:

$$\frac{2\pi eB}{c}n < A_k(E) < \frac{2\pi eB}{c}(n+1) \tag{36}$$

where $A_k(E)$ is the area in the first Brillouin zone between two open orbits with energy E, and is separated by gaps of order $\sim\omega_c(d/a_M)^2$ from the neighbouring bands.

The results of this analysis are in excellent agreement with the magneto-optical interband experiments of Belle et al. [77] in $GaAs-AlGaAs$ superlattices with $5nm$ period (as well as with their numerical results for the Kronig-Penney model). A series of sharp absorption peaks is observed in the region corresponding to transitions between the quasi-discrete energy levels of the valence and conduction subband. Above this energy range, no sharp structure is observed, corresponding to the broad 'bands' discussed above. The intraband results of Duffield et al. [78,79] in n-type samples are also in agreement with this general picture.

It is however important to point out that the semiclassical analysis cannot be invoked for the valence subbands, because of their origin from a degenerate bulk band. However, one would suspect that the identification of the k_x-bandwidth with the discrete portion of the spectrum should hold (it just says that a hole with that energy does not 'see' the barriers). This was confirmed by a full quantum mechanical numerical calculation [80].

Excitons in Quantum Wells

Excitons are bound electron-hole pairs and are the lowest electronic excited states of non-metallic crystals. They are easily detected in optical spectra, because they give rise to sharp line structures, in contrast to broad continuum transitions. In a semiconductor like *GaAs* bulk excitons have a binding energy of about $4meV$. The theoretical determination of this binding energy may at first seem an even more challenging problem than the acceptor one. In the bulk, however, an important simplification is possible [81], which makes a very good approximate calculation easy. In fact, the bulk effective mass Hamiltonian for excitons can be written:

$$H_{\mathrm{exc}} = H_e(\mathbf{p}_e) - H_h(\mathbf{p}_h) - e^2/\epsilon|\mathbf{r}_e - \mathbf{r}_h|. \tag{37}$$

Here the first term is simply $p_e^2/2m_e^*$, the second is the Luttinger Hamiltonian for the hole (i.e. equation (25) with $p_h = -i\partial/\partial r_h$ replacing $\mathbf{k}$). We can define a relative coordinate $r = r_e - r_h$, its conjugated momentum p, and a coordinate $R = 1/2(r_e + r_h)$, with conjugate momentum $P = p_e + p_h$. P is actually a constant of the motion, corresponding to the overall momentum of the pair (we avoid mentioning the 'center of mass' because, due to the structure of the valence band edge, there is no center of mass for the electron-hole pair [82]). Optical transitions can only create or destroy $P = 0$ excitons, for which the Hamiltonian becomes

$$H_{\mathrm{exc}} = H_e(\mathbf{p}) - H_h(\mathbf{p}) - e^2/\epsilon r. \tag{38}$$

Notice that this Hamiltonian is a 4×4 matrix. The electron term and the Coulomb term are assumed multiplied by the unit matrix, i.e. they are added to the diagonal terms of the Luttinger Hamiltonian. Now, apart from the Coulomb term, there are two kinds of terms: the terms proportional to $(\gamma_1 + 1/m_e^*)$, and other terms proportional to γ_2 or γ_3. For *GaAs* $m_e^* = 0.067$, $\gamma_1 = 6.85$, $\gamma_2 = 2.10$, $\gamma_3 = 2.90$. Therefore, $\gamma_1 + 1/m_e^* >> \gamma_2, \gamma_3$. Neglecting all γ_2, γ_3 terms in first approximation one finds four degenerate exciton states with binding energy

$$R^* = \frac{e^4}{2(\gamma_1 + 1/m_e^*)\hbar^2\epsilon^2}. \tag{39}$$

The off-diagonal terms can be included as a second-order perturbation [81], which does not remove the fourfold degeneracy, but provides a correction to the binding energy.

This procedure cannot be applied to acceptor states calculations, because the large value of $1/m_e^* (\approx 15)$ is essential to the argument. Unfortunately it is not easy to carry out this procedure for a quantum well either. First of all, we lose translation invariance in the z direction: we can, therefore, set the component of P along the interfaces, $P_{/\!/} = 0$, but P_z is no longer a good quantum number. Also, the boundary conditions depend separately on z_e and z_h, not on $z_e - z_h$. The two-particle nature of the problem actually gives rise to a situation opposite to the bulk case: the exciton problem is more complicated than the acceptor problem.

A correct treatment of the effects produced by the degeneracy of the valence band is necessary to ensure quantitative understanding of the spectra observed, especially in presence of external electric [83] or magnetic [84-87] fields; a qualitative general picture is on the other hand obtainable by simple considerations on the physics of a hydrogenic atom between two plane barriers which enhance the binding energy (with respect to the subband edge) by keeping the electron and the hole closer to each other. After some early attempts to quantify this kind of picture by more refined variational treatments of the kind hydrogenic problem, full scale calculations have recently been undertaken, which aim at a complete quantitative description of the experiments, with essentially no approximations except those intrinsic to the envelope-function method. Among the most recent and complete are those of Zhu and Huang [88,89], of Andreani and Pasquarello [90] and of Bauer and Ando [91,92], who went quite far in the inclusion of external fields. The results [91] for the transition energies in a magnetic field are in good agreement with experiments of Ossau et al. [93].

The problem of the exciton in quantum wells seems therefore well understood, at least at the level of extensive numerical diagonalizations. One problem raised by recent experiments is the doublet splitting of the lowest lying exciton state [94], which was claimed to be as large as $6meV$ for 10 A $GaAs-AlGaAs$ quantum wells. Explanations in term of the electron-hole exchange interaction [95], which in bulk $GaAs$ produces an effect of the order of $0.02meV$, do not seem convincing; the enhancement due to confinement in one direction can hardly produce such a large effect. More recent experimental [96] and theoretical [97] work in thin-layer $GaAs-AlGaAs$ multiple quantum well systems seems to suggest a much smaller splitting due to exchange effects, although more direct evidence would be desirable.

References

1. T. Ando, A.B. Fowler and F. Stern, *Rev. Mod. Phys.* **54**, 437 (1982)
2. J.N. Schulman and Y.C. Chang, *Phys. Rev.* **B24**, 4445 (1981).
3. J.N. Schulman and Y.C. Chang, *Phys. Rev.* **B27**, 2346 (1983).
4. Y.C. Chang, *Phys. Rev.* **B37**, 8215 (1988).
5. M. Jaros, K.B. Wong and M.A. Gell, *Phys. Rev.* **B31**, 1205 (1985).
6. M. Jaros, A. Zoryk and D. Ninno, *Phys. Rev.* **B35**, 8277 (1987).
7. C.G. van de Walle and R.M. Martin, *Phys. Rev.* **B24**, 5621 (1986).
8. L.J. Sham and M. Schlüter, *Phys. Rev.* **B32**, 3883 (1985).
9. W. Kohn, in: *Solid State Physics: Advances in Research and Applications*, F. Seitz and D. Turnbull editors, Academic, New York, 1957, vol. 5, p. 257.
10. M. Altarelli, in: *Heterojunctions and Semiconductor Superlattices*, G. Allan, G. Bastard, N. Boccara, M. Lannoo, M. Voos editors, Springer, Berlin, 1986, p. 12.
11. E.O. Kane, in: *Semiconductors and Semimetals*, R.K. Willardson and A.C. Beer, editors, Academic, New York, 1966, vol. 1, p. 75.
12. W.A. Harrison, *Phys. Rev.* **123**, 85 (1961).
13. D.J. Ben Daniel and C.B. Duke, *Phys. Rev.* **152**, 683 (1966).
14. See e.g. E.O. Kane, ref. 11.
15. L.M. Roth, B. Lax and S. Zwerdling, *Phys. Rev.* **114**, 90 (1959); L.M. Roth, in: *Handbook of Semiconductors*, W. Paul, editor, North-Holland, Amsterdam, 1982, vol. 1, p. 451.
16. J.M. Luttinger, *Phys. Rev.* **102**, 1030 (1956).
17. J.M. Luttinger and W. Kohn, *Phys. Rev.* **97**, 869 (1955).

18. M. Cardona, in: *Atomic Structure and Properties of Solids*, E. Burstein, editor, Academic, New York, 1972, p. 514.

19. M. Altarelli, in: *Applications of High Magnetic Fields in Semiconductor Physics*, G. Landwehr, editor, Springer, Berlin, 1983, p. 174.

20. M.F.H. Schuurmans and G.W. 't Hooft, *Phys. Rev.* **B31**, 8041 (1985).

21. C. Mailhiot and D.L. Smith, *Phys. Rev.* **B35**, 1242 (1987).

22. T. Ando, in: *Proceedings of the 3rd Brazilian School of Semiconductor Physics*, World Scientific, Singapore, 1987, p. 23.

23. Y.C. Chang, J.N. Schulman, G. Bastard, Y. Guldner and M. Voos, *Phys. Rev.* **B31**, 2557 (1985).

24. M.G. Burt, *Semicond. Sci. Tech.* **2**, 401 and 706 Erratum (1987).

25. M.G. Burt, *Semicond. Sci. Tech.* **3**, 739 (1988).

26. H. Kroemer, *Surf. Sci.* **174**, 299 (1986).

27. G. Duggan, *J. Vac. Sci. Tech.* **B3**, 1224 (1985).

28. J. Menéndez, A. Pinczuk, D.J. Werder, A.C. Gossard, J.H. English, T.H. Chiu and W.T. Tsang, *Superlattices and Microstructures* **3**, 163 (1987).

29. W.I. Wang, E.E. Mendez and F. Stern, *Appl. Phys. Letters* **45**, 639 (1984).

30. G. Duggan, H.I. Ralph and K.J. Moore, *Phys. Rev. B* **32**, 8395 (1985).

31. See e.g. K. Ploog, *Ann. Rev. Mat. Science* **12**, 123 (1982).

32. G. Bastard, *Phys. Rev. B* **24**, 5693 (1981).

33. E.O. Kane, in: *Handbook of Semiconductors*, W. Paul editor, Academic, New York, 1982, vol. 1, p. 193.

34. D.A. Broido and L.J. Sham, in: *Proc. of the 17th International Conferencere on the Physics of Semiconductors*, San Francisco (1984), J.D. Chadi and W.A. Harrison eds., Springer, New York, 1985, p. 337.

35. M. Altarelli, U. Ekenberg and A. Fasolino, *Phys. Rev. B* **32**, 5138 (1985).

36. L.C. Andreani, A. Pasquarello and F. Bassani, *Phys. Rev. B* **36**, 5887 (1987).

37. S.S. Nedorezov, *Soviet Phys. Sol. State* **12**, 1814 (1971).

38. See e.g. T. Ando, A.B. Fowler and F. Stern ref. [1].

39. M. Altarelli, in: *Festkörperprobleme*, P. Grosse ed., Vieweg, Braunschweig, 1985, vol. XXV, p. 381.

40. T. Ando, *J. Phys. Soc. Japan* **54**, 1528 (1985).

41. Y.C. Chang and G.D. Sanders, *Phys. Rev. B* **32**, 5521 (1985).

42. A. Pinczuk, D. Heiman, R. Sooryakumar, A.C. Gossard and W. Wiegmann, *Surf. Sci.* **170**, 573 (1986).

43. R. Sooryakumar, D.S. Chemla, A. Pinczuk, A. Gossard, W. Wiegmann and L.J. Sham, *J. Vac. Sci. Tech. B* **2**, 349 (1984).

44. See e.g. L.L. Chang, in: *Heterojunctions and Semiconductor Superlattices*, G. Allan, G. Bastard, N. Boccara, M. Lannoo and M. Voos editors, Springer, Berlin, 1986, p. 152.

45. J.C. Maan, Y. Guldner, J.P. Vieren, P. Voisin, M. Voos, L.L. Chang and L. Esaki, *Solid State Commun.* **39**, 683 (1981).

46. M. Altarelli, *Phys. Rev. B* **28**, 842 (1983).

47. E.E. Mendez, L. Esaki and L.L. Chang, *Phys. Rev. Lett.* **55**, 2216 (1985).

48. M. Altarelli, J.C. Maan, L.L. Chang and L. Esaki, *Phys. Rev. B* **36**, 9867 (1987).

49. Y. Guldner, G. Bastard, J.P. Vieren, M. Voos, J.P. Faurie and A. Million, *Phys. Rev. Lett.* **51** 907 (1983).

50. G.S. Boebinger, J.P. Berroir, Y. Guldner, J.P. Vieren, M. Voos and J.P. Faurie, in: Proceedings of the MSS III Conference, Montpellier (1987), *J. de Physique* **C5**, 301 (1987).

51. D.J. Olego, J.P. Faurie and P.M. Raccah, *Phys. Rev. Lett.* **55**, 328 (1985).

52. S.P. Kowalczyk, J.T. Cheung, E.A. Kraut and R.W. Grant, *Phys. Rev. Lett.* **56**, 2755 (1986).

53. Tran Minh Duc, C. Hsu and J.P. Faurie, *Phys. Rev. Lett.* **58**, 1127 (1987). See also ref. [50], pp. 307.

54. Y.R. Lin-Liu and L.J. Sham, *Phys. Rev. B* **32**, 5561 (1985).

55. M. Jaros, *Phys. Rev. Lett.* **60**, 2560 (1988).

56. G. Bastard, *Phys. Rev. Lett.* **60**, 2561 (1988).

57. N.F. Johnson, P.M. Hui and H. Ehrenreich, *Phys. Rev. Lett.* **61**, 1993 (1988).

58. R. Ferreira and G. Bastard, *Phys. Rev. B* **38**, 8406 (1988).

59. A. Fasolino and M. Altarelli, *Surf. Sci.* **142**, 322 (1984); and in: *Two-Dimensional Systems, Heterostructures and Superlattices*, G. Bauer, F. Kuchar and H. Heinrich eds., Springer, Berlin, 1984, p. 176.

60. G.D. Sanders and Y.C. Chiang, *Phys. Rev. B* **32**, 4282 (1985).

61. G. Platero and M. Altarelli, *Phys. Rev. B* **36**, 6591 (1987).

62. C. Mailhiot, Y.C. Chang and T.C. McGill, *Phys. Rev. B* **26**, 4449 (1982).

63. W.T. Masselink, Y.C. Chang and H. Morkoç, *Phys. Rev. B* **28**, 7373 (1984); *J. Vac. Sci. Tech. B* **2**, 376 (1984).

64. E.E. Mendez, F. Agullo-Rueda and J.M. Hong, *Phys. Rev. Lett.* **60**, 2426 (1988); P. Voisin, J. Bleuse, C. Bouche, S. Gaillard, C. Alibert and A. Regrenyi, *Phys. Rev. Lett.* **61**, 1639 (1988).

65. See e.g. N.W. Ashcroft and N.D. Mermin, *Solid State Physics*, Holt, Rinehardt and Winston, New York, 1976.

66. H.L. Störmer, Z. Schlesinger, A. Chang, D.C. Tsui, A.C. Gossard and W. Wiegmann, *Phys. Rev. Lett.* **51**, 126 (1983).

67. A. Broido and L.J. Sham, *Phys. Rev. B* **31**, 888 (1985).

68. U. Ekenberg and M. Altarelli, *Phys. Rev. B* **32**, 3712 (1985).

69. E. Bangert and G. Landwehr, *Superlattices and Microstructures* **1**, 363 (1985); *Surf. Sci.* **170**, 593 (1986).

70. U. Ekenberg, *Surf. Sci.* **170**, 601 (1986).

71. Y. Iwasa, N. Miura, S. Tarucha, H. Okamoto and T. Ando, *Surf. Sci.* **170**, 587 (1986).

72. D. Heiman, A. Pinczuk, A.C. Gossard, A. Fasolino and M. Altarelli, in: *Proceedings of the 18th International Conference on the Physics of Semiconductors, Stockholm, 1986*, E. Engström, editor, World Scientific, Singapore, 1987.

73. M.L. Claessen, J.C. Maan, M. Altarelli, P. Wyder, L.L. Chang and L. Esaki, *Phys. Rev. Lett.* **57**, 2556 (1986).

74. J.C. Maan, in: *Festkörperprobleme, vol. 27*, P. Grosse editor, Vieweg, Braunschweig (1987).

75. See e.g. J.C. Slater, *Insulators, Semiconductors and Metals*, McGraw-Hill, New York, 1976.

76. G.E. Zilberman, *Soviet Physics JETP* **5**, 208 (1957); *ibid* **6**, 299 (1958).

77. G. Belle, J.C. Maan and G. Weimann, *Solid State Commu.* **56**, 65 (1985).

78. T. Duffield, R. Bhat, M. Koza, F. De Rosa, D.M. Hwang, P. Grabble and S.J. Allen, Jr., *Phys. Rev. Lett.* **56**, 2724 (1986).

79. T. Duffield, R. Bhat, M. Koza, F. De Rosa, K.M. Rush and S.J. Allen, Jr., *Phys. Rev. Lett.* **59**, 2693 (1987).

80. A. Fasolino and M. Altarelli, in: *Proceedings of the Int. Conf. on the Physics of Semiconductors, Warsaw, 1988*, in press.

81. A. Baldereschi and N.O. Lipari, *Phys. Rev. B* **3**, 439 (1971).

82. M. Altarelli and N.O. Lipari, *Phys. Rev. B* **15**, 4898 (1977).

83. See e.g. L. Viña, R.T Collins, E.E. Mendez, W.I. Wang, L.L. Chang and L. Esaki, in: *Excitons in Confined Systems*, R. Del Sole, A. D'Andrea and A. Lapiccirella eds., Springer, Berlin, 1988, p. 230.

84. J.C. Maan, G. Belle, A. Fasolino and M. Altarelli, *Phys. Rev. B* **30**, 2253 (1984).

85. N. Miura, Y. Iwasa, S. Tarucha and H. Okamoto, in: *Proceedings of the 17th International Conference on the Physics of Semiconductors, San Francisco, 1984*, J.D. Chadi and W.A. Harrison, editors, Springer, New York, 1985, p. 359.

86. W. Ossau, B. Jäkel, E. Bangert, G. Landwehr and G. Weimann, *Surf. Sci.* **174**, 188 (1986).

87. D.C. Rogers, J. Singleton, R.J. Nicholas, C.T. Foxon, K. Woodbridge, *Phys. Rev. B* **34**, 4002 (1987).

88. B. Zhu and K. Huang, *Phys. Rev. B* **36**, 8102 (1987).

89. B. Zhu, *Phys. Rev. B* **37**, 4689 (1988).

90. L.C. Andreani and A. Pasquarello, *Europhysics Lett.* **6**, 259 (1988).

91. G.E.W. Bauer and T. Ando, *Phys. Rev. B* **37**, 3130 (1988).

92. G.E.W. Bauer and T. Ando, *Phys. Rev. B* **38**, 6015 (1988).

93. W. Ossau, B. Jäkel, E. Bangert and G. Weinmann, in: *The Basic Properties of Impurity States in Superlattice Semiconductors*, C.Y. Fong editor, Plenum, New York, 1988, in press.

94. R. Bauer, D. Bimberg, J. Christen, D. Oertel, D. Mars, J.N. Miller, T. Fukunaga and H. Nagashima, in: *18th International Conference on the Physics of Semiconductors*, O. Engström editor, World Scientific, Singapore, 1987, p. 525.

95. Y. Chen, B. Gil, P. Lefebvre, H. Mathieu, T. Fukunaga and H. Nagashima, in Ref. 83, p. 200.

96. H.W. van Kesteren, E.C. Cosman, F.J.A.M. Greidanus, P. Dawson, K.J. Moore and C.T. Foxon, *Phys. Rev. Lett.* **61** 129 (1988).

97. B.R. Salmassi and G.E.W. Bauer, *Phys. Rev. B* **39**, 1970 (1989).

ELECTRONIC STATES AND OPTICAL TRANSITIONS IN QUANTUM WELLS

GUENTHER HARBEKE

Paul Scherrer Institut
c/o Laboratories RCA
Zurich, Switzerland

1. Introduction

In recent years, there has been remarkable progress in the fabrication of layered semiconductor structures by molecular beam epitaxy (*MBE*) or metal-organic chemical vapour deposition (*MOCVD*). Artificial microstructures of controlled chemical composition and doping level, precisely defined to within interatomic distances, have become available. When the physical dimensions are comparable to or smaller than the de Broglie wavelength of electronic particles, quantum size effects occur which may strongly modify the behaviour of the bulk crystal.

Among these structures, the electronic and optical properties of quantum wells and superlattices have been extensively studied because of their interesting physical phenomena and possible device applications. The optical properties of the quantum well were first investigated by Dingle [1] in 1975. Since then the quantum well laser has received considerable attention because of its low threshold current and weak temperature dependence [2]. Both the spectroscopy and the laser emission have been concerned with transition between electronic states in the quantized electron and hole subbands, i.e. transitions across the band gap.

The term 'intersubband transitions', on the other hand is used to refer to subband-to-subband transitions within either the conduction or the hole subbands. This type of spectroscopy has first been demonstrated on electronic states in semiconductor space charge layers, notably on the two-dimensional electron system in silicon inversion layers [3]. The energy of these states can be changed by application of a voltage across the interface and thus bending the bands. Typically, the electron surface densities are in the $10^{12} cm^{-2}$ range and the energies of the transition lie in the 30 to $100 meV$ range.

Intersubband transitions in heterostructures have been investigated by resonant Raman scattering [4] and, more recently, by resonance infrared absorption between electron subbands in *GaAs* quantum wells [5]. The direct intersubband transitions are indeed of resonant character since they occur between bands of nearly the same dispersion whereas interband transitions involve bands of opposite curvature. By appropriate choice of well and barrier material and, in particular, the well thickness the resonance wavelength can be varied from about $3.5 \mu m$ to longer wavelengths over the entire infrared range. These transitions between the different quantized subbands are also of great interest for technological applications in the infrared spectrum such as detectors, modulators and lasers.

In this paper the field of electronic states and optical transitions in quantum well structures is reviewed with particular emphasis on conduction subband spectroscopy. It is not attempted to cover the large and rapidly growing amount of literature on this subject.

R. Girlanda et al. (eds.), Progress on Electron Properties of Solids, 373–383.

Instead, the basic principles and phenomena will be illustrated, often by way of example.

2. Bulk Semiconductors

The discussion of electronic states and optical transitions in semiconductor quantum structures requires a good deal of knowledge about the bulk material. A lot of progress in the calculation of the electronic band structure has been made in the sixties and was reviewed by Bassani and Pastori Parravicini [6]. One of the more successful concepts was the pseudopotential approximation but for a practical realization for a specific material it still required two or three input parameters from experiment such as the band gap and energy values at other critical points.

More recently, the local-density approximation (*LDA*) in density-functional theory [7] has become the method of choice to calculate exactly ground-state properties and the valence band structure of semiconductors. As far as electronic excitation energies are concerned, one has to be aware of the difference of calculating the quasiparticle energy, with which an electron can be added to the system and which is the experimental observable, as opposed to a set of eigenvalues of a set of noninteracting electrons. In other words, the band gaps in *LDA* are seriously in error, they come out systematically too low by a wide margin, although the dispersion of individual bands is generally well described. This has led to the situation where still the experimental band gap value and the use of the 'scissors operator' were required to obtain a realistic description.

The past few years have seen a renaissance of techniques which have originally been developed by Hedin [8]. In this approach the self-energies are calculated using the one-particle Green's function G and the screened Coulomb interaction W: the GW approximation. In this way it has now become possible to calculate quasiparticle band structures of semiconductors such as *Si* or *GaAs* in good agreement with experimental transition energies [9]. Next to energy values the properties of a semiconductor are determined by effective masses, g factors and nonparabolicities in the vicinity of extremum points in the band structure. It has been realized three decades ago that these parameters, in other words the $E(\mathbf{k})$ dispersion, can be calculated to high precision in a semiempirical theory, namely the $\mathbf{k}\cdot\mathbf{p}$ method [10]. It is a powerful subsidiary tool which uses interband transition energies and interband matrix elements at extremum points, known from experiment or from *a priori* calculations, as input parameters. The results will therefore depend on the quality of the input as well as on the number of bands included in the calculation. In the case of the conduction band of *GaAs*, e.g. it was shown [11] that a three-level model involving the Γ_6^c, Γ_7^v and Γ_8^v bands in Figure 1 does not correctly give the observed values of effective mass and g factor. A three-level model is suitable for small-gap semiconductors such as *InSb* in which this interaction is by far the dominating one. In *GaAs* a five-level model including the interaction with the Γ_7^c and Γ_8^c bands in Figure 1 is required since E_g is about $1.5\,eV$ and the separation between Γ_6^c and Γ_7^c, Γ_8^c is just about $2E_g$. The five-level model (14×14 Hamiltonian counting orbital degeneracies) has been shown to describe non-parabolicity, warping and spin-splitting of the conduction band quite accurately, even for energies higher than $50\,meV$ above the bottom of the band [12]. Further improvements have recently been achieved by Cardona, Christensen and Fasol [13] who introduced a six-level model (16×16 Hamiltonian) including the next higher conduction band $\Gamma_6'^c$ and also the previously omitted spin-orbit interaction between the p-like valence and conduction bands.

From the knowledge of the electronic band structure one can proceed to calculate the complex dielectric function $\epsilon(\omega)$ or the absorption coefficient $K(\omega)$ but the expected

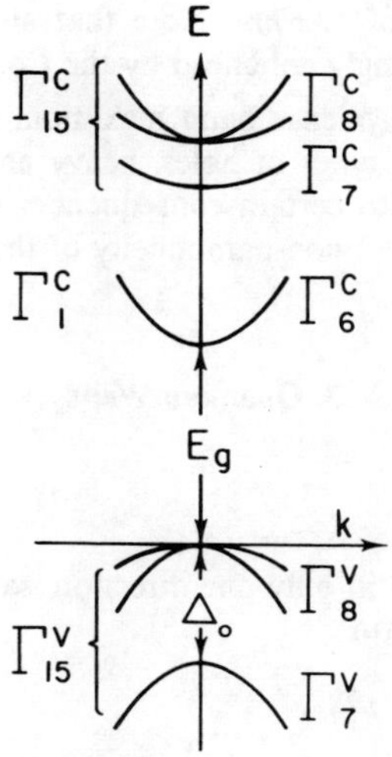

Figure 1. Schematics of the band structure near $k = 0$ in a diamond or zincblende structure, direct-gap semiconductor.

lineshape is often strongly modified by exciton effects. Excitons are electronic excitations consisting of electron-hole pairs correlated by attractive Coulomb interaction. Excitons associated with the band gap have therefore discrete bound pair states $n = 1,2,3, \cdots$ with energies $E_n = E_g - R^* / n^2$ which can dominate the absorption edge spectrum even for weakly bound excitons as in *GaAs* [14] in Figure 2.

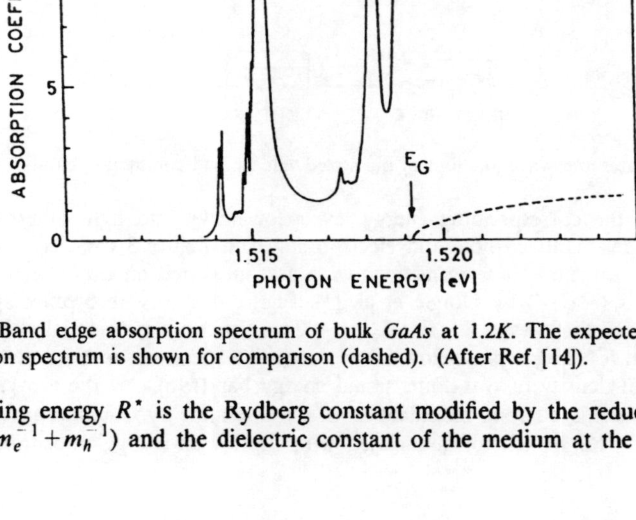

Figure 2. Band edge absorption spectrum of bulk *GaAs* at 1.2*K*. The expected one-electron approximation spectrum is shown for comparison (dashed). (After Ref. [14]).

The binding energy R^* is the Rydberg constant modified by the reduced exciton mass μ ($\mu^{-1} = m_e^{-1} + m_h^{-1}$) and the dielectric constant of the medium at the Rydberg energy.

The binding energy derived from the *GaAs* spectrum in Figure 2 is 4.2 *meV* corresponding to an exciton Bohr radius of 13.6 *nm*. Note that also in the continuum above the energy gap the absorption is strongly enhanced by the Coulomb interaction.

Due to the degeneracy of the valence band maximum at $k = 0$ in diamond and zinc-blende type semiconductors two types of holes, heavy and light, have to be included in the exciton problem. This leads to certain consequences such as the existence of 'heavy' and 'light' excitons, anisotropy and non-parabolicity of the exciton dispersion.

3. Quantum Wells

3.1. Interband Transitions

If the crystal electron is confined in only one direction, say z, by an infinitely high potential, the eigenvalues are of the form

$$E = E_n + \frac{\hbar^2}{2m_e^*} (k_x^2 + k_y^2) \tag{1}$$

with the discrete quantized states

$$E_n = \frac{\hbar^2}{2m_e^*} (\frac{n\pi}{L_z})^2; \; n = 1,2,3, \cdots \tag{2}$$

where L_z is the well width. The finite potential in real quantum well structures allows the wavefunction to extend slightly into the barrier material as indicated in Figure 3 and modifies the eigenvalue distribution.

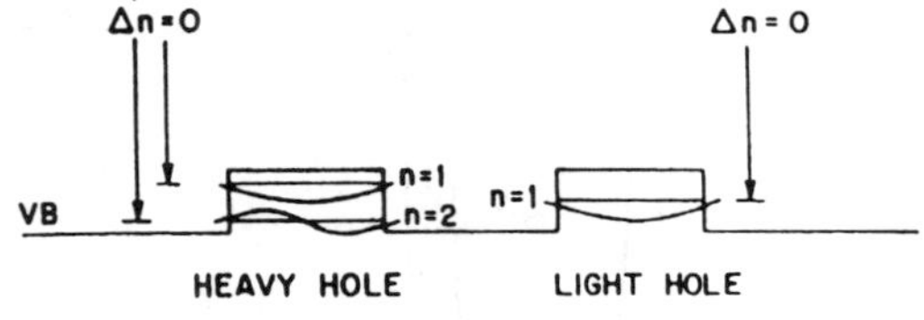

Figure 3. Bound states and wavefunctions of quantized valence and conduction bands.

Considering also the corresponding energy levels for heavy and light holes, interband transitions between quantized hole and electron states in Figure 3 can occur, subject to the $\Delta n = 0$ selection rule. These transitions were first measured on *GaAs/AlGaAs* multiple quantum wells (*MQW*) by Dingle et al. [1]. Figure 4 shows absorption spectra for several different well thicknesses L_z. For $L_z = 400\,nm$ the result is essentially the same as for bulk *GaAs* but for thicknesses below 50*nm* a step-like spectrum develops reflecting the piecewise constant density of two-dimensional energy bands [6]. At the low-energy side of each plateau there is a strong modification of the lineshape by Coulomb interaction of the $\Delta n = 0$ pair. The 1,1 exciton associated with the lowest-energy transition shifts to

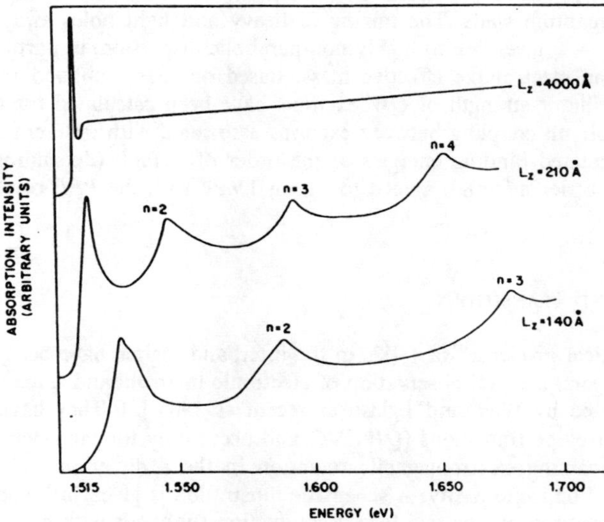

Figure 4. Absorption of multiple $GaAs/Al_{.2}Ga_{.8}As$ quantum well structures of different well thickness L_z, $T = 2K$. (After Ref. [1]).

higher energy with decreasing well thickness, approximately proportional to L_z^{-2} as given by eq. (1). The curve for $14nm$ shows already a splitting of this transition since the different confinement felt by heavy and light holes gives rise to heavy hole (HH) and light hole (LH) exciton transitions. Systematic studies of the spectra for a range of L_z indicated an increase of the HH exciton binding energy from $4.2meV$ in the bulk to about $9meV$ for $L_z < 10nm$ [1]. This trend towards the two-dimensional exciton with a binding energy four times that of a three-dimensional exciton is also apparent in the higher thermal stability of exciton features in the absorption coefficient [15].

Photoluminescence (PL) and photoluminescence excitation (PLE) spectroscopy are widely used for the study of the electronic states in quantum wells. From a lineshape analysis of the PL measurements on $GaAs/AlGaAs$ MQW structures from $12K$ to room temperature Chen et al. [16] found a 5% increase of the confinement energy. This effect was shown to be due to the temperature variation of electron and hole masses analog to that of the bulk masses. Dawson et al. [17] have observed the $2s$ state of both the $1,1HH$ and $1,1LH$ exciton of $GaAs/AlGaAs$ MQW structures by PLE. It is the equivalent of the $n = 2$ peak in Figure 2 for the bulk case except for the $HH - LH$ exciton splitting due to the quantum size effect. The accurate determination of the $1s - 2s$ splittings facilities the comparison with calculations of the binding energy of the QW excitons.

The electronic states of semiconductor quantum wells have been calculated by Bastard [18] and other authors in the envelope function scheme based on the effective-mass approximation. In this approach the quantum well functions are sums of products of Bloch wave functions at Γ, which are periodic with the host unit cells, and envelope functions, slowly varying on the scale of the unit cells. The procedure consists of first solving the effective-mass equation in the well and barrier material and then matching the linear combinations of bulk wave functions at the interfaces in order to find the subband eigenstates. The continuity conditions ask for the total wave function and the probability current to be continuous at the interface. Andreani, Pasquarello and Bassani [19] have obtained exact solutions of the effective-mass equation to calculate the hole subbands in *GaAs*/*AlGaAs* quantum wells. The mixing of heavy and light holes away from the in-plane vector $k_{/\!/} = 0$ gives rise to highly non-parabolic dispersion, in particular the $1, LH$ band has even an electronlike effective mass. Based on these subband results, binding energies and oscillator strength of *QW* excitons have been calculated for *GaAs*/*AlGaAs* [20]. Taking Coulomb coupling between excitons associated with different subbands into account the calculated binding energies of the order of $10\,meV$ (depending in detail on well width and barrier material) agreed to within $1\,meV$ with the *PLE* results of Dawson et al. [17].

3.2. Intersubband Transitions

Whereas the optical properties of *QW*'s in the interband regime have been intensely studied for many years, the first observation of electronic intersubband transitions in doped *QW*'s was reported by West and Eglash as recent as 1985 [5]. They have been termed quantum well envelope transitions (*QWEST*) and occur only for the electric polarization vector of the incoming electromagnetic radiation in the z direction and only between envelope states of opposite parity. A schematic illustration is given in Figure 5 where the full eigenvalue spectrum of the two lowest conduction subbands with dispersion in $\mathbf{k}_x, \mathbf{k}_y$ according to eq. (1) is shown. If the dispersion would be exactly the same for the two subbands, the joint density of states (*JDOS*) for transitions for all $\mathbf{k}$ up to $\mathbf{k}_F$ would be a delta function. Since, however, the wavefunction of the higher subband has a larger admixture from the wavefunction of the barrier material with the higher effective mass, its effective mass will always be higher. The *JDOS* at $T = 0K$ will thus have a box type distribution from ΔE at $\mathbf{k}_F$ to ΔE_0. At $T \neq 0K$ the Fermi function causes the asymmetric distribution with the low-energy tail as indicated in Figure 5.

The subband states can be populated by optical pumping, i.e. generation of electron-hole pairs by irradiation with above-band-gap-energy photons, or by doping. The dopant atoms can be placed directly in the well or, alternatively, in the barrier, in each case away from the interface. In the latter case, also called modulation doping, the carriers will fall into the lower-energy states in the well, thereby separating themselves from the ionized impurity centers and thus avoiding this specific scattering process. If the subband separation is very much larger than kT, only the lowest subband will be populated.

The first measurement of intersubband transitions has been achieved on modulation doped *MQW* structures [5], the results are reproduced in Figure 6. In order to optimize the coupling of the transitions to the incoming light, the measurements were done with p-polarized light under incidence at Brewster's angle of 73 degree. The internal electric field has still only a small component in the z direction because of the strong refraction. The peak positions for the two *MQW*'s show the expected relation to the well widths; the measurement also reveals well width inhomogeneities of the sample with $8.2nm$ average width. Also the other sample has a considerably larger halfwidth than that expected from an estimate of the *JDOS*. West and Eglash [5] also determined the dipole matrix element which is as large as $2e\ nm$.

Several other coupling schemes have subsequently been applied in order to increase the polarization component in the quantization direction. One is based on multiple internal reflection (MIR) where the incoming light is focussed onto a 45 degree polished edge of the wafer and repeatedly reflected in the crystal by total internal reflection [21]. The absorption has been enhanced by a factor of 100 or more over the single pass Brewster's angle configuration. The MIR technique has also been shown to yield *QWEST* absorption with good signal-to-noise ratio on single *QW*'s [22]. In this case the coupling can be

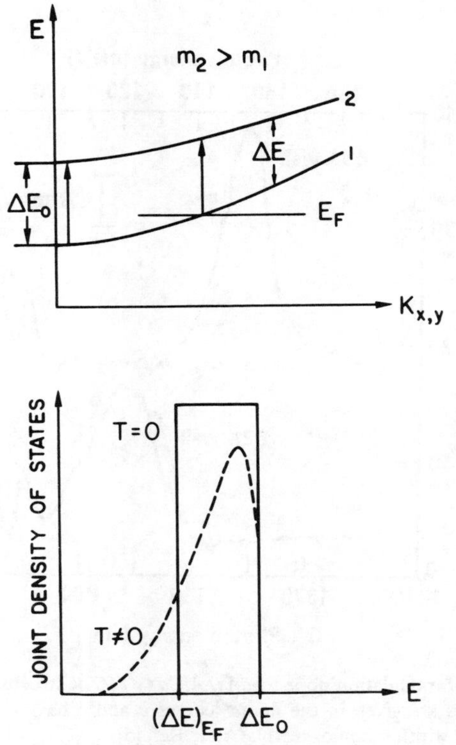

Figure 5. Schematics of conduction intersubband transitions in doped quantum wells and joint density of states.

further improved by depositing a metal layer on the surface which ensures that there is a large radiation field perpendicular to the well [23]. The metal layer can act as a Schottky contact to change the electron density in the QW and hence to modulate the intersubband absorption. Other coupling enhancements have been achieved by a special prism geometry [24] or by depositing metal gratings having a period equal to or smaller than the wavelength at which the subband transitions occur [25].

Peak absorption wavelengths have been varied from 8 to $14\mu m$ by well design in the $GaAs/AlGaAs$ system. There is no basic obstacle to go to longer wavelengths by increasing the well width and lowering the barrier height. A short-wavelength limit is, however given around $6\mu m$ by the maximum direct-gap conduction band offset of about $330\,meV$. Levine et al. [26] have therefore investigated lattice matched MQW structures of $In_{.53}Ga_{.47}As\,/\,In_{.52}Al_{.48}As$ on InP with a conduction band discontinuity of about $550\,meV$. The short-wavelength limit is expected to be $3.6\,\mu m$ and the authors have experimentally obtained a peak wavelength of $4.4\mu m$ for a well width of $5nm$.

The analysis of the absorption lineshape as a function of temperature, doping mechanism and doping concentration can give valuable information on the subband dispersion, optical transition times and electron-electron interaction. Figure 7 shows the comparison of the $4.2K$ lineshape of well and modulation doped $GaAs/Al_{.3}Ga_{.7}As$ MQW's of the same nominal silicon doping concentration of $n = 10^{18}cm^{-3}$ [27]. The fact that the linewidth difference, which is already pronounced at $4.2K$, increases further upon increasing the temperature [27,28] shows that it is caused by an additional genuine broadening

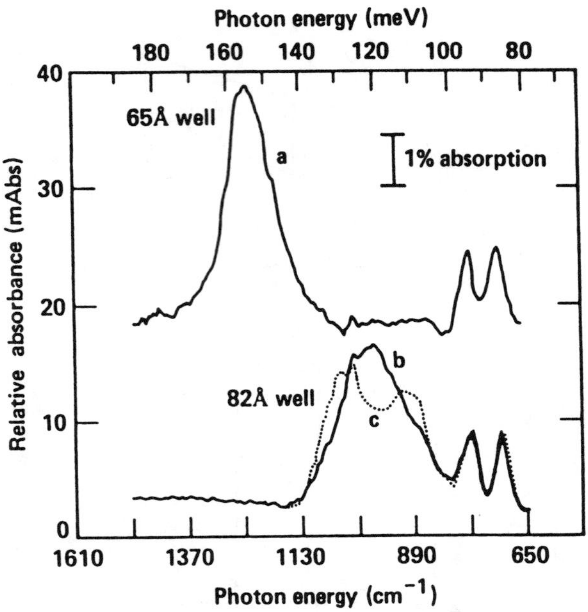

Figure 6. Absorbance of modulation doped *GaAs / AlGaAs MQW* structures with 50*QW*'s. The estimated well thicknesses are given in the figure; curves *b* and *c* have been obtained on the same sample showing its well width inhomogeneity. (After Ref. [5]).

mechanism rather than by well width fluctuations of individual samples. This broadening mechanism is obviously the scattering of the electrons at ionized impurity centres in the well reducing the optical transition time in the same way as the mobility scattering time [28].

The first requirement for a detailed lineshape analysis is to calculate the conduction subband dispersion with good precision. Von Allmen et al. [28] have presented results of a **k**·**p** calculation for a 6×6 Hamiltonian ($\Gamma_6^c, \Gamma_8^v LH$ only, Γ_7^v) and a 14×14 Hamiltonian corresponding to the **k**·**p** calculation for bulk *GaAs* by Roessler [12]. A comparison of effective masses and non-parabolicities gives marked differences between the two models, in particular for the higher subband and for energies at larger **k** values. The absorption lineshape in the absence of broadening is then given by the *JDOS* as illustrated in Figure 5. A phenomenological energy state linewidth is introduced by replacing the energy conserving delta function in the expression for the absorption coefficient by a Lorentzian. There is very little broadening in good modulation doped samples such as the one shown in Figure 7. The 14-band model gives a halfwidth of the *JDOS* of 4.2 *meV* and 5.3 *meV* for the two spin-polarized bands, whereas the experimental halfwidth is 5.6 *meV*. There is thus no detectable contribution from well width fluctuations, a deviation by half a monolayer of *GaAs* would cause a 5 to 6 *meV* difference in resonance energy. Even the energy state width resulting from deconvoluting the experimental lineshape with the calculated *JDOS* can only be determined to within an order of magnitude, i.e. between .3 and 3 psec, due to uncertainties in the input parameters for the calculations and the experimental resolution. Further refinements of the model have been introduced by adding an electrostatic potential solution of the Poisson equation to the Hamiltonian and by taking exchange and correlation effects in *LDA* into account [29]. These effects have no or very

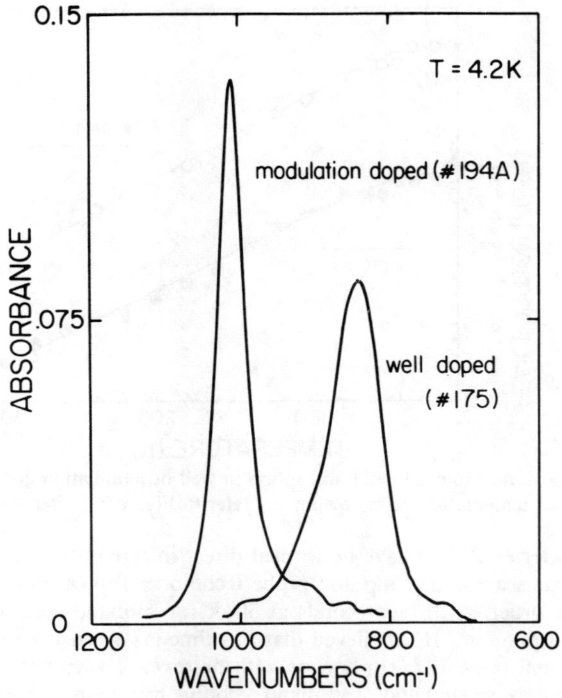

Figure 7. Intersubband absorbance in well or modulation doped $GaAs/Al_{.3}Ga_{.7}As$ *MQW* structures at $T = 4.2K$. (After Ref. [27]).

little influence on the dispersion but they are extremely important for the absolute peak position. The peak position is also influenced by the depolarization shift due to the polarization of the electron gas layer [4]. The depolarization field can be directly measured by Raman scattering. Spin density or single particle excitations observed in back scattering with crossed polarization of the incident and scattered light yield the bare subband spacing. Collective charge density excitations observed as polarized spectra are shifted in energy by the depolarization field. If all this is considered, the absolute positions at $4.2K$ in Figure 7 are in very good agreement with theory [29] (with the well width measured by scanning electron microscopy). This statement cannot be made for the negative temperature coefficient of the peak position shown in Figure 8. The resonance energy is influenced by the temperature dependence of the barrier height and of the effective masses. The former part has the temperature characteristics as in the experiment but the latter part is by far the stronger one and the decrease of the bulk conduction band mass with increasing temperature as input in the calculation yields a positive temperature coefficient for the resonance energy. This discrepancy raises the question whether the bulk parameters can be used also at finite temperature to calculate the subband dispersion. It is indeed well known that the electron-phonon interaction, which is responsible for the temperature variation of the band parameters, is different in a two- and three-dimensional system [30].

Some work has been done to investigate the carrier dynamics in quantum well structures. Oberli et al. [31] used picosecond Raman spectroscopy to measure the intersubband relaxation in wide quantum wells with subband splitting smaller than the optical-phonon energy. They found cooling rates of the order of several hundred picoseconds, probably caused by interaction of the excited electron with acoustic phonons and other

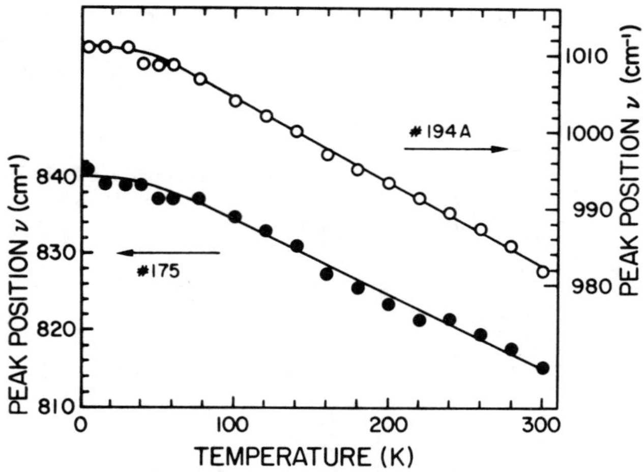

Figure 8. Peak position of intersubband absorption in well or modulation doped $GaAs/Al_{.3}Ga_{.7}As$ *MQW* structures *vs*. temperature. Sample numbers refer to Figure 7. (After Ref. [27]).

electrons. Seilmeier et al. [32] have performed direct infrared absorption spectroscopy on a picosecond time scale in a pump and probe technique. Typical intersubband relaxation times are of the order of 10 picoseconds at $300K$ for subband splittings larger than the energy of optical phonons. It is believed that this time is governed by emission or absorption of longitudinal optical (*LO*) phonons with finite wave vector to transfer the excited carrier into the lowest subband. Intraband cooling can then proceed much faster by efficient scattering by *LO* phonons of practically $\mathbf{k} = 0$ wave vector.

Summarizing the discussion on intersubband spectroscopy it can be said that it offers the possibility to study separately the conduction and hole subbands and the one-component carrier gas of either type. The physics of one component is not dominated by exciton effects or the complications of the subband dispersion of the other component. On the other hand, the knowledge gained on each of the two subsystems can be put to use in the discussion of the electron-hole system encountered in interband spectroscopy.

Acknowledgement

This contribution is dedicated to Franco Bassani on the occasion of his sixtieth birthday.

I am grateful to P. von Allmen, M. Berz, F.K. Reinhart and E. Meier for a close cooperation on intersubband spectroscopy.

References

1. R. Dingle, in: Festkörperprobleme, *Advances in Solids State Physics*, Vol. XV, H.J. Queisser ed., Pergamon/Vieweg, Braunschweig, 1975, pp. 21.
2. Y. Arakawa and A. Yariv, *IEEE J. Quantum Elect.* **QE-22**, 1887 (1986).
3. F. Schäffler and F. Koch, *Solid State Commun.* **37**, 365 (1981).
4. G. Abstreiter, M. Cardona and A. Pinczuk, in: Light Scattering in Solids IV, M. Cardona and G. Güntherodt eds., Springer, Berlin 1984, pp. 5.

5. L.C. West and S.J. Eglash, *Appl. Phys. Lett.* **46**, 1156 (1985); L.C. West, Ph. D. Thesis, Stanford University, 1985.

6. F. Bassani and G. Pastori Parravicini, Electronic States and Optical Transitions in Solids, Pergamon, Oxford, 1975.

7. W. Kohn and L.J. Sham, *Phys. Rev.* **140**, A1133 (1965).

8. L. Hedin, *Phys. Rev.* **139**, A796 (1965).

9. S.G. Louie, Proc. 18th Int. Conf. 'The Physics of Semiconductors', Stockholm, O. Engström ed., World Scientific, Singapore, 1987.

10. M. Luttinger and W. Kohn, *Phys. Rev.* **97**, 869 (1955); E.O. Kane, *J. Phys. Chem. Solids* **1**, 249 (1957).

11. C. Hermann and C. Weisbuch, *Phys. Rev. B* **15**, 823 (1977).

12. U. Rössler, *Solid State Commun.* **49**, 943 (1984).

13. M. Cardona, N.E. Christensen and G. Fasol, *Phys. Rev.* **38**, 1806 (1988).

14. R.G. Ulbrich, in: Festkörperprobleme, *Advances in Solid State Physics* Vol. XXV, P. Grosse ed., Vieweg, Braunschweig, 1985, pp. 299.

15. D.A.B. Miller, D.S. Chemla, D.J. Eilenberger, P.W. Smith, A.C. Gossard and W.T. Tang, *Appl. Phys. Lett.* **41**, 679 (1982).

16. Y. Chen, R. Cingolani, L.C. Andreani, F. Bassani and J. Massiers, *Il Nuovo Cimento* **10D**, 847 (1988).

17. P. Dawson, K.J. Moore, G. Duggan, H.I. Ralph and C.T.B. Foxon, *Phys. Rev. B* **34**, 6007 (1986).

18. C. Bastard, in: Molecular Beam Epitaxy and Heterostructures, L.L. Chang and K. Ploog, eds., M. Nijhoff, Dordrecht, 1985.

19. L. Andreani, A. Pasquarello and F. Bassani, *Phys. Rev. B* **36**, 5887 (1987).

20. L.C. Andreani and A. Pasquarello, *Europhys. Lett.* **6**, 259 (1988).

21. B.F. Levine, R.J. Malik, J. Walker, K.K. Choi, C.G. Bethea, D.A. Kleinmann and J.M. Vandenberg, *Appl. Phys. Lett.* **50**, 273 (1987).

22. J.Y. Andersson and G. Landgren, *J. Appl. Phys.* **64**, 4123 (1988).

23. M.J. Kane, M.T. Emeny, N. Apsley, C.R. Whitehouse and D. Lee, Superlattices and Microstructures, in print.

24. G. Abstreiter, T. Egeler, S. Beeck, A. Seilmeier, H.J. Huebner, G. Weimann and W. Schlapp, *Surface Science* **196**, 613 (1988).

25. M. Helm, E. Colas, P. England, F. DeRosa and S.J. Allen, Jr., *Appl. Phys. Lett.* **53**, 1714 (1988).

26. B.F. Levine, A.Y. Cho, J. Walker, R.J. Malik, D.A. Kleinmann and D.L. Sivco, *Appl. Phys. Lett.* **52**, 1481 (1988).

27. G. Harbeke, P. von Allmen, M. Berz and F.K. Reinhart, in: Proc. of the 14th Intern. Conf. on Infrared Physics, R. Kesselring and F.K. Kneubuehl eds., Zurich, 1988.

28. P. von Allmen, M. Berz, G. Petrocelli, F.K. Reinhart and G. Harbeke, *Semicond. Sci. and Technol.* **3**, 1211 (1988).

29. P. von Allmen, M. Berz, F.K. Reinhart and G. Harbeke, *Superlattices and Microstructures,* **5**, 259 (1989).

30. S. Das Sarma and A. Madhukar, *Phys. Rev. B* **22**, 2823 (1980).

31. D.Y. Oberli, D.R. Wake, M.V. Klein, J. Klem, T. Henderson and H. Morkoc, *Phys. Rev. Lett.* **59**, 696 (1987).

32. A. Seilmeier, H.J. Huebner, G. Abstreiter, G. Weimann and W. Schlapp, *Phys. Rev. Lett.* **59**, 1345 (1987).

A SELF ENERGY APPROACH FOR OPTICAL TRANSITION ENERGIES IN SEMICONDUCTORS AND SUPERLATTICES

M. SCHLÜTER AND M.S. HYBERTSEN

AT&T Bell Laboratories
Murray Hill, NJ 07974
USA

A quasiparticle self energy approach directly taking into account the many-body nature of the electron-electron interaction is described which gives an excellent account of the quasiparticle band energies in semiconductors and insulators. The calculated gaps are in excellent agreement with the absorption edge and higher energy features observed in optical probes of semiconductors e.g. *Si*, *Ge*, *GaAs* and *AlAs*. This first-principles approach can be extended with confidence to novel semiconductor materials such as superlattices which is illustrated here for the *Si*/*Ge* strained layer superlattices.

1. Introduction

The optical properties of many materials can be successfully interpreted in terms of a quasiparticle band structure. In particular, the empirical pseudopotential approach early on proved crucial for unraveling the optical spectra of tetrahedral semiconductors [1]. However, the theory of quasiparticle band energies in solids is generally a many body problem involving the electron-electron interaction and is therefore of considerable fundamental interest. From the practical point of view, a first principles approach is also highly desirable for extension to consider novel materials. Self consistent field theories, such as the local density functional approach (*LDA*), suffer from well known deficiencies when the eigenvalues are interpreted as quasipracticle energies. For example, the absorption edge (minimum gap) in semiconductors is underestimated by 30-100% in the *LDA* case [2]. In this paper, a conceptually well founded self energy approach to this problem is reviewed. It has proven quantitatively applicable to the calculation of quasiparticle energies in real materials systems including bulk semiconductors [3-5], semiconductor surfaces [6] and interfaces [7,8]. The self energy approach can be combined [2] with *LDA* ground state calculations (well founded and of proven accuracy [9]) to provide a predictive tool which may be applied to novel semiconductor systems.

2. The Self Energy Approach

The quasiparticle in a solid are described by a Schroedinger type equation [10]:

$$(T + V_{ext} + V_H)\psi_{nk}(\mathbf{r}) + \int d\mathbf{r}'\Sigma(\mathbf{r},\mathbf{r}';E_{nk})\psi_{nk}(\mathbf{r}') = E_{nk}\psi_{nk}(\mathbf{r}), \tag{1}$$

including the kinetic energy, the external potential due to the ion cores, the average electrostatic (Hartree) potential, and the non-local, energy dependent and in general complex

R. Girlanda et al. (eds.), Progress on Electron Properties of Solids, 385–393.

self energy operator. The self energy operator Σ describes the exchange and correlations among the electrons. The real part of the eigenvalue gives the energy of the quasiparticle while the imaginary part corresponds to the lifetime.

The physically important approximation is for the self energy operator, here described by the so-called GW approximation [10]:

$$\Sigma(\mathbf{r},\mathbf{r}';E) = i\int \frac{dE'}{2\pi} e^{-i\delta E'} G(\mathbf{r},\mathbf{r}';E-E')W(\mathbf{r},\mathbf{r}';E'), \qquad (2)$$

where $\delta = 0^+$. The full interacting Green's function G as well as the dynamically screened Coulomb interaction W enter explicitly. The latter is fully described by the dielectric matrix ϵ^{-1} which screens v, the bare Coulomb interaction:

$$W(\mathbf{r},\mathbf{r}';E) = \int d\mathbf{r}'' \epsilon^{-1}(\mathbf{r},\mathbf{r}'';E)v(\mathbf{r}''-\mathbf{r}'). \qquad (3)$$

For the Green's function, a quasiparticle approximation is used:

$$G(\mathbf{r},\mathbf{r}';E) = \sum_{n,k} \frac{\psi_{n\mathbf{k}}(\mathbf{r})\psi^*_{n\mathbf{k}}(\mathbf{r}')}{E - E_{n\mathbf{k}} - i\delta_{n\mathbf{k}}}. \qquad (4)$$

Here $\psi_{n\mathbf{k}}, E_{n\mathbf{k}}$ are quasiparticle wavefunctions and energies implicitly defined by Eq. (1), $\delta_{n\mathbf{k}} = 0^+$ for occupied states and $\delta_{n\mathbf{k}} = 0^-$ for empty states.

The full dynamical dielectric matrix is required. This represents a major difficulty for evaluating the self energy operator for a semiconductor. Due to the charge inhomogeneity (covalent bonding), the so-called local fields in the screening response are physically very important [11]. These local fields yield the variation in the screened Coulomb interaction around an added electron as it moves from regions of high density, e.g. bonds, to regions of low density, e.g. interstitial [12]. The local fields are described by the off-diagonal elements of the dielectric matrix in a reciprocal space representation, $\mathbf{q},\mathbf{G}$. Various approaches have been applied to obtain the full dynamical response. Hybertsen and Louie [3] have taken the problem in two steps. First, the static dielectric matrix is calculated as a ground state property from the *LDA* [13]. Second, the dielectric matrix is extended to finite frequency using a generalized plasmon pole model [3] which represents each component $\mathbf{q},\mathbf{G},\mathbf{G}'$ of the imaginary part of dielectric matrix by a delta function. No adjustable parameters enter. A variation on this approach has been introduced by von der Linden and Horsch [14] with similar results. Godby, Schlüter and Sham [4] directly calculate the dynamical dielectric matrix without the plasmon pole approximation but only for imaginary frequencies which is appropriate for integrated quantities and allows quasiparticle energies to be extracted only in the near gap region.

Clearly proper treatment of the dielectric response of semiconductors is a difficult and important component in the self energy approach. Some progress has been made in finding a model for the screening which contains the essential role of local fields. Since the plasmon pole model is found to be adequate, attention can be focussed on the static response function. One may observe (see e.g. Ref. [12]) that the important part of the local fields for the self energy problem is the variation in the local screening potential near the site of the added electron as the site is varied. This approximately follows the local charge density, as is intuitively clear: more electrons nearby lead to more complete screening locally. This is exploited in a model introduced by Hybertsen and Louie [15]. The model consists in taking the screening potential around in added electron at $\mathbf{r}'$ to be the same as for a homogeneous medium of the local density at $\mathbf{r}'$. The model for the homogeneous medium must include the finite dielectric constant. Thus ϵ_0 is the single parameter entering the model. With the necessary symmetry,

$$V_{scr}(\mathbf{r},\mathbf{r}') = \frac{1}{2}(V^{hom}_{scr}(\mathbf{r}-\mathbf{r}';r_s(\mathbf{r}'))+V^{hom}_{scr}(\mathbf{r}'-\mathbf{r};r_s(\mathbf{r})). \qquad (5)$$

Upon Fourier transformation, for a given **q**, each row of ϵ^{-1} can be obtained by a fast Fourier transform. It is quite easy to implement this calculation, and the computer time required is significantly less than for the full calculation of the dielectric matrix. This model quantitatively reproduces the screening potential around an added electron in bulk materials such as *Si* and *Ge*. When incorporated into the self energy calculation, the resulting quasipracticle energies are within about $0.1\,eV$ of those found from the complete calculation of the dielectric matrices.

A somewhat different model has been proposed by Gygi and Baldereschi [16]. The model consists in separating the self energy in the semiconductor into a long-range and short-range part. The latter is taken to be the corresponding expression for the homogeneous electron gas at the local density in the crystal. For the long-range part, the static screened exchange plus Coulomb hole approximation is made using a model (diagonal) dielectric function. In this fashion non-locality is preserved in the long-range part (also at intermediate length scales) and local field effects are included via the *LDA* framework. The accuracy of this approximation is about $0.2\,eV$ for a variety of group IV and III-V semiconductors.

Once the screening response has been established, the self energy must be evaluated and Eq. (1) solved. In the Green's function Eq. (4), the quasiparticle wavefunctions and spectrum are initially approximated by the solutions of the *LDA* Kohn-Sham equations. Then the spectrum is updated. The *LDA* wavefunctions are adequate [3,4]. Equation (1) can be solved perturbatively to first order in the difference $\Sigma - V_{xc}$ (V_{xc} is the *LDA* exchange-correlation potential.) Second order contributions are negligible. The calculation of the diagonal matrix elements $<n\mathbf{k}|\Sigma(E)|n\mathbf{k}>$ is significantly faster than the calculation of $\Sigma_{GG'}(\mathbf{k},E)$ required for direct solution of Eq. (1). For bulk semiconductors, numerical precision of $0.1\,eV$ can be typically achieved for band gaps and dispersions. The diverse approaches for handling the screening response above generally give final quasiparticle energies in good agreement with each other.

3. Calculation of Optical Transition Energies in Semiconductors

The optical properties are interpreted here in terms of an independent quasiparticle picture. Direct electron-hole interactions (excitonic effects) are neglected. Then the quasiparticle band structure calculated as described above can be directly compared to experiment.

Consider first the minimum absorption edge. The *LDA* typically yields a value too small in comparison to experiment. The quasiparticle gap calculated using the self energy approach is in excellent agreement with experiment [17] for a wide range of semiconductors and insulators, as illustrated in Table I.

The frequency dependent optical response of semiconductors depends on the transition energies at critical points in the Brillouin zone. These have been exhaustively analyzed for bulk semiconductors [1]. Here we utilize the previous assignment of various spectroscopic features with transitions in particular regions of the Brillouin zone e.g. E_1 features arise from transitions along Λ. The direct transition energies are calculated at those symmetry points and compared directly to the features identified in differential reflectivity measurements. Of these assignments, the E_2 case is probably least well defined introducing extra uncertainty into the comparison of the calculations to experiment in this case. The results for *Si* and *Ge* are summarized in Table II and those for *GaAs* and *AlAs* are given in Table III. As can be seen, the overall agreement with experiment is excellent. The only systematic deviation is found for the E_0 transition which is

TABLE I

Comparison of the minimum gap in the *LDA* spectrum to experiment and recent results of the self energy approach (*QP*).

E_g	*LDA*	*QP*	*Exp*[c]
diamond	3.9	5.6[a]	5.48
Si	0.52	1.29[a,b]	1.17
Ge	<0	0.75[a]	0.744
LiCl	6.0	9.1[a]	9.4
AlAs	1.18	1.99[b]	2.14
GaAs	0.56	1.48[b]	1.52

[a]Ref. [3] [b]Refs. [4,5]
[c]Ref. [16]

TABLE II

Comparison of the calculated direct transition energies from Ref. [3] to critical point energies derived from optical spectroscopy. The results for *Ge* include the spin-orbit interaction.

	Theory	Expt.[a]	
Si			
$\Gamma_{25'v} \rightarrow \Gamma_{2c}$	4.08	4.2	E_0
$\Gamma_{25'v} \rightarrow \Gamma_{15c}$	3.35	3.4	E_0'
$L_{3'v} \rightarrow L_{1c}$	3.54	3.45	E_1
$L_{3'v} \rightarrow L_{3c}$	5.51	5.50	E_1'
$X_{4v} \rightarrow X_{1c}$	4.43	4.44,4.60	E_2
Ge			
$\Gamma_{8v} \rightarrow \Gamma_{7c}$	0.71	0.887	E_0
$\Gamma_{7v} \rightarrow \Gamma_{7c}$	1.01	1.184	$E_0 + \Delta_0$
$\Gamma_{8v} \rightarrow \Gamma_{6c}$	3.04	3.006	E_0'
$\Gamma_{8v} \rightarrow \Gamma_{8c}$	3.26	3.206	$E_0' + \Delta_0'$
$L_{4,5v} \rightarrow L_{6c}$	2.18	2.25	E_1
$L_{4,5v} \rightarrow L_{4,5c}$	5.76	5.576	E_1'
$X_{5v} \rightarrow X_{5c}$	4.45	4.501	E_2

[a]Ref. [16]

from the valence band edge *p*-states to the conduction band *s*-state at the zone center, Γ. This can be traced to the core-valence exchange contribution to the self energy operator which is here subsumed in the ionic pseudopotential. Atomic estimates suggest that inclusion of this term would increase the E_0 by about the amount needed.

If one turns to the placement of the bands relative to the valence band edge e.g. probed by photoemission and inverse photoemission, similar agreement with experiment is found. Details are given in Refs. [3-5]. One particularly interesting case is the placement of the L_{1c} conduction band in *AlAs*. A discrepancy of 0.5 *eV* is found in comparison to experiment. As discussed in Ref. [5], this can be traced to a misinterpretation

TABLE III

Comparison of the calculated direct transition energies from Ref. [5] to critical point energies derived from optical spectroscopy. The results do include the spin-orbit interaction. The data for *AlAs* available in the literature are less complete than for *GaAs*.

	Theory	Expt.[a]	
GaAs			
$\Gamma_{8v}\rightarrow\Gamma_{6c}$	1.47	1.52	E_0
$\Gamma_{7v}\rightarrow\Gamma_{6c}$	1.81	1.86	$E_0+\Delta_0$
$\Gamma_{8v}\rightarrow\Gamma_{7c}$	4.40	4.49	E_0'
$\Gamma_{8v}\rightarrow\Gamma_{8c}$	4.57	4.66	$E_0'+\Delta_0'$
$L_{4,5v}\rightarrow L_{6c}$	2.93	3.02	E_1
$L_{6v}\rightarrow L_{6c}$	3.14	3.25	$E_1+\Delta_1$
$L_{4,5v}\rightarrow L_{6c}$	6.52	6.63	E_1'
$X_{7v}\rightarrow X_{6c}$	4.81	4.94	E_2
$X_{6v}\rightarrow X_{6c}$	4.88	5.01	$E_2+\Delta_2$
$X_{7v}\rightarrow X_{7c}$	5.03	5.34	$E_2+\Delta_2'$
$X_{6v}\rightarrow X_{7c}$	5.10	5.42	$E_2+\Delta_2+\Delta_2'$
AlAs			
$\Gamma_{8v}\rightarrow\Gamma_{6c}$	3.26	3.13	E_0
$\Gamma_{7v}\rightarrow\Gamma_{6c}$	3.54	3.41	$E_0+\Delta_0$
$\Gamma_{8v}\rightarrow\Gamma_{7c}$	5.05	-	E_0'
$L_{4,5v}\rightarrow L_{6c}$	3.91	3.80	E_1
$L_{6v}\rightarrow L_{6c}$	4.11	4.00	$E_1+\Delta_1$
$L_{4,5v}\rightarrow L_{6c}$	6.36	-	E_1'
$X_{7v}\rightarrow X_{6c}$	4.43	4.54	E_2
$X_{7v}\rightarrow X_{7c}$	5.33	-	E_2'

[a]Ref. [16]

of the sequence of alloy data used to extrapolate the position of this band. A correction of the 'experimental' value is proposed.

The essential physics behind the *GW* self energy approach to quasiparticle energies in semiconductors is three-fold. First, the self-energy operator $\Sigma(\mathbf{r},\mathbf{r}';E)$ is non-local, i.e. the 'potential' seen by a quasiparticle at site **r** depends on its own shape at **r**′. Most importantly, the range of this non-locality encompasses typical variations in crystal Bloch functions of order interatomic distances or longer. The self energy operator thus, for example, strongly discriminates between bonding (valence band) and anti-bonding (conduction band) states. Second, the screened Coulomb interaction which enters Σ includes the strong local field effects described above. Third, dynamical renormalization effects in Σ are retained. The picture of quasi particles renormalized by interaction with plasmons implicit in the plasmon pole model is accurate provided local field effects are taken into account. Including all these effects quantitatively in the GW approximation yields excellent results for semiconductors and insulators. Higher order diagrams (vertex corrections) are evidently less important. However, no direct calculations of these terms have yet been undertaken.

The next step in an *ab initio* theory of the optical response of semiconductors is to examine the self energy corrections to the band structure in more detail e.g. changes in the reduced effective mass at critical points. With a well founded theory for the quasiparticle energies, it would be interesting to return once again to the role of excitonic interactions in the dynamical optical response. Other topics related to optical properties which are of interest include the behaviour near the metal-insulator transition.

4. Si/Ge Strained Layer Superlattices

We turn now to a case where the self energy approach has been applied to a novel material and where the predictive power of the theory is important. Superlattices alternating a few monolayers of Si and Ge may be regarded as new semiconductors with an 'engineered' unit cell. Recent electroflectance (ER) measurements for the case of $Si_4Ge_4/Si(001)$ (alternating four layers of Si and Ge grown on Si) showed strong new features at low energy ($0.76\,eV$ and $1.1-1.25\,eV$) [18]. Also, a rich spectrum is found between $2\,eV$ and the onset of the E_1 feature in bulk Si (about $3.4\,eV$). Since the spectrum for $Si_{0.5}Ge_{0.5}$ alloys shows no features below about $2.5\,eV$, these results strongly suggest that interesting new optical transitions have been introduced by the ordered structure. Central to this process is the concept of zone folding. The real-space unit cell is illustrated in Figure 1a. For the case of an infinitely extended superlattice a new Brillouin zone results which is down-folded from the original diamond structure case (Figure 1b). The zone-edge X points along the growth direction (longitudinal) are folded back to the zone center. For this structure, the L points are folded back to $k_z = 0$, and split into two inequivalent classes due to the lower symmetry introduced by the heteropolar $Si-Ge$ bonds at the interfaces. Note that the transverse Δ minima (k_x, k_y) are unaffected. A crucial question is whether the artificial order leads to a direct gap material e.g. with the conduction band minimum now at the zone center. Furthermore, it is necessary that the superlattice potential lead to dipole allowed transitions to the zone-folded states.

An initial survey of the bandstructure was made using the LDA [7] since the overall band topology is expected to be correct. The LDA band structure for the infinite Si_4Ge_4 superlattice is shown in Figure 2. The conduction band minimum is in fact near the zone edge along k_x and k_y parallel to the interface. Due to the effect of strain, the Δ minima along k_z folded back near Γ are higher in energy. Thus the absorption edge is indirect. For quantitative comparison to experiment, the full quasiparticle self energy calculation was performed, including spin orbit splitting of the valence band edge. The absorption edge is calculated to be $0.85\,eV$. Transitions from the light hole band start at $0.95\,eV$. The lowest dipole allowed direct transitions are at 1.24, $1.34\,eV$ with matrix elements 100-1000 times weaker than a fully allowed transition. The lowest possible transition at the zone center is dipole forbidden. The calculated energy gaps strongly suggest that the lowest feature in the ER spectrum is due to the indirect absorption edge while the next feature arises from the new direct transition. These conclusions are substantiated by photocurrent measurements [19] which show absorption edges ($0.78, 0.90\,eV$) with energy dependence consistent with an indirect edge. The open question here is the apparent intensity of the low energy structure in the ER spectrum. The features observed around $2.5-3.0\,eV$ can be explained in terms of E_1 type transitions (derived from the folded L

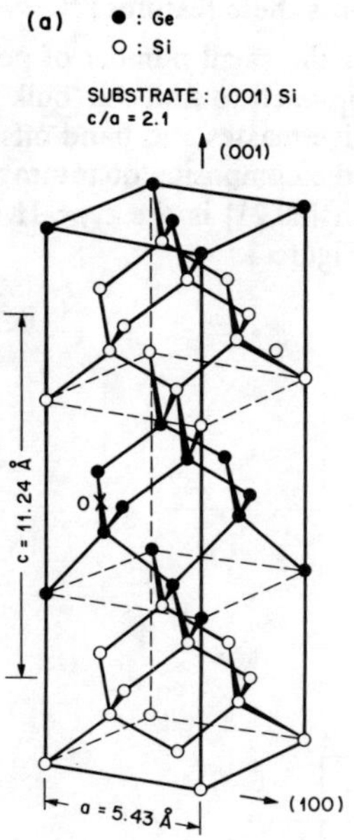

Figure 1a. The real space unit cell for the Si_4Ge_4 superlattice.

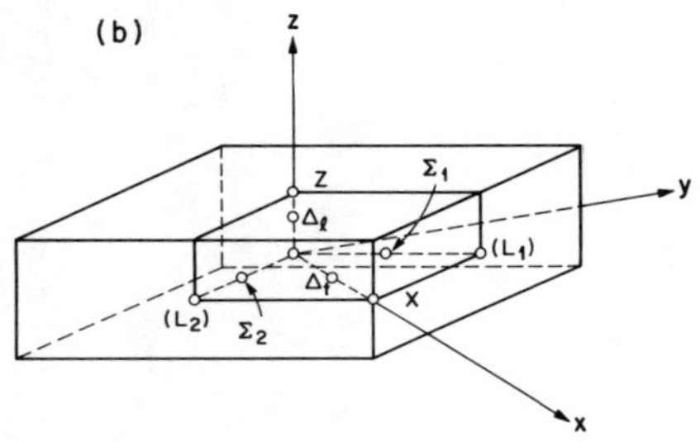

Figure 1b. The Brillouin zone for this structure.

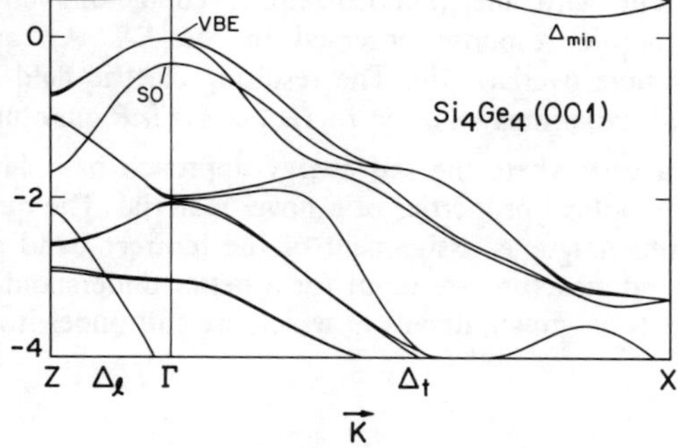

Figure 2. The *LDA* band structure for Si_4Ge_4 grown on a *Si*(001) substrate. The Brillouin zone directions refer to Figure 1b.

points) in the superlattice which are split by the combined effect of strain and spin-orbit splitting. In addition, the lowered symmetry further splits these features [7].

An important feature of these materials as grown is the small number of periods. This can be taken into account by using the self energy approach to find the 'bulk' properties of the Si_4Ge_4. The relevant band edge energies, effective masses and band offset are then used as input for an effective mass type treatment of the composite 'quantum well' structure composed of Si_4Ge_4 embedded in *Si*. The result [20,21] is the type II (staggered) alignment of the superlattice to bulk *Si* illustrated in Figure 3.

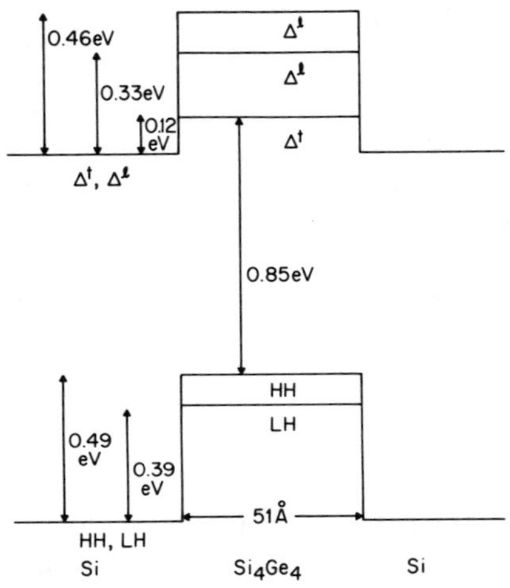

Figure 3. The energy band diagram for Si_4Ge_4 on *Si* showing the allignement of the valence band edge states and near conduction band edge states to those of bulk *Si*.

This has a number of consequences for optical properties of these materials. First, while holes are confined to the superlattice region, electrons are in scattering states in the bulk *Si* region. When the absorption edge is properly corrected for this, one finds 0.77,0.78 *eV* in good agreement with the photocurrent. Second, this structure may explain the enhanced electro-optic response observed in the *ER*. An applied electric field will enhance electron-hole overlap [20]. The resulting electric field enhanced absorption has been seen in static experiments on the $In_xGa_{1-x}As/InP$ quantum well structures [21].

This then is a case where the self energy approach has played an important role in understanding the optical properties of a novel material. The quantitative accuracy of the theory allowed unambiguous assignment of the indirect band gap in this material. The results for the band structure are input for a better understanding of the optical properties of the materials as grown, details of which are still under investigation.

acknowledgements

It is our great pleasure to dedicate this article to Professor Franco Bassani in honor of his sixtieth birthday. It is also a pleasure to acknowledge collaboration with Prof. S.G. Louie, Dr. R. Godby and Prof. L.J. Sham on various aspects of the work described in this paper.

References

1. See, for example, M.L. Cohen and J.R. Chelikowsky, *Electronic Structure and Optical Properties of Semiconductors*, Springer, New York (1988).
2. L.J. Sham and M. Schlüter, *Phys. Rev. Lett.* **51**, 1888 (1983); *Phys. Rev. B* **32**, 3883 (1985).
3. M.S. Hybertsen and S.G. Louie, *Phys. Rev. Lett.* **55**, 1418 (1985); *Phys. Rev. B* **34**, 5390 (1986).
4. R.W. Godby, M. Schlüter and L.J. Sham, *Phys. Rev. Lett.* **56**, 2415 (1986); *Phys. Rev. B* **37**, 10159 (1988).
5. R.W. Godby, M. Schlüter and L.J. Sham, *Phys. Rev. B* **35**, 4170 (1987).
6. M.S. Hybertsen and S.G. Louie, *Phys. Rev. Lett.* **58**, 1551 (1987); *Phys. Rev. B* **38**, 4033 (1988).
7. M.S. Hybertsen and M. Schlüter, *Phys. Rev. B* **36**, 9683 (1987).
8. S.B. Zhang, et al., *Solid State Comm.* **66**, 585 (1988).
9. See, for example, S.G. Louie in *Electronic Structure Dynamics and Quantum Structural Properties of Condensed Matter*, edited by J. Devreese and P. van Camp (Plenum, New York, 1985) pp. 335.
10. L. Hedin and S. Lundquist, *Solid State Phys.* **23**, 1 (1969).
11. R. Resta and A. Baldereschi, *Phys. Rev. B* **23**, 6615 (1981).
12. M.S. Hybertsen and S.G. Louie, *Phys. Rev. B* **35**, 5602 (1987).
13. M.S. Hybertsen and S.G. Louie, *Phys. Rev. B* **35**, 5585 (1987).
14. W. von der Linden and P. Horsch, *Phys. Rev. B* **37**, 8351 (1988).
15. M.S. Hybertsen and S.G. Louie, *Phys. Rev. B* **37**, 2733 (1988).
16. *Zahlenwerte und Funktionen aus Naturwissenschaften und Technik*, in Vol. III of Landolt-Bornstein (Springer, New York, 1982) pt. 17a; see also Refs. 3-5.
17. F. Gygi and A. Baldereschi, *Phys. Rev. Lett.* **62**, 2160 (1989).
18. T.P. Pearsall, et al., *Phys. Rev. Lett.* **58**, 729 (1987).
19. M.S. Hybertsen et al., *Phys. Rev. B* **37**, 10195 (1988).
20. M.S. Hybertsen and M. Schlüter, in *Epitaxy of Semiconductor Layered Structures*, edited by R.T. Tung, L.R. Dawson and R.L. Gunshor (Materials Research Society, Pittsburgh, 1988) pp. 413-418.
21. M.S. Hybertsen, P. Friedel and M. Schlüter, Proc. of the Nineteenth Int. Conf. on the Physics of Semiconductors, Warsaw, 1988, Institute of Physics, Polish Academy of Sciences, Warsaw, (1988) p. 491.
22. D. Gershoni, et al., *Phys. Rev. Lett.* **60**, 448 (1988).

THREE-WAVE MIXING VIA SUBBAND LEVELS IN INVERSION LAYERS AND IN QUANTUM WELLS

M.Y. JIANG AND E. BURSTEIN
Physics Department and Laboratory for Research on the Structure of Matter
University of Pennsylvania
Philadelphia, PA 19104

We discuss the underlying concepts and the microscopic electronic processes that contribute to three-wave mixing via subband levels in quantum wells. $3WM$ in an n-type inversion layer involves Franz-Keldysh type interband transitions between the valence band and the various subbands in the conduction band, intersubband transitions and intraband transitions, which are all optically allowed. $3WM$ in a symmetric quantum well is forbidden as the envelope functions of the confined electrons and holes, and the excitons they form, have well defined parity. $3WM$ is allowed in the presence of an electric field perpendicular (or parallel) to the walls. The field also leads to a Stark shift of the subband and exciton energies. $3WM$ in an asymmetric quantum well is allowed in the absence of an applied electric field, and is conceptually the same as in a symmetric well in the presence of a perpendicular electric field. The dependence of the $3WM$ signal on the polarization and frequency of the input EM waves can be used as a spectroscopic probe of the intersubband excitations of inversion layers and quantum well structures.

1. Introduction

A wide variety of linear and nonlinear optical phenomena have been used to study the electronic states and excitations of inversion layers, and quantum wells. They include infrared absorption [1], optical absorption [2], photoresponse [3], resonant Raman scattering [4], photoluminescence [5], photoreflectance [6], optical Stark effect [7] and the effects of electric field, magnetic field and stress on the interband and intersubband optical transitions that are involved. We have recently initiated a theoretical investigation of three-wave mixing ($3WM$) in inversion layers and quantum well structures, both for its intrinsic interest as a nonlinear optical phenomenon and for its potential as a spectroscopic probe of the 2-dimensional electronic states and excitations [8]. In this paper we discuss $3WM$ via subband levels as still another phenomenon that can be used to study the 2-dimensional electronic structure and excitations of inversion layers and quantum wells. Although $3WM$ via subband levels is forbidden in symmetric quantum wells, it can be induced by an externally applied electric field. An experimental investigation of $3WM$ in inversion layers and in single and multiple quantum wells structures, using a pulsed CO_2 laser as the source of the ω_2 input wave and a tunable pulsed dye laser in the visible as the source of the ω_1 input wave, is now under way in our laboratory. We expect that, as in the case of resonant Raman scattering by intersubband transitions [4,9], it should be readily possible to observe resonant $3WM$ via intersubband and intrasubband transitions.

R. Girlanda et al. (eds.), Progress on Electron Properties of Solids, 395–408.

2. General Considerations

Three-wave mixing phenomena, including second harmonic generation (*SHG*) and sum and difference frequency generation (*SFG* and *DFG*), which involve the coherent interaction of two 'input' *EM* waves, ω_1 and ω_2, and an 'output' *EM* wave ω_3 (i.e., $\omega_3 = \omega_1 + \omega_2$), have been formulated by Bloembergen and collaborators and experimentally investigated by them and by other investigators in centrosymmetric, as well as noncentrosymmetric, media [10]. In terms of the macroscopic fields of the input *EM* waves, the nonlinear polarization generated in a medium is written as:

$$\mathbf{P}^{NL}(\omega_3,\mathbf{k}_3) = \chi^{(2)}\mathbf{E}(\omega_1,\mathbf{k}_1)\mathbf{E}(\omega_2,\mathbf{k}_2)+\gamma\mathbf{E}(\omega_1,\mathbf{k}_1)\nabla\mathbf{E}(\omega_2,\mathbf{k}_2)$$
$$+ \ \xi\ \mathbf{E}(\omega_1,\mathbf{k}_1)\times\mathbf{H}(\omega_2,\mathbf{k}_2)\cdots$$

where the first term represents the 'electric dipole' contribution which exists in media lacking inversion symmetry; the second and third terms represent the 'quadrupole' and 'magnetic dipole' contributions which exist in centrosymmetric, as well as noncentrosymmetric media, and $\mathbf{k}_3 = \mathbf{k}_1 \pm \mathbf{k}_2$. Since the surface layers of centrosymmetric media lack a center of inversion, they also take part in $3WM$ via an 'electric dipole' contribution [11]. The recent investigations of *SHG* by the 2×1 and 7×7 reconstructed surfaces of $Si(111)$ by Heinz et al. [12] have been particularly valuable in providing information about the symmetry of the reconstructed surfaces.

$3WM$ can also take place in centrosymmetric media as a result of the change in symmetry that is induced by an electric field [13]. The space-charge electric field and associated band-bending that exist at a semiconductor surface have been shown to induce a sizeable resonant Raman scattering by *LO* phonons in both centrosymmetric and noncentrosymmetric semiconductors (i.e., III-V, IV-VI compound semiconductors) via a Franz-Keldysh two-band mechanism [14]. An analogous Franz-Keldysh two-band mechanism is also expected to play a role in space-charge electric field induced $3WM$ in semiconductors [15]. In fact, when the ω_2 input *EM* wave is a low frequency *p*-polarized *EM* wave which sets up a Coulomb field in the surface region, the microscopic processes which play a role in $3WM$ are fundamentally the same as those involved in the Raman scattering by *LO* phonons which is also a three-waving mixing phenomenon.

From the microscopic point of view of $3WM$ as three-photon mixing, the elementary $3WM$ processes in bulk semiconductors involve three virtual electronic transitions which are accompanied in any time order by the annihilation of an input photon of frequency ω_1; the annihilation, or creation, of an input photon of frequency ω_2; and the creation of an output photon of frequency ω_3 [16]. In the absence of free carriers, the first step involves an interband transition creating an 'interband' electron-hole pair; the second step involves either an interband transition, or an intraband transition, of the excited electron (or hole) in the first intermediate state; and the third step involves the recombination of the interband electron-hole pair in the second intermediate state, returning the system to its electronic ground state. Processes in which all three steps are interband transitions are termed three-band processes, and those in which one of the steps involves an intraband transition are termed two-band processes (Figures 1 and 2). In the case of $3WM$ via subband levels, the 3-level processes that involve two interband transitions and an intersubband transition are in effect two-band processes since the intersubband transitions only involve a change in the envelope part of the electron and hole subband wavefunctions. The three interband transitions in three-band $3WM$ processes are all electric dipole allowed only in media lacking a center of inversion. In centrosymmetric media, where the electronic wave functions have well-defined parity, only two of the

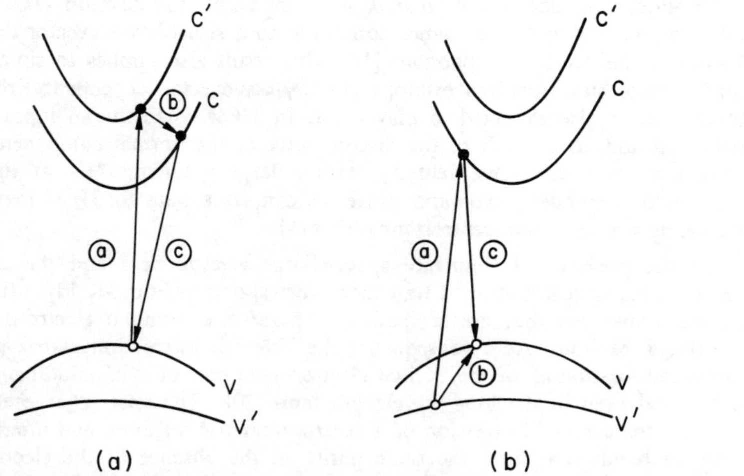

Figure 1. The sequence of three interband transitions that are involved in three-band $3WM$ processes in bulk media: a) the process in which the excited electron undergoes an interband transition in the second step, and b) the process in which the excited hole undergoes an interband transition in the second step.

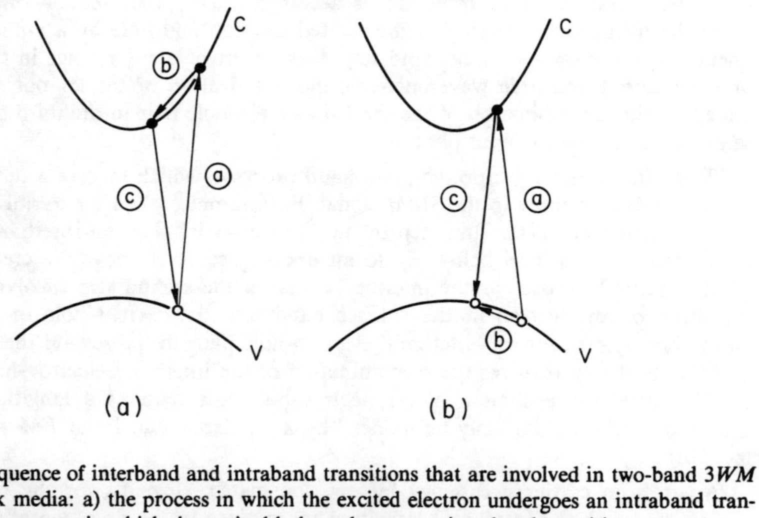

Figure 2. The sequence of interband and intraband transitions that are involved in two-band $3WM$ processes in bulk media: a) the process in which the excited electron undergoes an intraband transition, and b) the process in which the excited hole undergoes an intraband transition.

interband transitions can be electric dipole allowed. The third interband transition, which necessarily involves bands of the same parity, may be either quadrupole, or magnetic dipole, allowed.

In the case of Raman scattering by LO phonons via a two-band process, Loudon [17] has shown that in the long wavelength ($k_{LO} \approx 0$) limit, the relatively large Fröhlich (e.g., electron-longitudinal electric field) interaction matrix elements for the intraband scattering of the excited electron and the excited hole have the same magnitude, but opposite signs, and thier contributions cancel. On the other hand, the electron and hole contributions do not cancel when k_{LO} is finite and the electron and hole masses are different, leading under resonance conditions to a sizeable wavevector dependent contribution to the *RS* by *LO* phonons [18]. This result also applies to situations where the intermediate state involves excitons [19]. A wavevector dependent Fröhlich two-band mechanism is also expected to play a role in 3*WM* when the ω_2 input *EM* wave is *p*-polarized and, as a result of the discontinuity of the normal component of the electric field, sets up a Coulomb field E_C with a large gradient ∇E_C at the surface. The wavevector dependent two-band processes can contribute to 3*WM* in centrosymmetric media, as well as in non-centrosymmetric media.

In the presence of a surface space-charge electric field and the associated band-bending, the optical interband transitions correspond to Franz-Keldysh tunneling-assisted optical transitions that create spatially separated continuum electrons and holes and 'polarized' excitons. As a consequence the Fröhlich interaction matrix elements for the intraband 'scattering' of the excited electron-hole pair of a longitudinal electric field do not cancel even in the long wavelength limit [20]. The space-charge electric field also removes the center of inversion of a centrosymmetric medium, and interband transitions between bands that have the same parity in the absence of the electric field become electric-dipole allowed.

When ω_1 is in the vicinity of an optical energy band gap and $\omega_2 << \omega_1$, the dominant contribution to the nonlinear electric susceptibility are from the two-band processes in which the first (interband) transition is accompanied by the annihilation of an ω_1 photon, the second transition is accompanied by the annihilation (*SFG*) or creation (*DFG*) of an ω_2 photon, and the third transition is accompanied by the creation of an ω_3 photon. Since the intraband scattering of the excited electron and hole by a *p*-polarized low frequency ω_2 input wave in the second step does not involve any change in the orbital parts of the electron and hole wavefunctions, the polarization of the ω_3 output photon generated by the recombination of the excited electron-hole pair in the third step, will be the same as that of the ω_1 input photon.

When free carriers are present, two-band processes which involve a different sequence of steps also contribute to the 3*WM* signal. For example, when the Fermi level lies above the conduction band, the first step in such processes involves an interband excitation of an electron from a level below E_F to an unoccupied level above E_F creating an 'intraband electron-hole pair' as the intermediate state; the second step involves an interband transition of an electron in the valence band into the excited hole in the conduction band, thereby creating an 'interband electron-hole pair' in the second intermediate state; and the third step involves the recombination of the interband electron-hole pair (Figure 3). The intraband excitation of an electron-hole pair requires a longitudinal Coulomb field and can, therefore, only be induced by a *p*-polarized ω_2 input *EM* wave at the surface [21].

Apart from carrier occupancy factors, the matrix elements for the two-band 3*WM* process in which the intraband transition takes place in the first step are essentially the same as those for the 3*WM* processes in which the intraband transition takes place in the second step. However, the energy denominators of the dominant terms in the third-order time-dependent perturbation expression for the nonlinear susceptibility under resonance conditions (i.e., ω_1 is near an optical band gap) are different. In the 3*WM* process where the intraband transition occurs in the second step, the energy denominator has the form

$$(E_{a-0} - \hbar\omega_1 - i\gamma_{a-0})(E_{b-0} - \hbar\omega_3 - i\gamma_{b-0})$$

where E_{a-0} and E_{b-0} are the energies (relative to the ground state) of the interband electron-hole pairs in the first and second intermediate states, respectively, and γ_{a-0} and γ_{b-0} represent the corresponding energy widths of the two interband transitions. In the

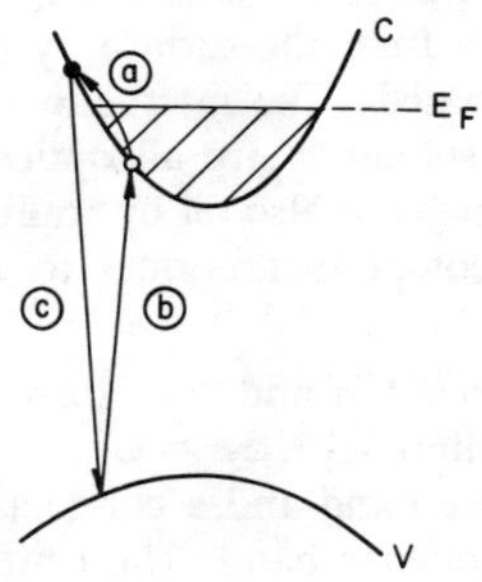

Figure 3. The sequence of interband and intraband transitions that are involved in two-band processes in a bulk medium in which the Fermi level lies above the conduction band edge.

3*WM* process where the intraband transition occurs as the first step, the energy denominator is given by

$$(E_{b-a}-\hbar\omega_2-i\gamma_{b-a})(E_{b-0}-\hbar\omega_3-i\gamma_{b-0})$$

where E_{b-a} is the energy of the intraband electron-hole pair in the first intermediate state and γ_{b-a} is the energy width of the intraband transition. In both situations the denominators are small when ω_1 is near an optical energy gap, and $\omega_2 << \omega_1$.

In the next two sections we will discuss 3*WM* via subband levels in inversion layers and in quantum wells and, in particular, 3-level processes which involve two interband transitions and an intersubband transition and 2-level processes which involve two interband transitions and an intrasubband transition.

3. Three-Wave Mixing via Subband Levels in Inversion Layers

In the strong band-bending region of a narrow inversion layer (taken to be *n*-type for convenience) at a surface, or at a heterojunction, the electron states at the conduction band edge are localized in the direction (z) normal to the layer forming 2-dimensional energy subbands. The corresponding hole states at the valence band edge are not localized and have a continuous distribution of energies. The envelope functions of the electron states in the *n*-th subband (c_n) of the conduction band have the form

$$|c_n(\mathbf{k}_{\|})> = \phi_n(z)\exp(i\mathbf{k}_{\|}\cdot\vec{\rho})$$

and the extended envelope functions of the hole states in the valence band have the form

$$|v(k_z,\mathbf{k}_{\|})> = \phi_{k_z}(z)\exp(i\mathbf{k}_{\|}\cdot\vec{\rho})$$

where $\phi_n(z)$ is the quantized z dependent wavefunction of the $1-D$ potential energy well; $\phi_{k_z}(z)$ is the extended $1-D$ Airy type wavefunction which at large z goes over to $\exp(ik_z z)$, and $\mathbf{k}_{\|}$ is the wavevector in the $x-y$ plane of the inversion layer. The space-charge electric field in the inversion layer controls the spacing between the $2-D$ subbands in the conduction band. It also modifies the orbital parts of the wavefunctions of the electron and hole states and, in the case of a centrosymmetric semiconductor, removes the center of inversion.

The interband transitions between the valence band and the conduction subbands are

Franz-Keldysh type transitions which generate spatially-separated electrons and holes. Although the electron-hole pairs are Coulomb correlated, they do not form 'free' excitons since the holes are accelerated away from the surface by the strong space-charge electric field i.e., free excitons are disassociated. The interband transitions between the valence band and the various conduction subbands are all optically allowed. The non-spin-flip transitions between different subbands are also all optically allowed. They are induced by photons having an electric field component normal to the inversion layer, i.e. by *p*-polarized photons [22].

We consider the contributions to $3WM$ and, for illustrative purposes, discuss the contributions to sum frequency generation, $\omega_1+\omega_2 = \omega_3$, from processes that involve interband transitions between the valence band and a conduction subband and an intraband transition within the subband, or valence band. The contributions from these processes can be enhanced by using an ω_1 input *EM* wave in the vicinity of the optical band gap and a low frequency (i.e., infrared) ω_2 input *EM* wave in the vicinity of the energy separation between the subbands. As a model system we discuss a $GaAs\,/\,Al_{1-x}Ga_xAs$ heterojunction, specifically one in which the Fermi level lies between the c_1 and c_2 conduction subband edges of the *GaAs* inversion layer.

There are several *SFG* processes of interest:

(i) a 3-level process that involves a 'direct' ($\Delta k_{\parallel} \approx 0$) non-spin flip $c_1 \rightarrow c_2$ intersubband transition of an electron in the c_1 subband to an unoccupied subband level in the c_2 subband, creating an 'intersubband electron-hole pair', as the first step; a Franz-Keldysh $v \rightarrow c_1$ interband transition of an electron in the valence band into the excited hole in the c_1 band, creating an interband electron hole pair as the second step; and a Franz-Keldysh $c_2 \rightarrow v$ interband recombination of the interband recombination of the interband electron-hole pair as the third step (Figure 4). Under resonance conditions, the dominant contribution to the nonlinear susceptibility from this process comes from the sequence in which the intersubband transition in the first step is accompanied by the annihilation of a *p*-polarized ω_2 photon; the interband transition in the second step is accompanied by the annihilation of an ω_1 photon and the interband (recombination) transition in the third step is accompanied by the creation of an ω_3 photon.

(ii) a 3-level process in which the participating subband levels lie above E_F that involves Franz-Keldysh $v \rightarrow c_1$ and $c_2 \rightarrow v$ interband transitions as the first and third steps; and an intersubband transition as the second step (Figure 5).

(iii) a 2-level process in which the participating subband levels lie above E_F that involves Franz-Keldysh $v \rightarrow c_1$ and $c_1 \rightarrow v$ interband transitions as the first and third steps; and intraband interaction of the excited electron and hole with the longitudinal electric field set up by a *p*-polarized photon in the second step.

The dominant contribution to the nonlinear susceptibility from type (ii) and type (iii) processes comes from the sequence in which the interband transition in the first step is accompanied by the annihilation of an ω_1 output photon; the intersubband and intraband transitions in the second step are accompanied by the annihilation of a low frequency *p*-polarized ω_2 photon; and the interband recombination of the interband electron-hole pair is accompanied by the emission of the ω_3 photon. Since the intrasubband transitions and the intraband transitions do not change the periodic parts of the electron and hole wavefunction, the two interband transitions in each of these processes will involve photons having the same polarization.

Apart from occupancy factors, the matrix elements for the two 3-level *SFG* processes that involve intersubband transition are the same. The difference between the type (i) and type (ii) processes appears in the form of the energy denominators of the dominant term

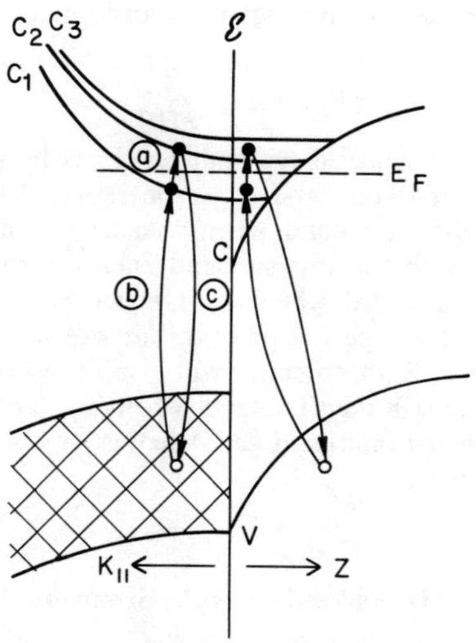

Figure 4. Schematic diagram of an n-type inversion layer showing the spatial dependence of the valence and conduction band edges and the conduction subbands to the right, and the E versus $k_{\|}$ dispersion of the valence band and the conduction subbands to the left. The sequence of interband and intersubband transitions that are involved in the $3WM$ process in which the $c_1 - c_2$ intersubband transition is the first step is shown in both the E versus z and E versus $k_{\|}$ parts of the diagram.

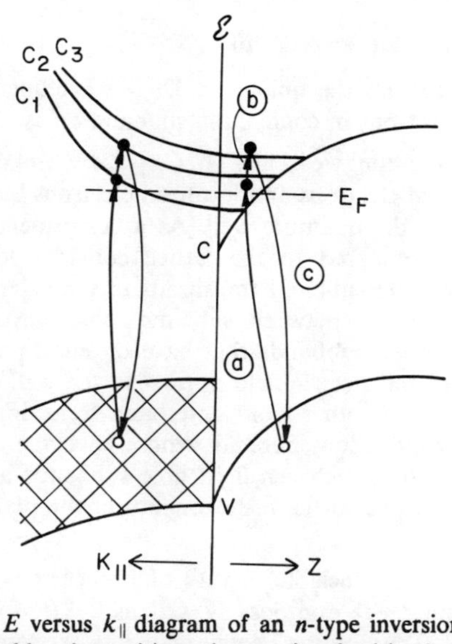

Figure 5. The E versus z and E versus $k_{\|}$ diagram of an n-type inversion layer showing the sequence of interband and intersubband transitions that are involved in the $3WM$ process in which the $c_1 - c_2$ intersubband transition is the second step.

in the time-dependent third order perturbation expression for their contribution to the nonlinear susceptibility, $\chi^{(2)}$. In the process that involves the intersubband transition as the first step, the energy denominator has the form

$$(E_{vc_1} - \hbar\omega_1 - i\gamma_{vc_1})(E_{vc_2} - \hbar\omega_3 - i\gamma_{vc_2}).$$

In the process that involves the intersubband transition as the second step, the denominator has the form

$$(E_{c_1c_2} - \hbar\omega_2 - i\gamma_{c_1c_2})(E_{c_2v} - \hbar\omega_3 - i\gamma_{c_2v}).$$

Double resonance can be achieved for both processes by tuning ω_1 to be resonant with the $v_1 \rightarrow c_1$ interband separation and ω_2 to be resonant with the $c_1 \rightarrow c_2$ intersubband energy separation. On the other hand, when ω_1 is away from the optical band gap, tuning ω_2 through resonance with the intersubband energy separation, will yield a resonance peak in the output *SFG* signal when ω_2 is equal to the non-spin-flip $c_1 \rightarrow c_2$ energy separation in the case of a type (iI) process, but will not yield a resonance peak in the case of a type (i) process. Both processes will exhibit resonance peaks which are appreciably broadened by Franz-Keldysh effects when ω_1 (and therefore $\omega_3 \approx \omega_1$) is tuned through resonance with the interband gap, whether or not ω_2 is resonant with the intersubband energy separation.

4. 3*WM* via Subband Levels in Symmetric Quantum Wells

In a quantum well both electron and hole states are localized by the potential energy barriers that exist at the interfaces. Moreover, the degeneracy of the heavy and light hole valence bands at $k_{\parallel} = 0$ is lifted [23]. The envelope functions of the hole states in the valence subbands and of the electron states in the conduction subbands have the form

$$|v_m(\mathbf{k}_{\parallel})> \; = \; \phi_m(z)\exp(i\mathbf{k}_{\parallel} \cdot \vec{\rho})$$

$$|c_n(\mathbf{k}_{\parallel})> \; = \; \phi_n(z)\exp(i\mathbf{k}_{\parallel} \cdot \vec{\rho})$$

where $\phi_{vm}(z)$ and $\phi_{cn}(z)$ are the quantized $1-D$ wavefunctions of the holes in valence subband v_m and the electrons in conduction subband c_n.

In a symmetric quantum well, i.e., $Al_xGa_{1-x}As/GaAs/Al_xGa_{1-x}As$, the envelope functions of the localized electrons and localized electrons have well defined parity due to the mirror symmetry of the quantum well. As a consequence, for quantum wells whose dimensions are large compared to the lattice constant for which the effective mass approximation is good, intersubband transitions in the valence band and in the conduction band are only allowed between subbands that have different parity [22]. The allowed transitions between subbands that have different parity are induced by photons having a component of the electric field normal to the wall of the quantum well, i.e., by *p*-polarized photons. On the other hand, interband transitions are optically allowed only between valence and conduction subbands whose envelope functions have the same parity. We note that transitions between light hole subbands and heavy hole subbands are electric-dipole forbidden due to the orthogonality of the periodic part of their wavefunctions.

In the absence of an appreciable density of free carriers, the optically allowed interband transitions create $2-D$ excitons, as well as $2-D$ Coulomb-correlated continuum electron-hole pairs [2]. The binding energy of the $1s$ exciton involving an electron in the c_1 subband and a heavy hole in the h_1 subband, designated by $1s(h_1c_1)$, is greater than that of a bulk $1s$ exciton in *GaAs*. The $1s$ exciton resonance peak in a *AlGaAs/GaAs/AlGaAs* quantum well has appreciable strength and is quite narrow. The envelope functions of the exciton levels, which are $2-D$ hydrogenic in character, are given by

$$\Phi_{vmcn}(\mathbf{r}_e, \mathbf{r}_h) = \phi_{vmcn}(z_e, z_h)\phi_H(|\vec{\rho}_e - \vec{\rho}_h|)$$

where $\phi_{v_m c_n}(z_e, z_h)$ is the Coulomb correlated z dependent envelope function of the exciton made up of a v_m hole and a c_n electron; and $\phi_H(\vec{\rho}_e - \vec{\rho}_h)$ is the corresponding $2-D$ hydrogenic wavefunction of the exciton in the $x-y$ plane, which is not sensitive to the specific v_m and c_n subbands involved. When the binding energy of the $1s$ exciton is small compared to the $c_1 \rightarrow c_2$ intersubband energy, which is the case in a narrow (100Å or less) quantum well, the z-dependent part of the wavefunction is given, to a good approximation, by the product of the separate electron and hole subband wavefunctions, i.e., $\phi_{v_m c_n}(z_e, z_h) = \phi_{v_m}(z_h)\phi_{c_n}(z_e)$.

The allowed interband transitions between the hole subbands and the electron subbands lead to $1s$ excitons, but not to $2p$ excitons. Because of the difference in the periodic part of the light and heavy hole wave functions, the polarization selection rules are different for the transitions leading to light and heavy hole excitons [23]. Thus, for *GaAs/AlGaAs* quantum wells grown in (100) direction, the $1s(h_1c_1)$ excitons can be created only by photons having E_x and E_y components, whereas the $1s(l_1c_1)$ excitons can be created by photons having an E_z component, as well as by photons having E_x and E_y components. A transition from the $1s(h_1c_1)$ level to the $2p(h_1c_1)$ level can be induced by photons having a component of the electric field parallel to the walls of the well, i.e., by either s- or p-polarized photons. A transition from a $1s(h_1c_1)$ exciton level to a $1s(h_1c_2)$ exciton can be induced by a p-polarized photon, but the radiative recombination of the $1s(h_1c_2)$ exciton is forbidden, i.e., c_2 and h_1 do not have the same parity. We note that an optical (i.e., one-photon) transition from a $1s(h_1c_1)$ exciton level to a $2p(h_1c_2)$ exciton level is also forbidden.

It follows from an inspection of the selection rules for the interband and intersubband transitions in a symmetric quantum well that $3WM$ via intersubband transitions and interexciton level transitions are dipole forbidden. For example, the allowed interband transitions yield excited electrons and holes whose envelope functions have the same parity which combine to form $1s$ excitons whose envelope functions have even parity. Since the allowed electron (or hole) intersubband transitions lead to a change in the parity of the subband envelope function, the interband transition that corresponds to the recombination of the electron and hole in the third step is forbidden.

$3WM$ via subband levels in a symmetric quantum well can be induced by the application of an electric field, either in the direction normal to, or parallel to, the walls of the quantum well. An electric field in the direction normal to the quantum well, E_z, polarizes the excited electron hole pair (or exciton) along z and removes the parity of the z part of the subband envelope functions. It also causes a Stark shift in the subband energy levels. The net effect of the E_z field is to induce 3-level processes via intersubband transitions and 2-level processes via intrasubband transitions. In this connection we note that $3WM$ in an asymmetrical quantum well is conceptually equivalent to $3WM$ in a symmetric quantum well in the presence of an externally applied E_z, and is allowed in the absence of an externally applied electric field. An electric field oriented parallel to the walls, $E_{\parallel}$, which polarizes the $2-D$ hydrogenic exciton states in the $x-y$ plane, induces 3-level processes via inter-excitonic level transitions and 2-level processes via intra-exciton transitions. It should be noted also, that E_z does not induce $3WM$ processes via inter-exciton and intra-exciton level transitions, and that $E_{\parallel}$ does not induce $3WM$ processes via intersubband and intrasubband transitions.

We consider here only the case of E_z induced $3WM$. The case of $E_{\parallel}$ induced $3WM$ via $2-D$ exciton levels involves the same conceptual framework as $E_{\parallel}$ induced $3WM$ via excitons in bulk semiconductors [24]. It will be treated in a future publication. For simplicity, we leave out the distinction between light and heavy hole subbands.

In absence of free carriers (i.e., E_F lies within the forbidden gap) the 3-step electronic processes of interest for *SFG* via subband levels in a symmetric quantum well are:

(i) a 3-level process which involves a $v_1 \rightarrow c_1$ interband transition as the first step; a $c_1 \rightarrow c_2$ (non-spin-flip) intersubband transition as the second step; and an electric field induced $c_2 \rightarrow v_1$ interband recombination as the third step (Figure 6).

(ii) a 3-level process which involves a $v_1 \rightarrow c_1$ interband transition as the first step; an electric field induced $c_1 \rightarrow c_3$ intersubband transition as the second step; and a $c_3 \rightarrow v_1$ interband recombination as the third step,

(iii) a 2-level process which involves a $v_1 \rightarrow c_1$ interband transition as the first step; intrasubband transitions of the spatially polarized electron-hole pair in the intermediate state as the second step; and a $c_1 \rightarrow v_1$ interband recombination as the third step. (We do not include here 3*WM* processes that involve intersubband transitions in the valence band, since they are simply the counterparts of the processes that involve intersubband transitions in the conduction band).

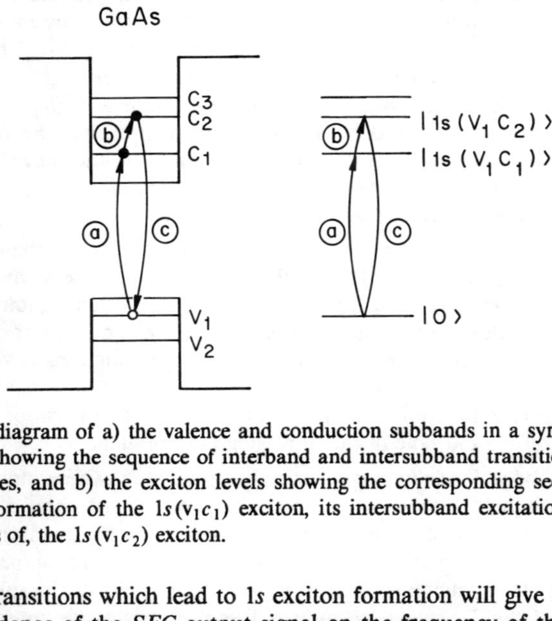

Figure 6. Schematic diagram of a) the valence and conduction subbands in a symmetric quantum well without carrier showing the sequence of interband and intersubband transitions of one of the 3-level 3*WM* processes, and b) the exciton levels showing the corresponding sequence of transitions involving the formation of the $1s(v_1c_1)$ exciton, its intersubband excitation to, and subsequent recombinations of, the $1s(v_1c_2)$ exciton.

The interband transitions which lead to $1s$ exciton formation will give rise to resonant peaks in the dependence of the *SFG* output signal on the frequency of the ω_1 input *EM* wave [2]. The resonant exciton contribution from process (i) to the non-linear susceptibility involves an interband matrix element for creating a virtual $1s(v_1c_1)$ exciton, a matrix element for the intersubband transition of the exciton creating a virtual $1s(v_1c_2)$ exciton, and an interband matrix element for the E_z-induced recombination of the virtual $1s(v_1c_2)$ exciton. The denominator has the form

$$(E_{1s(v_1c_1)} - \hbar\omega_1 - i\gamma_{1s(v_1c_1)})(E_{1s(v_1c_2)} - \hbar\omega_3 - i\gamma_{1s(v_1c_2)}).$$

Resonance peaks occur when either ω_1 matches the $1s(v_1c_1)$ exciton energy, or ω_3 matches the $1s(v_1c_1)$ exciton energy. When ω_2 corresponds to the difference in energy of the $1s(v_1c_1)$ and $1s(v_1c_2)$ excitons the peak will be doubly resonant when ω_1 matches the $1s(v_1c_1)$ energy and, therefore, ω_3 matches the energy of the $1s(v_1c_2)$. A major subsidiary benefit of double resonance is a greatly enhanced signal.

The corresponding resonant exciton contribution from process (ii) will involve a field induced matrix element for the intersubband transition of the $1s(v_1c_1)$ exciton to form a

$1s(v_1c_3)$ exciton and its recombination, and the contribution from the process (iii) will involve a field induced matrix element for the intrasubband scattering of the polarized $1s(c_1v_1)$ exciton and its recombination. Their energy denominators also involve two factors which are resonant when ω_1 and ω_3 match the $1s$ exciton energies.

When carriers are present in the c_1 subband, i.e., when E_F lies between the c_1 and c_2 subband edges, two other processes occur.

(iv) a 3-level process which involves a $c_1 \rightarrow c_2$ (non-spin-flip) intersubband transition as the first step; a $v_1 \rightarrow c_1$ interband transition as the second step which yields an interband electron-hole pair; and an electric field induced $c_2 \rightarrow v_1$ recombination of the interband electron-hole pair as the third step (Figure 7).

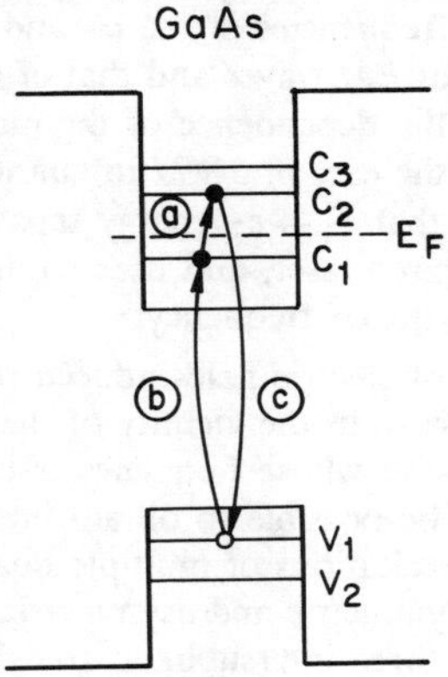

Figure 7. Schematic diagram of the valence and conduction subbands in a symmetric quantum well in which the E_F lies between the c_1 and c_2 subband edges showing the sequence of interband and intersubband transitions of the 3-level $3WM$ process in which the $c_1 - c_2$ intersubband transition is the first step.

(v) a 3-level process which involves an electric field induced $c_1 \rightarrow c_3$ intersubband transition as the first step; a $v_1 \rightarrow c_1$ interband transition as the second step; and a $c_3 \rightarrow v_1$ interband transition as the third step.

The presence of carriers in the c_1 subband suppresses the $1s(v_1c_1)$ exciton and, thereby, suppresses the resonance at the $1s(v_1c_1)$ exciton energy. However, a $1s(v_1c_2)$ exciton exists, providing the carrier density in the c_1 subband is not too large.

The contribution from processes (iv) (or (v)) to $\chi^{(2)}$ via $1s$ excitons involves a matrix element for the intersubband transition $c_1 \rightarrow c_2$ (or $c_1 \rightarrow c_3$), a matrix element for the interband transition $v_1 \rightarrow c_1$ which yields a $1s(v_1c_1)$ exciton (or a $1s(v_1c_3)$ exciton), and a matrix element for the field induced interband recombination of the exciton or continuum electron-hole pair. The denominator for the process involving, for example, the $1s(v_1c_2)$ exciton in the second step is given by

$$(E_{v_1c_2} - \hbar\omega_2 - i\gamma_{v_1c_2})(E_{1s(v_1c_2)} - \hbar\omega_3 - i\gamma_{1s(v_1c_2)}).$$

The process exhibits an exciton resonance peak when ω_3 matches the $1s(v_1c_2)$ exciton energy. When ω_2 corresponds to the non-spin-flip $c_1 \rightarrow c_2$ energy separation, the peak will be doubly resonant. Tuning ω_2 through the $c_1 \rightarrow c_2$ energy separation at fixed ω_1 (or varying E_z and thereby the $c_1 \rightarrow c_2$ energy separation) will yield a peak when ω_2 is resonant with the $c_1 \rightarrow c_2$ transition. Measurements of the dependence of the output signal on the frequency of the ω_2 input photon can, thus, be used as a spectroscopic probe of the intersubband excitations.

5. Concluding Remarks

It must be emphasized that the output $3WM$ signal is a coherent superposition of contributions from different spatial regions (i.e., surface, substrate, quantum well, etc.,) and from different $3WM$ processes (i.e., three-band processes, two-band processes, etc.). It does not give any information per se about the specific 3-step processes that are involved. Such information can, however, be obtained by carrying out measurement of the dependence of the output signal on the frequencies of the ω_1 and ω_2 input EM waves, and the polarizations of the ω_1 and ω_2 input EM waves and that of the ω_3 output EM waves, and by carrying out measurements of the dependence of the output signal on the magnitude of the applied E_z (or $E_{\parallel}$) field. In the case of $3WM$ in quantum wells, one ideally chooses the width of the quantum well so that $c_1 \rightarrow c_2$ energy separation is in the vicinity of the frequency (ω_2) of an available infrared laser, and uses an applied E_z (or $E_{\parallel}$) to tune the $c_1 \rightarrow c_2$ energy separation through the ω_1 frequency.

By carrying out measurements of electric field-induced resonant SFG (or DFG), using a tunable source of the ω_1 input wave in the vicinity of the interband gap and a tunable source of a p-polarized ω_2 input wave whose frequency is in the vicinity of the intersubband energy separation, it should be possible to obtain information about the interband excitations and the intersubband excitations of multiple quantum wells and superlattices, as well as of heterojunctions and symmetric and asymmetric quantum wells. We note that an electric field-induced SHG via three intersubband transitions has been observed in a symmetric $AlGaAs/GaAs/AlGaAs$ quantum well using a CO_2 laser beam as the fundamental EM wave [25].

Finally, we note that by using an intense ω_2 input EM wave (i.e., CO_2 laser) it should be possible to observe readily 4-wave mixing ($\omega_1+2\omega_2=\omega_3$) in inversion layers and quantum wells, via processes that involve two interband transitions and two intersubband transitions that are allowed in the absence of an applied dc electric field.

Acknowledgement

We are pleased to express our appreciation to Y.J. Chen, R.T. Collins, T.F. Heinz, G. Pajer, A. Pinczuk and L. Sham for valuable discussions.

Research supported in part by the NSF/MRL Program at the University of Pennsylvania through Grant no. DMR-8216718.

References

In some of the references listed here, we have made no attempt to be complete, but rather have cited typical papers on the subject.

1. P. Kneschaurek, A. Kamgar and J.F. Koch, *Phys. Rev. B* **14**, 1610 (1977); W. Beinvogel and J.F. Koch, *Solid State Commun.* **21**, 133 (1977); L.C. West and S.J. Eglash, *Appl. Phys. Lett.* **46**, 1156 (1095); A. Harwit and J.S. Harris, Jr., *Appl. Phys. Lett.* **50**, 685 (1987).

2. R.C. Miller, D.S. Kleinman, W.T. Tsang and A.C. Gossard, *Phys. Rev. B* **27**, 1134 (1981); D.A. Miller, D.S. Chemla, T.C. Damen, A.C. Gossard, T.H. Wood and C.A. Burns, *Phys. Rev. B* **32**, 1043 (1985); J.A. Brum and G. Bastard, *Phys. Rev. B* **31**, 3893 (1985).

3. R.T. Collins, K. von Klitzing and K. Ploog, *Phys. Rev. B* **33**, 4378 (1986); L. Viña, R.T. Collins, E.E. Mendez and W.I. Wang, *Phys. Rev. Lett.* **58**, 832 (1987); R.T. Collins, L. Viña, W.I. Wang, C. Mailhiot and D.L. Smith, *Proc. SPIE* **792**, 2 (1987).

4. G. Abstreiter, A. Pinczuk and M. Cardona in *Light Scattering in Solids IV* (Topics in Applied Physics) edit. by M. Cardona and G. Guntherodt (Springer, Berlin, Heidelberg, New York 1984), Vol. 54, p. 74; G. Abstreiter and K. Ploog, *Phys. Rev. Lett.* **42**, 1308 (1979); A. Pinczuk, H.L. Stormer, R. Dingle, J.M. Worlock, W. Wiegmann and A.C. Gossard, *Solid State Commun.* **32**, 1001 (1979); E. Burstein, A. Pinczuk and D.L. Mills, *Surf. Sci.* **98**, 451 (1981).

5. C. Weisbuck, R.C. Miller, R. Dingle, A.C. Gossard and W. Weigmann, *Solid State Commun.* **37**, 219 (1981); R.C. Miller, A.C. Gossard, G.D. Sanders, Y.C. Chang and J.N. Shulman, *Phys. Rev. B* **32**, 8452 (1985).

6. R.A. Hoepfel, J. Shah, A.C. Gossard and W. Weigmann, *Appl. Lett.* **47**, 163 (1985); H. Shen, P. Parayanthal, F.H. Pollak, M. Tomkiewicz, T.S. Drummond and W.T. Beard, *Appl. Phys. Lett.* **48**, 126 (1986); B.V. Shanabrook, O.J. Glembocki and W.T. Beard, *Phys. Rev. B* **35**, 2540 (1987); O. Glembocki, B.V. Shanabrook, N. Bottka, W.T. Beard and J. Comas, *Appl. Phys. Lett.* **46**, 970 (1985).

7. A. von Lehman, D.S. Chemla, J.E. Zucker and J.P. Heritage, *Optics Lett.* **11**, 609 (1986); D. Frölich, R. Wille, W. Schlapp and G. Weimann, *Phys. Rev. Lett.* **59**, 1748 (1987).

8. M.Y. Jiang, E. Burstein and A. Pinczuk, *Bull. Opt. Soc. Amer.* **22**, 40 (1987).

9. E. Burstein, A. Pinczuk and D.L. Mills, *Surf. Sci.* **98**, 451 (1981).

10. N. Bloembergen, *Nonlinear Optics* (Benjamin, New York, 1965) and *J. Opt. Soc. Amer.* **70**, 1429 (1980); N. Bloembergen, H.J. Simon and C.H. Lee, *Phys. Rev.* **181**, 1261 (1969); N. Bloembergen, R.K. Chang, S.S. Jha and C.H. Lee, *Phys. Rev.* **174**, 813 (1968); R.K. Chang, E. Yi Chen, F.G. Parsons and M. Selders, *Proc. Conf. on Polaritons*, Taormina, 1972, edit. by E. Burstein and F. De Martini (Pergamon Press, New York, 1974) p. 3119.

11. Y.R. Shen, *J. Vac. Sci. Tech. B* **3**, 1464 (1984).

12. T.F. Heinz, M.M. Loy and W.A. Thomson, *Phys. Rev. Lett.* **54**, 63 (1985).

13. R.W. Terhune, P.D. Maker and C.M. Savage, *Phys. Rev. Lett.* **8**, 404 (1962); C.H. Lee, R.K. Chang and N. Bloembergen, *Phys. Rev. Lett.* **18**, 167 (1967).

14. A. Pinczuk and E. Burstein, *Phys. Rev. Lett.* **27**, 808 (1971); S. Buchner, L.Y. Ching and E. Burstein, *Phys. Rev. B* **14**, 4459 (1976); L. Brillson and E. Burstein, *Phys. Rev. Lett.* **27**, 808 (1971).

15. G. Pajer, E. Burstein and A. Pinczuk, *Proc. Int. Conf. on Physics of Semiconductors*, Stockholm 1986, edit. by O. Engstrom (World Scientific Publ., Singapore, 1987) p. 1699.

16. P.M. Butcher and T.P. McLean, *Proc. Phys. Soc. (London)* **81**, 219 (1963); E. Burstein and A. Pinczuk in *Physics of Opto-Electronic Materials*, edit. by W. Albers (Plenum Press, New York, 1971) p. 33.

17. R. Loudon, *Proc. Royal Soc. A* **218**, 275 (1963).

18. D.C. Hamilton, *Phys. Rev.* **188**, 1221 (1969).

19. R.M. Martin and T.C. Damen, *Phys. Rev. Lett.* **26**, 86 (1971).

20. A. Pinczuk and E. Burstein, *Surface Science* **37**, 153 (1973); see also J.D. Gay, J.D. Dow, E. Burstein and A. Pinczuk, *Proc. Int. Conf. on Light Scattering in Solids*, edit. by M. Balkanski (Flammarion, Paris, 1971) p. 33 which deals with excitons.

21. P. Apell and S. Lundqvist, *J. de Physique Coll. C10* **44**, 305 (1983).

22. T. Ando, *Solid State Communications* **21**, 133 (1977); T. Ando and S. Mori, *J. Phys. Soc. Japan* **47**, 1518 (1979).

23. R. Sooryakumar, A. Pinczuk, A.C. Gossard, D.S. Chemla and L.S. Sham, *Phys. Rev. Lett.* **58**, 1150 (1987); G.E.W. Bauer and T. Ando, *Phys. Rev. B* **38**, 6015 (1988); for overview see R. Sooryakumar, *Comments on Condensed Matter Physics* **14**, No.1, 43 (1988).

24. G. Pajer, M.Y. Jiang and E. Burstein, *Bull. Amer. Phys. Soc.* **33**, 749 (1988).

25. S.J.B. Yoo, M.M. Fejer, A. Harwit, R.L. Byer and J.S. Harris, Jr., *Bull. Opt. Soc. Amer.* **22**, 40 (1987).

ELECTRONIC SPECTRA OF SOLIDS, IMPURITIES AND SUPERSTRUCTURES WITH THE RECURSION AND RENORMALIZATION METHODS

R.D. GRAFT[(*)], G. GROSSO[(**)], D.J. LOHRMANN[(***)], L. MARTINELLI[(**)], S. MORONI[(****)], G. PASTORI PARRAVICINI[(**)] AND L. RESCA[(***)]

[(*)] *Night Vision and Electro-Optics Laboratory*
Fort Belvoir, Virginia, 22060, USA

[(**)] *Dipartimento di Fisica*
Università di Pisa, INFM, CISM and GNSM
piazza Torricelli 2, 56126 Pisa, Italy

[(***)] *Department of Physics, The Catholic University of America*
Washington D.C. 20064, USA

[(****)] *Dipartimento di Fisica 'A. Volta'*
Università di Pavia, via A. Bassi 8, 27100 Pavia, Italy

The recursion and the renormalization methods have greatly enriched the traditional field of electronic state calculations of periodic and aperiodic materials. After discussing their formal relationship, we present problems where the mentioned iterative procedures provide significant breakthrough. We consider first the application of the recursion method in the reciprocal space for periodic structures, where the unprecedented number of recursions reached allows us to analyze directly the asymptotic region. The recursion method in real space is then discussed and extended to the treatment of large clusters, either regular or with built in foreign atoms or vacancies. Impurity states of arbitrary range can be studied and non-perturbative results concerning ionized substitutional atoms in silicon are here provided for the first time. Localized impurity atoms in the presence of a dynamical Jahn-Teller effect also benefit of the recursion method: several vibronic systems as well as infrared absorption spectrum of transition metal impurities in cubic semiconductors are discussed. Using a mixed real-space reciprocal-space representation, it is shown that layered structures (surfaces, heterostructures, quantum wells and multiple quantum wells) can be studied in a very elegant and economical way with the renormalization method. In particular, we interpret the intrinsic origin of the semiconductor-semimetal transition of $InAs - GaSb$ superlattices.

1. Introduction

The recursion and the renormalization methods have become extremely valuable tools for the study of electronic states in perfect crystals as well as in broken-symmetry solids. Among the most recent developments, we will outline in this paper the progress in the study of periodic crystals and their continued fraction asymptotic behaviour, impurity problems, dynamical Jahn-Teller effect in localized systems, electronic properties of surfaces and superlattices. The perspectives of the recursion and the renormalization methods extend to many other problems, such as fractal structures, quasicrystals, disordered systems, correlation effects in Auger spectra, where further advances are expected with the help of these special iterative procedures.

The recursion [1] and the renormalization [2] methods for electronic states calculations

R. Girlanda et al. (eds.), Progress on Electron Properties of Solids, 409–438.

have appeared in the literature following unrelated patterns, the former as a conceptual development of the Lanczos method [3], the latter as an implementation of the Wilson renormalization group method [4]. Their formal relationship has been pointed out only recently [5]. Both methods aim to provide a calculational procedure for the Green's function of an operator. All the relevant physical information becomes then accessible: for instance, the projected density of states, bound and resonant states, bond order parameters, whose knowledge opens the way to self-consistent calculations, and atomic forces from atomic energies via the Hellman-Feynman theorem. Although these methods have been used almost exclusively in connection with the tight-binding method, nothing prevents their use in connection with other formalisms, empirical or a priori, whenever the rank of the matrices to handle is very large and direct diagonalization impractical.

The basic feature of the recursion method is to provide a continued fraction expansion of the Green's function matrix elements. Given an operator (usually the one-electron crystal Hamiltonian) and a state of interest, the recursion method generates a new hierarchical set of orthonormal basis functions, in a chain-like sequence. This transformation is completely general and applies to any operator. However, it is particularly convenient for operators represented by sparse matrices. This occurs naturally in solids when localized orbitals are adopted as basis functions, because the orbitals on a given site interact only with a reasonably small number of nearby orbitals. The renormalization approach has entered the arena of the workable methods for electronic band structure calculations about a decade after the original formulation of Wilson in the theory of phase transitions. The renormalization approach consists of a systematic reduction of the degrees of freedom of a given system. Any reduction is of practical value only under the condition that the original and the reduced Hamiltonian have the same form, so that the renormalization process can be repeated over and over; then either a fixed point Hamiltonian is reached or the large but finite number of degrees of freedom of the system is exhausted. The form invariance condition is in general very restrictive; however, the method has been enriched with concepts and techniques peculiar to solid state physics and has become now a most promising tool in the study of electronic properties of superlattices and other multilayer structures.

In Section 2, we summarize the theoretical foundation of the recursion and the renormalization methods, and discuss their equivalence, on the basis of elementary properties of matrix algebra. The other Sections of the paper contain applications in several fields of relevant interest.

In Section 3 and 4 we consider the recursion method in **k**-space and in real-space, respectively. For periodic structures, we show that it is possible to exploit translational symmetry and achieve an unprecedented number of recursions, which permit to analyze the asymptotic behaviour of continued fractions. In real space, we show how to handle very large clusters of atoms by exploiting local symmetry and avoiding the construction of the map of neighbours.

Sections 5 and 6 deal with defect problems; the first one treats the electronic structure of impurities, the second one the phonon coupling to localized electronic states. In Section 5, we show that the recursion method provides a non-perturbative treatment for impurity potentials of any arbitrary range, and we present some new results for substitutional impurities in silicon. In Section 6 we discuss the recursion method in the treatment of strongly interacting vibronic systems, and the role of Jahn-Teller effect and spin-orbit coupling on the optical absorption spectrum.

Sections 7 and 8 illustrate the application of the renormalization method to multilayer structures. The perspectives of the method in the investigation of surfaces, heterostructures, quantum wells and superlattices are outlined. In addition, we present new results and interpretation on the challenging semiconductor-semimetal transition in $InAs-GaSb$ superlattice. Section 9 contains the conclusions.

2. Theoretical Foundation of the Recursion and Renormalization Methods

The recursion and renormalization methods are, in essence, physically motivated iterative procedures to obtain directly the Green's function of an operator H, avoiding the intermediate step of its explicit diagonalization. Both methods are precious tools to handle matrices (preferably in sparse form), whose rank is high (in the range, say, of thousand to millions). In addition, the renormalization method (whenever applicable) can also handle the 'thermodynamic' limit of infinite order matrices.

The most straightforward way [5] to derive these methods is to partition the space in which H is represented and to use elementary algebraic properties of matrices. Given a Hamiltonian H, let us represent it, quite generally, on an orthonormal basis $\{\phi_i\}$ as:

$$H = \sum_{ij} T_{ij} |\phi_i><\phi_j|. \tag{2.1}$$

We wish to calculate relevant matrix elements of the Green's function operator

$$G(E) = \frac{1}{E-H} \tag{2.2}$$

avoiding the explicit diagonalization of H.

For this purpose, we begin to separate the space S spanned by $\{\phi_i\}$ into two arbitrary complementary subspaces S_A and S_B, so that

$$P_A = \sum_{i \in S_A} |\phi_i><\phi_i|; \quad P_B = \sum_{i \in S_B} |\phi_i><\phi_i|$$

$$P_A + P_B = 1.$$

Using the projection operators P_A and P_B, and a proper arrangement of the corresponding basis sets, we separate H into four parts:

$$H = \begin{bmatrix} H_{AA} & H_{AB} \\ H_{BA} & H_{BB} \end{bmatrix}. \tag{2.3}$$

Similarly,

$$G(E) = \begin{bmatrix} G_{AA}(E) & G_{AB}(E) \\ G_{BA}(E) & G_{BB}(E) \end{bmatrix}. \tag{2.4}$$

From Eqs. (2.2) and (2.3) and straightforward matrix multiplication we obtain the expressions of $G_{AA}(E)$, $G_{AB}(E)$, $G_{BA}(E)$ and $G_{BB}(E)$. In particular, for $G_{AA}(E)$ we have:

$$G_{AA}(E) = \frac{1}{E - H_{AA} - H_{AB}\dfrac{1}{E-H_{BB}}H_{BA}}. \tag{2.5}$$

Eq. (2.5) basically means that the effective energy dependent Hamiltonian defined on the subspace S_A

$$H_{AA}^{\text{eff}}(E) = H_{AA} + \Sigma(E), \tag{2.6a}$$

$$\Sigma(E) = H_{AA} + H_{AB}(E - H_{BB})^{-1} H_{BA}, \tag{2.6b}$$

is exactly equivalent to the original Hamiltonian (2.1) for what concerns its Green's function matrix elements on the subspace S_A.

It is evident that, provided the reduced Hamiltonian has the same form as the initial one, the procedure can be repeated leading either to a fixed point Hamiltonian or to a rapid decrease, and eventual exhaustion, of the degrees of freedom: this is the basic strategy of the renormalization method.

A close inspection of Eq. (2.5) immediately leads to the recursion method too. In fact, let us include in the subset S_A just one (normalized) state $|f_0\rangle$, so that $P_A = P_0 = |f_0\rangle\langle f_0|$ and $P_B = 1-P_0$. Let us define

$$\begin{aligned} a_0 &= \langle f_0|H|f_0\rangle \\ |F_1\rangle &= (1-P_0)H|f_0\rangle = H|f_0\rangle - a_0|f_0\rangle \\ H_1 &= (1-P_0)H(1-P_0). \end{aligned}$$

In this case Eq. (2.5) yields:

$$G_{00}(E) \equiv \langle f_0|\frac{1}{E-H}|f_0\rangle = \cfrac{1}{E-a_0-\langle F_1|\frac{1}{E-H_1}|F_1\rangle}. \tag{2.7}$$

Let us indicate with b_1^2 the normalization of $|F_1\rangle$ and with $|f_1\rangle$ the corresponding normalized state. We can repeat for $|f_1\rangle$ and H_1 the procedure adopted for $|f_0\rangle$ and $H_0(\equiv H)$, and proceed iteratively to obtain with straightforward algebra the standard three-terms recurrence relations

$$\begin{aligned} |F_{n+1}\rangle &= H|f_n\rangle - a_n|f_n\rangle - b_n|f_{n-1}\rangle \\ b_{n+1}^2 &= \langle F_{n+1}|F_{n+1}\rangle \\ a_{n+1} &= \langle f_{n+1}|H|f_{n+1}\rangle. \end{aligned} \tag{2.8}$$

The continued fraction expression of the diagonal Green's function matrix element is then:

$$G_{00}(E) = \cfrac{1}{E-a_0-\cfrac{b_1^2}{E-a_1-\cfrac{b_2^2}{E-a_2-\cfrac{b_3^2}{\cdots}}}}. \tag{2.9}$$

In general, for an infinite system the chain of variables generated from a given initial state is also infinite. As a result, the accuracy and utility of the recursion method depends heavily on the number of exact recursions that can be performed. Once the maximum effort has been done in this direction, one has to determine the most fruitful use of the information obtained with the recursion process.

A first possibility exploits mathematical properties on the part of the chain that is actually known to set up rigorous bounds, which hold regardless of the (infinite) part of the chain that remains unaccessible. The well known mathematical bounds on the integrated density of states is such an example. In other cases one diagonalizes the tridiagonal chain at several steps and looks for convergence of relevant physical quantities; this is very useful to obtain information on states localized in the first steps of the chain. In other situations, simplified models are used to continue the recursive process, or to provide a reasonable description of the asymptotic behaviour. In any case, the treatment of the tail of the continued fractions is a nonroutine problem [6,7], whose solution, if possible at all, requires skilful use of mathematical properties and physical considerations. On the other hand, the renormalization method, when applicable, is very rapidly convergent and presents no problem of asymptotic behaviour.

Finally we refer to the literature [5,6] for the formal interconnections of the recursion and renormalization methods with other methods such as the method of moments [8], the subspace Hamiltonian technique [9], the equation of motion method [10], the transfer matrix method [11,12], as well as other techniques used in statistical physics [13] and many-body problems [14]. Notice, furthermore, that our unified treatment of the recursion and renormalization methods opens in principle the possibility of joined renormalization-recursion procedures; it is likely that the treatment of disordered systems will eventually benefit from such possibility.

3. Recursion Method in Reciprocal Space for Periodic Structures

As first application of the iterative procedures outlined above, we consider the implementation of the recursion method for periodic structures.

In this case it is possible to exploit the translational symmetry and the **k**-space representation. The one-electron crystal Hamiltonian can be diagonalized with standard methods of band theory, and indicating by $\psi_{n\mathbf{k}}$ the resulting Bloch functions, the Green's function is given by

$$G(E) = \sum_{n\mathbf{k}} \frac{|\psi_{n\mathbf{k}}><\psi_{n\mathbf{k}}|}{E - E_{n\mathbf{k}}}. \tag{3.1}$$

In the literature, the matrix elements of the Green's function operator are routinely obtained from Eq. (3.1) or appropriate manipulations. Thus, one may wonder why the recursion method should be employed at all. The answer is that the recursion method provides the Green's function with less effort than the standard **k**-space methods (Eq. (3.1)), which require sampling and processing a large number of **k** points in the Brillouin zone and repeat the time consuming procedure for all energies E of interest; this is avoided in the recursion method where the unitary transformation to a linear chain has to be performed just once, regardless of the trial energy E.

The usefulness of the recursion method in **k** space has been pointed out only recently [15,16]. Anlage et al. [15] have shown the convenience of **k**-space for taking into account the point group symmetry of the crystal and have embodied the special **k**-points technique [17] in the recursion method. Cordelli et al. [16] have shown how to fully exploit translational symmetry, thus reaching a very large number of recursions. We briefly summarize this procedure [16], and present new results for the cubium model crystal, and its asymptotic behaviour.

We start from a microscopic description of the crystal in terms of localized orbitals. The one-electron crystal Hamiltonian in real space can be represented in the form

$$H = \sum_{mn,\mu\nu} T_{mn,\mu\nu} \, |\phi_{m\mu}><\phi_{n\nu}| \tag{3.2}$$

where $\{\phi_{m\mu}\}$ indicates a set of orthonormal localized orbitals, centered on the position $\mathbf{d}_\mu$ within the cell τ_m. The indices m,n run over all couples of interacting atoms, while μ,ν run over the (small) number of independent orbitals at each site.

The first step to exploit translation symmetry consists in using the Bloch sum representation instead of the local orbital representation:

$$|\Phi_{\mathbf{k}\mu}\rangle = \frac{1}{\sqrt{\mathcal{N}}}\sum_{\vec{\tau}_m} e^{i\mathbf{k}\vec{\tau}_m}|\phi_{m\mu}\rangle, \tag{3.3}$$

where $\mathcal{N}$ is the number of **k** vectors in the Brillouin zone. In this new representation, the crystal Hamiltonian is given by

$$H = \sum_{\mathbf{k}\mu\nu} T_{\mu\nu}(\mathbf{k})|\Phi_{\mathbf{k}\mu}\rangle\langle\Phi_{\mathbf{k}\nu}|. \tag{3.4}$$

The iterative procedure should start in real space with a localized state, say $|\phi_{0\nu}\rangle$, and the Hamiltonian (3.2). Alternatively, and much more conveniently, one can implement the iterative procedure in **k**-space with the Hamiltonian (3.4) and the starting state

$$|f_0\rangle = \frac{1}{\sqrt{N^3}}\sum_{\mathbf{k}}^{N^3}|\Phi_{\mathbf{k}\nu}\rangle \tag{3.5}$$

where a large, but finite number of selected **k** vectors in the Brillouin zone has been chosen appropriately. For instance, one can make the choice:

$$\mathbf{k} = \frac{1}{2N}[(2m-1)\mathbf{h}_1 + (2m_2-1)\mathbf{h}_2 + (2m_3-1)\mathbf{h}_3] \tag{3.6}$$

$$1 \leqslant m_1,\ m_2,\ m_3 \leqslant N$$

where $\mathbf{h}_1$, $\mathbf{h}_2$, $\mathbf{h}_3$ are the fundamental vectors of the reciprocal lattice. With this choice it is possible to demonstrate that the first N recursions in reciprocal space using as seed state $|f_0\rangle$ are exact and coincide with the recursion in real space starting from $|\phi_{0\nu}\rangle$. With the special **k**-points technique, choices other than (3.6) are possible.

Eqs. (3.5) and (3.6) show that in order to achieve, say, hundred exact coefficients, we have correspondingly to consider about one million **k** vectors in the Brillouin zone. It is not easy to handle such a large number of degrees of freedom at the same time, due to computer storage limitation, but the new procedure [16] recently devised allows to overcome the problem. Notice, in fact, that the application of the Hamiltonian H to a Bloch function of vector **k** does not connect it to other Bloch functions of different wave vector; hence, the application of H to a set of Bloch functions within a given subregion of the Brillouin zone does not involve spreading to Bloch functions in other subregions. We can thus focus on an arbitrarily chosen subregion (the only limitation is that it can be handled by the computer); from the huge number of its degrees of freedom we generate an appropriate tridiagonal matrix, and all we have to store are the first N pairs of chain parameters. We have thus the full computer storage capabilities available to attack the next subregion, and so on until all subregions are processed. A final iteration is then performed on the chains so generated.

The recursion procedure in each subzone is very simple to implement. Let us indicate with M the number of **k** vectors in each given subzone S and consider

$$|f_{0S}\rangle = \frac{1}{\sqrt{M}}\sum_{\mathbf{k}}^{M}|\Phi_{\mathbf{k}\nu}\rangle. \tag{3.7}$$

Let us indicate with $c_{k\mu}^{(0)} \equiv 1/\sqrt{M}$ the expansion coefficients of the seed state (3.7). The expansion coefficients $c_{k\mu}^{(1)}(i=1,2,...)$ of the hierarchical set generated by the seed state (3.7) and the Hamiltonian (3.4) yield

$$C_{\mathbf{k}\mu}^{(i+1)} = \sum_{\nu} T_{\mu\nu}(\mathbf{k})c_{\mathbf{k}\nu}^{(i)} - a_i c_{\mathbf{k}\mu}^{(i)} - b_i c_{\mathbf{k}\mu}^{(i-1)}, \tag{3.8a}$$

with the next couples of chain parameters given by

$$b_{i+1}^2 = \sum_{\mathbf{k}\mu} C_{\mathbf{k}\mu}^{(i+1)*} C_{\mathbf{k}\mu}^{(i+1)}; \; c_{\mathbf{k}\mu}^{(i+1)} = \frac{1}{b_{i+1}} C_{\mathbf{k}\mu}^{(i+1)} \tag{3.8b}$$

$$a_{i+1} = \sum_{\mathbf{k}\mu\nu} c_{\mathbf{k}\mu}^{(i+1)*} T_{\mu\nu}(\mathbf{k}) c_{\mathbf{k}\nu}^{(i+1)}.$$

As a result of the unitary transformation from the Bloch functions for each subzone to the hierarchical set of functions which give the tridiagonal representation (within that subzone) of the Hamiltonian H, we obtain a set of independent linear chains. These chains can be handled following the same procedure and yield at the end N exact pairs of the continued fraction parameters.

The importance of a large number of exact parameters is not only relevant for the perfect crystal, but also in many other problems that use the Green's function of the perfect crystal as the basic ingredient to treat break of symmetry or disorder effects. Furthermore, the unprecedented number of exact recurrences obtained for the perfect crystal opens the road for gaining a new insight into the asymptotic behaviour of the continued fraction parameters.

It is worth noticing that the possibility of evaluating a large number of continued fraction parameters, up to the asymptotic region, is essential to the knowledge of the physical properties of a given system. For this purpose, numerical calculations in conjunction with analytic treatments have been pursued for several model densities of states [18,19]. In particular, the effect of anomalies in the density of states gives rise to well defined asymptotic behaviours of the continued fraction coefficients, and the presence of gaps produces undamped oscillations of these coefficients [20,21]; for a single connected band also the effect of van Hove critical points [22] has been fully analyzed [23,24]. This means that in the asymptotic behaviour of the continued fraction coefficients information is retained concerning the long range periodicity of the crystal, its dimensionality, and also position and type of the singularities within the spectrum.

Let us examine briefly the main analytic results on this last point, because we are now in the position to verify the theoretical predictions with numerical results obtained by means of the **k**-space recurrence procedure outlined above.

Let us consider a single connected band with edges E_1 and E_2. It is well known that the parameters of its associated continued fraction must reach the asymptotic values a_∞ and b_∞, where a_∞ is the midpoint of the band ($a_\infty = (E_1+E_2)/2$) and b_∞ is a quarter of the bandwidth ($b_\infty = (E_2-E_1)/4$). If we indicate with $\alpha_n = a_n - a_\infty$ and with $\beta_n = b_n - b_\infty$ the deviations of the continued fraction parameters from their asymptotic value and with $\phi_s = \arccos(E_s - a_\infty)/2b_\infty$ the angular position of the van Hove singularity E_s within the band, it is possible to deduce the results of Table I, through an appropriate use of the Dyson equation [24].

We can now verify the theoretical predictions in the case of cubium, a model crystal [16] with only s-type functions at the sites of a simple cubic lattice. The single connected band is

$$E(\mathbf{k}) = E_0 + 2t[\cos(k_x a) + \cos(k_y a) + \cos(k_z a)] \tag{3.9}$$

where a is the lattice parameter and t is the nearest neighbour interaction. It can be easily seen that the function (3.9) has two internal van Hove critical points (saddle points) at the energies $E_s = E_0 \pm 2t$. From Table I, indicating with ϕ_s the phase of the critical points, the theoretical predicted value of the b_n's in the asymptotic region is

$$(b_n - b_\infty) n^{3/2} = A \cos[(2n+1)\phi_s + \eta] \tag{3.10}$$

where the amplitude A and the phase η depend on the particular model.

TABLE I

	INTERNAL SINGULARITY	BOUNDARY SINGULARITY
3-Dimensional Crystal (++) Minimum M_0 (−−) Saddle M_1 (−+) Saddle M_1 (+−) Maximum M_3	$\beta_n = \pm \frac{k}{n^{3/2}} \cos[(2n+1)\phi_s + \delta_s \pm \frac{\pi}{4}]$	$\beta_n = \frac{k'}{n^3} \cos[(2n+1)\phi_s + \delta_s]$
2-Dimensional Crystal Minimum P_0 Saddle P_1 Maximum P_2	$\beta_n = \frac{c}{n} \cos[(2n+1)\phi_s + \delta_s]$ $\beta_n = \frac{c}{n} \sin[(2n+1)\phi_s + \delta_s]$ same as P_0	$\beta_n = \frac{c'}{n^2} \sin[(2n+1)\phi_s + \delta_s]$
1-Dimensional Crystal Minimum Q_0 Maximum Q_1	same as M_0 same as M_3	same as M_0 same as M_3

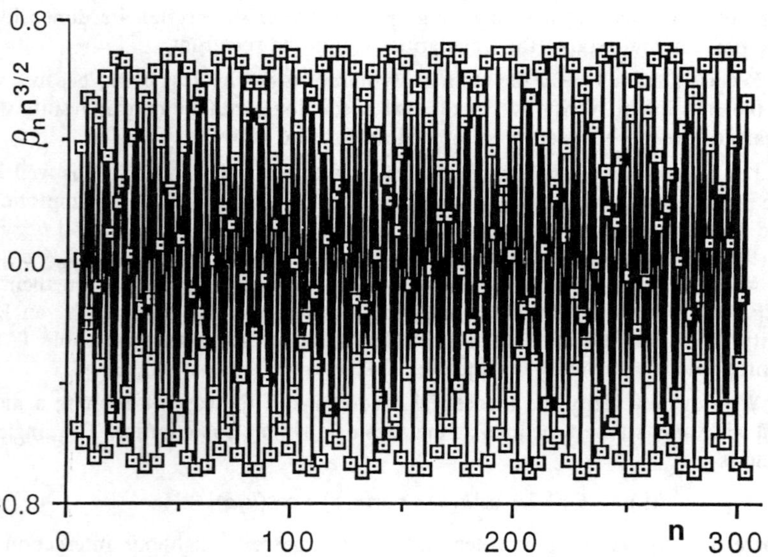

Figure 1. Values of $\beta_n n^{3/2} (\beta_n = b_n - b_\infty)$ plotted for as many as 300 values of n to provide an experimental study of the asymptotic behaviour in cubium.

We have evaluated as many as 300 coefficients for the cubium model. They are reported in Figure 1 (the hopping parameter t is taken as the unity of energy), and follow

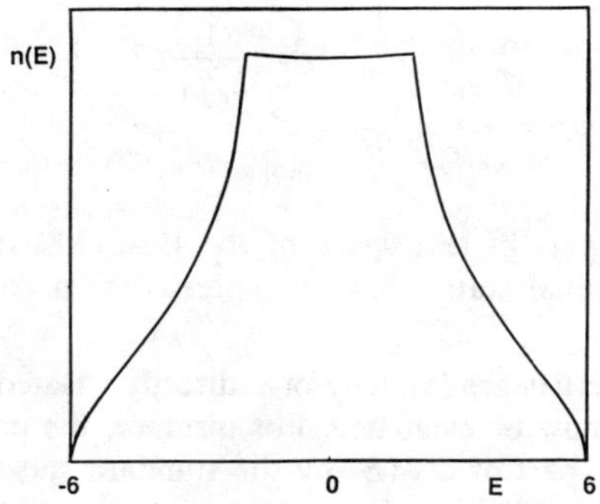

Figure 2. Density of states of the cubium model with 300 exact coefficients and square root terminator.

the expected asymptotic behaviour. From these coefficients the well known truncated bell-shape of the density of states can be reconstructed (Figure 2). It can thus be seen that spurious oscillations in the density of states near the critical points M_1, and M_2 disappear if 100 (or more) continued fraction parameters are exactly calculated before applying a square root terminator. The same order of coefficients is expected to be accessible also for realistic semiconductors with standard parametrization models [25], and intensive research is under way in this direction [16].

4. Recursion Method in Real Space for Periodic or Aperiodic Systems

In the previous section we have discussed the application of the recursion method in reciprocal space. The procedure given there was explicitly devised for periodic structures, being based on the **k**-space representation. In the case of non-periodic structures it is evident that we have to implement the recursion method in real space. In these cases, a cluster of atoms as large as possible, compatibly with the computer storage capabilities, is constructed. The large cluster is supplemented by appropriate 'boundary conditions' for the atoms at or near the boundaries, to mimic as accurately as possible the actual infinite system: free clusters, saturated clusters, or repeated clusters are examples used in the literature.

Let $\{\phi_{m\mu}\}$ denote the basis set of orthonormal orbitals for the cluster under consideration (m labels the sites in the cluster and μ the localized orbitals on each sites) and $T_{m\mu,n\nu}$ the matrix elements of the one-electron Hamiltonian. The recursion method begins, as usual, with the preliminary choice of a state of interest. Depending on physical considerations, the initial state can be a single localized orbital or an appropriate combination of them. Quite generally, we can write:

$$|f_0> = \sum_{m\mu} c^{(0)}_{m\mu} \, |\phi_{m\mu}>. \tag{4.1}$$

Once the choice (4.1) is made, the recursion method generates a unique hierarchy of states $|f_0>...|f_n>$ and chain parameters a_n, b_n. The recursion relationship for the expansion coefficients is

$$C^{(i+1)}_{m\mu} = \sum_{n\nu} T_{m\mu,n\nu} c^{(i)}_{n\nu} - a_i c^{(i)}_{m\mu} - b_i c^{(i-1)}_{m\mu}. \tag{4.2a}$$

The normalization coefficients b^2_{i+1} and the expectation values a_{i+1} are given by

$$b_{i+1}^2 = \sum_{m\mu} C_{m\mu}^{(i+1)*} C_{m\mu}^{(i+1)}; \quad c_{m\mu}^{(i+1)} = \frac{1}{b_{i+1}} C_{m\mu}^{(i+1)} \tag{4.2b}$$

$$a_{i+1} = \sum_{m\mu} c_{m\mu}^{(i+1)*} T_{m\mu,n\nu} c_{n\nu}^{(i+1)}. \tag{4.2c}$$

Eqs. (4.2) are the counterpart in real space of the Eqs. (3.8) in reciprocal space. The Green's function on the initial state $|f_0>$ is expressed as a continued fraction of type (2.9).

Relevant physical properties are (directly or indirectly) related to the Green's function matrix elements, and can now be evaluated. For instance, the projected density of states is related to the imaginary part of $G_{00}(E)$ by the standard spectral theorem result. The poles of the real part of $G_{00}(E)$ in the forbidden energy regions determines the localized levels. For self-consistent calculations, it is the density matrix that is of primary interest. Its expression in terms of a local basis representation is:

$$\rho = \sum_{ij,\mu\nu} \rho_{i\mu,j\nu} |\phi_{i\mu}><\phi_{j\nu}|. \tag{4.3}$$

The coefficients $\rho_{i\mu,j\nu}$ are called bond parameters and can be expressed in the form

$$\rho_{i\mu,j\nu} = -\frac{1}{\pi}\mathrm{Im} \int_{-\infty}^{E_F} G_{i\mu,j\nu}(E)dE, \tag{4.4}$$

E_F being the Fermi energy. In particular, the diagonal bond parameters $\rho_{i\mu,i\mu}$ provide the fractional occupation number of the orbital, while off-diagonal terms are related to the concepts of hybridization and charge transfer.

Let us examine more closely the recursion relation (4.2) from an operational point of view in the case of a general cluster (a cluster without any particular local symmetry, or in which the symmetry is not explicitly used). The standard way [26] to operate is to begin to label with a sequential index all the atoms of the cluster. Then one creates the map of all mutually interacting neighbour atoms. Finally, the interaction matrix for each couple of interacting atoms is evaluated; only at this stage the recursion process can be started.

It is evident that for a general cluster a large number of computer storage locations are required; if this infinite cluster simulates an infinite solid, the storage limitation poses a serious obstacle to the ability of achieving more than a handful of exact recursions (ten, twenty or so). In some calculations, 'over-recursions' are performed without increasing the cluster size, but reliability is doubtful due to the possible border effects which are not physical in nature.

The recursion method become much more effective, and many more recursions can be performed, if the cluster under consideration is 'regular' and the local symmetry is exploited. By a regular cluster we mean that all the atoms (or most of them, except possibly a few at or near disrupted regions) occupy a finite volume of a translationally invariant lattice. When the geometry of the atomic positions in the cluster is regular, it is possible to avoid the time and storage consuming construction of the map of neighbours [27,28]. Furthermore, the possible pairs and orientation of nearest neighbours is drastically reduced, and consequenlty a very small number of independent interaction matrices need to be considered. In recent calculations in clusters with the zincblende structure [27,28] (and five orbitals per site) it has been possible to obtain about 50 exact recursions, and considerably more, easily a factor 2 or better, if real space group symmetry is employed to maximum advantage. This opens new perspectives in a number of problems, but the full potentialities of the procedure are still to be explored.

5. Impurity Problems of Arbitrary Range Potentials with the Recursion Method

5.1. Green's function approach to the impurity problem

The study of the electronic structure of defects in solids has a long tradition, and we refer to the literature [29-32] for a description of the numerous lines of approach devised to treat the challenging effects brought about by the breaking of the translational symmetry. Some defects can be treated within the envelope function approximation, whose strategy essentially consists in avoiding a detailed account of the host perfect crystal, but rather in describing it with relevant bulk parameters (usually concerning energy band extrema in the Brillouin zone). In most actual situations, however, a detailed 'microscopic approximation' of the impurity is needed. We consider here the role that the recursion method is assuming in the latter line of attack.

The recursion method can be fruitfully discussed if compared with the defect Green's function method [33,34], widely applied for impurity problems of short range nature. Let us picture the whole perturbed system as a perfect crystal hosting a (small) impurity region. The Hamiltonian can be written as

$$H = H^0 + V \tag{5.1}$$

where H^0 is the Hamiltonian of the perfect host crystal, and V is the potential associated to the impurity. By expanding both H^0 and V in terms of the localized orbitals $\{\phi_{m\mu}\}$ of the perfect crystal, we obtain:

$$H^0 = \sum_{m\mu,n\nu} T_{m\mu,n\nu} |\phi_{m\mu}><\phi_{n\nu}| \tag{5.2}$$

and

$$V = \sum_{m\mu,n\nu} \gamma_{m\mu,n\nu} |\phi_{m\mu}><\phi_{n\nu}|. \tag{5.3}$$

The reason of splitting H into a perfect crystal part H^0 and a perturbation V is that the Green's function $G^0(E)$ of H^0 is (reasonably) easy to be obtained; the Green's function $G(E)$ of H is then obtained from the knowledge of $G^0(E)$ and V by means of the Dyson equation:

$$G(E) = G^0(E) + G^0(E)VG(E). \tag{5.4}$$

In particular, the bound impurity states within the gap (if any) are obtained by solving the determinantal equation:

$$\|1 - G^0(E)V\| = 0. \tag{5.5}$$

The solution of the secular equation (5.5) requires the knowledge of all the matrix elements of $G^0(E)$ on the impurity space; thus the defect Green's function method is workable only if V involves a small number of orbitals (impurity of short-range nature).

The matrix elements of $G^0(E)$ in the impurity subspace are the ingredients to set up and solve the impurity problem. In Section 3 and Section 4 we have shown the advantage of the recursion method in calculating $G^0(E)$, and we have seen how to evaluate the Green's function matrix elements of interest for the perfect crystal Hamiltonian H^0, operating either in real space or in reciprocal space. Thus the recursion method is an invaluable auxiliary tool in the defect Green's function method.

This remarkable advantage, however, is not the most important one; indeed what is even more important is the fact that the recursion method opens new perspectives in the impurity problems of arbitrary range, including extended impurities with Coulomb tails, a problem of major importance in the design of semiconductor devices. Alternative theoretical treatments are of limited value for a quantitative account of the complicate interplay of short and long range effects, especially in multi-valley semiconductors.

The recursion method can treat the problem of extended impurities on a microscopic basis [6,35,36]. The idea is to avoid the traditional split of the crystal plus impurity system into a periodic Hamiltonian and whatever left, but to treat on the same footing the host crystal and the guest impurity, setting up clusters as big as possible. The Hamiltonian to be considered is directly

$$H = \sum_{m\mu,n\nu} (T_{m\mu,n\nu} + \gamma_{m\mu,n\nu}) |\phi_{m\mu}><\phi_{n\nu}| \tag{5.6}$$

with no restrictions on the extension of the impurity subspace. In Section 4 we have seen how to calculate a large number of recursions in real space. From the poles of the corresponding Green's function in the energy gap region the impurity energy levels can be obtained with high accuracy.

The recursion method is emerging as a precious tool in the study of electronic structure of crystals with impurities of arbitrary range, as shown by the recent analysis of impurities [27,28] in cubium and actual multiband semiconductors. Although a more satisfactory treatment of the asymptotic behaviour is desirable, it is already possible to perform significant calculations and to predict further advances in this field.

5.2. Substitutional Impurities with Localized and Extended Potentials in Silicon

As an example of the concepts discussed in Section 5.1, we present here new results concerning localized and extended impurities in silicon. For the microscopic description of the silicon crystal, we use the local orbital representation [25] with five orbitals per site (s,p_x,p_y,p_z,s^*) and nearest neighbour interactions.

We begin by approximating a perfect infinite *Si* solid with a finite *Si* cluster. The cluster we consider is a cube with dimensions N_1a, N_2a, N_3a, $(N_1=N_2=N_3=N)$ along the translation vectors $\vec{\tau}_1 = a(1,0,0)$, $\vec{\tau}_2 = a(0,1,0)$, $\vec{\tau}_3 = a(0,0,1)$, and eight atoms in the conventional cubic cell. We typically take $N = 8$, i.e. a cluster with 4096 atoms and 20480 localized orbitals, although occasionally clusters of higher size are studied for convergence tests. Periodic boundary conditions are employed to connect bonds extending past the cluster borders to atoms on the opposite sides of the cluster. With $N = 8$, a seed state placed at the center of the cluster yields 16 exact recursions before the cluster borders are encountered. Actually recursions up to a few hundreds have been performed for the *Si* cluster, and we verified that the number of steps was more than sufficient to assure that no spurious state occurred in the gap. In fact, the cyclic boundary conditions correspond to selecting out of the band structure dispersion curves of *Si* those energy levels that fold back at the $\mathbf{k} = 0$ vector of the Brillouin zone of the 'repeated cluster'; of course no such energy level can appear outside the band edges. At this point we are ready to place the impurity into the cluster and evaluate Green's function matrix elements. Whenever necessary the cluster size is increased to include up to 32768 atoms and 163840 orbitals in order to verify that cluster size effects or impurity band broadening are negligible.

We begin by calculating the Green's function matrix elements of interest. The Green's function $G^0_{ss}(E)$ on the orbital s for the perfect cluster is

$$G_{ss}^0(E) = \cfrac{1}{E - a_0 - \cfrac{b_1^2}{E - a_1 - \cfrac{b_2^2}{E - a_2 - \cfrac{b_3^2}{\cdots}}}} = \frac{1}{E - E_s^{(Si)} - \Sigma_s^{(Si)}(E)}. \quad (5.7)$$

Similarly for $G_{pp}^0(E)$ we obtain:

$$G_{pp}^0(E) = \frac{1}{E - E_p^{(Si)} - \Sigma_p^{(Si)}(E)}. \quad (5.8)$$

In Figure 3 we report the s orbital projected density of states in silicon calculated in our repeated cluster with 370 recursions. The termination scheme is here simply an imaginary part given to E. This is equivalent to broaden with a Lorentzian curve the palisade of 371 delta-like functions which would appear by straightforward truncation.

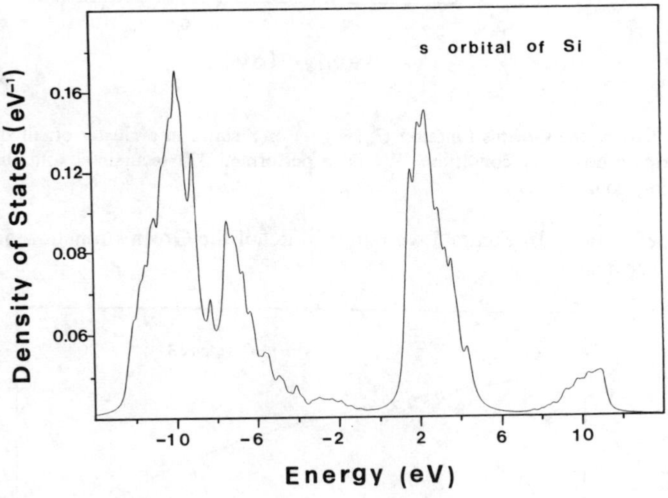

Figure 3. Projected density of states ($-1/\pi \operatorname{Im} G_{ss}^0(E+i\epsilon)$) on s states in a cluster of silicon with 4096 atoms and periodic boundary conditions. We have performed 370 recursions, and the imaginary part of the energy equals $0.15\,eV$.

The valence band density of states presents two broad bands separated by a dip at $-8\,eV$, which corresponds to the X_1 point energy in the valence band structure (the zero of energy is taken at the top of the valence band). The conduction band density of states also presents two broad peaks, separated by a small gap; however, this small gap at $\simeq 6.5\,eV$ is an artifact of the parametrization used, and has no physical relevance. From inspection of Figure 3 and comparison with the nominally exact results obtained with Gilat-Raubenheimer integration [15,37] it is evident that undesired oscillations are present in the allowed energy regions.

In Figure 4 we show the real part of the Green's function on the s orbital (the energy E has been given a small but finite imaginary part equal to $0.15\,eV$). We note the regular monotonically decreasing behaviour in the forbidden energy regions ($E < -12.5\,eV$, $0 < E < E_g = 1.17\,eV$ and $E > 11.38\,eV$), but spurious oscillations are still present in the allowed energy bands. In the forbidden energy regions, on the contrary, we have verified that the Green's function is essentially independent of the cluster size and of the number

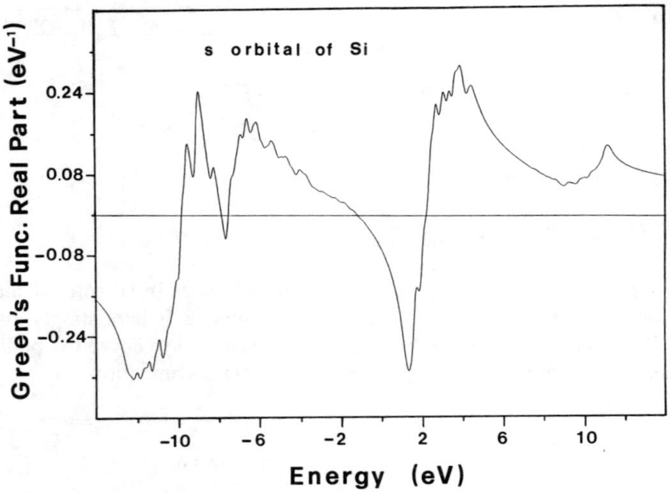

Figure 4. Real part of the Green's function $G^0_{ss}(E+i\epsilon)$ on s states in a cluster of silicon with 4096 atoms and periodic boundary conditions. We have performed 370 recursions, and the imaginary part of the energy equals $0.15\,eV$.

of recursion performed. In Figure 5 we report in detail the Green's function on s orbitals in the energy gap region.

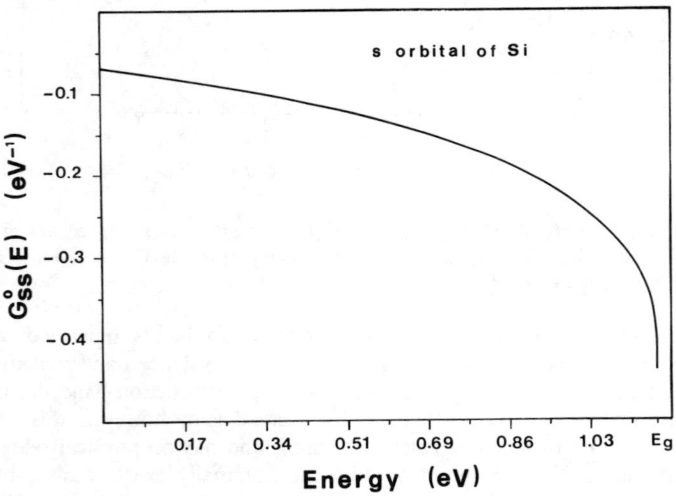

Figure 5. Real part of the Green's function $G^0_{ss}(E)$ for real E in the energy gap in the silicon crystal. For central cell impurity potentials, the bound s state (if any) is determined by the equation $G^0_{ss}(E) = 1/\Delta^{(i)}_s$.

We study now the effect of a foreign localized impurity in the silicon cluster. The impurity potential is first assumed to influence only the s and p orbitals on a single site (the s^* orbital is introduced in the parametrization [25] just to reproduce the indirect energy gap, and in some sense is more related to the host crystal than to the actual impurity). The short range impurity potential is thus of the type

$$V = (E_s^{(i)} - E_s^{(Si)})\,|\phi_{0s}><\phi_{0s}| + \tag{5.9}$$
$$+(E_p^{(i)} - E_p^{(Si)})[\,|\phi_{0x}><\phi_{0x}| + |\phi_{0y}><\phi_{0y}| + |\phi_{0z}><\phi_{0z}|\,].$$

The quantities $\Delta_s^{(i)} = E_s^{(i)} - E_s^{(Si)}$ and $\Delta_p^{(i)} = E_p^{(i)} - E_p^{(Si)}$ are often assumed to be related to the difference between the ionization potentials of the impurity and perfect crystal atoms; the schematization (5.9) has enjoyed a broad success in predicting the overall trends of impurity levels [38,39].

Because of the local tetrahedral symmetry, the state ϕ_{0s} never mixes in the recursion with the p-like states on the same lattice point. We have thus that the Green's functions of the perturbed crystal on the s orbital becomes simply:

$$G_{ss}(E) = \frac{1}{E - E_s^{(i)} - \Sigma_s^{(Si)}(E)}. \tag{5.10}$$

In the energy gaps the impurity levels of symmetry A_1 are given by the poles of $G_{ss}(E)$ or, equvalently, by the condition

$$G_{ss}^0(E) = \frac{1}{\Delta_s^{(i)}}. \tag{5.11}$$

Similarly, for central cell corrections on p states, T_2 impurity states are determined by the condition $G_{pp}^0(E) = 1/\Delta_p^{(i)}$.

Let us discuss Eq. (5.11), with $G_{ss}^0(E)$ given in Figure 5. Since $G_{ss}^0(E)$ for E in the energy gap is negative and monotonically decreasing, it is possible to have bound states only if $\Delta_s^{(i)}$ is negative and below the threshold value. When the perturbation parameter $\Delta_s^{(i)}$ is less than $-14\,eV$, a resonance develops in the valence band. The resonance seems to be pinned at about $-0.68\,eV$, that compares favorably with the value $-0.79\,eV$ for the unrelaxed A_1 vacancy resonance in silicon [40]. The energies of bound and resonant A_1 impurity states are shown in Figure 6. For completeness, we report in Figure 6 also the hyperdeep impurity level that arises below the bottom of the valence bands; below threshold, it soon reaches a simple linear behaviour without pinning.

It is worthwhile to discuss briefly the identification of the resonant states in the allowed energy regions. We notice that the crystal plus impurity system has wavefunctions which exhibit two extreme behaviours: itinerant states delocalized on the whole system and localized states (if any) surrounding the impurity region, with resonant states (if any) having intermediate nature.

Led by the above considerations, we have devised a very effective method to identify bound and resonant states of impurities in crystals. After performing as many recursions as possible (say N), we diagonalize the tridiagonal matrix (of order of $N+1$, if the states $|f_0>\ldots|f_N>$ have been determined). Let $E_i (i = 1,\ldots N+1)$ indicate the eigenvalues of the truncated chain, and $c_{i,j}$ the coefficients of the corresponding eigenvectors. Let us consider in particular the quantities $W(E_i,f_0) = |c_{i,0}|^2$ and $W(E_i,f_N) = |c_{i,N}|^2$; it is evident that;

$$\sum_{i=1}^{N+1} W(E_i,f_0) = 1 \tag{5.12a}$$

$$\sum_{i=1}^{N+1} W(E_i,f_N) = 1. \tag{5.12b}$$

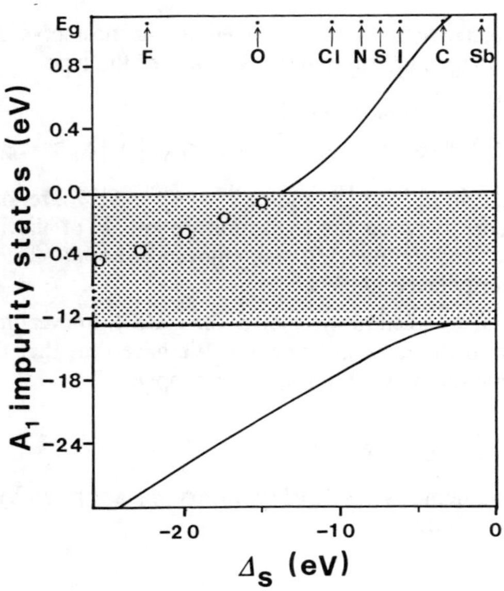

Figure 6. Energy of the bound states and resonances of A_1 symmetry versus central cell potential in silicon crystals. The shaded region is the valence band region; notice the change of vertical energy scale therein. The hyperdeep impurity state is also reported. Some $\Delta_s^{(i)}$ for impurity atoms are indicated.

It is also clear that $W(E_i,f_0)$ and $W(E_i,f_N)$ can be used to give an indication of the degree of localization of the wavefunction corresponding to E_i. It is found by careful analysis that for N sufficiently large, bound states in the gap are identified by the fact that $W(E_i,f_0)$ is rapidly convergent, and $W(E_i,f_N)$ is orders of magnitude smaller than average. Also, most resonances can clearly be identified by the increase of the oscillator strength $W(E_i,f_0)$ on a level E_i (or on a few adjacent ones) of at least one order of magnitude with respect to the average intensity of nearby eigenvalues.

The analysis so far illustrated for A_1 impurity states can be done for scattering states of T_2 symmetry, as well. In Figure 7 we show the Green's function $G_{pp}^0(E)$ in the energy gap region. The real part of $G_{pp}^0(E)$ now changes its sign at $E \simeq 0.51\,eV$, so that we have the possibility of impurity levels for both sufficiently attractive and sufficiently repulsive central cell potentials. The pinning energy $\simeq 0.51\,eV$ corresponds to the dangling bond of a neutral unrelaxed vacancy level in the gap, and is again in good agreement with the value of $0.87\,eV$ found in Ref. [40].

Until now, we have discussed short-range perturbations in silicon with the recursion method. But the full potentiality of the method becomes evident in discussing the modified electronic structure when short-range and long-range potential contributions are both present with comparable importance. The potential operator due to a foreign impurity at the origin can be approximated in the form

$$V = \sum_{\mu} \gamma_\mu \,|\phi_{0\mu}\rangle\langle\phi_{0\mu}| - \sum_{m\mu}{}' \frac{Ze^2}{\epsilon R_{m\mu}} \,|\phi_{m\mu}\rangle\langle\phi_{m\mu}| \qquad (5.13)$$

where the last term represents a screened Coulomb tail. The recursion method is now applied considering the total Hamiltonian of the perfect crystal plus the perturbating potential (5.13). In Figure 8, we report the A_1 impurity states (bound states in the energy gap region and resonances near the top of the valence band) in the presence of an

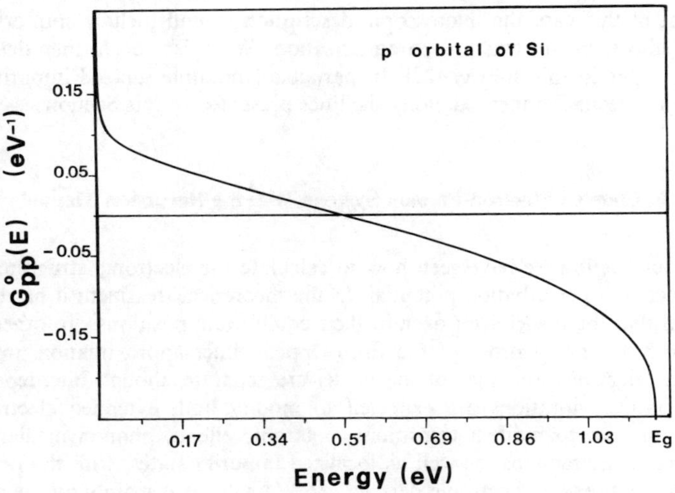

Figure 7. Green's function $G^0_{pp}(E)$ for real E in the energy gap in silicon crystals. For central cell impurity potentials, the bound T_2 state (if any) is determined by the equation $G^0_{pp}(E) = 1/\Delta^{(i)}_p$.

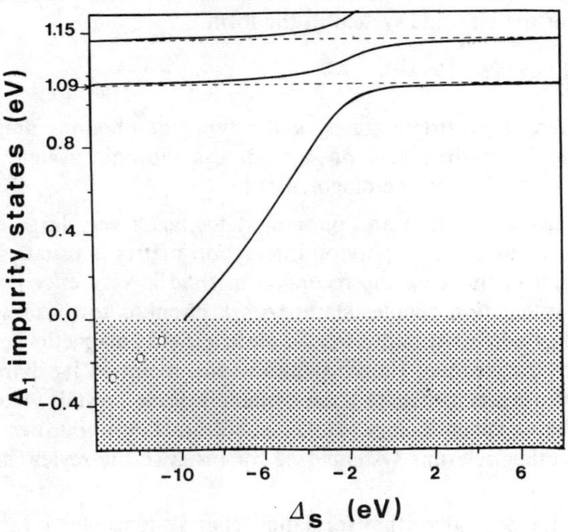

Figure 8. Energy of the bound states and resonances of A_1 symmetry versus central cell potential in the presence of an attractive Coulombic tail ($Z = 1$) in silicon. The shaded region is the valence band region. Notice the change of vertical energy scale between the $n = 1$ pinning level and the bottom of the conduction band.

attractive coulombic tail ($Z = 1$). We report only the lowest Rydberg-like series terms; notice that the levels $n = 2, 3$ are rather insensitive to the value of Δ_s and are pinned to the positively charged vacancy states. On the other hand, it is evident from Figure 8 that the lowest state is very sensitive to the interplay of short-range potential and long-range Coulombic tail, as predicted on the basis of the intervalley interference effects [41].

Finally, we mention that the structure of the acceptor states could also be worked out similarly, but in this case the microscopic description should include spin-orbit interaction in the silicon band structure parametrization. We refer for further details to the forthcoming paper in this subject [27]. In particular, multiple ionized impurities can be studied with the recursion method along the lines presented in this Section.

6. Coupled Electron-Phonon Systems with the Recursion Method

In the previous Section we have seen how to calculate the electronic structure of a solid in the presence of a perturbation potential. In the theoretical treatment it has been implicitly assumed that the nuclei are frozen in their equilibrium positions. In other words we were accepting the lowest order of the Born-Oppenheimer approximation, in which the motion of the electrons and that of the nuclei are separate, though interdependent. In general, the lattice vibrations are expected to modify both extended electronic states (leading for instance to indirect transitions, polaronic effects, phonon mediate electron-electron effective interactions) as well as localized impurity states, with the possibility of total or partial removal of electronic degeneracy. We will focus mainly on the classic subject of dynamical Jahn-Teller effect in localized electron states; but the concepts and the procedures that we illustrate should find a broad application in other coupled electron-boson systems.

The direct way to solve the problem of localized electrons coupled to phonons is to expand the states of the vibronic system in the form:

$$\psi = \sum_{in\alpha} c_{in\alpha} |\psi_i\rangle |\phi_{n\alpha}\rangle, \tag{6.1}$$

where i labels different electronic states, α the types of phonons and n is the phonon occupation number. One then acts on ψ with the vibronic system Hamiltonian and obtains, in principle, a matrix to be diagonalized.

The product space of electron and phonon states has a very large number of degrees of freedom; however, the electron-phonon interaction matrix is usually rather sparse, and it is well known that in this case the recursion method is very effective. For instance, a linear Jahn-Teller interaction couples states with n phonons to those with $n \pm 1$ phonons only; similarly other perturbations, such as electric field, magnetic field, spin-orbit and strain, add terms to the vibronic Hamiltonian but still maintain its sparse form. Examples of classic Jahn-Teller systems include color centers in ionic crystals [42,43], ions of transition metals in cubic semiconductors [44,45], simple metals impurities in rare gas solids [46,47], and many other relevant systems (see for instance the review articles in Ref. [48] and Ref. [49]).

The use of the Lanczos algorithm for Jahn-Teller systems was initited by Muramatsu and Sakamoto [50], who succeeded in obtaining for some simple systems analytic expression of the continued fraction coefficients. For example, the vibronic system constituted by a doubly degenerate electronic state of symmetry E in a cubic crystal interacting with a two-dimensional phonon of symmetry ϵ (elsewhere referred as Γ_3 or E) leads to

$$a_n = E_e + n\hbar\omega \quad n = 0,1,...; \quad b_n^2 = V_E^2 I(\frac{n+1}{2}) \quad n = 1,2,... \tag{6.2}$$

where I stands for integer part, V_E is the coupling constant in the linear Jahn-Teller interaction, and E_e is the orbital energy (the initial state chosen for the recursion procedure is a zero phonon vibronic state). Then it is straightforward to obtain the absorption spectrum by diagonalizing the tridiagonal matrix and by taking the projection (modulus square) of the corresponding eigenfunctions on the initial state. The absorption spectrum shows the characteristic two-bands shape produced by the two minima of the adiabatic potential surface.

Also for the $T \oplus \epsilon$ systems (this is a triply degenerate p-like electronic state, interacting in a cubic crystal with two-dimensional phonons of ϵ symmetry) one can obtain analytic expressions for a_n and b_n. Choosing as initial state a zero phonon vibronic state one obtains:

$$\begin{aligned} a_n &= E_p + (n+1)\hbar\omega \qquad n = 0,1,\ldots \\ b_n^2 &= nV_T^2 \qquad\qquad\qquad\; 1,2,\ldots \end{aligned} \tag{6.3}$$

where V_T is the coupling constant in the linear Jahn Teller interaction, and E_p is the orbital energy of the p-like states.

The vibronic system $T \oplus \epsilon$, in the presence of spin-orbit interaction, can be easily treated numerically with the recursion method and a large number (say hundred or more) of parameters of the continued fraction can be evaluated. The parameters show very complicated oscillations, responsible for the three bands in the absorption spectrum. We have noticed an interesting and useful fact: these oscillations are of decreasing amplitude around the straight line defined by Eq. (6.3).

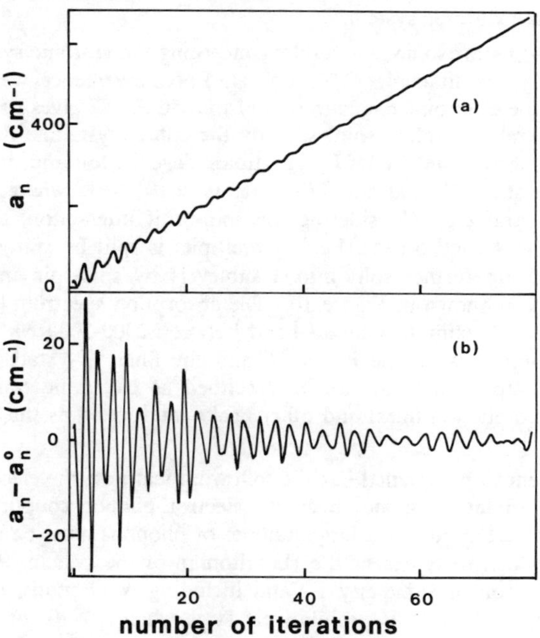

Figure 9. (a) Behaviour of the coefficients a_n in the case of the vibronic model $T \otimes \epsilon$ and spin-orbit interaction. (b) Oscillatory behaviour of $a_n - a_n^0$, where a_n^0 are the coefficients of the $T \otimes \epsilon$ model without spin-orbit interaction.

This is clearly seen in Figure 9, where the vibronic model has been specified for Ag impurities in argon matrix [47]. We thus find an asymptotic behaviour very different from that discussed for electronic states in Section 3; a comparative systematic study is desirable and should open further perspectives in the field of vibronic systems. An asymptotic behaviour of type (6.3) has been exploited to improve the accuracy of calculations on the vibronic system $T\oplus\tau$ (the phonons involved are three-dimensional) in the presence of spin-orbit interaction; the complicated oscillations of the a_n and b_n parameters occur also in this case around a straight line, whose coefficients can be obtained by a least square fitting procedure [47].

The recursion method has been gradually extended to systems of increasing computational difficulty. Examples are multi-mode systems such as $E\oplus(\epsilon_1+\epsilon_2)$ and $T\oplus(\epsilon+\tau)$ with spin-orbit coupling. A review of physical and numerical aspects for Jahn-Teller and pseudo Jahn-Teller systems is presented in Ref. [51].

More recently, further pursuing the advances from both computational and physical point of view, we have been able to successfully discuss the Jahn-Teller effect in the complicated problem of transition metal ions in cubic semiconductors. From a technical point of view, two elegant improvements should be mentioned. The first one concerns the technique of making convolutions with a continued fraction approach [47], and the second one concerns the concepts and properties of the so-called [45] 'dipole carrying state'. Such a state is the particular linear combination of dipole allowed states, with coefficients proportional to the matrix elements of the dipole operator. When such a state is chosen as the initial one, all the states generated by the recursion procedure are dipole free. Thus the Green's function projected on the initial state gives directly the absorption band of the vibronic system.

As an example, we report some new results concerning the vibronic system constituted by a Fe^{++} impurity ion in cubic $CdTe$ crystal. For convenience we recall that the ground state $3d^6$ of the electronic configuration of the ion Fe^{++} gives rise to a ^5D multiplet (25 times degenerate including spin) split by the cubic crystalline field [22] into the levels ^5E (10 times degenerate) and $^5T_2$ (15 times degenerate); the energy separation between the ground state ^5E and the $^5T_2$ state is $\simeq 10|Dq|$, where Dq is the one-electron cubic-field parameter. Considering the spin-orbit interaction, the ^5E multiplet splits in 5 sublevels to second order. The $^5T_2$ multiplet is split by spin-orbit interaction in 3 sublevels (these are further split into 6 sublevels by spin-spin interaction). The corresponding scheme is shown in Figure 10. The absorption spectrum [52] of Fe^{++} in $CdTe$ at low temperature exhibits a broad band between 2200-3000 cm^{-1}, arising from an electronic transition between the initial ^5E and the final $^5T_2$ states of Fe^{++}. The spectrum contains sharp peaks that can be described as electronic transitions without phonon emission (zero-phonon lines) and other peaks interpreted as the phonon assisted transitions.

The vibronic system to be handled has the following features: the electronic part is 15 times degenerate (multiplet $^5T_2$); moreover, the electron-phonon coupling is strong (the Huang-Rhys factor is $\simeq 6$). Hence, a large number of phonons must be included to construct the basis on which to represent the Hamiltonian of the system. Even considering only one type of phonon of symmetry Γ_3 and including N phonons, one arrives at a matrix whose dimension $15\times N^2$ is prohibitively large when $N\geq 10$ for standard diagonalization techniques. Hence, it is apparent the usefulness of the recursion procedure, taking as initial state a carrying dipole state. In Table II we report for Fe^{++} ion in $CdTe$ the calculated energies and intensities of the lowest transition lines, compared with the experimental ones, as well as the quantities $W(E_i,f_N)$ (defined in Section 5 as the magnitude square of the projection $<\psi_i|f_N> = c_{iN}$ of the chain eigenfunction ψ_i on the chain state f_N). The quantities $W(E_i,f_N)$ give an indication of the localization and accuracy of the eigenvalue E_i after N recursions. Notice that in the literature the 'accuracy parameter' is often provided [51] via the alternative quantity $\rho(E_i,f_N)$, called the 'residual vector' and defined as $\rho(E_i,f_N) = |c_{iN}\cdot b_{N+1}|$. The results of Table II are obtained with the spin-orbit coupling $\lambda = -100cm^{-1}$ as in the free Fe^{++} ion, the phonon energy

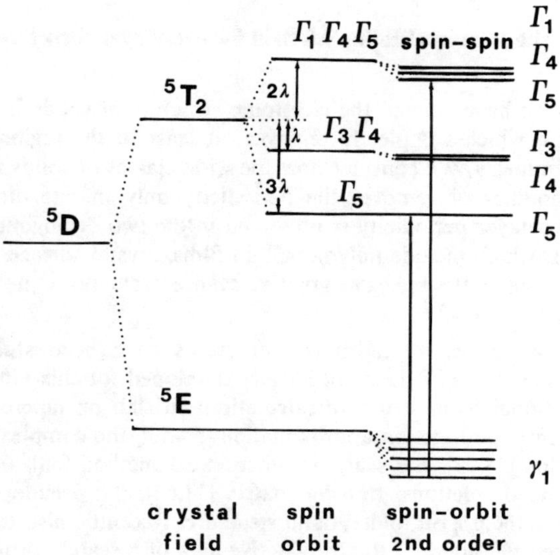

Figure 10. Schematic indication of the energy levels of the atomic multiplet ^5D of Fe^{++} in a crystal field of T_d symmetry, including spin-orbit and spin-spin interactions.

TABLE II.

Energies E_i and intensities $W(E_i,f_0)$ of the dipole allowed transitions from the ground γ_1 level to Γ_5 vibronic levels for $CdTe{:}Fe^{++}$. Experimental data are also shown for comparison. The convergence parameters $W(E_i,f_N)$ for $N = 50$, 70 and 100 iterations are also reported. The theoretical calculations are made with the following choice of the vibronic model parameters: $\gamma = -100cm^{-1}$, $\hbar\omega = 36cm^{-1}$, $E_{JT} = 232cm^{-1}$ and $|Dq| = 258.5cm^{-1}$.

Transition calcul.	energy (cm^{-1}) experim. [52]	$W(E_i,f_0)$	$W(E_i,f_N)$ $N=50$	$N=70$	$N=100$
2282	2282	0.106	6.5×10^{-11}	1.6×10^{-15}	9.3×10^{-22}
2295	2294	0.112	1.9×10^{-8}	2.5×10^{-10}	5.3×10^{-18}
2306	2309	0.081	6.7×10^{-7}	3.2×10^{-8}	2.8×10^{-15}
2320	2318	0.040	1.7×10^{-5}	1.3×10^{-7}	9.6×10^{-12}
2334	2334	0.029	1.9×10^{-4}	3.6×10^{-6}	7.9×10^{-9}

$\hbar\omega = 36cm^{-1}$ and the Jahn-Teller energy $E_{JT} = 232cm^{-1}$. From the detailed calculations we see that 50 iterations or more are necessary to obtain a good description of the absorption peaks in the initial part of the infrared spectrum.

The recursion method has greatly extended the richness and sophistication of the vibronic systems that can be handled theoretically, thus reducing the gap between the wealth of available experimental data and their quantitative interpretations. The recursion method, with the mathematical and physical developments which are accompanying its deployment in this field [47,53] is bringing renewed interest and stimulus in the comprehension of the Jahn-Teller effect in localized systems, and, more generally, in the physics of coupled electron-boson systems.

7. The Renormalization Method for Multilayer Structures

In Section 5 and 6 we have studied the electronic structure of solids in the presence of a perturbing potential, which completely removes (at least in the region of the impurity) the translational symmetry. We consider now the wide classes of solids in which a perturbation potential modifies or removes the periodicity only in one direction (say the z direction), while intralayer periodicity is preserved in the two orthogonal (x and y) directions. Such systems, which include polymers, thin films, crystal surfaces, heterostructures, quantum wells and superlattices, have great relevance from both the fundamental and technological point of view.

The electronic properties of multilayer structures have been studied with several theoretical approaches, most of them specifically developed for this kind of materials. In addition to conventional band structure calculations in slab or supercell geometries, we may mention the envelope function approximation [54-56], the complex-**k** vector [57] and complex **k·p** methods [58,59], the scattering-theoretical method [60], ordinary [61,62] or matrix [5,63] continued fractions, transfer matrix [11,64], the pseudopotential approach [65] or other schemes [66,67] on folded band structure. Recently, also the renormalization approach has entered the arena of this very active area of research, providing for surfaces [68-70], interfaces [69] and superlattices [71,72] a procedure which is at the same time physically transparent and computationally relatively simple. In this Section, we describe the common peculiar features of the multilayer structures that make them so suitable for the renormalization approach; in the next Section we will discuss the specific problem of the semiconductor-semimetal transition in indium-arsenide gallium-antimonide superlattices.

The prototype layer structure of physical relevance is graphite. In the two-dimensional approximation it consists of a single plane of carbon atoms at the corners of an array of hexagons. Pioneering studies [73] were performed adopting a microscopic description with s and p localized orbitals, and using the two-dimensional Bloch sums (known in the current literature as 'layer orbitals') as basis functions, because these automatically account for the intralayer symmetry properties. While early calculations considered a single plane only, later calculations included successfully the interactions among planes stacked along the anisotropic axis, and an increasing number [74-76] of layered structures and politype effects were interpreted.

In the usual layer structures [76] the stacking of planes in the z direction is periodically repeating with reasonably short lattice constant; thus, the full **k**-space representation and direct diagonalization of the Hamiltonian at **k** points of interest can be done with standard procedures. On the other hand, in the case of surfaces or superlattices, the periodicity in the z-direction is removed or may exhibit a very long periodicity, respectively. In these cases the most economical way to proceed is to use a mixed representation (**q**-space representation for intralayer symmetry and real space representation for the orthogonal direction). The renormalization procedure can then be applied.

It is well known from elementary crystallography that any solid, with or without surfaces, can be resolved into a family of parallel atomic planes. The microscopic description of the crystal can then conveniently be done in terms of two-dimensional Bloch sums $|\Phi_{\mathbf{q}l\lambda}\rangle$, called layer orbitals, where $\mathbf{q}$ is a vector of the two-dimensional Brillouin zone, and λ labels the independent orbitals on each layer. Atomic planes can always be grouped into layers so that interlayer interactions are restricted to nearest neighbour in the layer index 1, even if interactions extend to higher order neighbours in real space. The index λ is most important because, as we will see, it determines the order of the matrices to be handled in the renormalization. For instance, in the case of Si (001)

surface with five localized orbitals per atom and nearest neighbour interactions, each layer is actually composed of a single atomic plane, and n_λ is 5.

In the layer orbital representation, the crystal Hamiltonian of any multilayer structure takes a peculiar block-tridiagonal form; for instance, for surfaces we have the semi-infinite form:

$$H = \begin{bmatrix} A_1 & B_1 & 0 & 0 & 0 \\ C_1 & A_2 & B_2 & 0 & 0 \\ 0 & C_2 & A_3 & B_3 & 0 \\ & & \cdots & \cdots & \cdots \end{bmatrix}, \tag{7.1}$$

where all matrices A, B, C have dimension $n_\lambda \times n_\lambda$ and depend only on **q**.

For films the dimension of the block matrix (7.1) is finite. For heterostructures, the block tridiagonal form is infinite, instead of semi-infinite. For superlattices with supercell periodicity d, the dimension is finite, and the interaction matrices $B_N e^{-ik_z d}$ and $C_N e^{ik_z d}$ are to be added at the upper right and lower left corners. All these systems are described by a (finite, semi-infinite or infinite) linear chain with n_λ orbitals per site, and nearest neighbours interactions only; thus they can be handled with the renormalization method with a simple and exact iterative procedure. We shall discuss the method as applied to (7.1); minor and straightforward modifications are needed to treat all the other layered structures.

The calculation of the Green's function on a given layer is performed by successive decimation (Section 2) of all the other layers. For instance, let us eliminate the nth layer in Eq. (7.1); the renormalized matrices on the $(n-1)$th and $(n+1)$th layers and their renormalized interactions are given by:

$$\tilde{A}_{n-1}(E) = A_{n-1} + B_{n-1} \frac{1}{E - A_n} C_{n-1} \tag{7.2}$$

$$\tilde{A}_{n+1}(E) = A_{n+1} + C_n \frac{1}{E - A_n} B_n$$

$$\tilde{B}_{n-1}(E) = B_{n-1} \frac{1}{E - A_n} B_n$$

$$\tilde{C}_{n-1}(E) = C_n \frac{1}{E - A_n} C_{n-1}.$$

The numerical calculation of Eqs. (7.2) only requires the inversion of the small matrix $(E - A_n)$ of order n_λ, besides straightforward multiplication and addition of matrices.

Notice that the block tridiagonal form is preserved upon the transformation (7.2), and the procedure can therefore be iterated without the problem of undesired induced interactions. Notice also that successive single layer elimination is formally analogous to repeated transfers of wavefunction using the transfer matrix [11,64]; however, the renormalization procedure remains stable independently of the number of iterations which are performed.

A nice feature of the renormalization method is that the local symmetry can be exploited to substantially reduce the number of decimations in bulk-like regions. Consider, for instance, a standard III-V compound with zincblende structure, sampled along the (001) direction. Since all alternate layers in the bulk crystal are equivalent, all even (or odd) layers can be eliminated in a single stroke. The procedure can then be repeated and a fixed point Hamiltonian is rapidly reached. Slightly different considerations also apply to superlattices, heterostructures and other layered structures. In all cases, a limited amount of renormalizations allow to obtain the Green's function on a given layer.

The knowledge of the energy levels can be supplemented, whenever necessary, by the calculation of the wavefunctions. One may simply repeat the calculation by preserving in turn successive layers. An alternative and simpler procedure is just to propagate the Green's function, known on any given pair of layers, to adjacent layers, by inverting formulas such as Eqs. (7.2); the multilayer structure wavefunction can thus be reconstructed site by site.

To conclude this section, we remark that also the standard recursion method can handle an Hamiltonian of type (7.1), which is essentially a multi-orbital linear chain Hamiltonian; for a linear chain, hundreds (or thousands) coefficients of the continued fraction expansion of a diagonal matrix element of the Green function can be readily calculated. Both the problem of the computer storage limitations and the difficulty of the continued fraction termination are thus overcome. In fact, the recursion method has been proposed for the calculation of superlattices electronic structures [61], and actually employed in the study of surface vibrational modes [62]. However, since this approach does not explicitly exploit the local symmetry properties, it is not as rapidly convergent as the renormalization method, but probably it becomes competitive with it in the presence of large non-bulk-like zones, for instance depletion layers in doped heterostructures.

8. Electronic States in Superlattices with the Renormalization Approach

As indicated in the previous Section, the renormalization method is very appealing both from the formal and computational point of view. It unifies the treatment of different situations, including strained or disrupted regions and allows self-consistent calculations to be performed. Each step of the renormalization procedure has a clear physical meaning and defines the effective Hamiltonian acting on the preserved layers.

Recent applications of the renormalization method include the widely studied electronic structure of *Si* (100)-(2$\times$1) reconstructed surface [70], and the effect of a super-periodic potential in *Si* [72]. In particular, the method has allowed to reveal the transition from indirect to direct gap materials, as well as the multivalley origin of the conduction band energy doublets in silicon superlattices [72].

To further illustrate the potentialities of the renormalization method, we consider the electronic structure of $InAs-GaSb$ superlattice and we focus on the challenging problem of its semiconductor-semimetal transition, that has defied until now a theoretical interpretation.

The microscopic description of *InAs* and *GaSb* is based on a sp^3s^*, tight-binding model [25], with nearest neighbour interactions only; spin-orbit coupling [77] is included and the parametrization was modified by us to accurately reproduce the effective masses of the valence bands and conduction bands edges [78]. The independent parameters of the model and further details will be published elsewhere [79].

The composite $InAs-GaSb$ system [80] is the prototype of kind II superlattices: the bottom of the *InAs* conduction band is lower in energy than the top of the *GaSb* conduction band. Despite the peculiar band alignment, small period superlattices are semiconducting; only for large enough unit supercells the system becomes semimetallic. The transition takes place at a *InAs* thickness of about 100A.

This well known experimental behaviour [81] is due to quantum size effects, and can be qualitatively understood in terms of simple models [64]. However, it is not easy to provide a quantitative account and predict the electronic features of the system: for instance, we find that the critical supercell width for the semiconductor-semimetal transition is very sensitive to the valence band offset, whose theoretical calculation is still uncertain.

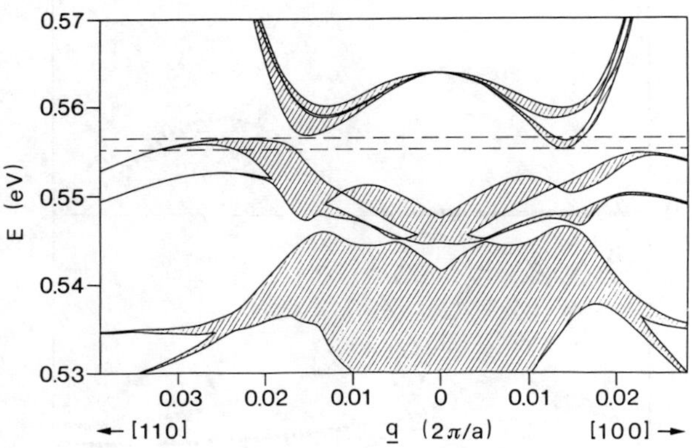

Figure 11. Energy bands of the *InAs* − *GaSb* (001) superlattice with 80 + 80 atomic planes. The parallel wave vector **q** is taken along the [100] and [110] directions; the dispersion along k_z is projected and originates shaded zones. Dashed lines indicate the top of the valence band and the bottom of the conduction band.

The nature itself of the semimetallic regime is not obvious, and requires a detailed knowledge of the superlattice band structure. In the envelope function approximation [55], the high carrier concentration is attributed to extrinsic effects. Our present calculations with the renormalization method show the occurrence of negative indirect gaps and the possibility of intrinsic semimetallic behaviour. We have computed the superlattice band structure near the Fermi energy for several unit cells containing up to 140 + 140 atomic planes. In Figure 11, we report the energy bands of a superlattice with 80 + 80 atomic planes per unit cell. The superlattice energy bands have been projected on the plane $k_z = 0$ of the superlattice Brillouin zone; this means that for any given **q** vector we consider the energies for all k_z from $-\pi/d$ to π/d, and we plot in Figure 11 the spread of energy at that given **q**. Notice the extreme simplicity by which k_z projected bands can be obtained with the renormalization method. In discussing Eq. (7.1) and its modifications for superlattices, we pointed out that only the corner elements contain the k_z dependence, and that only in the last renormalization are these elements varied. This allows to obtain, with negligeable additional effort, the band energies for any k_z component and hence the projected energy bands.

The main features of Figure 11 are high non-parabolicity of the bands, mixing and anticrossing behaviour (even for $\mathbf{q} = (0,0)$), and quasi-two-dimensional regime (small k_z dispersion) for large enough values of **q**. The conduction band minimum is lower than the valence band maximum by $1.3\,meV$, showing intrinsic semimetallic behaviour (even larger negative indirect gaps are found for other supercell thickness). Notice that the spin-orbit splitting for $\mathbf{q} \neq (0,0)$ and the anisotropy of the bands with respect to the direction of **q** have an important role on the negative indirect gap size.

Figure 12a displays the localization of the conduction band state at $0.564\,eV$ for $\mathbf{k} = (0,0,0)$. The wave function is confined in *GaSb*; since it is almost entirely of *p* character, and virtually dispersionless in the *z* direction, it is mostly derived from a bulk heavy-hole state of *GaSb*. The remarkable asymmetry of the wave function amplitude

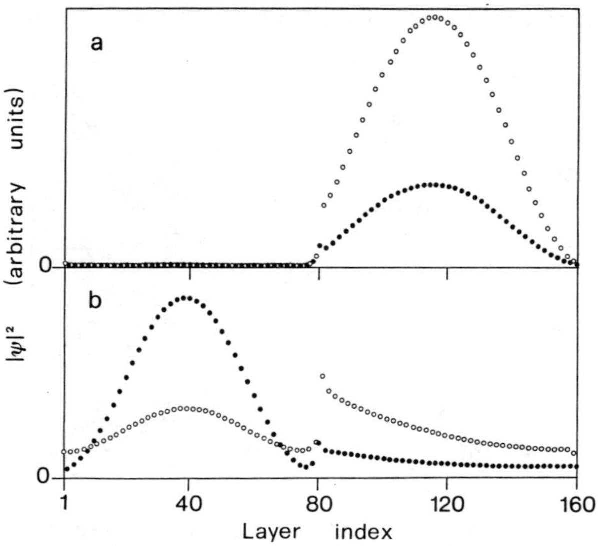

Figure 12. Wavefunction squared amplitude of (a) the state at $\mathbf{k} = (0,0,0)$ with $E = 0.564\,eV$, and (b) the state $\mathbf{k} = (0.02,0,0)2\pi/a$ with $E = 0.569\,eV$. Atomic planes 1 to 80 are *InAs*, 81 to 160 are *GaSb*. Amplitudes on anions and cations are indicated by white and black circles, respectively.

with respect to the *GaSb* slab center is a consequence of the difference between the two interfaces (one is of kind $In-Sb$, the other $Ga-As$). This effect is rather surprising, in view of the considerable thickness of the slabs of materials: it shows the importance of microscopic features of the interfaces, even for large period superlattices. In Figure 12b a state in the same subband is considered for a different $\mathbf{k}$ vector ($\mathbf{q}$ is now $0.02(2\pi/a)$ in the [100] direction, and $k_z = 0$): the mixing with the conduction band of *InAs* is evident, since the wavefunction now is mainly concentrated in *InAs*, and its orbital character is mostly *s*-like. Away from the mixing regions, the superlattice subbands quickly recover their bulk-like (*InAs* electron-like or *GaSb* hole-like) character.

Some quantitative aspects of the reported results, could be influenced to some extent by refinements in the model Hamiltonian employed, or by inclusion of charge transfer in a self-consistent way. However, from Figure 11 it can be realized that the formation of the negative indirect gap is due to spin-orbit splitting and band anisotropy; neither these effects, nor the basic quantum size effect mentioned above, are expected to heavily depend on details of the chosen parametrization or on the introduction of self-consistent charge transfer. The onset of intrinsic semimetallic regime has thus an actual physical meaning; further detailed calculations are in progress [79].

The ability of the renormalization method to provide a microscopic description of the not simple $InAs-GaSb$ system, while keeping the computational effort to a manageable level, opens interesting perspectives in the study of electronic structure and transport properties of superlattices and other multilayer materials.

9. Conclusions

In this paper we have reviewed the important role which the recursion and renormalization methods are gaining in the study of the electronic structure of periodic and aperiodic solids. For perfect crystals, it is now possible to reach many recursions and very accurate results. This encourages further investigation in the analytic or semianalytic treatment of disorder and localization in aperiodic solids, too. In the problem of the electronic structure of impurities with both short-range and long-range potential contributions, the recursion method has opened new perspectives. Also for impurities strongly coupled to lattice vibrations a major breakthrough has been achieved with iterative procedures. In the case the break of symmetry occurs only in one direction, the fastly convergent renormalization approach can be applied. In multilayer structures, the renormalization approach can play a substantial role in the theoretical understanding of the vast phenomenology that is enriching the very active area of heterostructure and superlattices research. The subjects discussed in this paper do not exhaust by any means the range of problems which could benefit from systematic application of the renormalization or recursion methods. Interesting new perspectives concern quasicrystals, fractal structures, as well as resonant tunneling and ballistic transport; these and other fields of active research are adding new flavour and interest in the iterative procedures in solid state physics.

Acknowledgements

This work has been partially supported by Commissione delle Comunità Europee through contract number ST2J-0254-2-I.

References

1. R. Haydock, V. Heine and M.J. Kelly, *J. Phys. C* **5**, 2845 (1972); *J. Phys. C* **8**, 2591 (1975). See also D.W. Bullett, R. Haydock, V. Heine and M.J. Kelly, *Solid State Physics* **35** (1980).
2. H. Aoki, *J. Phys. C* **13**, 3369 (1980); *J. Phys. C* **14**, 2771 (1981).
3. C. Lanczos, *J. Res. Nat. Bur. Standards* **45**, 255 (1950); **49**, 33 (1952); Applied Analysis, Prentice-Hall, Englewood Cliffs, N.J. 1956.
4. K. Wilson, *Phys. Rev. B* **4**, 3174, 3184 (1971).
5. P. Giannozzi, G. Grosso, S. Moroni and G. Pastori Parravicini, *Applied Numerical Mathematics* **4**, 273 (1988).
6. G. Grosso and G. Pastori Parravicini, *Adv. Chem. Phys.* **62**, 81 (1985); **62**, 133 (1985).
7. D.J. Pettifor and D.L. Weaire eds., Recursion Method and its Applications, *Solid State Science Series* **58**, Springer, Berlin, 1985.
8. F. Cyrot-Lackmann, *Adv. Phys.* **16**, 393 (1967); J.P. Gaspard and F. Cyrot-Lackmann, *J. Phys. C* **6**, 3077 (1973).
9. R.E. Allen and M. Menon, *Phys. Rev. B* **33**, 5611 (1986).
10. J.W. Halley and H.B. Shore, *Phys. Rev. B* **36**, 6640 (1987).
11. D.H. Lee and J.D. Joannopoulos, *Phys. Rev. B* **23**, 4988 (1981); **B23**, 4997 (1981).
12. M.P. López Zancho, J.M. López Sancho and J. Rubio, *J. Phys. F* **14**, 1205 (1984); *J. Phys. F* **15**, 851 (1985); L.R. Ram-Mohan, K.H. Yoo and R.L. Aggarwal, *Phys. Rev. B* **38**, 6151

(1988).

13. M. Evans, P. Grigolini and G. Pastori Parravicini eds, Memory Function Approaches to Stochastic Problems in Condensed Matter, *Adv. Chem. Phys.* **62**, Wiley, New York, 1985.
14. M.H. Lee, J.Hong and J. Florencio, Jr., in Proceedings of the 7th General Conference of the Condensed Matter Division of the European Physical Society, eds. F. Bassani, G. Grosso, G. Pastori Parravicini and M. Tosi, *Physica Scripta T* **19**, 498 (1987).
15. S.M. Anlage and D.L. Smith, *Phys. Rev. B* **34**, 2336 (1986); D.M. Woodruff, S.M. Anlage and D.L. Smith, *Phys. Rev. B* **36**, 1725 (1987).
16. A. Cordelli, G. Grosso and G. Pastori Parravicini, *Phys. Rev. B* **38**, 2154 (1988); and to be published.
17. A. Baldereschi, *Phys. Rev. B* **7**, 5212 (1973); D.J. Chadi and M.L. Cohen, *Phys. Rev. B* **8**, 5747 (1973); N.J. Monkhorst and J.D. Pack, *Phys. Rev. B* **13**, 5188 (1976).
18. G. Allan, *J. Phys. C* **17**, 3945 (1984); R. Haydock and C.M.N. Nex, *J. Phys. C* **18**, 2235 (1985); R. Haydock, *Phil. Mag. B* **53**, 545 (1986); M.U. Luchini and C.M.M. Nex, *J. Phys. C* **20**, 3125 (1987).
19. P. Giannozzi, G. Grosso and G. Pastori Parravicini, *Phys. Stat. Solidi B* **128**, 643 (1985).
20. A. Magnus, in Padé Approximation and its Applications, *Lecture Notes in Mathematics*, 765, L. Wuytack ed., Springer, Berlin, 1979, p. 150; A. Magnus, in Padé Approximation and its Applications, *Lecture Notes in Mathematics*, 1071, H. Werner and H.J. Bünger eds., Springer, Berlin, 1984, p. 213.
21. P. Turchi, F. Ducastelle and G. Treglia, *J. Phys. C* **15**, 2891 (1982).
22. F. Bassani and G. Pastori Parravicini, *Electronic States and Optical Transitions in Solids*, Pergamon Press, Oxford 1975; Nauka, Moscow 1982.
23. H.H. Hodges, *J. Phys. Lett.* **38**, L187 (1977).
24. G. Grosso, P. Pastori Parravicini and A. Testa, *Phys. Rev. B* **32**, 627 (1985).
25. P. Vogel, H.P. Hjalmarson and J.D. Dow, *J. Phys. Chem. Solids* **44**, 365 (1983).
26. C.M.M. Nex, Cambridge Recursion Method Library, cited in *Solid State Physics* **35**, (1980) p. 78.
27. D.J. Lohrmann, L. Resca, G. Pastori Parravicini and R.D. Graft, *Phys. Rev. B* (1989) in the press.
28. D.J. Lohrmann, Thesis submitted for his Ph.D. degree, Catholic University of America, Washington D.C., December 1988.
29. F. Bassani, G. Iadonisi and B. Preziosi, *Rep. Progr. Phys.* **37**, 1099 (1974).
30. S.T. Pantelides, *Rev. Mod. Phys.* **50**, 797 (1978).
31. M. Altarelli and F. Bassani in Handbook on Semiconductors, Vol. 1, ed. W. Paul, North Holland, Amsterdam, 1982.
32. M. Lannoo and J. Bourgain, Point Defects in Semiconductors, Vol. I Theoretical Aspects, Vol. II Experimental Aspects, Springer Verlag, Berlin Vol. I 1981; Vol. II 1983); M. Jaros, *Rep. Progr. Phys.* **48**, 1091 (1985).
33. G.F. Koster and J.C. Slater, *Phys. Rev.* **95**, 1167 (1954); **96**, 1208 (1954).
34. G.A. Baraff and M. Schlüter, *Phys. Rev. Lett.* **41**, 892 (1978); J. Bernholc, N.O. Lipari and S.T. Pantelides, *Phys. Rev. Lett.* **41**, 895 (1978); G.B. Bachelet, G.A. Baraff and M. Schlüter, *Phys. Rev. B* **24**, 915 (1981); B**24**, 4736 (1981); M. Schlüter in Highlights of Condensed Matter Theory, edited by F. Bassani, F. Fumi and M.P. Tosi, North-Holland, Amsterdam 1985, p. 495.
35. D.M. Bylander and J.J. Rehr, *J. Phys. C* **13**, 4157 (1980).
36. R.D. Graft, G. Grosso, G. Pastori Parravicini and L. Resca, *Solid State Commun.* **51**, 247 (1986).
37. G. Gilat and L.J. Raubenheimer, *Phys. Rev.* **144**, 390 (1966).

38. P. Vogl in Festkorperprobleme, Vol. XXI of Advances in Solid State Physics, edited by J. Treusch, Vieweg, Braunschweig 1981, p. 191.

39. H.P. Hjalmarson, P. Vogl, D.J. Wolford and J.D. Dow, *Phys. Rev. Lett.* **44**, 810 (1980); H.P. Hjalmarson, H. Buttner and J.D. Dow, *Phys. Rev.* **B24**, 6010 (1981); S.Y. Ren, R.E. Allen, J.D. Dow and I. Lefkoreitz, *Phys. Rev. B* **25**, 1205 (1982).

40. W. Yong-Liang and U. Lindefelt, *Phys. Rev. B* **37**, 1320 (1988); see also *Phys. Rev. B* **38**, 4107 (1988).

41. L. Resca, *Phys. Rev. B* **26**, 3238 (1981); **B29**, 866 (1984).

42. F.S. Ham, *Phys. Rev. B* **8**, 2926 (1973); F.S. Ham and U. Grevsmühl, *Phys. Rev.* **B8**, 2945 (1973).

43. L. Martinelli, G. Pastori Parravicini and P.L. Soriani, *Phys. Rev. B* **32**, 4106 (1985).

44. J. Rivera-Iratchet, M.A. De Orúe and E.E. Vogel, *Phys. Rev. B* **34**, 3992 (1986); *Phys. Rev. B* **38**, 3556 (1988).

45. L. Martinelli, M. Passaro and G. Pastori Parravicini, *Phys. Rev.* **B** (1989), in the press.

46. M.C.M. O'Brien and S.N. Evangelou, *J. Phys. C* **13**, 611 (1980); *Solid State Commun.* **36**, 29 (1980); M.C.M. O'Brien, *J. Phys. C* **16**, 85 (1983); **16**, 6345 (1983); **18**, 4963 (1985).

47. L. Martinelli and G. Pastori Parravicini, *Phys. Rev. B* **37**, 10612 (1988).

48. M.D. Sturge, Solid State Physics, edited by F. Seitz, D. Turnbull and H. Ehrenreich, Academic, New York, 1967, vol. 20.

49 Yu.E. Perlin and M. Wagner eds., The Dynamical Jahn-Teller Effect in Localized Systems, North-Holland, Amsterdam, 1984.

50. S. Muramatsu and N. Sakamoto, *J. Phys. Soc. Japan* **44**, 1640 (1978); **46**, 1273 (1979); *Phys. Rev. B* **17**, 868 (1978).

51. D.R. Pooler in The Dynamical Jahn-Teller Effect in Localized Systems, edited by Yu.E. Perlin and M. Wagner, North-Holland, Amsterdam, 1984, p. 199.

52. G.A. Slack, F.S. Ham and R.M. Chrenko, *Phys. Rev.* **152**, 376 (1966).

53. M. Cini and A. D'Andrea, *J. Phys. C* **21**, 193 (1988).

54. G. Bastard, *Phys. Rev. B* **24**, 5693 (1981); **B25**, 7584 (1982); *IEEE J. Quant. Electr.* **22**, 1625 (1986).

55. M. Altarelli, *Phys. Rev. B* **28**, 842 (1983).

56. M. Altarelli, V. Ekenberg and A. Fasolino, *Phys. Rev. B* **32**, 5138 (1985); G. Platero and M. Altarelli, *Phys. Rev. B* **36**, 6591 (1987); F. Ancillotto, A. Fasolino and J.C. Maan, *Phys. Rev. B* **38**, 1788 (1988).

57. J.N. Schulman and Yia-Chung Chang, *Phys. Rev. B* **24**, 4445 (1981); **B27**, 2346 (1983); **B31**, 2056, 2069 (1985); **B33**, 2594 (1986).

58. C. Mailhiot and D.L. Smith, *Phys. Rev. B* **33**, 8345 (1986); **B33**, 8360 (1986); **B38**, 5520 (1988).

59. L.C. Andreani, A. Pasquarello and F. Bassani, *Phys. Rev. B* **36**, 5887 (1987).

60. J. Pollmann and S.T. Pantelides, *Phys. Rev. B* **18**, 5524 (1978); M. Schmeits, A. Mazur and J. Pollmann, *Phys. Rev. B* **27**, 5012 (1983); J. Pollmann, P. Kruger and A. Mazur, *J. Vac. Sci. Technol. B* **5**, 945 (1987).

61. R.D. Graft, G. Pastori Parravicini and L. Resca, *Solid State Commun.* **54**, 115 (1985).

62. S. Tiersten, S.C. Ying and T.L. Reinecke, *Phys. Rev. B* **33**, 4062 (1986).

63. O. Fromm and J. Koutecký, in ed., Catalysis in Chemistry and Biochemistry: Theory and Experiment, edited by B. Pullmann, Reidel, Dordrecht, 1979, p. 335; K.S. Dy, Shi-Yu Wu and T. Spratlin, *Phys. Rev. B* **20**, 4237 (1979); J.D. Brasher and K.S. Dy, *Phys. Rev. B* **22**, 4868 (1980).

64. G.A. Sai-Halasz, L. Esaki and W.A. Harrison, *Phys. Rev. B* **18**, 2812 (1978).

65. I. Morrison, M. Jaros and B. Wong, *Phys. Rev. B* **35**, 9693 (1987); L.D.L. Brown, M. Jaros and D. Ninno, *Phys. Rev. B* **36**, 2935 (1987); R.J. Turton, M. Jaros and I. Morrison, *Phys. Rev. B* **38**, 8397 (1988).

66. C. Tejedor, M. Calleja, F. Messeguer, E.E. Mendez, C.-A. Chang and L. Esaki, *Phys. Rev. B* **32**, 5303 (1985).

67. A. Elçi, *Phys. Rev. B* **34** 8616 (1986).

68. G.W. Bryant, *Phys. Rev. B* **31**, 5166 (1985); B**35**, 5547 (1987).

69. F. Flores, J. Durán and A. Muñoz, in Proceedings of the 7th General Conference of the Condensed Matter Division of the European Physical Society, edited by F. Bassani, G. Grosso, G. Pastori Parravicini and M.P. Tosi, *Physica Scripta T* **19**A, 102 (1987).

70. G. Grosso, S. Moroni and G. Pastori Parravicini, *Physica Scripta* **37**, 930 (1988); *Physica Scripta*, **T25**, 316 (1989).

71. R.D. Graft, G. Pastori Parravicini and L. Resca, *Solid State Commun.* **57**, 699 (1986).

72. R.D. Graft, D.J. Lohrmann, G. Pastori Parravicini and L. Resca, *Phys. Rev. B* **36**, 4782 (1987).

73. F. Bassani and G. Pastori Parravicini, *Il Nuovo Cimento* **50**, 95 (1967).

74. E. Doni and G. Pastori Parravicini, *Il Nuovo Cimento A* **63**, 117 (1969).

75. E. Doni, G. Grosso, G. Harbeke, E. Meier and E. Tosatti, *Phys. Stat. Sol. B* **68**, 569 (1979); E. Doni, R. Girlanda, V. Grasso, A. Balzarotti and M. Piacentini, *Il Nuovo Cimento B* **51**, 154 (1979); B**54**, 269 (1979).

76. See for instance the review article E. Doni and R. Girlanda in Electronic Structure and Electronic Transitions in Layered Materials, edited by V. Grasso, Reidel, Dordrecht, 1986, p. 1-171.

77. D.J. Chadi, *Phys. Rev. B* **16**, 790 (1977).

78. P. Lawaetz, *Phys. Rev. B* **4**, 3460 (1971).

79. G. Grosso, S. Moroni and G. Pastori Parravicini, *Phys. Rev. B* in the press; *Helvetica Physica Acta*, in the press.

80. G.A. Sai-Halasz, R. Tsu and L. Esaki, *Appl. Phys. Lett.* **30**, 651 (1977); L. Esaki, *IEEE J. Quant. Electr.* **22**, 1611 (1986).

81. L.L. Chang, N.J. Kawai, G.A. Sai-Halasz, R. Ludecke and L. Esaki, *Appl. Phys. Lett.* **35**, 939 (1979); L.L. Chang, N.J. Kawai, E.E. Mendez, C.-A. Chang and L. Esaki, *Appl. Lett.* **38**, 30 (1981); Y. Guldner, J.P. Vieren, P. Voisin, M. Voos, L.L. Chang and L. Esaki, *Phys. Rev. Lett.* **45**, 1719 (1980); J.C. Maan, Y. Guldner, J.P. Vieren, P. Voisin, M. Voos, L.L. Chang and L. Esaki, *Solid State Commun.* **39**, 683 (1981).

CORE EXCITONS IN STRAINED-LAYER SUPERLATTICES

JOHN D. DOW AND JUN SHEN

Department of Physics
University of Notre Dame
Notre Dame, Indiana 46556 USA

SHANG YUAN REN

Department of Physics
University of Notre Dame, Indiana, USA
and
University of Science and Technology of China
Hefei, Anhui, People's Republic of China

The physics of core excitons in semiconductors is reviewed, with emphasis on the fact that Hjalmarson-Frenkel 'deep' core excitons are observed, and co-exist with Wannier-Mott 'shallow' excitons which are not normally resolved experimentally. The theory of Hjalmarson-Frenkel excitons is extended to excitons in superlattices, and the *Ga3d* core exciton in $GaAs_{1-x}P_x/GaP$ strained-layer superlattices is predicted to change from a resonance in the conduction band (with apparent negative binding energy) to a bound state in the gap (positive binding energy), as the $GaAs_{1-x}P_x$ layer thickness decreases.

1. Introduction

Franco Bassani pioneered the theory of core excitons and was one of the first theorists to call attention to the fact that the *Si 2p* core exciton exhibits a binding energy significantly different from the shallow donor binding energy in *Si* [1]. Since he was also one of the first theorists interested in artificial superlattices, we shall honor him by discussing the physics of core excitons in superlattices.

When a soft *x*-ray excites an atom in a semiconductor, it creates a core hole of very small radius and an electron with the same wave vector (see Figures 1 and 2). Because the core orbital has such a small radius, it does not overlap the corresponding orbital on an adjacent site, and so the core band is flat, with infinite mass. As a result, the hole is immobile, and is a fixed point charge around which the electron can revolve. One expects the electron's orbit to be hydrogenic, with an envelope wave function ψ obeying the effective-mass Schrödinger equation

$$[(-\hbar^2/2m^*)\nabla^2 - e^2/\epsilon r]\psi(\mathbf{r}) = E\psi(\mathbf{r}).$$

Here E is the energy of the orbiting electron with respect to the conduction band edge, ϵ is the dielectric constant of the semiconductor, and m^* is the conduction band's effective mass [2] (we have assumed an isotropic, non-degenerate effective mass, for simplicity). The ground state of such an electron in the presence of its hole should be the hydrogenic 1*s* state, with a binding energy relative to the conduction band minimum of

$$E_B = (13.6\,eV)(m^*/m_0\epsilon^2),$$

R. Girlanda et al. (eds.), Progress on Electron Properties of Solids, 439–449.

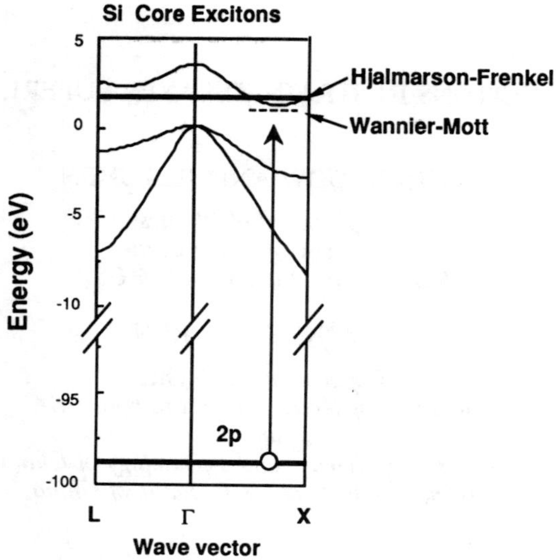

Figure 1. Energy versus wave vector: illustration of a *Si* 2*p* core excitation in *Si*. The core hole band is flat (lower heavy line), and the excited electron is either (i) associated with the conduction band minima near *X* for the shallow Wannier-Mott exciton (dashed line), or (ii) in the antibonding *s*-like A_1 deep level resonant with the conduction band for the Hjalmarson-Frenkel core exciton (upper heavy line).

typically tens of *meV*. Such a correlated electron-hole pair is a Wannier-Mott exciton [3].

Analogous impurity levels to the Wannier-Mott core excitons have been thoroughly studied and were thought to be completely understood until recently [4]: the shallow hydrogenic donor states [2]. The electrons of these states obey the same effective-mass Schrödinger equation as the electrons of the core excitons, and exhibit the predicted binding energies. In the case of *P* substituting for a *Si* atom in bulk *Si*, the physics of shallow donor levels has been well-established for decades. Therefore one expected that the *Si* 2*p* core exciton would exhibit the same binding energy as the *P* donor. It did not, being in many experiments two orders of magnitude larger: 0.1 to 0.9 *eV* [5-9].

The failure of the *Si* 2*p* core exciton binding energy to match the *P* donor binding energy was particularly perplexing because, to an adequate approximation (the $Z+1$ rule [1,10] or the optical alchemy approximation [11]), the core hole has zero radius and the same charge distribution as a proton. In the case of core-excited *Si*, the excited atom with its 'proton' and electron is '*P*' (similarly core-excited *Ga* and *In* are '*Ge*' and '*Sn*', respectively).

Hence in the case of the *Si* 2*p* core exciton, one expected an absorption spectrum that reflected the spectrum of the '*P* donor' impurity on *Si*, with its small hydrogenic effective-mass binding energy: $13.6\,eV\,(m^*/m_0\epsilon^2)$. The observation of much larger *Si* 2*p* core exciton binding energies ranging from 0.1 to 0.9 *eV* [5] became known as the '*Si* core exciton problem' - a problem whose explanation was complicated by the fact that the experimental results apparently did not agree. Indeed, the experiments calling for a binding energy almost the size of the *Si* band gap were particularly difficult to accept in the light of theories existing at that time. Bassani [1] was one of the first theorists to recognize that these observations called for a new theory of core excitons.

The differences between *P* in *Si* and a core-excited *Si* atom, Si^*, are small: (i) because

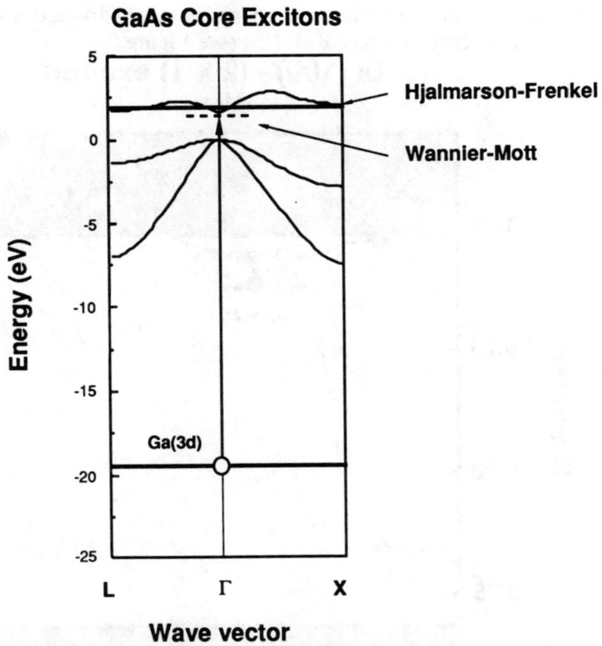

Figure 2. Illustration of the *Ga* 3*d* core excitons of *GaAs*. The shallow Wannier-Mott state lies only 2 *meV* below the conduction band edge. The deep Hjalmarson-Frenkel state lies 0.4 *eV* above the edge and is a T_2 resonance [13].

of the Franck-Condon principle [12], there is no lattice relaxation surrounding the core hole, whereas the lattice certainly relaxes around a *P* impurity - an effect that has always been thought to be negligible, and (ii) the differences in the core hole and proton charge distributions have always been believed to be negligible. Hence a successful theory of the *Si* core exciton problem would have either relied on dynamical effects such as time-dependent screening or explicitly embraced the sameness of Si^* and *P* - and ascribed the apparent differences in binding energies to the fact that experiments probe different aspects of the same defect [1]. The latter approach now provides a simple and natural solution to the *Si* core exciton problem, but had been rejected by many theorists in the 1970's because it implied that the then-current understanding of *P* in *Si* had been incomplete.

2. Co-Existence of 'Deep' and 'Shallow' States

The many apparently contradictory facts about the *Si* 2*p* exciton can be understood once one recognizes that impurities such as *P* in *Si* have *both* 'deep' and 'shallow' states [4] - and that there are corresponding deep or 'Hjalmarson-Frenkel' [13] and shallow or 'Wannier-Mott' excitons [3]. Infrared and transport experiments are sensitive to the shallow impurity levels while core exciton experiments resolve the 'deep' Hjalmarson-Frenkel excitons rather than the shallow Wannier-Mott states. For *P* in bulk *Si* the deep level lies slightly *above* the conduction band edge, so that the corresponding exciton has an apparent *negative* binding energy [4,14] (see Figure 1). However, if the Hjalmarson-

Frenkel $Si\,2p$ exciton is near a surface, its energy is rather dramatically perturbed, so that its apparent binding energy can be $0.8\,eV$ [15] (see Figure 3).

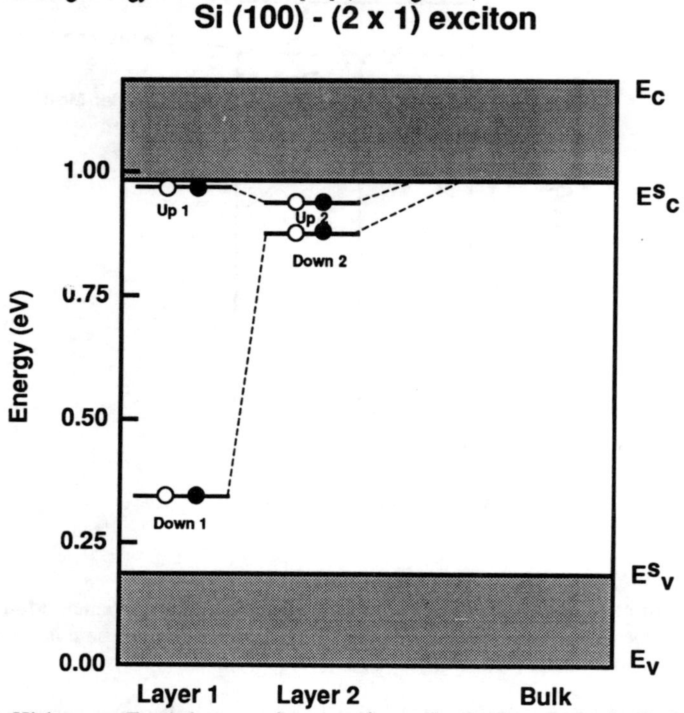

Figure 3. $Si\,2p$ Hjalmarson-Frenkel core exciton energies at the (2×1) asymmetric-dimer (100) surface of Si, after Ref. [15]. 'Up' ('down') denotes the upper (lower) surface site (assuming the Si occupies the lower half plane), and results are given for the surface layer (Layer 1) and the uppermost sub-surface layer (Layer 2). These levels coalesce into a resonance in the conduction band for excitons in the bulk (layer ∞).

Thus the explanation of the wide range of observed exciton energies is that the different measurements had different surface sensitivities, and probed core exciton binding energies in Si, ranging from slightly negative values characteristic of the bulk to large positive values (nearly the size of the band gap) at the surface.

A central element of this picture for the $Si\,2p$ core exciton was the prediction that the Hjalmarson-Frenkel exciton lies slightly *above* the conduction band edge in Si, with an apparent negative binding energy - a prediction that could be tested by measuring its energy in Si_xGe_{1-x} alloys as a function of alloy composition x. The theory [14] predicted that the Hjalmarson-Frenkel state would descend into the gap and assume a positive binding energy for $x\approx0.27$, and then re-enter the conduction band for larger x. The composition $x\approx0.27$ [16] corresponds to a cross-over in the conduction band minima of Si_xGe_{1-x} from being near the (100) X-point of the Brillouin zone of Si for $x = 1$ to the (111) L-point of Ge. Bunker et al. [16] performed an experiment to test these ideas, and concluded that the picture is correct for the $Si\,2p$ core exciton, but that the predicted energies of the Hjalmarson-Frenkel exciton were slightly lower than observed (see Figure 4). The resulting $Si\,2p$ core exciton binding energy is about $-0.06\,eV$.

The Hjalmarson-Frenkel core exciton level slightly above the band gap in bulk Si

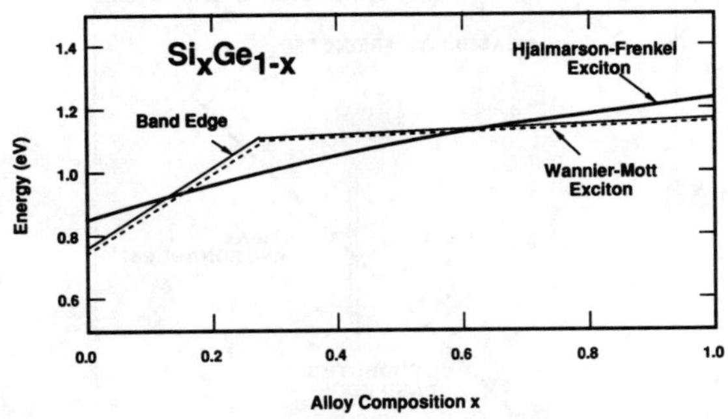

Figure 4. Energies (in *eV*) versus alloy composition x in Si_xGe_{1-x}, after Ref. [16]: conduction band edges (light solid line), (2) Wannier-Mott exciton levels (dashed line), and experimental conclusions [16] (heavy solid line), obtained by analyzing data with an adjusted version of the theory of Ref. [14].

corresponds to a 'deep' level of P in Si that is resonant with the conduction band. This deep level had not been a part of the theory of the P impurity, and yet was demanded by a fact that was well-known, but not fully appreciated: The central-cell defect potential of a P substitutional impurity in Si must be greater than $4\,eV$ deep, because the difference between P and Si s-state atomic energies is $4\,eV$. In a solid, it is impossible to have a $4\,eV$ perturbation potential that does not alter the electronic structure on a $4\,eV$ energy scale - and the old shallow donor theory indicated alterations of the electronic structure only on the tens of meV scale of the donor binding energy. Thus, in retrospect [4], P in Si must have, in addition to the shallow donor levels, four deep levels that lie above the conduction band edge and are resonances: one s-like level that corresponds to the Hjalmarson-Frenkel core exciton and a triply degenerate p-like level at higher energy. These four levels are localized in space and arise from the four perturbed bonds formed by the P impurity when it replaces Si.

The same theoretical picture of Hjalmarson-Frenkel core excitons in Si described Ga 3d core excitons in the bulk [13,17,18] (Figure 5) and Ga 3d and In 4d excitons at surfaces (Figures 6 and 7) [17,19,20] in $GaAs$, $GaSb$, $InAs$, $InSb$, and InP - a dramatic success because the surface exciton energies differ markedly from those in the bulk, yet generally agree with the theory.

The major difference between the Si 2p core exciton on the one hand and the Ga 3d and In 4d excitons on the other is symmetry. Dipole selection rules imply that the core-excited Ga and In generate electrons in odd-parity states and therefore have the energies of the p-like T_2 deep levels of Ge and Sn, respectively. In Si, the core-excited electron occupies an s-like A_1 level.

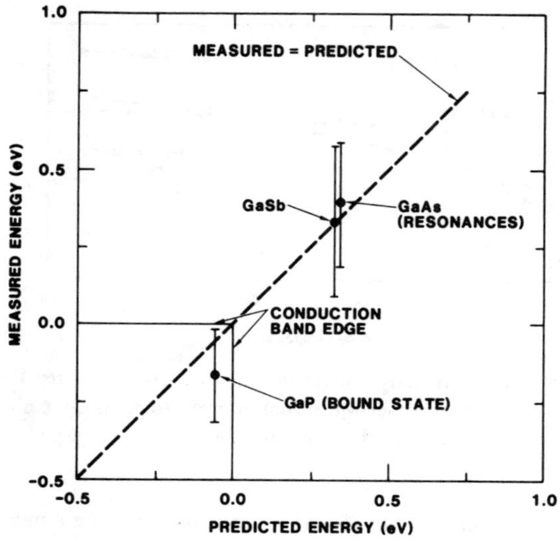

Figure 5. Comparison of measured [18] and predicted energies (in *eV*) of Hjalmarson-Frenkel *Ga* 3*d* core excitons in *GaP*, *GaSb*, and *GaAs*, after Ref. [13]. The zero of energy is the conduction band edge.

3. Core Excitons in Superlattices

Superlattices are particularly interesting materials for studying core excitons because the superlattice band edges are very sensitive to the layer thicknesses, while deep impurity levels are not. Therefore a *Ga* 3*d* core excitation in a *GaAs* quantum well, for example, would have its core excitation energy (i.e., the *Ge* impurity T_2 level) remain almost independent of the width of the *GaAs* well while the conduction band edge would dramatically increase in energy as the well became thinner (due to quantum confinement). As a result, the apparent core exciton binding energy (i.e., the energy of the *Ge* impurity level with respect to the conduction band edge) increases markedly as the well-width decreases. The physics is even more complex when the superlattice is not lattice-matched, and strain is important.

In this paper we apply these ideas to *Ga* 3*d* core excitons in $GaAs_{0.6}P_{0.4}/GaP$ strained-layer superlattices. We believe that these are the first theoretical results for this type of Hjalmarson-Frenkel core exciton in III-V superlattices. Our goal is to show that a Hjalmarson-Frenkel *Ga* 3*d* core exciton at a site near the center of a $GaAs_{0.6}P_{0.4}$ layer can emerge from the conduction band into the fundamental band gap of the superlattice as the layer thickness decreases. That is, its apparent binding energy with respect to the conduction band edge changes from negative to positive with decreasing layer thickness (see Figure 8).

The theory is identical to a theory for a *Ge* impurity at a central *Ga* site of a $GaAs_{0.6}P_{0.4}/GaP$ superlattice. The theoretical formalism has been published for such a defect in strain-free $GaAs/Al_xGa_{1-x}As$ superlattices [21-24], and can be modified in a straight-forward manner to treat strained-layer superlattices. The strain energy is

$$E_s = (a^3/64)(\Sigma_{ij}[6c_{11}+12c_{12}]([d_{ij}-d_0]/d_0)^2 + \Sigma_{ijk}[c_{11}-c_{12}](\delta\theta_{ijk})^2)$$

where the sums are over the bonds (of length d_{ij}) connecting atoms i and j and over the

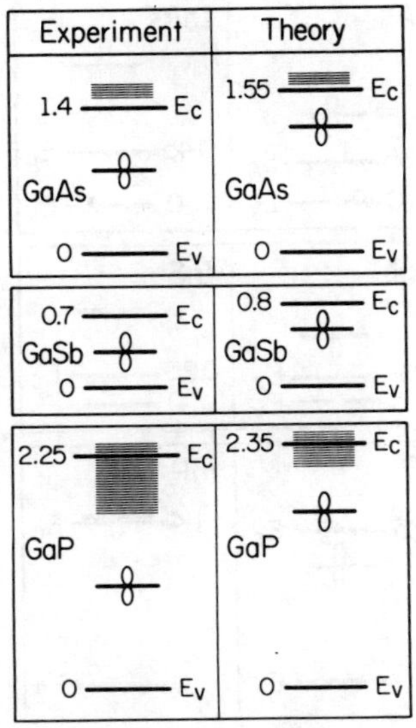

Figure 6. Experimental and theoretical *Ga* 3*d* Hjalmarson-Frenkel core exciton energies (propellers) at the (110) surfaces of *GaAs*, *GaSb*, and *GaP*, after Ref. [19]. The valence and conduction band edges are denoted E_v and E_c. The horizontal lines denote the bottom of the surface state band.

distortions θ_{ijk} of bond angles defined by atoms *i,j*, and *k*. Here c_{11}, c_{12}, d_0, and a are the elastic constants, perfect-crystal bond lengths, and perfect-crystal lattice constant of the relevant material. There are three types of contributions to the strain energy: (i) from *GaP*, (ii) from $GaAs_{1-x}P_x$, and (iii) from the interfaces. Three parameters define the geometry of the strained superlattice: (i) a_{perp}, the lattice constant in the direction perpendicular to the growth or z direction (this is the same in both $GaAs_{1-x}P_x$ and *GaP* layers), (ii) $c(GaP)$, the strained lattice constant in the z or growth direction in a *GaP* layer, and (iii) $c(GaAs_{1-x}P_x)$, the corresponding lattice constant in $GaAs_{1-x}P_x$. The strain energy is minimized with respect to these three parameters, and a_{perp}, $c(GaP)$, and $c(GaAs_{1-x}P_x)$ are determined. Then the off-diagonal (only) matrix elements of the Hamiltonian are scaled according to the rules of Harrison (i.e., as d^{-2} [25]) and Slater and Koster (i.e., the appropriate angle dependences [25,26]). Finally the superlattice electronic structure and Green's functions are calculated, in the usual fashion [21-24].

The predictions of the theory are contained in the *p*-like T_2 '*Ge* impurity' states, because the 3*d* core hole is associated with T_2 electron states as a consequence of the dipole selection rules for optical absorption. The predicted band gaps for [001] $GaAs_{0.6}P_{0.4}/GaP$ superlattices, are expected to lie slightly below [27,28] the experimental gap, but within about $\approx 0.1\,eV$ of it. The band edge of the bulk materials, both unstrained and strained as in the superlattice, are compared with the superlattice band edges in Figure 9 (we have assumed a valence band offset for a strained $GaAs_{0.6}P_{0.4}/GaP$ heterojunction of 60% of $0.33\,eV$ [29]). The corresponding 'binding

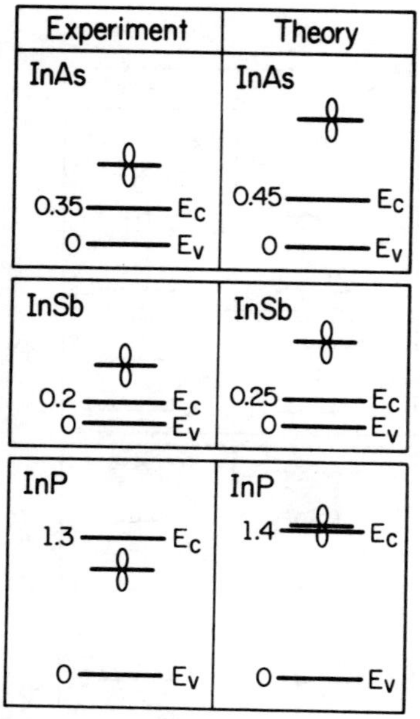

Figure 7. Experimental and theoretical *In* 4*d* Hjalmarson-Frenkel core exciton energies (propellers) at the (110) surfaces of *InAs*, *InSb*, and *InP*, after Ref. [19].

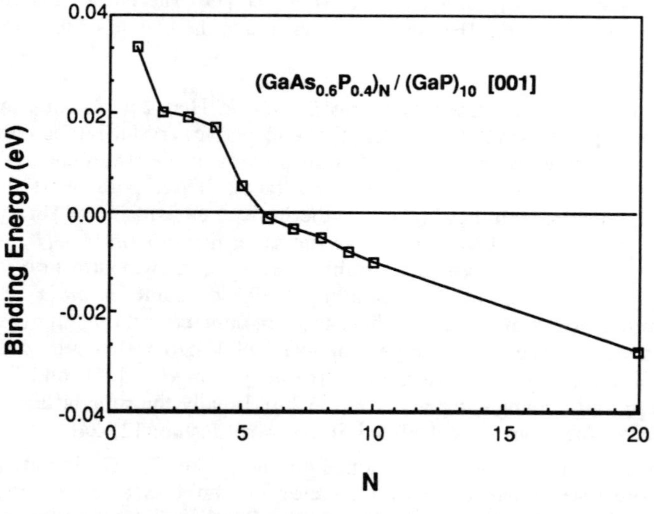

Figure 8. Predicted apparent binding energy of a *Ga* 3*d* core exciton p_z-like level at the center of a $GaAs_{0.6}P_{0.4}$ layer in an $N \times 10$ $GaAs_{0.6}P_{0.4}/GaP$ superlattice as a function of *N*. Note that for small (large) *N* the binding energy is positive (negative).

energies' of the Hjalmarson-Frenkel excitons are displayed in Figure 8 [28]. As the $GaAs_{0.6}P_{0.4}$ layer becomes thinner, the conduction band edge moves up in energy with

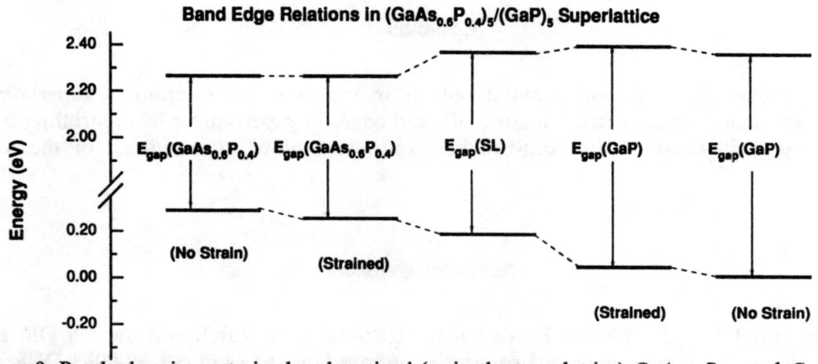

Figure 9. Band edges in unstrained and strained (as in the superlattice) $GaAs_{0.6}P_{0.4}$ and GaP, and in the 5×5 $GaAs_{0.6}P_{0.4}/GaP$ superlattice.

respect to the core exciton, until the core-exciton level descends into the gap. For simplicity of presentation in Figure 8, we have plotted only the p_z-like T_2 core exciton level for an exciton far from an interface. In the superlattice, and especially near interfaces, the T_2 levels (which are degenerate in the bulk) split, with the splittings being largest near the interfaces.

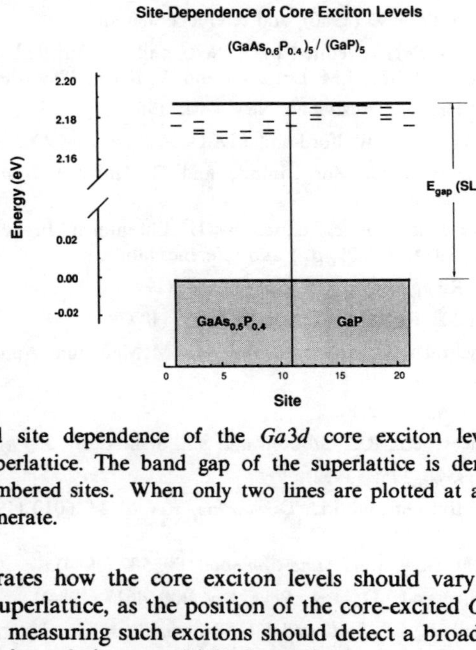

Figure 10. Predicted site dependence of the $Ga3d$ core exciton levels (in eV) in a 5×5 $GaAs_{0.6}P_{0.4}/GaP$ superlattice. The band gap of the superlattice is denoted $E_{gap}(SL)$. The Ga atoms are at odd-numbered sites. When only two lines are plotted at a site, two of the exciton levels are nearly degenerate.

Figure 10 illustrates how the core exciton levels should vary in energy for a 5×5 $GaAs_{0.6}P_{0.4}/GaP$ superlattice, as the position of the core-excited Ga atom is changed. Of course, experiments measuring such excitons should detect a broadened line whose shape reflects the distribution of sites at which the excitons are created and the splittings at each site.

4. Summary

We hope that this work will stimulate efforts to measure core excitons in superlattices and to show that suitable manipulation of band edges in superlattices by controlling layer thicknesses will cause Hjalmarson-Frenkel excitons to move into and out of the band gap.

Acknowledgements

We are grateful to the US Air Force Office of Scientific Research and the US Office of Naval Research for their generous support (Contract Nos. AF-AFOSR-89-0063-DEF and N00014-89-J-1136). We have benefitted from conversations with B. Bunker and K. Newman.

References

1. F. Bassani, *Appl. Optics* **19**, 4093 (1980), and reference therein.
2. W. Kohn, in *Solid State Physics* (edited by F. Seitz and D. Turnbull, Academic Press, New York, 1957), Vol. 5, pp. 258-321; J.M. Luttinger and W. Kohn, *Phys. Rev.* **97**, 969 (1955).
3. R.S. Knox, *Theory of Excitons*, Academic, New York, 1963.
4. H.P. Hjalmarson, P. Vogl, D.J. Wolford and J.D. Dow, *Phys. Rev. Lett.* **44**, 810 (1980).
5. A. Quattropani, F. Bassani, G. Margaritondo and G. Tinivella, *Nuovo Cimento* **51B**, 335 (1979), and reference therein.
6. F.C. Brown, in *Solid State Physics*, edited by H. Ehrenreich, F. Seitz and D. Turnbull, Academic, New York, 1974, Vol. 29, p. 1 and references therein.
7. F.C. Brown and O.P. Rustgi, *Phys. Rev. Lett.* **28**, 497 (1972).
8. G. Margaritondo and J.E. Rowe, *Phys. Lett.* **59A**, 464 (1977).
9. R.S. Bauer, R.Z. Bachrach, D.E. Aspnes and J.C. McMenamin, *Nuovo Cimento* **B39**, 409 (1977).
10. M. Altarelli, *J. Phys. C* **4**, 95 (1978).
11. J.D. Dow, D.R. Franceschetti, P.C. Gibbons and S.E. Schnatterly, *J. Phys. F* **5**, L211 (1975).
12. M. Lax, *J. Chem. Phys.* **20**, 1752 (1952).
13. H.P. Hjalmarson, H. Büttner and J.D. Dow, *Phys. Rev. B* **24**, 6010 (1981); *Phys. Lett.* **85A**, 293 (1981).
14. K.E. Newman and J.D. Dow, *Solid State Commun.* **50**, 587 (1984).
15. M.A. Bowen, R.E. Allen and J.D. Dow, *Phys. Rev.* **B30**, 4617 (1984).
16. B.A. Bunker, S.L. Hulbert, J.P. Stott and F.C. Brown, *Phys. Rev. Lett.* **53**, 2157 (1984). The cross-over of conduction band edges occurs between $x = 0.1$ and $x = 0.3$, depending on temperature. Figure 4 corresponds to a low temperature.
17. M. Skibowski, G. Sprüssel and V. Saile, in *Proceedings of the Fourteenth International Conference on the Physics of Semiconductors*, Edinburgh, 1978, edited by B.L.H. Wilson, Institute of Physics, Bristol, 1979, p. 1359.
18. D.E. Aspnes, C.G. Olson and D.W. Lynch, in *Proc. XIII-th Intern. Conf. Phys. Semiconductors* (Rome, 1976), edited by F.G. Fumi (North-Holland, Amsterdam, 1976) pp. 1000; *Phys.*

Rev. **B12**, 2527 (1975); **B14**, 2534, 4450 (1976); D.E. Aspnes, M. Cardona, V. Saile and G. Sprüssel, *Solid State Commun.* **31**, 99 (1979).

19. R.E. Allen and J.D. Dow, *Phys. Rev.* **B24**, 911 (1981).
20. D.E. Eastman and J.L. Freeouf, *Phys. Rev. Lett.* **33**, 1601 (1974); **34**, 1624 (1975); W. Gudat and D.E. Eastman, *J. Vac. Sci. Technol.* **13**, 831 (1976); D.E. Eastman, T.-C. Chiang, P. Heimann and F.J. Himpsel, *Phys. Rev. Lett.* **45**, 6546 (1980).
21. S.Y. Ren, J.D. Dow and J. Shen, *Phys. Rev.* **B38**, 10677 (1988).
22. S.Y. Ren and J.D. Dow, *J. Appl. Phys.* **65**, 1987 (1989).
23. R.-D. Hong, D.W. Jenkins, S.Y. Ren and J.D. Dow, *Mater. Res. Soc. Symp. Proc.* **77**, 545 (1987), in *Interfaces, Superlattices, and Thin Films*, ed. J.D. Dow and I.K. Schuller.
24. J.D. Dow, S.Y. Ren and J. Shen, NATO Advanced Science Institutes Series **B183**: *Properties of Impurity States in Superlattice Semiconductors*, ed. by C.Y. Fong, I.P. Batra and S. Ciraci (Plenum Press, New York, 1988), p. 175.
25. J.C. Slater and G.F. Koster, *Phys. Rev.* **94**, 1498 (1954).
26. W.A. Harrison, *Electronic Structure and the Properties of Solids*, W.H. Freeman, San Francisco, 1980, p. 481.
27. P.Vogl, H.P. Hjalmarson and J.D. Dow, *J. Phys. Chem. Solids* **44**, 365 (1983).
28. The electronic structure model employed [27] is an sp^3s^* empirical tight-binding model whose matrix elements exhibit chemical trends depending on atomic energies and bond lengths. The matrix elements were obtained by simultaneously fitting the band structures at the Γ and X points of sixteen semiconductors. In the case of $GaAs_{1-x}P_x$, the L conduction minima are slightly lower than they should be, especially for $x \approx 0.5$, where they can lie slightly below the Γ and X minima. Thus the quantitative conclusions of the present work for the band gap and the exciton binding energy have theoretical uncertainties of order $0.2\,eV$, and the core exciton binding energies are, if anything, *underestimated* by about this amount. Hence the binding energies we predict are likely to be even larger in Nature.
29. C. Delerue, M. Lannoo and J.M. Langer, *Phys. Rev. Lett.* **61**, 199 (1988).

QUANTUM HARMONIC OSCILLATOR IN A MAGNETIC FIELD: AN EXAMPLE OF HOLOMORPHIC REPRESENTATION

V. MARIGLIANO RAMAGLIA, B. PREZIOSI, A. TAGLIACOZZO AND F. VENTRIGLIA

Dipartimento di Scienze Fisiche
Universita' di Napoli e GNSM (CNR)
Mostra d'Oltremare Pad. 19
I-80125 Napoli
Italy

The propagator of an harmonic oscillator in a constant magnetic field is given in closed holomorphic form. The analytical expression of the unitary map is evaluated, which converts the holomorphic representation into the coordinate one. This is applied to obtain a closed form of the Feynman's propagator in polar coordinates. The procedure is of general validity and can be extended to many non interacting oscillators.

1. Introduction

There is a revival of interest in the classical problem of a quantum particle or an oscillator in an external magnetic field [1,2].

A complete new ground for investigations is in fact provided by the, so called, 'small systems' in a magnetic field [3,4,5]. The size of the samples is so small that the mean free path for the electron-electron scattering is much larger than the sample itself, so that breaking of the phase coherence in the wavefunction of the single carriers is unlikely. Therefore, experiments can reveal how the phase of the single particle wavefunctions is affected by the magnetic field.

In particular, an Aharonov-Bohm $(A - B)$ solenoid threading a small ring gives rise to those interference effects of a particle with itself, which are by now well known [4,5,6].

In this respect, a promising theoretical tool is the Feynman's propagator for the motion in a plane orthogonal to the field, what is in fact the contents of this work.

It is straightforward to lead the problem of evaluating the propagator for a particle back to that of an oscillator, as the former is the limit case of the latter when the frequency of the oscillator ω_0 goes to zero.

However, we will not deal with the problems that arise in the definition of a propagator for geometries of the sample which are not simply connected [7] (e.g. rings). In that case, paths have to be divided into homotopy classes labeled by the winding number and separate propagators for these classes superimpose with phases that are dictated by the $A - B$ flux threading the holes in the sample [8].

Nonetheless, the closed analytic form for the propagator of an oscillator in a magnetic field has appeared in the literature only recently [9,10,11,12] and its derivation up to now follows rather complicated routes.

We show in this work, how it is possible to obtain the same result by choosing a representation which is most suitable for harmonic oscillators, that is the holomorphic representation [13].

Among the other advantages, this representation allows for direct generalization to the

R. Girlanda et al. (eds.), Progress on Electron Properties of Solids, 451–459.

evaluation of Green's function and Bloch density matrices of many non interacting particles, both bosons and fermions. Therefore our formalism can be a good starting point to discuss two-dimensional quantum electron systems which are increasingly popular thanks to the high quality *GaAs/GaAlAs* heterojunctions [14] that are being constructed by experimentalists.

It is well known that many-body correlation in a $2-d$ electron gas in presence of a magnetic field induce very relevant phenomena as the Integral Quantum Hall Effect and its fractional counterpart at higher fields [15]. It has been reported quite recently that at fields that are even higher a magnetically induced Wigner solid has been observed [16].

In this work, special care will be used in dealing with the degeneracy of the Landau levels: this is studied in terms of the principle quantum number n and the magnetic quantum number m, which are appropriate for a polar coordinate representation.

In section 2 we collect some results about the solution of the Schroedinger equation in polar coordinates which will be used in the following.

In section 3 we introduce the holomorphic representation and apply it to the derivation of the propagator $\hat{U}(T)$ in the space of the holomorphic functions.

In section 4 we deal with the problem of transferring the result for $\hat{U}(T)$ onto the space of configuration in polar coordinates.

To conclude with, we spend fwe words on the advantages of our procedure in section 5.

2. Eigenvalues and Eigenfunctions in Polar Coordinates

The motion of a non relativistic particle of mass M and charge $-e(e>0)$ in an harmonic potential of frequency ω_0 with an external homogeneous magnetic field B, is ruled by the hamiltonian.

$$H = H_z + \frac{1}{2M}(p_x^2 + p_y^2) + \frac{\omega_c}{2}L_z + \frac{1}{2}M\omega^2(x^2+y^2) = H_z + H_t \qquad (1)$$

where $\omega_c = eB/Mc$ is the cyclotron frequency and $\omega^2 = \omega_0^2 + \omega_c^2/4$. The magnetic field is applied along the z-axis, and we choose the vector potential $A = \frac{1}{2}B(-y,x,0)$. Hereafter the z-component of the motion will be dropped, because it is separable. The angular momentum L_z is an integral of motion, so that in polar coordinates (r,θ) the Schroedinger equation for our system is readily solved, yielding the eigenvalues:

$$E_{n,m} = (2n + |m| + 1)\hbar\omega + m\hbar\omega_c/2. \qquad (2)$$

Here $m = 0, \pm 1, ...$ is the magnetic quantum number which labels the eigenfunctions of L_z, while $n = 0, 1, ...$ is the principal quantum number of the radial wavefunction:

$$R_{n,m}(r) = \frac{1}{\Lambda}\left[\frac{n!}{(n+|m|)!}\right]^{\frac{1}{2}} e^{-\xi/2}\xi^{|m|/2}L_n^{(|m|)}(\xi). \qquad (3)$$

In eq. (3) the generalised Laguerre polynomials $L_n^{(\alpha)}$ appear and $\xi = r^2/2\Lambda^2$ is an adimensional variable, obtained from the length $\Lambda = (\hbar/2M\omega)^{\frac{1}{2}}$ that characterizes our problem. These functions are normalized according to $\int_0^\infty R^2(r)rdr = 1$. Finally the eigenfunctions, common to the complete set of commuting observables L_z and H_t, are:

$$\Psi_{n,m}(r,\theta) = \frac{e^{im\theta}}{\sqrt{2\pi}} R_{n,m}(r). \tag{4}$$

When $\omega_0 \to 0$, the harmonic potential disappears and the Landau's result [17] follows, for a charged particle in an homogeneous magnetic field. Its eigenvalues are $\epsilon_{n,m} = (2n + |m| + m + 1)\hbar\omega_c/2$, and the eigenfunctions are still given by $\Psi_{n,m}$, if the radial wavefunction of eq. (13) is rewritten with $\Lambda \to \lambda = (\hbar/M\omega_c)^{\frac{1}{2}}$.

We observe that classically the energy has to be independent of the angular momentum. Besides, for a particle with a negative charge, only non positive values of L_z can occur. On the contrary, in the quantum case, the positive values of m are allowed, and the energy contains m as above.

Back to the oscillator, its partition function

$$Z = \mathrm{Tr}\hat{U}(-i\hbar\beta) = \sum_{n,m} e^{-\beta E_{n,m}}$$

(here β^{-1} is the temperature) was easily evaluated long ago [2]. On the contrary, it is difficult to perform a calculation of the propagator in real time $\hat{U}(T) = \exp(-iTH_t/\hbar)$. In fact, in the coordinate representation, the kernel of $\hat{U}$ is written as

$$U(r\theta, r'\theta'; T) = \sum_{n,m} \Psi_{n,m}(r,\theta)\Psi^*_{n,m}(r',\theta')e^{-iTE_{n,m}/\hbar}, \tag{5}$$

so that it is a formidable task to obtain, by direct summation, a closed form for it.

In the next section we show how this can be done in the holomorphic representation.

3. The Holomorphic Representation

Following Faddeev [18], we use a simplified version of the holomorphic representation, and begin with the simple case of a one-dimensional harmonic oscillator with hamiltonian

$$\hat{h} = \omega(\hat{a}^\dagger \hat{a} + \frac{1}{2}). \tag{6}$$

Here and after we put $\hbar = 1$.

The usual commutation rule $[\hat{a}, \hat{a}^\dagger] = 1$, satisfied by the $\hat{a}$-operators, has a representation

$$\hat{a}^\dagger := a^*, \quad \hat{a} := \frac{d}{da^*}$$

in the space of the analytic functions $f(a^*)$ of a complex variable a^*, i.e. linear combinations of the monomials

$$\varphi_n(a^*) = \frac{(a^*)^n}{\sqrt{n!}} \tag{7}$$

orthonormalized with the scalar product

$$<\varphi_k | \varphi_j> = \int \varphi_k^*(a^*)\varphi_j(a^*)e^{-a^*a}\frac{da^*da}{2\pi i} = \delta_{k,j}.$$

An arbitrary operator $\hat{A}$ is represented by an integral operator with kernel $A(a^*, a)$:

$$(\hat{A}f)(a^*) = \int A(a^*, z)f(z^*)e^{-z^*z}\frac{dz^*dz}{2\pi i}. \tag{8}$$

Here the kernel $A = e^{a^* a}\tilde{A}$ is obtained from the function $\tilde{A}(a^*,a)$ associated to the normally ordered form of $\hat{A}$ (that is, a polynomial in variables a^* and a preserving the form of the normally ordered polynomial in operators $\hat{a}^\dagger$ and $\hat{a}$).

If the operator $\hat{A}$ appearing in eq. (8) is the identity operator $\hat{I}$, whose kernel is $e^{a^* a}$, we obtain the analogous of Dirac's δ-function:

$$f(a^*) = \int e^{a^* z} f(z^*) e^{-z^* z} \frac{dz^* dz}{2\pi i}. \tag{9}$$

For a small time interval ΔT, the kernel of the evolution operator $\hat{u}(\Delta T)$ for the hamiltonian of eq. (6) is

$$u(a^*,a;\Delta T) = \exp\{a^* a - i\tilde{h}(a^*,a)\Delta T\} \tag{10}$$

where $\tilde{h}(a^*,a) = \omega(a^* a + \frac{1}{2})$. Now, as the convolution of the kernels corresponds to the product of the operators $\hat{A}$ and $\hat{B}$:

$$(AB)(a^*,a) = \int A(a^*,a_1)B(a_1^*,a)d\mu_1$$

(here and after, $d\mu_s := e^{-a_s^* a_s} da_s^* da_s / 2\pi i$ is the measure), the propagator is expressed as a path integral, as soon as the time interval T is fractioned into N steps of width $\epsilon = T/N$:

$$u(a^*,a';T) = \lim_{N\to\infty} \int \dots \int \prod_{k=0}^{N-1} e^{-i\epsilon\omega/2} e^{a_{k+1}^* a_k (1-i\epsilon\omega)} d\mu_1 \dots d\mu_{N-1}.$$

Here $a_N^* = a^*$, $a_0 = a'$ are independent variables.

This is nothing but a set of integrals of the form of eq. (9), that is a convolution of δ-functions, so that we may carry out the integration on one variable after the other. After the integrations have been completed, the limit may be taken. The result is:

$$u(a^*,a';T) = \exp\{a^* e^{-i\omega T} a' - \frac{1}{2} i\omega T\}. \tag{11}$$

This shows that the holomorphic representation is the natural one for the harmonic oscillator, just like the coordinate representation is for the free particle [19].

This is because the basis functions φ_n are the eigenfunctions of $\hat{h}$:

$$\int u(a^*,a_1;T)\varphi_n(a_1^*)d\mu_1 = e^{-i\omega T/2}\varphi_n(a^* e^{-i\omega T}) = \varphi_n(a^*)e^{-iE_n T}. \tag{12}$$

In other words, if $|n>$ is the eigenket of $\hat{h}$ belonging to the eigenvalue $E_n = (n+\frac{1}{2})\omega$, then $<a^*|n> = \varphi_n(a^*)$.

To apply this formalism to our hamiltonian H_t of eq. (1), we introduce the usual creation (annihilation) operators $\hat{a}_1^\dagger, \hat{a}_2^\dagger (\hat{a}_1, \hat{a}_2)$ corresponding to the $x-$ and y-component of the motion respectively, and the number operators $N = N_1 + N_2$ so that $L_z = i(\hat{a}_1^\dagger \hat{a}_2^\dagger - \hat{a}_1 \hat{a}_2)$ and

$$H_t = \omega(N+1) + \frac{\omega_c}{2} L_z. \tag{13}$$

Now, as N and L_z form a complete set of compatible observables, the hamiltonian is diagonalized by the canonical transformation

$$\hat{a}_\pm^\dagger := \frac{1}{\sqrt{2}}(\hat{a}_1^\dagger \pm i\hat{a}_2^\dagger)$$

which preserves the commutation rules [20].

In terms of these new operators, $N = N_+ + N_-$ and $L_z = N_+ - N_-$, so that the hamiltonian is diagonal:

$$H_t = \omega(N_+ + N_- + 1) + \frac{\omega_c}{2}(N_+ - N_-) = \omega_+ N_+ + \omega_- N_- + \omega, \tag{14}$$

with $\omega_\pm = \omega \pm \omega_c/2$.

The operators N_+ and N_- both have as their spectrum the sequence of non-negative integers $n_\pm = 0, 1, \ldots$ and the complete orthonormal set common to them is formed by the vectors:

$$|n_+, n_- > = (n_+! n_-!)^{-\frac{1}{2}} (\hat{a}_+^\dagger)^{n_+} (\hat{a}_-^\dagger)^{n_-} |0,0> \tag{15}$$

where $|0,0>$ is the same vacuum of $N = N_1 + N_2$.

The evaluation operator $\hat{U}(T)$ is obtained in the holomorphic representation at once, if constructed on the space spanned by the direct product of the basis functions $\varphi_{n_+}(a_+^*)$ and $\varphi_{n_-}(a_-^*)$, each given by eq. (7), with a_+^* and a_-^* independent complex variables:

$$\Phi_{n_+, n_-}(a_+^*, a_-^*) = \varphi_{n_+}(a_+^*) \varphi_{n_-}(a_-^*). \tag{16}$$

In fact, being H_t diagonal, the kernel of $\hat{U}(T)$ has separate dependence on a_+^* and a_-^* and therefore becomes the product of two one-dimensional kernels, each of the form eq. (11):

$$U(a_+^* a_-^*, a'_+ a'_-; T) = \exp\{a_+^* e^{-i\omega_+ T} a'_+ + a_-^* e^{-i\omega_- T} a'_- - i\omega T\}. \tag{17}$$

Applying eq. (12) to this case, it is readily seen that the basis functions $\Phi_{n_+, n_-}(a_+^*, a_-^*)$ are just the eigenfunctions $<a_+^* a_-^* | n_+ n_- >$ of the hamiltonian, belonging to the eigenvalues $E_{n_+, n_-} = n_+ \omega_+ + n_- \omega_- + \omega$.

4. The Propagator in the Configuration Space

An invertible map $\hat{K}$, that converts the holomorphic representation into the coordinate one, allows us to obtain a closed form for the kernel $U(r\theta, r'\theta'; T)$ of eq. (5) from the one of $U(a_+^* a_-^*, a'_+ a'_-; T)$ of eq. (17).

In the case of the previous one-dimensional harmonic oscillator, the kernel of the map is written at once as:

$$K(x,a) = \sum_n <x|n><n|a^*> = \sum_n \psi_n(x) \varphi_n^*(a^*)$$

where the eigenfunctions $\psi_n(x)$ and $\varphi_n^*(a^*)$ correspond to the same label n, thus belonging to the same eigenvalue. In fact

$$\psi_j(x) = \int K(x,a) \varphi_j(a^*) d\mu.$$

In our problem, however, we have two sets of quantum numbers running over different ranges, i.e. n and m which label $\Psi_{n,m}(r,\theta)$, while n_+ and n_- label $\Phi_{n_+, n_-}(a_+^*, a_-^*)$. Therefore, it is a preliminary necessity to put them into a one-to-one correspondence, and this is accomplished by realizing that

$$|n_+, n_- >= \begin{cases} |n+m, n> & \text{if } n_+ > n_- \\ |n, n+|m|> & \text{if } n_+ \leqslant n_- \end{cases}. \tag{18}$$

Thus $n = \min\{n_+, n_-\}$ and $m = n_+ - n_-$. Once the correspondence has been established, the unitary map $K(r,\theta,a_+,a_-)$ can be constructed, which allows for changing from the holomorphic representation to that of the polar coordinates:

$$K(r,\theta;a_+,a_-) = \tag{19}$$
$$\sum_{n=0}^{\infty}\left\{\sum_{m=1}^{\infty}\Psi_{n,m}(r,\theta)\Phi^*_{n+m,n}(a^*_+,a^*_-) + \sum_{m=0}^{\infty}\Psi_{n,-m}(r,\theta)\Phi^*_{n,n+m}(a^*_+,a^*_-)\right\}.$$

This expression can be reduced to closed form, as shown in the Appendix, to give:

$$K(r,\theta,a_+,a_-) = \frac{1}{\Lambda\sqrt{2\pi}}\exp\left[-\frac{1}{2}\xi + a_+a_- + \sqrt{\xi}(e^{i\theta}a_+ - e^{-i\theta}a_-)\right] \tag{20}$$
$$= K^*(r,-\theta,a^*_+,a^*_-)$$

where Λ and ξ have been defined after eq. (3).

The map K, together with the operator U of eq. (17), is the main result of this work.

We are able now to evaluate:

$$U(r\theta,r'\theta';T) =$$
$$\int K(r\theta,a_+a_-)U(a^*_+a^*_-,a'_+a'_-;T)K^*(r'\theta',a'_+a'_-)d\mu_+d\mu_-d\mu'_+d\mu'_-.$$

The integrations on $d\mu'_+d\mu'_-$ can be performed according to eq. (9). In fact the kernel U acts like a Dirac's δ, so changing the variables of K^* from $(a'_\pm)^*$ into $a^*_\pm e^{-i\omega_\pm T}$.

Thanks to the exponential form of K given by eq. (20), a third integration (e.g. on $d\mu_-$) is of the same kind. A quadratic form is eventually obtained in the variables a^*_+, a_+: thus the last integral is gaussian and can be evaluated exactly.

Restoring $\hbar$, the final result is:

$$U(r\theta,r'\theta';T) = \tag{21}$$
$$\frac{M\omega}{2\pi\hbar i\sin\omega T}\exp\left\{i\frac{M\omega}{2\hbar\sin\omega T}\left[(r^2+r'^2)\cos\omega T - 2rr'\cos(\theta-\theta'-\frac{\omega_c}{2}T)\right]\right\}.$$

This is the result that was expected. Its periodic dependence on time follows from the shape of the hamiltonian given by eq. (1). The periodicity is well defined as long as the frequencies ω and $\omega_c/2$ are commensurate. In the case of a free particle in a magnetic field the propagator, or the causal Green function can be evaluated also by direct methods in cartesian coordinates, yielding

$$g(x,y,x',y';t) = -\frac{M\omega_c}{4\pi\hbar^2\sin\frac{\omega_c t}{2}}\exp\left[i\frac{M\omega}{2\hbar}(yx'-xy')\right]$$
$$\exp\left\{-i\frac{M\omega_c}{4\hbar}[(x-x')^2+(y-y')^2]\cot\frac{\omega_c t}{2}\right\}.$$

This Green function coincides with the previous one when $\omega = \omega_c/2$ apart from a factor $i\hbar$.

5. Conclusive Remarks

The evaluation of the propagator of eq. (21) has quite a long history. The efforts to evaluate the Feynman's path integral directly in cartesian coordinates immediately face a quite unaffordable problem, although it became clear that a closed form for the solution could be achieved [9]. This is due to the angular momentum term appearing in the hamiltonian of eq. (1), which gives rise to a non local term in the action after one of the variables has been integrated out.

A complete solution was given by means of other approaches [10,11], until it was recognized that the polar coordinate representation was the most suitable for handling the problem [12]. However, it is important to note that in this case an effective Lagrangian has to be considered, which includes a centrifugal term proportional to $\hbar^2$, what calls for justification [21].

The route here reported allows, for the first time, to start with the hamiltonian of eq. (13), which doesn't strictly depend on the choice of the coordinates, ending with the representation of $\hat{U}(T)$ in polar coordinates, via the holomorphic functional method. The steps we go through provide the unitary $\hat{K}$ between the holomorphic functions space and the configuration space, what is a result of wider applicability.

Few remarks are here in order.

A crucial requirement of the method is to establish the right correspondence between the labels of the holomorphic basis set and those of the eigenfunctions of the hamiltonian in polar coordinates, given by eq. (18).

We think that more insight in the formulation of the problem in cartesian coordinates could be gained with our method, whose roots can be found even in the Pauli's book [22].

Within what stated above, the holomorphic representation is particularly suitable to discuss problems in which one or more oscillators are involved and magnetic fields. We have shown that handling the problem in the holomorphic functions space is straightforward due to the fact that the basis set of the representation is just the set of the eigenfunctions of the hamiltonian.

In the case of a non interacting many body system the theory has also been applied to fermions via Grassmann variables.

For many particles, the propagator of eq. (17) can be extended to give the generating functional for the partition function, for the Green's functions and the correlation functions. As traces are independent of the representation, they can be more simply evaluated in the holomorphic functions space. In particular, in dealing with the $n-$ body Green's function, transformation back to the space of configurations, by means of the map $\hat{\mathbf{K}}$, is only required for those n fields on which the trace is not performed.

Discussions with G.P. Zucchelli and G. Morandi are gratefully acknowledged.

Appendix

We report here the calculation of the kernel $K(r\theta;a_+ a_-)$ of eq. (19).

The explicit form of the functions Ψ given by eqs. (4,3) and Φ given by eqs. (16,7) is substituted into eq. (19). This yields:

$$K(r,\theta;a_+,a_-) = \frac{e^{-\xi/2}}{\Lambda\sqrt{2\pi}} \sum_{n=0}^{\infty} \left\{ \sum_{m=1}^{\infty} \frac{1}{(n+m)!} e^{im\theta} \xi^{m/2} L_n^{(m)}(\xi) a_+^{n+m} a_-^n + \right.$$

$$\sum_{m=0}^{\infty} \frac{1}{(n+m)!} e^{-im\theta} \xi^{m/2} L_n^{(m)}(\xi) a_+^n a_-^{n+m} \Big\}.$$

Putting $z = a_+ a_-$, the sum on the label n gives the generating function of the generalized Laguerre polynomials [23], expressed in terms of the Bessel function of first kind J_m:

$$e^z |\xi z|^{-m/2} J_m(2\sqrt{\xi z}).$$

This result is:

$$K(r,\theta;a_+,a_-) = \frac{e^{z-\xi/2}}{\Lambda\sqrt{2\pi}} \Big\{ J_0(x) + \sum_{m=1}^{\infty} (t_+^m + t_-^{-m}) J_m(x) \Big\},$$

with $x = 2\sqrt{\xi z}$, and $t_\pm = \pm e^{i\theta}\sqrt{a_+/a_-}$.

Inspection of the r.h.s of the previous equation shows that the curly bracket is just the generating function of the Bessel functions of integer order, so that:

$$K(r,\theta;a_+,a_-) = \frac{e^{z-\xi/2}}{\Lambda\sqrt{2\pi}} \exp\frac{x}{2}(t_+ - t_+^{-1})$$

from which eq. (20) follows.

References

1. L.D. Landau, *Zeits. f. Physik* **64**, 629 (1930).
2. G.C. Darwin, *Proc. Camb. Phil. Soc.* **27**, 86 (1931).
3. Y. Imry, in Percolation, localization and superconductivity, p. 189, eds. A.M. Goldman, S.A. Wolf, Plenum, N.Y. 1984.
4. M. Buttiker and Y. Imry, *J. Phis. C: Sol. St.* **18**, L467 (1985).
5. Y. Gefen and D.J. Thouless, *Phys. Rev. Lett.* **59**, 1752 (1987).
6. Y. Aharonov and Bohm, *Phys. Rev.* **115**, 485 (1959).
7. P.A. Horvathy, G. Morandi, E.C.G. Sundarshan, *Il Nuovo Cimento D* **11**, 201 (1989), National Conference on Feynman's Quantum Mechanics 40 Years after Its Proposal, Eds. A. Ranfagni, D. Mugnai, P. Moretti and M. Cetica, Napoli, 16-18 June 1988.
8. G. Morandi and E. Menossi, *Eur. J. Phys.* **5**, 49 (1984).
9. B.K. Cheng, *J. Phys. A: Math. Gen.* **17**, 819 (1984).
10. N.H. March and M.P. Tosi, *J. Phys. A: Math. Gen.* **18**, L643 (1985).
11. I.M. Davies, *J. Phys. A: Math. Gen.* **18**, 2737 (1985).
12. J.M. Manoyan, *J. Phys. A: Math. Gen.* **19**, 3013 (1986).
13. F.A. Berezin, The Method of Second Quantization, Academic, NY, 1966.
14. L. Esaki, *IEEE J. of Quantum Electr.*, Vol. QE-22 **9**, 1611 (1986).
15. R.E. Prange and S.M. Girvin (eds), The Quantum Hall Effect, Springer, Berlin, 1987.
16. E.Y. Andrei, G. Deville, D.C. Glattly, F.I.B. Williams, E. Paris and B. Etienne, *Phys. Rev. Lett.* **60**, 2765 (1988).
17. L.D. Landau and E.M. Lifshitz, Quantum Mechanics, sect 111, Pergamon, Oxford, 1985.
18. L.D. Faddeev and A.A. Slavnov, Gauge Fields, sect. 2.2, Benjamin-Cummings, Reading Mass, 1980.
19. R.P. Feynman and A.R. Hibbs, Quantum Mechanics and Path Integrals, sect. 3.1, McGraw-Hill, NY, 1965.

20. A. Messiah, Quantum Mechanics, Chap. 12 sect. 14, North-Holland, Amsterdam, 1969.

21. A.M. Arthurs, *Proc. R. Soc.*, London, **A313**, 445 (1969).

22. W. Pauli, Wave Mechanics, sect. 44, C.P. Enz ed., the MIT Press, Cambridge Mass, 1973.

23. M. Abramowitz and I.A. Stegun eds., Handbook of Mathematical Functions, National Bureau of Standard, Washington, 1965.